R. Margesin · F. Schinner (Eds.) *Cold-Adapted Organisms*

Ecology, Physiology, Enzymology and Molecular Biology

Springer

Berlin
Heidelberg
New York
Barcelona
Hong Kong
London
Milan
Paris
Singapore
Tokyo

Rosa Margesin · Franz Schinner (Eds.)

Cold-Adapted Organisms

Ecology, Physiology, Enzymology and Molecular Biology

With 100 Figures and 26 Tables

Springer

Professor Rosa Margesin
Professor Franz Schinner

Institute of Microbiology
University of Innsbruck
Technikerstraße 25
A-6020 Innsbruck
Austria

rosa.margesin@uibk.ac.at
franz.schinner@uibk.ac.at

ISBN 3-540-64973-5 Springer-Verlag Berlin Heidelberg New York

Library of Congress Cataloging-in-Publication Data

Cold-adapted organisms : ecology, physiology, enzymology, and molecular biology / Rosa Margesin, Franz Schinner (eds.)
p. cm.
Includes bibliographical references (p.) and index.
ISBN 3-540-64973-5
1. Cold adaptation. I. Margesin, Rosa, 1962- . II. Schinner, Franz, 1947- .
QP82.2.C6C62 1999
571.4'64--dc21 99-13852

Cover design: Design & Production, Heidelberg
Typesetting: Camera ready by editors
SPIN 10690996 31/3136 - 5 4 3 2 1 0 - Printed on acid-free paper

Preface

Cold ecosystems cover the major part on earth and are colonized by cold-adapted microorganisms, plants and animals. These organisms have adapted to their environment in such a way that metabolic processes, reproduction and survival strategies are optimal. They are responsible for nutrient turnover as well as for production and decomposition of biomass in terrestrial and aquatic cold ecosystems.

Despite their important role, cold-adapted organisms have received little attention so far both in basic and applied research. In this book, prominent authors present fundamental and new knowledge and concepts regarding cold-adapted microorganisms, plants and animals in the fields of ecology, physiology, enzymology and molecular biology.

Beside of the ecological importance of cold-adapted organisms, their large potential for biotechnological applications has been recognized only recently. The interested reader is referred to our book "Biotechnological Applications of Cold-Adapted Organisms" (Springer-Verlag).

We are most grateful to the authors for their excellent contributions. This book has been produced within the Concerted Action "Eurocold" under the coordination of Professor N.J. Russell. We are also indebted to the publishers, especially to Dr. D. Czeschlik, for their cooperation, and to B. Marschall for skilful preparation of the text layout.

Innsbruck, January 1999 R. Margesin, F. Schinner

Contents

Ecology and Physiology 1

Cold-adapted microorganisms

Microbial life in permanently cold soils
A. M. Gounot 3
Lake ice microbial communities in alpine and antarctic lakes
R. Psenner et al. 17
Physiology of cold-adapted microorganisms
A. M. Gounot and N. J. Russell 33

Cold-adapted plants

Ecological aspects of cold-adapted plants with a special emphasis on environmental control of cold hardening and dehardening
O. Junttila and R. Robberecht 57
Plant responses to low temperature: signaling pathways involved in plant acclimation
A. Kacperska 79
Ice nucleation and deep supercooling in plants: new insights using infrared thermography
M. Wisniewski and M. Fuller 105
Modeling sequential responses of plant cells to freezing and thawing
M. J. T. Reaney and L. V. Gusta 119

Cold-adapted animals

Cold adaptations in terrestrial invertebrates
L. Sømme 137
Nival invertebrate animals in the East Alps: a faunistic overview
K. Thaler 165
High alpine streams: cold habitats for insect larvae
L. Füreder 181
Ecophysiological and morphological features of glacier-dwelling Collembola
G. Eisenbeis and E. Meyer 197
Ecological features of antarctic fishes
M. Vacchi et al. 219
Structure and function of hemoglobins from antarctic organisms: the search for correlations with adaptive evolution
G. di Prisco et al. 239

Enzymology **255**

Cold enzymes: a hot topic
C. Gerday et al. 257
Psychrophilic enzymes: insights into cold adaptation and catalysis from the first high resolution crystal structures
N. Aghajari and R. Haser 277
Temperature adaptation in enzymes of antarctic fishes
M. A. Ciardiello et al. 297
Aspartate aminotransferase from *Moraxella* TAC125: an unusual psychrophilic enzyme
M. L. Tutino et al. 305

Molecular Biology **317**

Genes and enzymes involved in arginine and pyrimidine biosynthesis in psychrophilic *Vibrio* strains from the deep sea
Y. Xu et al. 319
Plasmids from antarctic bacteria
M. L. Tutino et al. 335
Molecular responses in cold-adapted plants
M. Hiilovaara-Teijo and E. T. Palva 349
Gene expression and cold hardiness in animals
K. B. Storey and J. M. Storey 385

Subject index 409

Ecology and Physiology

Microbial life in permanently cold soils

A. M. Gounot

Laboratory of Microbial Ecology, UMR 5557, University Claude Bernard Lyon 1, Bat 405, 43 boulevard du 11 Novembre 1918, F-69622 Villeurbanne-Cedex, France; e-mail: gounot@biomserv.univ-lyon1.fr

1 Introduction

The earth contains many low-temperature environments. Over 80% of the earth's biosphere is permanently cold: 90% of the volume of oceans - which occupy 71% of the earth's surface - is colder than 5°C. Various aquatic and terrestrial continental environments are also cold, such as polar regions (14% of the earth surface), high mountains, and deep lakes.[1] These continental environments are characterized by stable or unstable temperatures. Many organisms are able to develop in cold habitats, but cold adaptation can be a tolerance or a preference. These microorganisms are called psychrophiles or psychrotrophs (more exactly psychrotolerants) depending on their maximum growth: psychrophiles do not grow at 20°C and above, psychrotrophs grow better at temperatures above 20°C.[2] Consequently, psychrophilic microorganisms, being heat-sensitive, can live only in permanently cold habitats, while psychrotrophic microorganisms are found in both permanently or temporarily cold habitats. Most continental habitats encounter large variations of temperature. Even temperate soils contain high numbers of psychrotolerant microorganisms which can be isolated after incubation at a temperature below 5°C but they grow faster in spring and summer, while psychrophiles are excluded. However, some continental habitats are permanently cold and can be favorable to true psychrophiles.

General reviews were published on microbial ecology of cold environments,[1,3] especially of Antarctica.[4,5] This chapter gives an overview of the diversity of terrestrial cold habitats and the particular features of their microbial communities, particularly those found in arctic and alpine terrestrial environments such as moraines of alpine glaciers, neves and cold caves where microorganisms encounter extreme conditions.

2
Methods

Various methods of soil microbiology were used for studying cold environments.[6,7] Direct light and electronic microscopy, and fluorescence microscopy after staining with acridine orange or 4,6-diamino-2-phenylindole (DAPI) give total cell numbers. The proportion of living bacteria is obtained by using a tetrazolium-reducing dye. Biochemical analyses of the sample, directly or after incubation with radioisotope-labeled substrates, give information on *in-situ* metabolic activities. Extraction and amplification of total DNA allow identification of microorganisms present in a soil sample.[8] However, the discovery, in a cold habitat, of intact microbial cells by acridine orange epifluorescence microscopy, or the successful amplification of bacterial DNA, does not indicate if there are viable and cold-adapted microorganisms in this habitat. These analyses must be completed by cultural methods (e.g. plate counts, enrichment cultures) which give the number of living (more exactly, culturable) microorganisms. They provide information on physiology, potential metabolic activities, and microbial diversity, after isolation and characterization of the isolates by phenotypic and molecular methods, such as 16S rDNA sequencing.[9,10]

3
Microbial types and functions

A temperature near 0°C is not very selective and a large variety of microorganisms, including bacteria, yeasts, fungi and microalgae, can be found in cold habitats.[1,3,11,12] These microorganisms are free-living in soil, fresh and saline waters, or associated with plants and cold-blooded animals. Some are pathogenic for plants or animals. Among bacteria, almost all types of bacteria were recognized either after isolation or detection of their metabolic products. Unlike hyperthermophiles, they do not belong to new phyla, but they are new taxa or species, or perhaps genera. The most frequently isolated bacteria are *Pseudomonas* spp., *Vibrio* spp. and other Gram-negative bacteria, Gram-positive bacteria of the coryneform group (e.g. *Arthrobacter* spp.) and *Micrococcus* spp., and also yellow-pigmented bacteria of the Flavobacteriaceae family (*Flexibacter-Bacteroides-Cytophaga* phylum). Spore-forming bacteria are less often found. Since Gram-negative bacteria are predominant in marine waters, which were more investigated, it is not surprising that most authors consider that the majority of psychrophilic bacteria are Gram-negative. Cyanobacteria are dominant in some cold environments. Yeasts, especially *Cryptococcus* species, are frequently isolated from cold soils.

The metabolic activities of almost all functional groups were detected in cold soils: photosynthesis as well as degradation of natural or xenobiotic organic molecules, such as cellulose, pectin, lignin,[1] methane,[13] hydrocarbons,[14] chlorophenol[15] and surfactants.[16] Although less studied, anaerobic respiration occurs.[6] Methane production at low temperature (10°C or below) was detected.[17–20] *In-situ* nitrification was rarely demonstrated.[1,6] Dinitrogen fixation was poor[1] or absent.[4] Psychrotrophic thiobacteria were isolated.[21] Archaebacteria were also detected by rRNA probe analysis,[22] but this did not demonstrate their ability to grow at low temperature.

4 Environmental conditions in cold habitats

In addition to temperature, the composition of microbial communities depends on several other environmental factors: salinity and osmotic pressure, especially in marine environments and saline lakes; dessication which increases saline stress; high pressure in deep sea; solar UV radiation and cosmic ionizing radiation in bare surface areas; long-term earth radiation in old ice and sediments; oxidative stress; darkness in deep waters, subsurface sediments and caves, which excludes phototrophic organisms; and nutrient availability. Unlike cooled food, cold waters and soils are often oligotrophic; amounts of organic carbon, nitrogen and phosphorus are very low in deep oceans and some lakes, and in moraines and subsurface sediments. The most extreme condition selects microorganisms which are the best adapted to it, and the species diversity decreases at the extremes.

It is also very important to consider whether these extreme conditions are stable or changing. Variations of temperature, particularly freeze-thawing cycles, result in drastic changes of other conditions. Cells can be damaged by two mechanisms, i.e. by the formation of ice crystals inside the cell, and by the increase in solute concentration due to water crystallization. In addition, thawing results in a transition from anoxia to full aeration.

Finally, cold areas offer a wide variety of habitats which have selected for different types of cold-adapted communities. The most extreme conditions are found in Antarctica which instigated many microbiological investigations in continental or marine areas.[4,5] However, typical cold-adapted communities are also found in alpine and arctic continental areas.

5 Cold deserts

In antarctic cold deserts, extreme conditions are dryness and drastic variations of temperature (-55 to +10°C). In soil, xero- and cryotolerant yeasts and cyanobacterial mats depend on water availability. Vishniac and Klinger[23] consider that psychrophilic yeasts, particularly *Cryptococcus* species, are the most important life form in antarctic desert soils. Pigmentation and intrinsic protection from harmful UV radiation, increased by the ozone hole over Antarctica, may partly account for the frequency of *Micrococcus* and *Deinoccus* isolates at the surface of soils.[4]

A protected habitat affording moisture and adequate illumination for primary production is found in porous rock, where the endolithic communities are constituted of xero- and cryotolerant microbes adapted morphologically to their habitat: cyanobacteria, green algae and Xanthophyceae are dominant, associated with filamentous fungi, yeasts, lichens and heterotrophic bacteria, especially Gram-positive bacteria (coryneforms, *Micrococcus* spp., *Deinococcus* spp.).[24]

6
Soils covered with snow in winter

In maritime Antarctica and in boreal tundra, soils are more hospitable because they are wet and cryptogams produce organic matter, but they withstand freeze-thaw cycles in the presence of free water. Water availability, rather than temperature itself, appears to be the limiting factor. Cryptogams release soluble organic matter, predominantly polyols, that are potential compatible solutes which confer a degree of freezing and dessiccation tolerance to cryptogams and microorganisms after release.[4] It is possible that this substantial supply of polyols and sugar not only acts as an energy source for the heterotrophic microbiota but may also provide cryoprotection.

The ability of yeasts to grow at low temperature, to accumulate cryoprotectants and to supercool, combined with frequent capsulation which acts as a moisture-regulating mechanism, may partly account for their large numbers in moss communities.[4] Algae and cyanobacteria also show frost tolerance. It has been suggested that sugar monophosphosphates accumulated in *Nostoc commune* may offer cryoprotection.[4] In these cold habitats, mucilaginous capsules or sheaths, produced by many bacteria, cyanobacteria and algae, can be effective in stabilizing the soil.[4] Phylogenetic diversity, studied by small subunit rDNA genes amplification, indicates that these bacteria are very diverse in phylogeny and that many probably reflect new genera or families.[8]

7
Permafrost sediments

The conditions are quite different in permanently frozen sediments (permafrost) that have not thawed for thousands or millions of years. These were studied in Siberia, North America and Antarctica.[6,25,26] The age of sediments can be determined from radiocarbon, paleomagnetic, palynological and paleontological analyses. In Siberia[26] they have 400-900 m depth and date back to the second half of the Pliocene (3-5 million years). The average temperatures of subsurface layers are -25 to -30°C in Antarctica and -10 to -12°C in arctic cryogenic regions. Total microbial counts give high numbers of microorganisms [10^7 to 10^9 colony-forming units (cfu) g^{-1} dry weight] while viable cell numbers are much lower (from <10 to 10^7 cfu g^{-1} dry weight). The library of isolated bacterial strains comprised more than 30 genera.[9] Characterization of 29 isolates by 16S rDNA sequencing and phylogenic analysis revealed that they fell into high-GC Gram-positive bacteria, β- and γ-proteobacteria, and low-GC Gram-positive bacteria.[9] They were mostly psychrotrophic and mesophilic, some strains were psychrophilic. Field analysis revealed the reduced state of most of the sediments. Culturable anaerobic bacteria exist in most permafrost samples in quantities comparable to aerobic bacteria. Metabolic activities were detected after thawing, including sulfide and methane production. The eukaryotic cells were less able to survive long-term cryoconservation. They were found only in recent sediments and their number decreased with depth.

The very low temperatures prevent growth of all microorganisms because of unavailability of liquid water, but they must remain alive as deep resting cells for a long time in "anabiosis".[6] After death, the lysis of cells is very low and components such as DNA can be preserved for a long time. Consequently, the discovery of intact microbial cells by acridine orange epifluorescence microscopy, or the successful amplification of bacterial DNA, does not indicate if they were viable and cold-adapted.

Microorganisms found in permafrost were living in soil before freezing and were adapted to the previous climate of the region. Some may also have been brought from far areas by wind or precipitated with snow. It is not surprising to find psychrotolerant rather than psychrophilic microorganisms. The survival depends on frost resistance and not on cold adaptation. Electron microscopy *in-situ* showed cell aggregates in an extracellular matrix, which may be an additional survival mechanism. Soil itself may have a protective effect. However, the possibility of active growth in a liquid film at -10°C is questionable.[26] In permafrost (and cryptoendolithic habitats), extremely long generation time slows down evolutionary processes. This explains the apparent absence of successful adaptation in these cold terrestrial environments.[6]

8 Ice sheet and glaciers

Ice contains much fewer microorganisms than permafrost. The conditions are more drastic and survival is shorter. Unlike sea ice, where the incidence of microorganisms has been well documented,[4,10] inland ice was much less investigated. In Antarctica, the ice sheet of the polar plateau is permanently too cold to permit the occurrence of free water. However, deep within the ice sheet viable microorganisms have been detected.[27] Microorganisms transported on terrestrial dust and precipitation with snow become embedded in the ice formed from falling snow. Ice contains also spores, pollen and other organic material of biological origin. They may have been brought from far areas by wind and remain viable in anabiosis, but the conditions are more drastic than in permafrost and survival is shorter. Abyzov[27] studied the frequency of viable microorganisms at various depths in ice sheet cores from the Vostok Station. The frequencies of viable microorganisms in 500-4000 ml of subsamples decreased with depth:

- 20% in the samples from the superfical ice sheet (100 m depth, age 0–3000 years),
- 14 and 20% at depths of 200 and 300 m, respectively (approximate age 7400–12500 years),
- 6% between 300 and 1500 m,
- 3% between 1500 and 2405 m (age 100,000–200,000 years).

Scanning electron microscopy of filters through which sample water had been passed revealed microbial cells even in samples in which bacterial growth was not found. Viable yeasts were found only in the 0–100 m samples, fungi and bacteria (Gram-negative and Gram-positive, actinomycetes) were found till 320 m depth, spore-forming bacteria (mainly of the genus *Bacillus*) being the only microorganisms found in the deepest samples. Most of these microorganisms were not cold-adapted but mainly produced a thick capsule that can protect from freezing shock. In addition, *Pseudomonas* strains synthesize a pseudomelanine black pigment, and several actinomycetes and fungi produce melanine. This could be a protection from radiation and free radicals. Finally, the distribution of viable microorganisms depends on the depth and on the resistance of the organisms to unfavorable factors in air and in ice, and not on cold adaptation.

In contrast, glaciers flowing from the antarctic plateau in sheltered valleys and at lower altitude occasionally bear localized pockets of cryoconite pools where melt water is produced by the black-body radiation of superficial wind-blown debris.[4]

In alpine glaciers, ice is not so old because it is renewed. For instance, the glacier of Bossons near Chamonix (France) flows down at a mean rate of 200 m in one year.[28] Glaciers are covered with snow during winter. The upper part is an accumulation area where the snowfield turns into ice that flows into the valley, where the ice progressively melts. However, in summer, ice warms up more slowly than the surrounding soil because it reflects solar radiation and katabatic currents bring cold air. Microbial activity is confined in the surface layer where meltwater is supplied. Green algae and cyanobacteria are abundant on permanently melting snowfields in mid and late summer.

"Snow algae" are among the best cold-adapted microorganisms with very low optimum temperature for growth.[1] They concentrate on the surface giving a red coloration. Their spores give them resistance to warmth and dryness, and allow transport and new colonization of snowfields in winter. Their red pigment may have a protective function against excess of light and UV radiation. Yoshimura et al.[29] studied the distribution of chlorophytes and cyanobacteria on a Himalayan glacier. These organisms are more abundant in the surface snow than in deeper ice. The lower (ablation) area made of ice and water and receiving direct light is more favorable to photosynthetic microorganisms than the upper (accumulation) area where the freezing period is longer and algae receive diffuse light through snow and have less inorganic nutrients. Heterotrophic microorganisms associated with these photosynthetic communities were rarely studied.

Few bacteria and fungal spores are found in the deeper ice. They are only survivors and are not really cold-adapted.

9
Sediments of glaciers

The silt which accumulates in the front of the glacier after ice thawing provides a favorable environment for strictly psychrophilic bacteria (A.M. Gounot, unpubl. data; Table 1).[31,32] Close to the glacier, katabatic currents and melting ice maintain the soil of the moraine at a temperature near 0°C. Temperature never rises above 1°C, the silt is wet but freezes every night during summer and all the time during other seasons. Ice contains organic detritus such as pollen, spores, pieces of wood and leaves, which accumulate at the front of the glacier, especially if the glacier flows down far in the valley. Soluble organic matter is released. Aerobic heterotrophic bacteria were found which were oligotrophic. They often belonged to the coryneform family (*Arthrobacter*, *Corynebacterium*) and to the yellow-pigmented *Flexibacter-Bacteroides-Cytophaga* phylum. They grow slowly even on rich media but give an appreciable biomass on very diluted media. They produce high amounts of extracellular polysaccharides that retain nutrients and can be a protection from freezing and thawing damage. Some have an optimum growth temperature as low as 6°C, others grow well at 2 and 20°C, and grow not at 28°C, as shown in Table 1 and confirmed by screenings of isolated strains. Many strains produce carotenoid pigments. In some corynebacteria these pigments are photoinduced, and may be a protection from oxidation and radiation damage.

At some distance from the glacier, the soil of the moraine warms up and true psychrophiles are excluded. Most bacteria grow better at 20 than at 2°C. Gram-negative psychrotolerant bacteria, such as *Pseudomonas* spp., that are absent below the glacier, are frequently isolated. A soil recently covered by a neve (e.g. Plate desert; Table 1), rich in organic matter, contained high numbers of bacteria which gave colonies at 20 and 28°C but very slowly, and not at all at 2°C. They were probably non-active cells trapped in ice. On the contrary, living cells from the moraine (e.g. Svartis, Columbia) gave colonies more rapidly on the agar plates: they were recolonizing the soil.

The analyses of alpine or arctic samples help to understand the evolution of cold-adapted microorganisms. In the Plate desert, the neve was in a phase of transformation into a new small glacier. In sediment below the ice, microorganisms were numerous and psychrotrophic, still typical of an alpine soil. On the contrary, at the foot of old glaciers all microorganisms previously entrapped in snow and ice have died. Authochtonous psychrophilic microorganisms have developed, which are well adapted to this habitat because they are in the active phase. Then, after the thawing of the glacier, low numbers of bacteria are found in sediments until they are colonized by vegetation which enriches the organic nutrients, promoting growth of psychrotrophic bacteria.[32] Psychrophiles are eliminated by both competition and warmer temperatures.

10
Cold cave sediments

A particular cold environment is found in caves situated in cold areas.[33–35] The temperature of a cave is constant and equal to the mean value of temperatures outside the cave. Caves situated in subarctic or mountain areas have a constant low temperature near 0°C, without freeze-thawing cycles. This is favorable to the selection of cold-adapted bacteria. However the caves often receive rain, melted snow or stream water, which bring organic matter and many microorganisms from outside.

Table 1. Enumeration of viable microorganisms in sediment samples from cold continental environments
Colonies on soil extract agar plates were enumerated after incubation at 28°C (1 week), 20°C (3 weeks) and 2°C (5 or 6 weeks) under aerobic conditions (after A.M. Gounot, unpubl. data, and refs. 30,33)

Origin	Altitude (m a.s.l.)	Characteristics	Temperature (°C)	Number of microorganisms (millions g^{-1} dry weight)		
				2°C	20°C	28°C
Lapland (Norway)						
Svartisen icefield	200	Sediment below melting ice	0.5	3.9	4.2	0.5
		Sediment defrost 2 years ago	Variable	2.0	2.6	1.0
Reingardslivatn						
Laphullet Cave	400	Silt in a small cave not liable to flooding	2.5	0.6	2.0	0.6
Larshullet Cave	400	Sand in a cave liable to flooding by an ice-covered lake	2.9	5.0	6.0	3.5
Surface soil near Lapphullet cave	400	Sandy soil below humus	Variable	4.0	20	10
Alpes (France)						
Chamonix: glacier des Bossons	1250	Sediment below melting ice	0.5	0.2	0.7	0.1
Desert de Plate						
Neve	2227	Mud below melting snow and ice	1.0	1.0	54	27
Karstic hole (10m)	2050	Sand below snow	1.5	4.3	11	3.0
Vassieux-en-Vercors						
Grotte des Beaumettes	1050	Silt in a galery liable to flooding by melting snow	6.0	0.4	3.9	4.1
		Silt in a galery not liable to flooding	6.0	0.5	2.5	0.4
Pyrenees (France)						
Seintein: Grotte des Corneilles	2400	Clay not liable to flooding	0	1.5	1.8	0.3

Table 1 – continuation

Origin	Altitude (m a.s.l.)	Characteristics	Temperature (°C)	Number of microorganisms (millions g^{-1} dry weight)		
				2°C	20°C	28°C
Karpates (Romania)						
Scarisoara Cave	1150	Red clay	2.2	0.2	3.0	0.3
		Lumpy clay	3.7	0.04	1.0	0.06
		Yellow clay	3.6	0.3	2.0	0.3
		Calcitic sand	-0.8	0.4	1.0	0.1
Rocky Mountains (Canada)						
Columbia glacier (Alberta)	2130	Sediment below melting ice	1.0	0.06	0.1	0.03
		Lithosol below Dryas	Variable	0.8	10	3.6
Front Range (Alberta)						
Ice Cave, Plateau Moutain	2316	Mondmilch (degraded calcite)	1.0	0.03	0.1	0.03
Cougar Mountain (B.C.)						
Nakimu Cave	1676	Mondmilch not liable to flooding	3.0	0.1	0.2	0.06
		Silt not liable to flooding	5.5	0.2	0.4	0.07

On the contrary, pristine caves, without pollution by draughts or water flowing down from the surface, are extremely oligotrophic because the lack of light excludes primary production by photosynthetic organisms. Pristine groundwaters contain less than 1 mg dissolved organic carbon l^{-1}. Fungi cannot develop under these oligotrophic conditions. Most microorganisms isolated from pristine subsurface sediments are chemoorganotrophic bacteria. Heterotrophic life in these ecosystems is supported by organic compounds that filter down from the soil above. Readily metabolized compounds are consumed by the surface microorganisms. Subsurface bacteria are oligotrophic and able to metabolize the remaining more recalcitrant molecules, such as humic substances, hydrocarbons, or phenolic compounds. Appendages may increase the cell surface for the sorption of nutrients. Some bacteria produce a sheath and, like bacteria from moraine sediments, many of them produce extracellular polysaccharides.

The activity of chemolithotrophic bacteria is restricted by low temperature and low availability of their substrate, such as ammonia for nitrifying bacteria.The iron-oxidizing bacteria of the genus *Gallionella* are frequently observed in cold groundwaters. Autotrophic thiobacteria are usually found in hot thermomineral waters. However, high numbers of sulfur- and thiosulfate-oxidizers were found in samples from two cold caves of the Canadian Rocky Mountains after incubation at low temperature (A.M. Gounot, unpubl. data).

When the cave temperature was around 10°C, bacteria were found to be psychrotolerant.[33] Psychrophilic bacteria were isolated in colder caves in alpine and arctic areas (Table 1).[33–35] According to the temperature of incubation, it is possible to distinguish autochthonous microorganisms from those brought by flooding water. For instance, in the cold alpine cave of Beaumettes (temperature 5.5°C), flooded in spring by the melted snow, it was observed that in the flooded area of the cave the silt contained many bacteria growing at 28°C and not at 4°C, only few bacteria grew at 4°C. In the not flooded galery, the bacteria were less abundant. They grew at 4 and 20°C, but slightly or not at all at 28°C. The autochthonous cold-adapted community of the cave was hinded by the allochthonous bacteria in the flooded silt.

Cold adaptation of microbial communities was studied in two Lapland caves and in the surface soil close to them. The superficial soil contained many bacteria which probably developed during summer and were not cold-adapted. In the small Laphullet Cave the bacteria were psychrotolerant and resembled phenotypically those of the alpine cave with many coryneform bacteria. The Larshullet Cave had a typically cold-adapted microbial community. It was observed that most of isolated strains were psychrophilic, growing better at 4 than at 20°C. Most of them were Gram-negative yellow-pigmented bacteria usually found in aquatic environments. This was not surprising because the cave is seasonally flooded by water from an ice-covered lake.

On the contrary, many coryneform bacteria are found in cave sediments and they are not pigmented. Among 515 *Arthrobacter* strains isolated from caves, less than 1% was pigmented.[33]

In many caves the classical definition of psychrophiles and psychrotrophs is questionable. The number of colonies developed at 20°C is much higher than that at 2 and 28°C. The best example was found in the Scarisoara Cave (e.g. lumpy clay; Table 1). After isolation, bacterial strains were actually stenothermic, growing at 10 and 20°C, and not at all growing at 2 and 28°C. It seems that microbial evolution in old caves results in the selection of stenotherms. Their ability to grow at moderate temperatures is lost rapidly but capacity to grow at low temperature is acquired much more slowly. Consequently, their optimum temperature for growth is higher than the temperature of their habitat. In cold springs associated with caves in Southern Indiana (temperature of 10–12°C) Brock et al.[36] found that these isolates had growth optima of 25–30°C. The authors suggested that evolution of cold-adapted bacteria does not take place readily and that either longtime periods are necessary, or that cold habitats of wide extent, such as are found in alpine and polar regions, are necessary.

11 Conclusions

Most continental (terrestrial) habitats encounter large variations of temperature unfavorable to strictly psychrophiles. As temperature rises to 10°C or more in summer, psychrotrophic bacteria prevail over psychrophiles, especially if frost prevents microbial growth during winter. Thermal flexibility of psychrotrophs and rapid restarting of activity after thawing improve their competitivness. High nutrient availability favors psychrotolerant microorganisms even at low temperatures. It is not necessary for a microbe to function at optimal rates as long as it can compete effectively in its particular environment.[11]

Microorganisms trapped in ice or frozen sediments are in anabiosis. They cannot evolve because the selection of better-adapted mutants occurs only if they are growing. They remain alive if they are preadapted to frozen conditions. Bacteria living in cold soils submitted to seasonal freezing-thawing cycles are actually resistant to freezing. Many of them produce protective mucus. They are preadapted for survival in permafrost and ice where mucoid microorganisms are also found. This property is not necessarily correlated with the capacity for growth at low temperature.

However, true psychrophiles are found in some particular terrestrial habitats, such as sediment in cold caves, at the foot of glaciers, and cryoconite on melting ice, where low but not freezing temperatures and water availibity remain compatible with growth, permanently or for a long enough time. They become dominant when the conditions are more limiting: permanent low temperature and low nutrient avaibility in caves not liable to flooding, and, in addition, freeze-thaw cycles in moraine sediments at the foot of glaciers.

12 References

1. Baross JA, Morita RY. Microbial life at low temperatures: Ecological aspects. In: Kushner DJ, ed. Microbial Life in Extreme Environments. London: Academic Press, 1978:9-71.

2. Morita RY. Psychrophilic bacteria. Bacteriol Rev 1975; 39:144-167.
3. Herbert RA. The ecology and physiology of psychrophilic microorganisms. In: Herbert RA, Codd GA, eds. Microbes in Extreme Environments. London: Academic Press, 1986:1-23.
4. Wynn-Williams DD. Ecological aspects of Antarctic microbiology. Adv Microb Ecol 1990; 11:71-146.
5. Friedmann EI, ed. Antarctic Microbiology. New-York: Wiley-Liss, 1993.
6. Gilinchisky D, ed. Viable Microorganisms in Permafrost. Puschino: Russ Acad Sci, 1994:1-115.
7. Delille D. Biodiversity and function of bacteria in the Southern Ocean. Biodiversity Conservation 1996; 5:1505-1523.
8. Zhou J, Davey ME, Figueras JB, Rivnika E, Gilichinsky D, Tiedje JM. Phylogenetic diversity of a bacterial community determined from Siberian tundra soil DNA. Microbiology 1997; 143:3913-3919.
9. Shi T, Reeves RH, Gilinchisky D, Friedmann EI. Characterization of viable bacteria from siberian permafrost by 16S rDNA sequencing. Microb Ecol 1997; 33:169-179.
10. Bowman JP, McCammon SA, Brown MV, Nichols DS, McMeekin TA. Diversity and association of psychrophilic bacteria in Antarctic sea ice. Appl Environ Microbiol 1997; 63:3068-3078.
11. Russell NJ. Cold adaptation of microorganisms. Phil Trans Roy Soc Lond 1990; 326:595-611.
12. Gounot AM. Bacterial life at low temperature: physiological aspects and biotechnological implications. A Review. J Appl Bacteriol 1991; 71:386-397.
13. Omelchenko MV, Vasilyeva LV, Zavarzin GA. A psychrophilic methanotroph from tundra soil. Curr Microbiol 1993; 27:255-259.
14. Margesin R, Schinner F. Efficiency of indigenous and inoculated cold-adapted soil microorganisms for biodegradation of diesel-oil in alpine soils. Appl Environ Microbiol 1997; 63:2660-2664.
15. Mohn WW, Westerberg K, Cullen WR, Reimer KJ. Aerobic biodegradation of biphenyl and polychlorinated biphenyls by Arctic soil microorganisms. Appl Environ Microbiol 1997; 63:3378-3384.
16. Margesin R, Schinner F. Biodegradation of the anionic surfactant sodium dodecyl sulfate at low temperatures. Int Biodeterioration Biodegradation 1998; 41:139-143.
17. Conrad R, Bak F, Seitz HJ, Thebrath B, Mayer HP, Schütz H. Hydrogen turnover by psychrotrophic homoacetogenic and mesophilic methanogenic bacteria in anoxic paddy soil and lake sediment. FEMS Microbiol Ecol 1989; 62:285-294.
18. Kotsyurbenko OR, Nozhevnikova AN, Zavarzin GA. Methanogenic degradation of organic matter by anaerobic bacteria at low temperature. Chemosphere 1993; 27:1745-1761.
19. Kotsyurbenko OR, Simankova MV, Nozhevnikova AN, Zhilina TN, Bolotina NP, Lysenko AM, Osipov GA. New species of psychrophilic acetogens: *Acetobacterium bakii* sp. nov., *A. paludosum* sp. nov., *A. fimetarium* sp. nov. Arch Microbiol 1995; 163:29-34.
20. Wagner D, Pfeiffer EM. Two temperature optima of methane production in a typical soil of the Elbe river marshland. FEMS Microbiol Ecol 1997; 22:145-153.
21. Ferroni GD, Leduc LG, Todd M. Isolation and temperature characterization of psychrotrophic strains of *Thiobacillus ferrooxidans* from the environment of a uranium mine. J Gen Appl Microbiol 1986; 32:169-175.
22. DeLong EF, Wu KY, Prézelin BB, Jovine RVM. High abundance of Archaea in Antarctic marine picoplankton. Nature 1994; 371:695-697.
23. Vishniac HS, Klinger J. Yeasts in the Antarctic deserts. In: Megusar F, Gantar M, eds. Perspectives in Microbial Ecology. Proceedings of the 4th ISME, 1986:46-51.
24. Friedman EI. Endolithic microorganisms in the Antarctic cold desert. Science 1982; 215:1045-1053.
25. Cameron RE, Morelli FA. Viable microorganisms from ancient Ross Island and Taylor Valley drill cores. Antarct J US 1974; 9:113-115.

26. Vorobyova E, Soina V, Gorlenko M, Minkovskaya N, Zalinova N, Mamukelashvili A, Gilichinsky D, Rivkina E, Vishnivetskaya T. The deep cold biosphere: facts and hypothesis. FEMS Microbiol Rev 1997; 20:277-290.
27. Abyzov SS. Microorganisms in the Antarctic ice. In: Friedmann EI, ed. Antarctic Microbiology. New York: Wiley-Liss, 1993:265-295.
28. Vivian R. Fiches des glaciers Français. Glaciers des Bossons et de Taconna. Rev Geogr Alp 1969; 57:871-874.
29. Yoshimura Y, Kohshima S, Ohtani S. A community of snow algae on a Himalayan glacier: change of algal biomass and community structure with altitude. Arctic Alpine Res 1997; 29:126-137.
30. Gounot AM. Etude microbiologique de boues glaciaires arctiques. CR Acad Sci Paris 1968; (D)266:1437-1438.
31. Moiroud A, Gounot AM. Sur une bactérie psychrophile obligatoire isolée de limons glaciaires. CR Acad Sci Paris 1969; (D)269:2150-2152.
32. Moiroud A. Etude microbiologique des limons de deux glaciers alpins. I. Activité biologique à basse température. Rev Ecol Biol Sol 1970; 7:457-464.
33. Gounot AM. Recherches sur les bactéries cavernicoles. CR 96e Congr Nat Soc Sav Toulouse 1971, Paris. Bibliothèque Nationale 1973, Sci 3:257-265.
34. Gounot AM. Microbial ecology of groundwaters. In: Gibert J, Danielopol DL, Stanford J, eds. Groundwater Ecology. San Diego: Academic Press, 1994:189-215.
35. Gounot AM. Bacteria. In: Decu V, Juberthie C, eds. Encyclopédie Biospéologique. Moulis, Bucarest: Soc Biospéol, 1994; 1:359-370.
36. Brock TD, Passman F, Yoder I. Absence of obligately psychrophilic bacteria in constantly cold springs associated with caves in Southern Indiana. Am Midland Nature 1973; 90:240-246.

Lake ice microbial communities in alpine and antarctic lakes

R. Psenner[1*], B. Sattler[1], A. Wille[1], C. H. Fritsen[2], J. C. Priscu[2], M. Felip[3] and J. Catalan[3]

[1] Institute of Zoology and Limnology, University of Innsbruck, Technikerstrasse 25, A-6020 Innsbruck, Austria
[2] Department of Biological Sciences, Montana State University, Bozeman, Montana 59717, USA
[3] Department of Ecology, Faculty of Biology, University of Barcelona, Avgda. Diagonal 645, E-08028 Barcelona, Spain

1 Introduction

The observation that metabolic processes are reduced or completely inhibited at freezing temperatures has influenced microbial ecologists and limnologists, who have commonly concentrated on warm seasons and temperate habitats. High altitude and high latitude sites are not easily accessible and conducting research at low temperatures is not a trivial matter. However, methods for *in-situ* studies of microbial processes such as bacterial growth and production have been improved to a degree that allows the study of very oligotrophic systems at low temperatures. Furthermore, the construction of observatories and field stations has increased accessibility to high alpine and antarctic ecosystems, which has largely extended our knowledge of microbial life in cold environments.

Recent interest in cold ecosystems such as snow and ice has been stimulated by the recognition that global change will strongly influence these systems, and by the notion that ice dynamics in the oceans and inland waters can largely influence the emission of greenhouse gases.[1] The study of lake ice microbial communities, although their contribution to climate change is likely to be marginal, may foster the understanding of microbial communities in extreme environments, their functional role in the cycling of nutrients and organic matter, and their sensitivity to climate variations. The ice cover plays a major role in determining the environmental properties of lakes. It reduces heat exchange and prevents wind-generated mixing of the water column; this means that transport processes in the water column will eventually depend only on diffusion, thus ice influences not only the transfer of energy and matter between the atmosphere and the water, but within the lake

* Corresponding author; e-mail: roland.psenner@uibk.ac.at

itself.[2] Light transmission is severely limited by ice and especially by snow layers (by up to 99.9%) and atmospheric gases are trapped in ice-covered lakes which can result in severe oxygen deficiency[3] and strong oversaturation of CO_2,[4] N_2 and O_2.[5] Nutrient input from the atmosphere is also inhibited by the ice cover and sedimentation fluxes can become very low. Thus, the formation of an ice cover triggers a switch in a lake's physicochemical dynamics and may be the major factor limiting productivity in the pelagic zone of ecosystems with long-lived or perennial ice covers.

In addition to their role as physical barriers and limiting factors for water column processes, ice covers on alpine and antarctic lakes contain active microbial communities which will be at the center of interest in this chapter.

2 Formation and characteristics of lake ice

2.1 Seasonal ice covers on high mountain lakes

Ice sheets on high mountain lakes in the Alps and the Pyrenees, lasting for half a year or longer, are generally characterized by slush layers located between solid ice created through dynamic flood-freeze cycles. The seasonal ice cover starts in general in a cold autumn night with the formation of black ice, which is optically clear and hard with nearly no inclusion of air bubbles. Snowfall on the black ice pushes it downward and lake water infiltrates the base of the snow cover – which is an excellent thermic isolator – through cracks. Freezing of the infiltrated water can create a solid layer consisting of snow crystals and frozen lake water (called white ice) on the surface of the black ice. Incomplete freezing of the infiltrated snow and subsequent flooding and freezing creates a sandwich-like structure consisting of alternating layers of snow, white ice, and slush on top of black ice (Fig. 1). Slush may be best compared to a sandy sediment but with grains consisting of ice crystals. The dynamic changes in the physical structure of these seasonal ice covers are driven by snow deposition, melting, freezing and flooding[6] which is similar to those processes creating productive microbial habitats on the surface of sea ice in the Southern Ocean.[7,8] The seasonal ice cover becomes thinner and structurally weaker during spring. This and the formation and widening of cracks enhances gas exchange with the atmosphere and increases sedimentation and light transmission. The increase in transmissivity can enhance photosynthesis of phytoplankton and, consequently, bacterial activity in the water column. The influence of the ice cover on pelagic life may, however, not cease with the complete melting, because the microbial communities developing in the ice column may provide an inoculum to the water column and thus influence composition and development of microbial communities in the open lake as suggested by Andreatta[9] for lakes which are not very deep.

It is remarkable that the physical and chemical composition of the ice cover, though resembling a rigid structure, is constantly changing. Associated with physical transformations, the different layers undergo chemical changes that can lead to

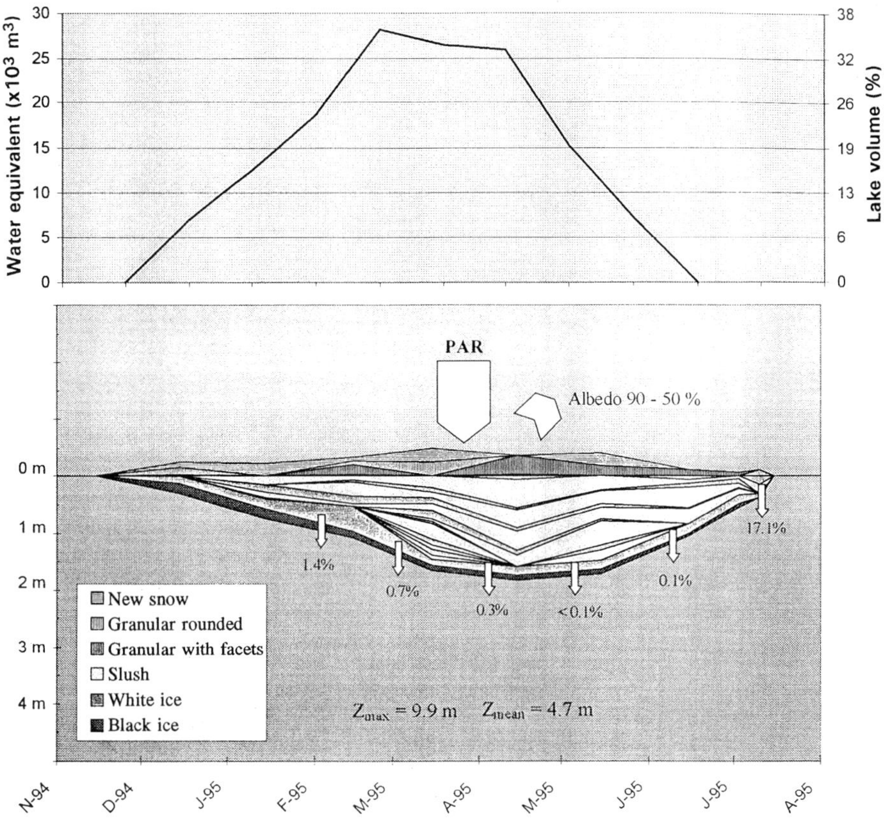

Fig. 1. Winter cover of a high mountain lake in the Alps (Gössenköllesee, Austria) (**Upper panel**) Water equivalent compared to the percentage of total lake volume (season 1994/95); (**lower panel**) Schematic representation of winter cover from ice formation until ablation with included slush layers. *Arrows* on top of the cover depict incoming photosynthetically active radiation (PAR) and outgoing albedo. Downlying arrows show percentage of radiation after passing snow and ice layers; Z_{max} maximum depth; Z_{mean} mean depth

either concentration or dilution of dissolved and particulate matter by partial melting, washout or entrainment of material from the atmosphere, the catchment and the lake. As a result, slush layers enriched in nutrients (phosphorus, ammonium and nitrate) are common.[6,10] At the end of winter, when light attenuation by snow decreases with thawing, light levels in the upper slush layers can be sufficient for net photosynthesis. In addition, seepage water from melting snowfields in the drainage area and atmospheric deposition can deliver organic matter and nutrients to the cover. The atmospheric input of pollen, conidia, leaves and stem debris, and of amorphous organic matter can become very high in springtime, when the lakes are still frozen but low altitude sites are free of snow. All these facts indicate that the ice and snow cover of lakes could support a microbial community, although the low temperatures – 0°C throughout the winter – may prevent them from reaching measurable activities.

2.2 Permanent ice covers on antarctic lakes

Aeolian deposition of sediments on the surface of the permanent ice covers is responsible for sorting and transporting lithogenic matter (primarily sand- and clay-sized particles) from the surrounding desert environments (soils, ephemeral streams, glaciers, etc.) onto the ice covers[11] that range in thickness from 3 to 20 m. These ice covers are cold (average temperature -10°C) and impermeable to materials for about 9 months out of the year. During the remaining 3–4 months of the annual cycle, the ice covers become relatively warm and isothermal at 0°C. Liquid water (a highly limiting factor for microbial proliferation in the polar desert) is generated near sedimentary aggregates in the ice due to the trapping of radiant energy by the relatively dark lithogenic and pigmented biogenic matter during this time.[8,12] Small-scale dynamics of the sediment aggregation and localized melting processes have not been investigated in detail, yet observations and preliminary analysis at scales of meters suggest that the 3 to 5-m-thick ice covers attain substantial liquid water contents[8] that makes them permeable to lithogenic matter and gases during the austral summer. The sedimentary aggregates are thought to remain near 2 m depth in the ice by the downward melting and migration during the summer months that is opposed by the relative movement of the ice upward (caused by ablation at the surface and freezing on the bottom).[13] The combination of these dynamic geophysical processes is responsible for the existence of seasonally transient liquid-water habitats that are characterized by the continual presence of ice surfaces, lithogenic sediments, low temperatures (0°C), annual freeze-thaw cycles, and extreme seasonal variability in solar radiation.

2.3 Comparison of lake ice structures

A comparison of mountain lake and antarctic lake ice structures is shown on Figure 2, which highlights similarities and differences in the habitats: the most obvious distinction is the layered structure of the ice covers in mountain lakes of the Alps and the Pyrenees, with a duration from 6 to 9 months, and the patchy distribution of sediments and water pockets in the Mc Murdo Dry Valley lakes within its permanent ice cover. These distinctions are primarily determined by the lower annual temperatures in the antarctic setting as well as the low amount of precipitation (below 100 mm) in the polar desert.

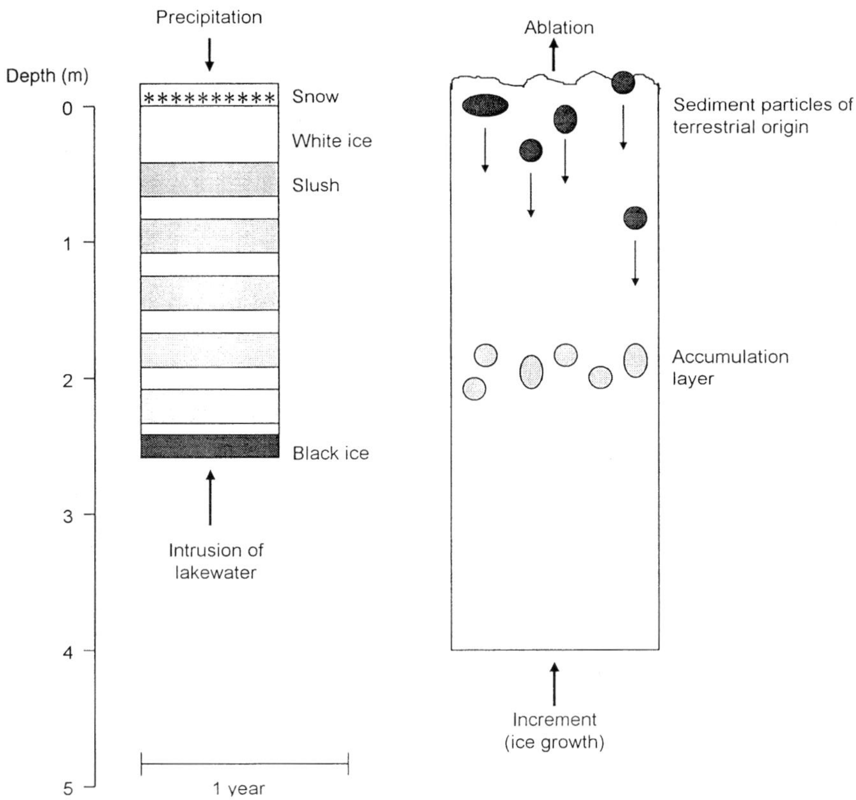

Fig. 2. Winter cover of high mountain lakes in the Alps and the Pyrenees (*left*) and permanent ice cover of antarctic lakes in the Mc Murdo Dry Valley region (*right*)

3
Composition, origin, morphology and activity of lake ice microbial communities

3.1
Alpine lake ice microbial communities

The infiltration of lake water into the ice cover transports organisms, inorganic nutrients and other lake water constituents into the snow and ice, where they often encounter more light than below the ice cover. Microorganisms found in slush layers, however, originate also from atmospheric and terrestrial sources. During the study of lake ice microbial communities (LIMCOs) we realized that fresh snow and supercooled cloud droplets collected at high altitudes contain active bacteria

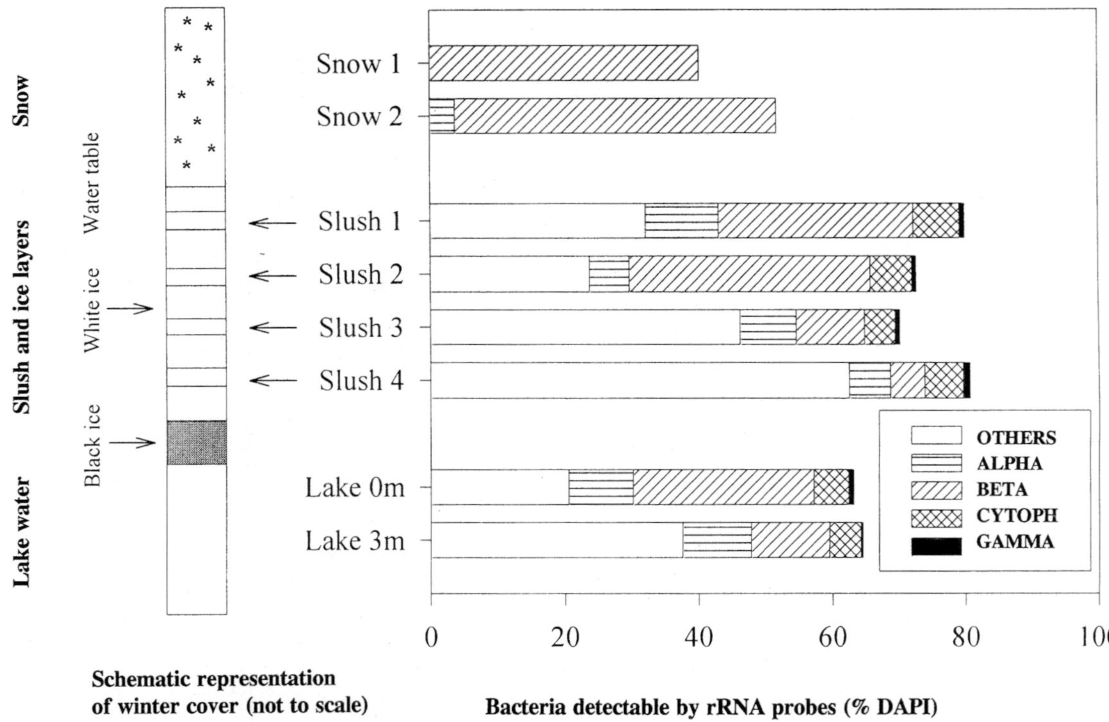

Fig. 3. Bacterial assemblages in snow and slush layers of the winter cover and two pelagic layers of a high mountain lake. Determination after *in-situ* hybridization with group-specific oligonucleotid probes (redrawn after ref. 14)
ALPHA complementary to the 16S rRNA (positions 19–35) of α-subclass Proteobacteria; *BETA* β-subclass Proteobacteria (23S rRNA, positions 1027–1043); *CYTOPH Cytophaga-Flavobacterium* group (16S RNA, positions 319–336); GAMMA γ-subclass Proteobacteria (23s rRNA, positions 1027–1043); *DAPI* staining with 4,6-diamidino-2-phenylindole

(unpubl. data), and Alfreider et al.[14] showed that the taxonomic composition of snow bacteria differed considerably from those found in slush layers or lake water (Fig. 3). These findings have led us to the assumption that microbial processes might not be restricted to lake ice habitats but that bacteria are metabolically active before the deposition of rain and snow. This hypothesis is supported by the findings of Puxbaum (unpubl. data), who showed that bacterial metabolites can be found in cloud droplets which may persist for several hours or days at temperatures below 0°C in a supercooled state. Thus, airborne bacteria may be the first organisms colonizing snow fields and may have an influence on the ecology of snow-covered remote environments such as high-altitude lakes and soils.

Bacterial communities are also morphologically diverse in slush layers and lake water. Water column bacteria are generally small free-living cells, mostly short rods and cocci (about 0.05–0.1 μm^3). In slush layers, however, together with these small forms, long thin filaments appear, sometimes longer that 100 μm, which are the dominant organisms regarding bacterial biomass. Long branched filaments and stalked bacteria are not uncommon, and in a few cases *Ancalomicrobium*-like morphologies are found. *In-situ* hybridization with fluorescent labeled rRNA-targeted oligonucleotid probes was carried out by using probes specific for all members of the domain bacteria, for the α-, β- and γ-subclasses of the class Proteobacteria, and for the *Cytophaga-Flavobacterium* group to obtain information concerning bacterial community structure. Results showed a distinct bacterial community composition in the different snow, slush and pelagic layers (Fig. 3).

Ciliates typical in high alpine lake water include *Askenasia chlorelligera*, *Urotricha* sp. and *Balanion planctonicum*, whereas slush layers are characterized by nonplanktonic genera such as *Urosoma*, *Lacrymaria* and *Dileptus* and some unidentified species. The only autotrophic species exclusive to the cover are those found in surface pools (*Chlamydomonas nivalis*, an unidentified *Chlamydomonas* sp., and an unidentified *Chromulina* sp.). Nonflagellates seldom appear in the slush layers. Other autotrophs found in slush are *Rhodomonas minuta*, *Gymnodinium* sp. and *Chromulina* sp. which are the dominant algae (Büsing, unpubl. data), but appear also in the pelagic zone of the lake. Felip et al.[15] have shown that the species composition (including phytoplankton, heterotrophic flagellates, and ciliates) of slush layers, surface pools and lake water in Estany Redó in the Pyrenees is clearly distinct in each case, and this difference persists in different years.

Bacterial production and respiration in slush layers generally increase with growth and development of the winter cover and range from values characteristic for ultraoligotrophic to eutropic lakes. Organisms living at low temperatures have higher nutrient requirements for sustenance of metabolic activities. Pomeroy et al.[16] and Wiebe et al.[17] found that marine bacteria living at 0°C can be substrate-limited as their growth was stimulated by the addition of nutrients and organic compounds. Felip et al.[18] pointed out that high growth rates can be observed at low temperatures if the amount of substrate is elevated. Therefore, high nutrient concentrations may be necessary in cold-water environments in order to compensate for the low temperature constraints on metabolic processes and growth. Experiments to determine temperature optima of water column and ice bacteria indicate that bacteria living in slush layers are adapted to temperatures from 0 to 5°C; pro-

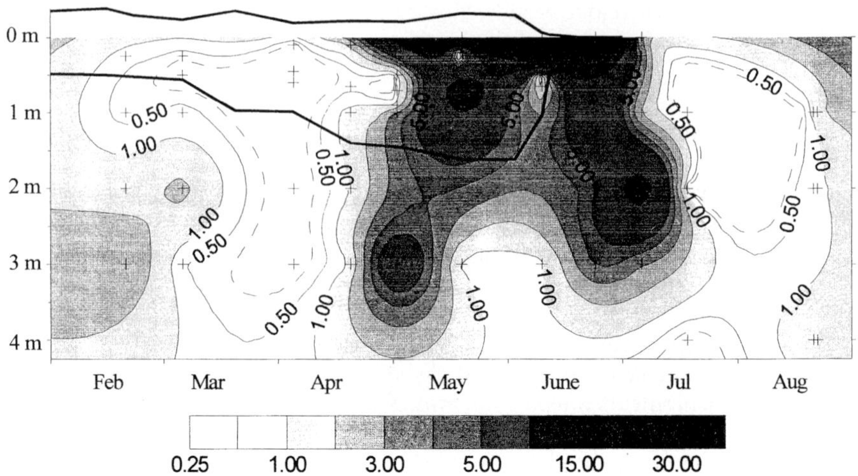

Fig. 4. Ratio of net primary production to bacterial production in slush layers of the winter cover (*bold line*) down to 4 m of the free water column in Gössenköllesee (Austria)

duction was decreased dramatically with rising temperatures. Lake water bacteria show an opposite trend, their preferred temperature range is near 15°C. Experiments indicate there is a taxonomic and physiological difference between the bacterial assemblages in the winter ice cover and those in the underlying lakewater (Fig. 3).

High nutrient concentrations in slush layers could lead to a stimulation of bacterial growth. Slush layers are inhabited by algae which support bacteria through extracellular release of carbon during high photosynthetic activity. From the formation of the ice cover to its decay we observed a shift from a system dominated by heterotrophs to an assemblage dominated by autotrophic organisms. This progression has consequences for the water column after ice break-up (Fig. 4).

Another effect favoring high microbial biomass inside slush can be the low density of grazers which may exert little grazing pressure. During such periods, microbial interactions in the ice cover are dominated mainly by bacteria and algae, thus being more similar to the antarctic lake ice than to the lake water itself. In a Pyrenean lake, however, Felip (unpubl. data) found higher numbers of heterotrophic nanoflagellates and ciliates, and also mixotrophic algae which may efficiently graze on bacteria. One possible reason for the low numbers of protistan grazers found in a lake in the Alps might be the reduced mobility due to the sediment-like structure of slush layers which can prevent effective grazing, while it is not known if the low temperature is a limiting factor for protistan activity. Sandwich-like structures and ice grains may also limit the access to metazoa, which were only occasionally observed in the winter cover of lakes in the Alps and the Pyrenees. The virtual absence of metazoa seems to be a common feature in antarctic lake ice but a remarkable contrast to the importance and densities of crustaceans in sea ice[19] even at very low temperatures. As mentioned before, ciliate species typical of sediments occur in the slush, which has led us to the assumption that the slush habitat resembles conditions in benthic strata or sandy sediments rather than in the pelagic zone.

3.2 Antarctic lake ice microbial communities

The community composition in the permanent ice covers on antarctic lakes also differs from the microbial communities found in the underlying water columns. The plankton within the liquid water column of permanently ice-covered antarctic lakes are typically composed of autotrophic eukaryotes (predominantly Chrysophytes, Cryptophytes, and Chlorophytes),[20,21] but also of cyanobacteria, bacteria[20,22] as well as heterotrophic flagellates and ciliates.[23] The phytoplankton are dominated by eukaryotic species that develop highly stratified distributions believed to be a result of preferential growth and survival of the different species along strong physicochemical gradients.[24] In contrast to the water column, lake-ice communities contain a mixture of heterotrophic bacteria, fungi, cyanobacteria, autotrophic flagellates and diatoms. Despite these ecologically functional members, the antarctic ice covers, the autotrophic biomass and activities are largely dominated by the filamentous cyanobacteria.[25–27] The filamentous cyanobacteria assemblages in the permanent ice covers can potentially originate from pelagic, benthic, atmospheric and terrestrial sources. However, they more closely resemble those genera (e.g. *Phormidium* and *Nostoc*) found in ephemeral streams and ponds, wetted soils in the surrounding desert, and on the lake's benthos. Cyanobacterial mats in these environments have been shown to be extremely resistant to desiccation and freezing[28,29] and are presumed to be capable of colonizing and growing within the extreme *in-situ* conditions of the permanent ice covers (e.g. long periods of freezing, freeze-thaw cycles, cold temperatures). Aeolian transport and deposition of sediment-associated cyanobacterial assemblages is likely to be the mode of introducing these assemblages into the ice covers.

Activities of microbes from several permanent ice covers has been examined at the time of ice melting. Active metabolic incorporation of radiolabeled substrates (e.g. ^{3}H-thymidine and ^{14}C-bicarbonate) was detectable within minutes to hours of encountering liquid water under laboratory conditions (i.e. irradiance >100 µmol photons m^{-2} s^{-1} and temperatures of 0–2°C) and enhanced bacterial and photoautotrophic biomass and activities were always found in ice meltwater containing ice-bound sediments.[30–32] Ice covers lacking substantial sediments (Lakes Vanda and Morning) had 10- to 1000-fold lower heterotrophic and photoautotrophic activities (Table 1). Note that the rates reported in Table 1 depict the depth-integrated values. However, the activity is presumed to have been restricted to the aggregates that were only 1–10 cm in diameter and were scattered throughout the ice covers. Therefore rates on a volumetric basis should be scaled upwards severalfold; this illustrates that microbial processes in these environments are likely to be highly concentrated.

The low ratio of bacterial production to primary production in the ice covers with sedimentary aggregates (Table 1) suggests that microbial assemblages in these ice covers possess the immediate capability for net community growth based on

Table 1. Standing stocks of antarctic lake-ice constituents (sediments, chlorophyll a and bacterial cells) and rates of primary production and secondary bacterial production in ice core meltwater from permanent ice covers on lakes in the McMurdo Dry Valleys, Antarctica

Lake	Sediment wt ($kg\ m^{-2}$)	Chl a ($mg\ m^{-2}$)	Bacteria[a] ($x10^{12}\ m^{-2}$)	BP ($\mu g\ C\ m^{-2}\ d^{-1}$)	PPR ($\mu g\ C\ m^{-2}\ d^{-1}$)	BP:PPR
East lobe Bonney	8.3	3.51	2.28	51.4	193	0.27
West lobe Bonney	1.4	1.33	1.11	27.8	27	1.02
Fryxell	8.9	7.42	6.00	504.1	1954	0.26
Hoare	8.8	5.25	2.12	235.3	2737	0.09
Joyce	7.2	5.08	0.76	56.5	580	0.10
Miers	11.4	22.9	7.43	989.2	33927	0.03
Morning	0.9	ND	3.66	43.3	ND	---
Vanda	0	1.20	0.62	5.9	2.9	2.07
Vida	22.2	5.05	10.0	138.9	1165	0.12

[a] Determined via epifluorescence.

Wt weight; *Chl a* chlorophyll a (determined fluorometrically)[28]; *BP* bacterial production (estimated from the incorporation of ^{3}H-thymidine into trichloroacetic-acid-insoluble material)[22]; *PPR* primary production (determined by the uptake of ^{14}C into particulate organic carbon)[26]; *ND* below detection autotrophic production of new biomass when solar radiation and liquid water becomes available within the ice.

Temperature records have shown that liquid water is not likely to exist in the ice covers until late November or early December.[8] Irradiances in the ice at this time are near their seasonal maximum and are well above the irradiance levels that saturate photosynthesis by the ice cyanobacteria. Therefore, community metabolism is likely to be supported by *in-situ* photoautotrophic production during the initial stages of the summer growth season. The seasonal progression of production and population dynamics of these communities throughout the summer growth season are unknown at this point in time.

4
Comparison between alpine and antarctic lake ice

A comparison between alpine and antarctic lake is given in Table 2 which summarizes the most important factors and processes leading to the formation of lake ice structures and microbial communities. Although antarctic lake ice is a permanent structure – in contrast to alpine lakes where the ice normally waxes and wanes in the course of 1 year – both systems are active sites of life for a period of only 5–7 months $year^{-1}$. Large differences exist also in the composition and the origin of organisms.

5
Conclusions and open questions

Is lake ice an ecosystem, an ecotone, or a habitat? In both alpine and antarctic lakes, it contains organisms which originate from the lake water, the atmosphere and the terrestrial catchment. These organisms form new assemblages in a physically complex and structured environment. The winter cover of mountain lakes can best be compared to a seasonally recurring habitat or a periodic ecotone. The occurrence of ciliates which are normally found only in benthic or terrestrial environments supports the idea of an ecotone linking atmospheric, aquatic and terrestrial systems. The relatively strict separation into different layers and the clear distinction of lake ice communities from those in surface pools and the pelagic zone[15] confirms the hypothesis that they are at least very distinct from plankton, although the first settlers in the ice cover immigrate from the pelagic zone. In the end, i.e. when the ice cover melts, they merge to form new pelagic biota. This does not normally happen in the ice cover on antarctic lakes, which can nonetheless be considered an intermittent habitat, though consisting of permanent ice, because liquid water and measurable life processes occur only during the austral summer. Both systems are microbial worlds. Whether antarctic lake ice is an ecotone or not is more debatable, but it allows bacterial life of mostly terrestrial origin to develop in a new environment, which would otherwise not happen because of the dry and cold conditions in the surrounding soils.

Table 2. Characteristics of alpine and antarctic lake ice

	Winter cover of alpine lakes	Antarctic lake ice
Location	High mountain lakes (Alps, Pyrenees, etc.)	Lakes in the Mc Murdo Dry Valley
Formation	Lake water penetrates through the ice cover; snow, rain and melting water trickles down; Annual precipition: 1,000–2,000 mm	Freeze-out of lake water; Annual precipitation: <100 mm
Thickness	1.5–3 m	3–6 m
Structure	Mixture of lake water, meltwater and rain water with snow and ice crystals; Sandwich-like structure	Uniformly hard lake ice with soil and rock particles surrounded by liquid water during summer
Duration	6–9 months: November through July	Permanent, but liquid water only in summer
Liquid water	10–30%; consists of lake water, rain and meltwater	About 10%, but only in summer around particles
Diffusion constant	Reduced: interstitial water, high tortuosity	Changing: melt-freeze intervals, sand clusters in water pockets
Radiation	Strong light gradient, from very high to <0.1%; UV-B radiation ca. 50% higher than at sea level	Strong light gradient, +/- constant radiation in summer; Low UV-B except during ozone hole events
Origin of organisms	Lake water, airborne, soils, littoral sediment	In-blown microorganisms from surrounding soils and long-range transport
Animals	No	No
Main feature	Intermittent (about 9 months); constantly at 0°C; Sandwich structure; "Sediment" of snow-ice particles; High bacterial production compared to lake water but reduced predation pressure; Microbial world: LIMCO (pro- and eukaryotes) originating from different sources	Permanent, but most time frozen: liquid water and microbial activity only in summer; Activity restricted to water pockets (patchy distribution); Microbial world consisting predominantly of cyanobacteria and heterotrophic bacteria; Eukaryotes are rare; Predation is almost absent

What our results clearly show is that temperatures at the freezing point do not inhibit the proliferation of microbial communities. Rather, active and diverse microbial communities develop in ice covers of distinct geographical regions of the world. The role of these communities in their respective ecosystems is still under study. In alpine lakes where the volume of the ice cover comprises a large portion of the total lake volume, the ice microbes may produce a substantial fraction of the ecosystem's biomass during extended periods of the year. In permanently ice-covered lakes in Antarctica, the contribution of the ice communities to the overall productivity of the lake ecosystem is relatively small.

The increase in UV-B radiation in the Antarctic as well as in the northern hemisphere is a new challenge for microorganisms in highly exposed environments. While pigmented organisms seem to be rather insensitive to UV radiation[34] it is still an open question how bacteria will react upon an increase of UV radiation. Therefore, while still at the beginning of the exploration of microbial communities in snow and ice,[10,15,24] we face a rapid change in those factors which are crucial for the formation, duration and physicochemical conditions of ice habitats, e.g. global warming, atmospheric deposition of acids and dust, and UV radiation.[35] We believe that it will be rewarding to investigate these habitats more intensively, for both a better understanding of ecological interactions under extreme environmental conditions and also for the investigation of microbial diversity and potential utilization of interesting traits.

Acknowledgements. The study of LIMCOs in alpine lakes is supported by the Austrian Science Foundation (FWF P-11685-MOB), the Acciones Integradas Hispano-Austriacas, and the EU project MOLAR (ENV4-CT95-0007). Investigations in Antarctica were funded by the US National Science Foundation (OPP 94-19423 and OPP 92-11773).

6 References

1. Arrigo KR, Worthen DL, Lizotte MP, Dixon P, Dieckmann G. Primary production in antarctic sea ice. Science 1997; 276:394-397.
2. Wharton RA, McKay CP, Clow GD, Anderson DT. Perennial ice covers and their influence on Antarctic lake ecosystems. In: Green WJ, Friedmann EI, eds. Physical and Biogeochemical Processes in Antarctic Lakes. Antarctic Res Ser, vol 59. Washington DC: Am Geophys Union, 1993: 53-70.
3. Psenner R. Alkalinity generation in a soft-water lake: Watershed and in-lake processes. Limnol Oceanogr 1988; 33:1463-1475.
4. Wögrath S, Psenner R. Seasonal, annual and long-term variability in the water chemistry of a remote high mountain lake: acid rain versus natural changes. Water, Air Soil Pollution 1995; 85:359-364.
5. Wharton RA, Jr, McKay CP, Simmons GM, Jr, Parker BC. Oxygen budget of a perennially ice-covered Antarctic lake. Limnol Oceanogr 1986; 31:437-443.
6. Catalan J. The winter cover of a high-mountain lake (Estany Redó, Pyrenees). Water Resour Res 1989; 25:519-527.
7. Meguro H. Plankton ice in the Antarctic Ocean. Antarctic Rec 1962; 14:192-199.
8. Fritsen CH, Adams EE, McKay CP, Priscu JC. Permanent ice covers of the McMurdo Dry Valley Lakes, Antarctica: Liquid water content. In: Priscu JC, ed. Ecosystem Dynamics in a Polar Desert: The McMurdo Dry Valleys, Antarctica. Antarctic Res Ser, vol 72. Washington DC: Am Geophys Union, 1998:269-280.
9. Andreatta S. Respiratoric activity of bacterial cells in a high mountain lake (Gossenköllesee, Tirol) during icebreak: quantification of tetrazolium-reduction image analysis. University of Innsbruck:Msc Thesis, 1997:1-53 (in German).
10. Jones HG, Quellet M. Mecanismes de translocation de matiére chimique et microbiologique dans la couverture de glace de quelques lacs. Eau Qué 1983; 16:71-80.

11. Squyres SW, Anderson DW, Neddell SS, Wharton RA Jr. Lake Hoare, Antarctica: Sedimentation through a thick perennial ice cover. Sedimentology 1991; 38:363-379.
12. Adams EE, Priscu JC, Fritsen CH, Smith SR, Brackman SL. Permanent ice covers of the McMurdo Dry Valley Lakes, Antarctica: bubble formation and metamorphism. In: Priscu JC, ed. Ecosystem Dynamics in a Polar Desert: The McMurdo Dry Valleys, Antarctica. Antarctic Res Ser, vol 72. Washington DC: Am Geophys Union, 1998:281-296.
13. Clow GD, McKay CP, Simmons RA. Climatological observations and predicted sublimation rates at Lake Hoare, Antarctica. J Climate 1988; 7:715-728.
14. Alfreider A, Pernthaler J, Amann R, Sattler B, Glöckner FO, Wille A, Psenner R. Commnity analysis of the bacterial assemblages in the winter cover and pelagic layers of a high mountain lake by in situ hybridization. Appl Environ Microbiol 1996; 62:2138-2144.
15. Felip M, Sattler B, Psenner R, Catalan J. Highly acticve microbial communities in the ice and snow cover of high mountain lakes. Appl Environ Microbiol 1995; 61:2394-2401.
16. Pomeroy LR, Wiebe WJ, Deibel D, Thompson RJ, Rowe GT, Pakulski JD. Bacterial responses to temperature and substrate concentration during the Newfoundland spring bloom. Mar Ecol Prog Ser 1991; 75:143-159.
17. Wiebe WJ, Sheldon Jr. WM, Pomeroy LR. Bacterial growth in the cold: evidence for an enhanced substrate requirement. Appl Environ Microbiol 1992; 58:359-365.
18. Felip M, Pace ML, Cole JJ. Regulation of planktonic bacterial growth rates: the effect of temperature and resources. Microb Ecol 1996; 3:15-28.
19. Kaufmann RS, Smith KL, Jr, Baldwin RJ, Glatts RC, Robison BH, Reisenbichler KR. Effects of seasonal pack ice on the distribution of macrozooplankton and micronekton in the northwestern Weddell Sea. Mar Biol 1995; 124:387-397.
20. Simmons GM, Vestal JR, Wharton RA. Environmental regulators of microbial activity in continental Antartic lakes. In: Green WJ, Friedmann, EI, eds. Physical and Biogeochemical Processes in Antarctic Lakes. Antarctic Res Ser, vol 59. Washington DC: Am Geophys Union, Antarctic Res Ser, 1993:165-195
21. Lizotte MP, Priscu JC. Pigment analsysis of the distribution, succession, and fate of phytoplankton in the McMurdo Dry Valley Lakes of Antarctica. In: Priscu JC, ed. Ecosystem Dynamics in a Polar Desert. The McMurdo Dry Valleys, Antarctica. Antarctic Res Ser, vol 72. Washington DC: Am Geophys Union, 1998:229-240.
22. Takacs CD, Priscu JC. Bacterioplankton dynamics in the McMurdo Dry Valley Lakes, Antarctica: Production and biomass loss over four seasons. Microb Ecol 1998 (in press).
23. James MR, Hall JA, Laybourn-Perry. Protozooplankton and microzooplankton ecology on lakes of the Dry Valleys, Southern Victoria Land. In: Priscu JC, ed. Ecosystem Dynamics in a Polar Desert. The McMurdo Dry Valleys, Antarctica. Antarctic Res Ser, vol 72. Washington DC: Am Geophys Union, 1998:255-267.
24. Priscu JC. Phytoplankton nutrient deficiency in lakes of the McMurdo Dry Valleys, Antarctica. Freshwater Biol 1995; 34:215-227.
25. Wing K, Priscu JC. 1993. Microbial communities in the permanent ice cap of lake Bonney, Antarctica: Relationships among chlorophyll-a, gravel, and nutrients. Antarctic J US 1993; 28:247-249.
26. Fritsen CH, Priscu JC. Cyanobacterial assemblages in permanent ice covers of Antarctic lakes: Distribution, growth rate and temperature response of photosynthesis. J Phycol 1998; 34:587-597.
27. Paerl HW, Priscu JC. Microbial phototrophic, heterotrophic and diazotrophic activities associated with aggregates in the permanent ice cover of Lake Bonney. Microb Ecol 1998 (in press).
28. Holm-Hansen O. Viability of blue-green andgreen algae after freezing. Physiol Plant 1963; 16:530-540.
29. Hawes I, Howard-Williams C, Vincent W. Dessication and recovery of Antarctic cyanobacterial mats. Polar Biol 1992; 12:587-594.

30. Fritsen CH, Priscu JC. Photosynthetic characteristics of cyanobacteria in permanent ice-covers on lakes in the McMurdo Dry Valleys, Antarctica. Antarctic J US 1996; 31:216-218.
31. Priscu JC, Fritsen CH. Antarctic lake ice microbial consortia: origin, distribution and growth physiology. Antarctic J US 1996; 31:223-224.
32. Priscu JC, Fritsen CH, Adams EE, Giovannoni SJ, Paerl HW, McKay CP, Doran PT, Gordon DA, Lanoil BD, Pinckey JC. Perennial Antarctic lake ice: An oasis for life in a polar desert. Science 1998; 280:2095-2098.
33. Hobbie JE, Daley RJ, Jasper S. Use of nuclepore filters for counting bacteria by fluorescence microscopy. Appl Environ Microbiol 1977; 33:1225-1228.
34. Sommaruga R, Obernosterer I, Herndl GJ, Psenner R. Inhibitory effect of solar radiation on thymidine and leucine incorporation by freshwater and marine bacterioplankton. Appl Environ Microbiol 1997; 63:4178-4184.
35. Sommaruga-Wögrath S, Koinig KA, Schmidt R, Sommaruga R, Tessadri R, Psenner R. Temperature effects on the acidity of remote alpine lakes. Nature 1997; 387:64-67.

Physiology of cold-adapted microorganisms

A. M. Gounot[1*] and N. J. Russell[2]

[1] Laboratory of Microbial Ecology, UMR 5557, University Claude Bernard Lyon 1, Bat 405, 43 boulevard du 11 Novembre 1918, F-69622 Villeurbanne-Cedex, France
[2] Department of Biological Sciences, Wye College, University of London, Wye, Ashford, Kent TN25 5AH, United Kingdom

1 Introduction

Microorganisms able to grow at low temperature, that is near 0°C, have been known for over 100 years in natural and manmade cold environments such as marine and continental waters, soils and foods. Until recently, the molecular basis of cold adaptation was poorly investigated. Most studies have been pragmatic, dealing with food microorganisms (mainly bacteria) with the aim of preventing their growth. During the late 1960s and 1970s, many papers were published on the ecology as well as the physiology of cold-adapted microorganisms. Most are now seldom referred to or even considered by those working in the field of cold adaptation. Except for lipid composition, more fundamental research was restricted by the lack of tools for such investigations. Moreover, most studies dealt with the heat sensitivity of cold-adapted microorganisms, which, although being of interest, is a consequence not the cause of cold adaptation. Others considered the effect of chilling on mesophiles such as *Escherichia coli*, which is mesophilic and cannot grow at 0°C, and cannot be considered as a universal model for organisms growing at low temperatures.

Recently, it has been appreciated that cold-adapted microorganisms could be useful in biotechnology, because they represent a valuable source of enzymes in processes operated at low temperatures.[1–7] Cold-adapted microorganisms are also models for investigating how biomolecules remain active at low temperature, particularly those which are strictly adapted to cold environments and can be considered as extremophiles. New research tools of molecular biology are now available for investigations on the physiology of these organisms, which together with the realization of their biotechnological potential has led to a stimulation of interest in and work on cold-adapted bacteria in particular.

* Corresponding author; e-mail: gounot@biomserv.univ-lyon1.fr

The ability to grow near 0°C is a characteristic of both psychrotrophic and psychrophilic microorganisms. Psychrotolerants (usually referred to as psychrotrophs in the food industry) usually only grow at temperatures approaching 0°C and grow optimally at temperatures around 20–25°C and may have upper limits as high as 40°C. In contrast, true psychrophiles are specifically adapted to low-temperature growth, usually below 0°C and have optimal growth temperatures below 15°C and upper limits below 20°C.[8] Species of snow algae are among the microorganisms reported to have the lowest optimum temperatures for growth based on cell counts to measure growth (but see discussion of "optimum" below). Although their growth rates are relatively slow compared with cold-adapted bacteria, some snow algae have apparent optima of 1–10°C and upper limits of 10°C.[9] Growth of yeast has been reported at subzero temperatures as low as -18°C.[10]

Recent general reviews on the physiology of cold-adapted microorganims are available.[4,5,11,12] In this chapter we discuss those factors which can preclude bacterial growth near 0°C (chilling but not freezing low temperature) in mesophiles and how psychrotolerant and psychrophilic bacteria are adapted for permanent growth at low temperatures near 0°C.

2
Effects of temperature on growth rate

The effect of temperature on bacterial growth is well described by the Arrhenius equation in which the log of growth rate is plotted as the reaction rate constant versus the reciprocal of the absolute temperature (i.e. degrees Kelvin). However, at temperatures near the upper or lower limits, such plots deviate from linearity, and below a minimum, no growth occurs at all. The linear range of the Arrhenius plot corresponds to a physiologically normal temperature for growth. A shift from a normal to a subnormal or supranormal temperature results in a lag phase (when growth ceases) and then growth re-starts at a lower rate than the theoretical. Temperatures outside the linear range of Arrhenius plot are stress-inducing temperatures. For psychrophiles, Arrhenius plots remain linear down to 0°C, whereas in psychrotolerants they deviate from linearity between 5 and 10°C, and in mesophiles at about 20°C.[13–16] At low temperatures, the growth rate of psychrophiles and psychrotolerants remains higher than that of mesophiles. It was suggested that the activation energy of the reaction is lower in psychrophiles;[17] however, this was not observed in several bacteria[8] and it is not regarded now as being diagnostic of psychrophiles.

The observed growth rate of a bacterial culture results from the conjunction of many complex, interacting mechanisms at the molecular level. Studies on *E. coli* and some other mesophilic bacteria have provided an understanding of what prevents growth at low but nonfreezing temperatures. In *E. coli*, a sudden down-shift of growing cells from 37 to 3°C results in the release of endogenous solutes and a rapid loss of viability.[18] The growth inhibition and decline of viability was attributed to a change in membrane permeability, correlated with the decrease of lipid fluidity[19] and a block of protein synthesis, specifically the initiation of translation.[20,21]

3 Metabolic activity

At low temperatures, the rate of enzymic reactions and the affinity of uptake and transport systems decrease more rapidly in mesophiles than in homologous psychrotolerant or psychrophilic strains. Many observations have been reported showing that cold-adapted bacteria express their activities, such as the production of exoenzymes, at low temperature. Proteinases in particular have been studied extensively in bacteria isolated from antarctic sites[22] and glaciers.[23] There have also been many studies of refrigerated food where the production of proteinases by psychrotolerant bacteria has drastic consequences for the quality of chilled foods (e.g. see ref. 24).

In psychrotolerant bacteria, the synthesis and/or activity of enzymes is often maximal near 17–20°C, i.e. at a temperature below the optimum for growth, which is about 30°C. This was observed for the production of proteinase in a *Pseudomonas* sp.[25–27] and in many other bacteria,[23,24] of a lipase in *Moraxella* sp.,[28] a lipase and two acid phosphatases in *Pseudomonas fluorescens*,[29] an amylase in a deep-sea psychrotolerant *Vibrio* sp.[30] and in *Alteromonas haloplanctis*,[31] as well as for enzymes involved in electron transport and coupled phosphorylation in a *Pseudomonas* sp.[32]

However, the definition of the "optimum" temperature for growth is not a physiological one. It has usually been defined as that temperature giving the highest growth rate, while less commonly growth yield is used. Moreover, it is generally measured for a short period only, regardless of the physiological state of the cells at the end of the experiment. More careful observations show that this supposed optimum occurs above the linear part of the Arrhenius plot.[33] This means that the cells are not in their optimal physiological state.[34] In various bacteria isolated from antarctic sites and tested for different exoenzyme activities, Feller et al.[22] observed that temperatures higher than those experienced in the natural environment (-2 to 4°C) gave faster growth rates, but concomitantly also markedly reduced culture growth of most strains, as measured by viable counts at the stationary phase. This has been also observed in a psychrophilic *Arthrobacter glacialis* (now redesignated as *Corynebacterium*; A.M. Gounot, unpubl. data).

In fact, the measurement of enzymic activity *in vivo* depends on protein synthesis and its regulation, enzyme maturation and stability, enzyme secretion (for exoenzymes) or substrate uptake (for intracellular enzymes). In a study by Feller et al.,[35] the yield of an extracellular proteinase at 4°C was almost double the amount at 25°C, despite the slower growth rate at the lower temperature. The production of inducible enzymes may be linked to the transport of the inducer or its precursor into the cell, and Orange[36] and Dé et al.[37] reported that the induction of β-lactamase in *Pseudomonas fluorescens* is modulated by temperature-dependent permeation of the inducer through the outer membrane.

The thermal effects on all these concomitant processes, involving both protein flexibility and lipid fluidity, result in an apparent optimal activity of enzymes that is about 20°C lower than that of their mesophilic counterparts. This point is discussed further below, in relation to cold-active enzymes.

4
Lipids and membrane stability

When considering the roles of lipids in membrane stability, there are two essential facets: the fluidity and the phase of the lipid continuum within the membrane. Both are important for the maintenance of such membrane functions as passive permeability and electron transport[38] including the control of proton permeability for which the term homeo-proton permeability adaptation has been proposed recently.[39] Membrane fluidity has received almost excessive attention, whereas membrane phase is usually ignored! An explanation may be that the normal phase is a bilayer (lamellar), although the propensity to form nonbilayer phases may be important for helping to drive membrane growth and segregation during cell division. Lipid fluidity is most influenced by the fatty acyl moieties, whereas lipid phase depends more on the nature of the head-group of the membrane lipids.[38] The gel to liquid-crystalline transition as well as the bilayer (lamellar) to nonbilayer phase transition are influenced by (growth) temperature, which usually triggers greater changes in the fatty-acyl composition than the head-group composition of membrane lipids. However, apart from the study by Finegold et al.[40] on the behavior of lipids extracted from an endolithic lichen community living beneath rock surfaces in Antarctica, there are no reports of lipid phase behavior specifically in relation to cold adaptation. Therefore, the following discussion focusses on membrane fluidity and its regulation via alterations in temperature-dependent changes of fatty acid composition.

The most frequent changes recorded are in the extent of unsaturation (saturated, monounsaturated or polyunsaturated), methyl branching and acyl chain length of the fatty acyl components of membrane lipids; there are also differences in the type of branching (iso or anteiso) and double bond (cis or trans). All of these changes influence the way in which the membrane lipids pack within the membrane by introducing steric constraints and modifying the degrees of freedom for molecular motion, i.e. they affect both the order and viscosity of the lipid bilayer, which is summarized in the combined terminology membrane fluidity.[38] For simplicity, this trivial term is used in the following discussion of thermal effects on bacterial membrane lipids, which are summarized in Table 1. It was proposed that the fatty acid changes after a temperature shift restored membrane fluidity to its pre-shift value, the "homeoviscous adaptation theory".[41] However, whilst membrane fluidity is important, it is probably unnecessary for exact fluidity levels to be maintained for bacterial growth to continue; indeed, *E. coli* can grow with 20–55% of its lipids in the gel state.[42]

When the growth temperature of a bacterial culture is altered, it has two effects with regard to lipid composition. One is a direct one on the physical properties of

Table 1. Fatty acid changes which increase or decrease membrane fluidity in bacteria

Increase fluidity		Decrease fluidity
Unsaturation	↔	Saturation
Cis double bond	↔	Trans double bond
Chain shortening	↔	Chain lengthening
sn-1-long, sn-2-short	↔	*sn*-1-short, *sn*-2-long[a]
Methyl branching	↔	Straight chain
Cis-unsaturation	↔	Cyclopropane

[a] Acyl chain "tailoring" (i.e. switching residues between the *sn*-1/2 positions of the lipid glycerol backbone) is included for completeness but, although "*Micrococcus cryophilus*" shows isomeric preference,[5,43] there are no examples in cold-adapted bacteria of this mechanism being used in cold adaptation

the lipid bilayer, making it more or less fluid for increases or decreases in temperature respectively; the other is on the enzymes of fatty acid biosynthesis, which leads to alterations in either the acyl chains of existing membrane lipids (fast changes) or in the pattern of fatty acid synthetase products for subsequent incorporation into new membrane lipids (slow changes; for a discussion see ref. 43). The basis of fast and slow changes lies in their metabolic origins, which to some extent depend upon the bacterial type.

In most bacteria, fatty acids are synthesized by a dissociable multienzyme complex [a Type II fatty acid synthetase (FAS), see ref. 44] in which the proteins are arranged around a central acyl carrier protein (ACP) molecule to which intermediates are covalently bonded; some bacteria have a so-called Type I FAS composed of large multifunctional protein to which ACP is covalently bound. However, for both systems the primary products are saturated fatty acids which are either straight-chain or branched-chain depending upon the type of primer molecule which is used.[44] This system does not synthesize unsaturated fatty acids which instead are made as secondary products of an aerobic desaturation system.[43] The desaturase enzymes are part of a complex within the membrane and the substrate for desaturation is probably the preformed acyl chains of membrane lipids.[43]

Some bacteria use another mechanism to make fatty acids, the so-called anaerobic system in which a modified fatty acid synthetase system produces a mixture of saturated and monounsaturated fatty acids. This is a soluble cytoplasmic multienzyme complex. Until recently, it was believed that the two pathways were mutually exclusive, but some exceptions have been discovered recently (e.g. see ref. 45). One class of bacteria which live at low temperatures are those inhabiting sea ice and other cold pelagic habitats, which are characterized by having not only the more usual complement of saturated and monounsaturated, but also polyunsaturated fatty acids,[46,47] which are regulated by temperature.[48,49] It appears that these organisms also possess two biosynthetic mechanisms, because they make the monounsaturated fatty acids using an anaerobic system but the polyunsaturated fatty acids by an aerobic desaturase mechanism.[50] Some antarctic and deep-sea bacteria, as well as a range of other bacteria, also contain trans-unsaturated fatty acids.[51] These are formed by the direct isomerization of cis-unsaturated substrates, probably membrane lipids.[52]

Temperature influences these biosynthetic processes differently: some are regulated by (de)activation of the enzyme protein(s), others by enzyme induction.[43] It is the nature of the thermosensor (or "molecular thermometer") and the biochemical mechanism of the change which together determine whether the response is fast or slow. Temperature-triggered changes that are mediated via desaturases can occur quickly because the substrates are usually existing membrane lipids; the same is true of formation of cyclopropane fatty acids (although there is little evidence for them being involved in temperature regulation) and for the conversion of cis- to trans-unsaturated fatty acids. In contrast, alterations in branching or unsaturation by the anaerobic pathway must be mediated by *de novo* fatty acid (and acyl lipid) synthesis, followed by integration into the membrane. The same is probably true of acyl chain shortening/elongation, which even if it occurred by "tailoring" of existing acyl chains would still involve their removal from the membrane lipid since the C_2-units are added/removed at the end of the molecule attached to the glycerol backbone of the membrane lipid; unless there is lipid turnover or an increase in the membrane lipid/protein ratio, the process would be a slower one. Evidence in several bacteria of increased membrane acyl-chain turnover following temperature down-shift has not been found (N.J. Russell, unpubl. data). Despite the large body of work on membrane lipid adaptation to temperature, it is still unclear just how the alterations in pattern of fatty acids synthesized and their incorporation in membrane lipids are linked to the regulation of membrane fluidity.[53] For example, the feedback mechanisms which modulate alterations in fatty acid synthetase activity once membrane fluidity has been restored are quite unknown; nor has the regulation of membrane enzymes such as desaturase been identified other than such vague comments that it is mediated via changes in lateral pressure of the membrane. It is a reasonable guess that two-component regulatory sytems will be involved and that there will be a link to the cold-shock and cold-acclimation systems (see below). In this respect it is significant that recently the product of a gene which is induced by cold shock in *Bacillus subtilis* has been identified as a membrane desaturase.[54] The same is probably true of the desaturase in other bacilli (e.g. see ref. 55).

5
Protein flexibility and cold-adapted enzymes

5.1
Effects of cold on proteins

Cell survival and activity depend on the maintenance of the structural and functional integrity of proteins, which in turn is derived from the sum of a many weak, noncovalent polar/ionic and hydrophobic interactions between the carbon skeleton of the polypeptide and the side-chains of amino acids.[56] For soluble proteins in aqueous solution, most hydrophobic regions are found as α-helices and/or β-sheets within the protein core, which has low hydration and high density. Although there are polar (including ionic) interactions on the surface, it is internal hydrophobic interactions which are mainly responsible for protein stability. Other hydrophobic amino acids are involved in domains that may be exposed to solvent or are shielded via quaternary interactions between polypeptide subunits of multimeric enzymes.[57] When the temperature is lowered, the strength of polar/ionic interactions is heightened but that of hydrophobic interactions decreases. Since it is the latter which are mainly responsible for stability, this results in the breaking of the compact structure and dissociation of the protein subunits, so that the nonpolar regions of the enzyme may become exposed to water. This can affect all cell proteins: enzymes and protein factors involved in macromolecular syntheses and regulation at the transcriptional and translational levels, metabolic enzymes and structural proteins.

In mesophilic bacteria, the inhibition of growth by cold can be due to a conformational change resulting in an inactivation or even a denaturation of enzymic or structural cell proteins. For instance, storage at 3°C noticeably diminishes the proteinase activity of a culture of *Streptococcus lactis*, probably due to the formation of an inactive trimer of the enzyme, but this does not affect bacterial viablity as it is not an essential component as long as other growth substrates are present.[58] The NAD^+-specific glutamate dehydrogenase of *Bacillus cereus* DSM31 is cold-labile and loss of activity in the cold was found to be due to dissociation of the holoenzyme into catalytically inactive subunits, which was reversible under certain conditions.[59]

The phenomenon of cold denaturation of proteins has been discussed fully by Privalov[60] and the proteins of cold-adapted bacteria must be folded in such a way as to not only retain activity at low temperatures but also to avoid denaturation. Thus, there is a balance between achieving conformational flexibility whilst retaining stability at low temperatures.

5.2 Cold-adapted enzymes

The overall catalytic cycle of an enzyme can be divided simplistically into three main phases: (1) recognition and binding of substrate, mediated by a number of amino acids within the enzyme; (2) conformational stresses induced by changes in the active site, leading to transition states and the formation of product; and (3) the release of product from the enzyme. Each of these phases involves weak bonding interactions that are sensitive to temperature.[61] Given these complexities, it is not surprising that different enzymes should have evolved with different answers to achieving low temperature activity and stability. The thermal characteristics of different enzymes within a bacterial cell may vary considerably and some will be more thermolabile than others. Some enzymes are very thermolabile and are probably a major determinant of the upper growth temperature limit of an organism. The block of protein synthesis in *Micrococcus cryophilus* (now redesignated as *Psychrobacter uratovorans*) results from the inactivation of certain thermolabile amino-acyl-tRNA synthetases which occurs in 10 min at 30°C.[62] The purified lactate dehydrogenase of the psychrophilic *Vibrio marinus* has apparent maximal activity at 10–15°C, but no activity at 40°C,[63] whilst the enzyme from an antarctic soil bacterium which has apparent maximal activity at 20°C loses its activity within 20 min at 50°C (J. Diez-Aguirre and N.J. Russell, unpubl. data). The isocitrate dehydrogenase of a psychrophilic *Vibrio* sp. contains a dimeric isoenzyme I which is relatively thermostable, and a monomeric isoenzyme II, which denatures above 15°C but is reactivated at 0°C.[64]

Compared with those from mesophiles, the purified enzymes obtained from cold-adapted bacteria are more active at low temperatures, retaining significant activity even down to 0°C and having a lower temperature optimum for activity. Although the latter parameter has no formal place in enzyme kinetics, it is a useful measurement for practical, comparative purposes. A lower activation energy was required for substrate hydrolysis compared with the corresponding enzymes of mesophiles. For instance the purified amylase of a psychrotolerant *Vibrio* sp. has an apparent optimum activity at 35°C, and retains 40 and 10% of its activity at 10 and 0°C.[65] A similar observation was made for the α-amylase of *Alteromonas haloplanctis,*[31] and the lactate dehydrogenase of an antarctic bacterium which retains 47% of its activity at 1°C (J. Diez-Aguirre and N.J. Russell, unpubl. data). Only recently has "true" (i.e. not modelled from gene sequences) three-dimensional (3D) structural information on cold-active enzymes from psychrophiles been available with the crystallization of first α-amylase,[66] then citrate synthase[67] and triosephosphate isomerase.[68] The molecular basis of cold activity and other aspects of enzymes from psychrophiles which function in the cold are dealt with more fully elsewhere in this book (see Chapter by Gerday et al.).

6
Heat sensitivity and heat-shock proteins

Cold-adaptation generally results in heat-sensitivity, and an increase in temperature above the relatively low growth optima of psychrotolerant and psychrophilic bacteria affects them in various ways. For instance, there is a block in protein synthesis, a leakage of intracellular substances and often a rapid loss of viability above 20°C in psychrophilic bacteria (e.g. *Pseudomonas* sp.),[14] *Vibrio marinus*,[8] and above 25 or 30°C in psychrotolerant bacteria (e.g. "*Micrococcus cryophilus*"[62] and *Arthrobacter* SI55[15]). The permeability changes have been attributed to a breakdown in barrier function of the membrane lipids which become hyperfluid at supramaximal temperatures, resulting in the indiscriminant leakage of intracellular molecules (for a discussion see ref. 12). The synthesis of cell wall components is also disturbed and changes in cell wall structure, leading to filamentation, have been described in *Bacillus psychrophilus, Aquaspirillum*[69] and *Arthrobacter globiformis*;[70] the formation of clumps of coccoid cells was observed in "*Arthrobacter glacialis*".[70] A number of metabolic changes may contribute to cell death, including the activation of some catabolic and respiratory enzymes that exhausts cellular reserves and leads to loss of viability. In some bacteria, the enhancement of autolytic enzyme activity also contributes to rapid death.

Like the enzymes from temperate organisms, those from cold-adapted bacteria usually remain active at temperatures well above the upper limit for growth and have apparent optima for activity (i.e. often between 30 and 40°C) that may be as much as 20°C higher than the optimum for growth.[22] However the enzymes from psychrotolerant and psychrophilic bacteria have reduced stability compared with their temperate counterparts, their cold-adapted conformations making them relatively heat sensitive. For example, the fructose-1,6-biphosphate aldolase of *Vibrio marinus*,[71] lactate dehydrogenase from the psychrotolerant *Bacillus psychrosaccharolyticus*,[56] β-galactosidase from a soil psychrotolerant *Arthrobacter* sp.,[72] proteinases from glacier bacteria,[73] and many enzymes from antarctic bacteria including alkaline phosphatase and ribonucleases,[74–76] triosephosphate isomerase,[77] α-amylase, lipase, exoproteinase and β-lactamase[22] and dehydrogenases for lactate (J. Diez-Aguirre, unpubl. data), alcohol (M. Stibor, B. Králová and N.J. Russell, unpubl. data) and glutamate (L. Camardella and G. di Prisco, unpubl. data) all exhibit marked heat-sensitivity compared with the corresponding enzymes from mesophiles.

When the cells of all animals, plants and microorganisms are exposed to temperatures above the normal for growth, they display a transient synthesis at high rates of proteins that are known as heat-shock-proteins (Hsps).[78] The primary structure of most Hsps appears to have been highly conserved during evolution, suggesting that they serve similar functions in all organisms. Hsps are normally present inside cells where they mediate such functions as molecular chaperones, mediating the correct folding and assembly of polypeptides, and destroying (by proteolysis) unfolded/denatured proteins. When cells are heat shocked there is a greater requirement for both molecular chaperoning to maintain the correct folding of nascent proteins and for the removal of heat-denatured mature proteins. Therefore, after heat shock there is rapid induction and increased synthesis of Hsps to cope with increased damage to cellular proteins, which in organismal terms is observed as an increased viability of cultures exposed to a second subsequent heat-stress. In *E. coli*, production of most of the Hsps is under the control of a single gene, *htpR*, which encodes a new sigma factor that confers specificity on RNA polymerase leading to increased transcription of heat-shock genes. There are also changes in the stability of the heat-shock sigma factor and its mRNA. Overall there is a complex set of transcriptional and translational controls that mediate the increased production of Hsps at the expense of so-called housekeeping gene products. More detailed discussion is outside the scope of this chapter and readers are referred to several excellent recent reviews on heat shock.[79,80]

It is not surprising that in psychrotolerant bacteria the heat-shock response is induced at rather low temperatures, as observed in *Bacillus psychrophilus*,[81] *Arthrobacter globiformis* SI55[82] and a *Rhizobium* sp.[83] Heat shock occurs at even lower temperatures in the psychrophilic *Vibrio* ANT-300[84] and *Aquaspirillum arcticum*.[85] In the psychrotolerant *A. globiformis* SI55, Hsps are synthesized between 28 and 32°C when intracellular proteolysis increases drastically.[82] Partial characterization of the proteolytic activities revealed two distinct proteolytic systems. The first proteolytic system is activated after a shift to temperatures above the maximum for growth, and may degrade to short peptides those intracellular proteins altered by the sudden increase in temperature. The second system is inactivated at ≥32°C and may degrade the peptides produced by the first system, because when it is inhibited, cell growth is blocked by accumulation of peptides as no amino acids are available for new protein synthesis.[82] Proteolytic activity in *Vibrio* ANT-300 is also stimulated by a heat shock.[86]

Although it is generally known as the heat-shock response, the synthesis of Hsps is induced by many other stresses, including toxic metals, ethanol and surfactants, and strictly speaking it is a general stress response that interlinks with a number of two-component regulatory systems within the bacterial cell.[87] It is noteworthy, therefore (despite the potential for cold-denaturation of proteins: see above), that sudden exposure of bacteria to low temperatures does not induce the heat-shock response. Instead, a unique response, the cold-shock response, is induced.

7
Protein synthesis at low temperature and cold-induced proteins

7.1
The cold-shock response in mesophilic bacteria

The limitation of growth of mesophilic bacteria at low temperature has been found to reside in the ribosomal fraction of the cell.[20] Shifting *E. coli* from 37 to 5°C results in an accumulation of 70S monosomes with a concomitant decrease in the number of polysomes. This has been interpreted as resulting from a reversible block *in vivo* in the initiation process.[20,21,88,89] In contrast, in a psychrotolerant *Pseudomonas* sp. ribosomes retain their ability to form polysomes after a shift-down to 5°C.[21] However, in *E. coli* after a shift from 37°C to 10 or 15°C the accumulation of 70S monosomes (and 30S and 50S subunits) is transient, and after a lag of 4 h at 10°C, the inhibition of translation initiation ceases and cell growth resumes at the new, slower rate. During the lag period, the cold-shock response is induced and a group of cold-shock proteins (Csps) are produced whereas the synthesis of other (housekeeping) proteins is inhibited.[90] These Csps are quite different from other stress proteins.

Most of what we know about their role comes from work with *E. coli*, although much remains to be elucidated.[91,92] About 18 Csps have been described at the present time in *E. coli* and putative functions have been proposed for 14 Csps (Table 2). Class I proteins are expressed at an extremely low level at 37°C and are drastically induced (>10-fold) after a shift to lower temperature. In contrast, Class II Csps are present at a certain level at 37°C and are less strongly induced after a cold shock. Some of these cold-induced proteins are essential for the bacteria to resume growth at the low temperature and have been shown *in vitro* to function at both the transcriptional and translational levels. Unlike heat-shock sigma factors, cold-shock inducible sigma factors have never been found.[95] The presence of a Hsp70 homologue (Hsc66) was found in *E. coli* after a cold shock,[94] which is surprising because the cold-shock and heat-shock responses are mutually exclusive in *E. coli* and *Bacillus subtilis*.[95] The Hsp70 was not induced after a heat shock and its role as protein-chaperone was not demonstrated.

Table 2. Putative functions of cold-shock proteins (Csps) in *Escherichia coli* (data taken from refs. 92–94)

Protein	Putative function
Class I (>10 fold induction)	
CspA family	
CspA	RNA chaperone, DNA-transcriptional activator (?)
CspB	RNA/DNA chaperone (?)
CspG	RNA/DNA chaperone (?)
CsdA	30S ribosomal subunit RNA-unwinding activity
RbfA	30S ribosomal binding factor
NusA	Involved in termination and anti-termination
Polynucleotide phosphorylase	Ribonuclease (mRNA degradation)
Class II (<10 fold induction)	
H-NS	Nucleoid-associated DNA binding protein DNA packaging, regulation of gene expression
RecA	DNA repair, recombination
GyrA	DNA (negative) supercoiling
IF-2	Translation (initiation factor)
Dihydrolipoamide acetyltransferase	Pyruvate decarboxylation
Pyruvate dehydrogenase	Pyruvate decarboxylation
Hsc66	Molecular chaperone (?)

CspA shows the highest induction after a cold shock in *E. coli*, its rate of production reaching 13% of the total protein synthesis within 1–1.5 h after a shift to 10°C.[96] This major Csp belongs to a family of structurally-related proteins of approximately 7 kDa. Nine members of this family (CspA to CspI) have been described but only three (CspA, CspB and CspG) are cold inducible.[97] The 3D structure of CspA, as determined by X-ray crystallography and nuclear magnetic resonance spectroscopy, consists of a five-stranded antiparallel beta-barrel.[98] Sequence similarity was found between the CspA of *E. coli* and a family of eukaryotic DNA-binding proteins, called Y-box transcription factors, which influence the rate of transcription by binding to an ATTGG motif in the promoter of target genes. However, whilst this motif is found in several cold-shock genes of *E. coli*, including the promoter of CspA,[90] the Y-box proteins of eukaryotic organisms are not cold inducible. Since activation of CspA expression precedes that of other cold-shock genes, it has been suggested that CspA is a transcriptional enhancer for several of the Csps in *E. coli*, especially H-NS and gyrA.[95,99] However, an ATTGG recognition motif is not essential for the cold-shock induction of some genes in *E. coli*, nor in other bacteria.[91,100]

The essential role of CspA is now thought to be at the translational level.[99,101] In fact, it had been suspected from as early as 1970 that a protein factor acting on the ribosomes was produced at low temperature in cold-adapted bacteria,[102] but the roles of Csps in the initiation of translation and the regulation of Csp induction have been only recently elucidated. CspA harbors two RNA-binding sequence motifs, known as HNP1 and RNP2. It is assumed that CspA binds to single-stranded RNA and functions as an RNA chaperone for efficient translation, either by preventing the formation of secondary structure in mRNA that would make it inaccessible for translation at low temperature or by protecting the message from enzymic degradation.[92,95] By whatever means it is accomplished, this function is complemented by CspB and CspG, as shown in a CspA-deletion mutant. In addition, CspB and CspG function as DNA chaperones, assisting RNA transcription and DNA replication at low temperature. The mRNA for CspA is produced constitutively at 37°C but cannot be translated, probably because of its extreme instability.[103,104] Therefore at low temperatures, mRNA stabilization plays a major role in maintaining the continuity of translation. It is not known at present how the mRNA is stabilized: possibly, upon cold shock, ribosomal function becomes impaired and its intrinsic mRNA-degradative machinery is dissociated and/or cold-induced changes in the secondary structure of rRNAs may increase their susceptibility to RNAse attack.

The preferential synthesis of Csps at low temperatures indicates that cold-shock mRNAs, unlike most other cellular mRNAs, possess a mechanism to form the translation initiation factors. The mRNAs of *E. coli* class I Csps are equipped with an extra ribosome-binding site called the downstream box in the coding region of mRNA.[92,105] The expression of CspA is complex and is regulated at numerous levels including transcription, mRNA stability and translation. Posttranscriptional regulation also seems to play the major role in the induction of Csps.[104] The autoregulation of CspA production is due to a unique sequence, called the cold box, in the long 5'-upstream translated region of the CspA mRNA, a sequence which is highly conserved in other cold-shock genes.[92]

Among the other Csps of *E. coli*, three ribosome-associated proteins are considered to be involved in translation. These are IF2 for the binding of the initiation tRNA to the 30S subunit; CsdA, a member of the so-called DEAD-box family of helicases; and RbfA, a 30S ribosome-associated protein.[92] It is postulated that these Csps associate with the subunits and the 70S ribosome to convert the cold-sensitive nontranslatable ribosome into a cold-resistant translatable state.[92]

Csps, in particular CspA homologues, have been found in a range of mesophilic, thermophilic and psychrotolerant Gram-positive and Gram-negative bacteria, including *Bacillus subtilis*,[91,98,106,107] *Streptomyces clavuligerus*,[108] *Vibrio vulnificus*[109] and lactic acid bacteria.[110–112] In contrast to *E. coli*, the cold-shock response of *B. subtilis* is more complex and balanced. After a cold shock, growth continues and 36 cold-induced proteins can be detected and 75 housekeeping proteins are still produced, but the synthesis of general stress proteins and Hsps is inhibited.[107] The CspB of *B. subtilis* has been studied extensively and its 3D structure has been described.[113] A cold-regulated gene cluster, which encodes a RNA-binding protein, has also been isolated from the cyanobacterium *Anabaena variabilis*.[114]

7.2 Cold shock and cold-acclimation proteins in cold-adapted bacteria

The electrophoretic patterns of cellular proteins produced after a temperature downshift or during continuous growth at low temperatures have been analyzed in the psychrotolerant bacteria *Bacillus psychrophilus*,[115] *Bacillus cereus*,[116] *Pseudo-*

monas fragi,[117,118] arctic strains of *Rhizobium*,[83] *Listeria* spp.[119,120] and *Arthrobacter globiformis*,[121,122] and in the psychrophilic *Vibrio* sp. ANT-300,[84] *Aquaspirillum arcticum*[123] and "*Arthrobacter glacialis*" (A.M. Gounot, unpubl. data); (Table 3). The synthesis of new proteins in response to a cold shock was also observed in the psychrotolerant yeast *Trichosporon pullulans*.[124]

The cold-shock response in cold-adapted bacteria is slightly different from that of mesophilic bacteria in several respects: the synthesis of domestic proteins is not inhibited; the number of Csps is higher than in mesophiles and it increases with the severity of the shock; and the level of Csp expression is weaker. Notably in relation to cold adaptation, the synthesis of one particular class of proteins is permanently induced during continuous growth at low temperature, which suggests that these proteins may be fundamental to life in the cold. These proteins are known as cold-acclimation proteins (Caps). Such Caps have been observed during cold adaptation in the psychrophilic *Aquaspirillum arcticum*,[123] and in the psychrotolerant *Bacillus psychrophilus*[115] and *A. globiformis* SI55.[121]

Some Csps and Caps of cold-adapted bacteria have been identified by immunoblotting of proteins or by cloning and sequencing of their genes. Homologies with the CspA of *E. coli* were found in psychrotolerant *Pseudomonas fragi*[117] and *A. globiformis* SI55.[121] In the psychrotolerant *A. globiformis* SI55, the CspA-like protein is also a Cap. It is produced very rapidly following cold shock but, unlike its mesophilic counterpart, it is still expressed during prolonged growth at low temperature. Like the CspA of *E. coli*, the synthesis of CapA in *A. globiformis* SI55 is regulated at the translational level, and growth resumption following a temperature

Table 3. Cold-shock proteins (Csps) and cold-acclimation proteins (Caps) in psychrotolerant and psychrophilic bacteria (data taken taken from refs. 83,84,115,118,120-123 and A.M. Gounot, unpubl. data)

Organisms	Temperature shift (°C)	Number of Csps	Growth temperature (°C)	Number of Caps[a]
Psychrotolerants				
Bacillus psychrophilus	20 → 0	9	10	4 (2 Csps)
	20 → 5	7	5	10 (7 Csps)
	20 → 10	5	0	11 (8 Csps)
Arthrobacter globiformis	25 → 4	23	4	18
Arctic *Rhizobium* sp.	25 → -2	7–24		
	25 → -5	7–12		
	25 → -10	5–14		
Listeria monocytogenes	37 → 5	12	5	4
Pseudomonas fragi	30 → 5	24		
	20 → 5	15	5	8 (6 Csps)
Psychrophiles				
Vibrio sp. ANT-300	13 → 0	7	0	13
Aquaspirillum arcticum	10 → 0	10	0	8 (5 Csps)
	5 → 0	9		
	10 → 5	7		
"*Arthrobacter glacialis*"	12 → 0	9	0	16 (4 Csps)

[a] The numbers in brackets indicate "Csp-like" Caps.

downshift is correlated with CapA expression. As for the CspA of *E. coli*, it is also induced at the optimum temperature (25°C in *A. globiformis*) by chloramphenicol, suggesting that it may have a more general role in cell metabolism. In *A. globiformis* SI55 a gene homologous to *cspA* of *E. coli* was cloned and sequenced.[122] The deduced protein sequence has a high level of identity with the sequences of CspA-related proteins from various sources, and no particular residue or domain that could be specific to cold-adapted microorganisms emerged.[122] A cold-shock gene homologue of the *cspA* in *E. coli* was also found to be present and expressed in antarctic psychrotolerant bacteria.[125]

Csps are "stress proteins" and they function to increase viability after a stress. However, although they are necessary, they are not sufficient alone for the resumption of growth at the new lower temperature, as the discovery of Caps has shown. However, we do not as yet have a picture of what are the functions of Caps. Most studies have used psychrotolerant strains for which a temperature near 0°C is not within the normal (linear) range of temperature for growth, and it is not clear in many published reports whether the downshifted bacterial cultures are truly in a state of permanent cold stress requiring the synthesis of Caps as well as Csps. In particular, it is not possible at present to state whether in psychrophiles there are specific protein factors for transcription or translation at low temperatures. Neither is the link between the well-characterized membrane lipid adaptations and Csps/Caps clear. Recently, it has been demonstrated that the desaturase of *Bacillus subtilis* which is induced following temperature downshift is a Csp.[126] More investigations must be done on true psychrophilic bacteria to clarify the function of Caps in particular and the relationship between Csps and Caps, and the integration of cellular events within the cytoplasm and those occurring in the membranes.

7.3 Sensors for low temperature in bacteria

Three suspected thermosensors have been suggested as playing a role in the induction of the cold-shock response: nucleic acids, ribosomes and the cytoplasmic membrane.[127]

It has been proposed that the state of the ribosome signals the induction of the temperature response,[91,95,128] and that the concentration of charged tRNA species within the cell may be the signal. After a sudden increase in temperature, the speed of translation exceeds that of the instantaneous supply of amino acyl-tRNA, so that the A-site of the ribosome would be empty. Certain antibiotics including kanamycin act in the same manner and induce the heat-shock response. Conversely a cold shock results in reduced translational capacity, whereupon the concentration of amino acyl-tRNA would be too high, thereby blocking the A-site in the same manner as certain antibiotics (e.g. chloramplenicol) which induce the cold-shock response.[91]

In a second theory for the cold sensor, it was proposed that the concentration of (p)ppGpp directly or indirectly affects the magnitude of the cold-shock response.[95,129] The guanosine nucleotides could also provide the link between intracellular events and those occurring in the membrane, since they are involved in regulation of lipid synthesis. In this respect it is relevant to note that Ray et al.[130] observed temperature-dependent phosphorylation and dephosphorylation of membrane and cytosol proteins in several bacteria from Antarctica, and it was suggested that cyclic (de)phosphorylation was involved in thermal sensing. Temperature-dependent phosphorylation of lipopolysaccharides has also been implicated in cold adaptation.[131]

A third theory implicates changes in DNA supercoiling in the sensing of temperature fluctuations, since it was demonstrated that negative supercoiling of plasmid DNA in *E. coli* decreases transiently when cultures are exposed to heat.[132] Conversely, supercoiling increases upon exposure to cold shock.[133] In *Yersinia enterocolitica*, thermoregulation is coincident with changes in DNA supercoiling, resulting in modulation of toxin and flagella synthesis.[134] Grau et al.[135] have proposed that DNA supercoiling may regulate fatty acid synthesis in *Bacillus subtilis*.

8 Antifreeze protection

In their natural habitats, cold-adapted bacteria generally encounter freezing temperatures and are subjected to freezing and thawing cycles. It may be supposed, therefore, that they are resistant to freezing. It is evident that a multiplicity of factors affect the response of microbial cells to chilling and freezing. The response depends on the physiological status, the rate of cooling, freezing and thawing, and the chemical composition of the freezing menstrum.[18,136] Growing cells are generally more susceptible than resting cells, and freezing at -20°C (that is the case in natural environments) is more damaging than at -196°C (the temperature of liquid nitrogen that is used for cryopreservation). It has often been observed that the resistance to freezing of bacteria harvested from the exponential phase of batch cultures is greater if they are preincubated at low but not freezing temperatures. When *E. coli* cultures grown at 37°C were cold shocked at 10°C for 6 h and then frozen, bacterial viability was significantly higher compared with cultures frozen without the cold shock.[96] Likewise, cold-shocked cultures of *Bacillus subtilis* and *Lactococcus lactis* subsp. *lactis* were significantly more tolerant to freezing than were non-cold-shocked cultures.[137–139] Moreover, increasing the incubation period at low temperature increased cryotolerance.[139] However, this is not a general finding. For instance, in an extensive study, it was shown that the freeze tolerance (at moderately low temperatures) of 259 bacterial strains belonging to 32 genera and 135 species is species- and genus-specific and that certain psychrophilic bacteria were very susceptible to frozen storage.[140] Strains of *Thiobacillus ferroxidans* which were cold shocked from 25 to 5°C were not protected against freezing at -15°C.[141] In lactic acid bacteria a cold shock at 10°C improved the viability of some strains but not that of others.[111] In passing it should be mentioned that the CspA of *E. coli* and CspB of *B. subtilis* were postulated initially to function as cellular antifreeze agents,[96,137] but this hypothesis was not confirmed later.

Antifreeze proteins are well known in cold-adapted fishes, insect larvae and other arthropods, and plants, but to date antifreeze proteins homologues of these in eukaryotic organisms have not been found in bacteria. Duman and Olsen[142] found thermal hysteresis activity in two psychrotolerant bacteria (*Rhodococcus erythropolis* and "*Micrococcus cryophilus*"). In spite of the fact that the two bacteria grew fast at 3°C, acclimation at 3°C for 3 or 4 weeks was required to induce measurable lowering of the freezing point. Treatment with proteinase eliminated the antifreeze activity, demonstrating that the activity was due to the presence of thermal hysteresis proteins, but these proteins were not identified. The claimed production of antifreeze protein by *Pseudomonas putida*[143] is not supported by evidence for the lowering of cellular freezing point.

The production of ice-nucleation proteins by bacteria is well documented.[144,145] Bacterial ice nucleation is attributed to the presence of a protein, comprised of highly repetitive sequences of amino acids, in the bacterial outer membrane that serves as a template for ice crystallization. Beside its nutritional role, one hypothesis for the significance of ice-nucleating agents in bacteria is that they provide resistance to desiccation by sequestering a reserve of water close to the cell surface as well as efficiently dispersing ice particles from the cell and thereby preventing damage by intracellular ice formation.[146]

Other molecules such as polysaccharides which are frequently produced by antarctic and arctic microorganisms (see Chapter by Gounot, in this book) may protect the cell. Freezing also results in osmotic shock, which can be prevented by the accumulation of intracellular compatible solutes such as polyols and sugars, as observed in antarctic yeast and algae.[147] Bacteria tend to accumulate a number of amino acids instead in response to osmotic stress, but there is no evidence for them playing a role in the freeze tolerance of bacteria. It may be that cultures need to be grown at low temperatures for a very long period of time for such an effect to be observed.

9
Conclusions

Bacteria are small and unable to insulate themselves against the effects of cold, so it follows that all of the cellular components of a cold-adapted bacterium must be adapted if the organism is to grow and divide successfully. The two major cellular components in this respect are proteins and lipids, providing for the continuance of enzyme activity and the functional integrity of membranes at low temperatures. Thus, the focus of this chapter has been on the cold adaptation of enzymes (proteins) and lipids (membranes). For enzymes, it is only in the past year or so have we been able to obtain molecular information about protein folding. For lipids, there is abundant long-standing compositional data and plenty of evidence for its relevance to membrane structure and function, and yet only comparatively recently have molecular techniques of gene cloning and transfer been successfully used to show directly, for example, the requirement for desaturase activity in the photosynthetic function[148] and chilling tolerance[149] of cyanobacteria. The aim of studies on these two broad fronts is to describe the molecular basis of cold adaptation, the "secret of psychrophily". It is not simply the opposite of adaptation to heat, as the work on cold shock and cold acclimation clearly demonstrates that there is a unique set of controls for cold adaptation; in this respect it will be particularly interesting to know what are the functions of cold-acclimation proteins that enable such a large proportion of the organisms on our planet to grow at low temperatures.

10 References

1. Sharp RJ, Munster MJ. Biotechnological implications for microorganisms from extreme environments. In: Herbert RA, Codd GA, eds. Microbes in Extreme Environments. London: Academic Press, 1989:215-295.
2. Gounot AM. Bacterial life at low temperature: physiological aspects and biotechnological implications. A review. J Appl Bacteriol 1991; 71:686-397.
3. Herbert RA. A perspective on the biotechnological potential of extremophiles. Trends Biotechnol 1992; 10:395-401.
4. Russell NJ. Physiology and molecular biology of psychrophilic microorganisms. In: Herbert RA, Sharp RJ, eds. Molecular Biology and Biotechnology of Extremophiles. London: Blackie, 1992:203-224.
5. Russell NJ. Molecular adaptations in psychrophilic bacteria: potential for biotechnological applications. Adv Biochem Eng/Biotechnol 1998; 61:1-22.
6. Margesin R, Schinner F. Properties of cold-adapted microorganisms and their potential role in biotechnology. J Biotechnol 1994; 33:1-14.
7. Brenchley JE. Psychrophilic microorganisms and their cold-active enzymes. J Ind Microbiology 1996; 17:432-437.
8. Morita RY. Psychrophilic bacteria. Bacteriol Rev 1975; 39:144-167.
9. Hoham RW. Optimum temperatures and temperature ranges for growth of snow algae. Arctic Alpine Res 1975; 7:13-24.
10. Collins MA, Buick RK. Effect of temperature on the spoilage of stored peas by *Rhodotorula glutinis*. Food Microbiol 1989; 6:135-141.
11. Herbert RA. The ecology and physiology of psychrophilic microorganisms. In: Herbert RA, Codd GA, eds. Microbes in Extreme Environments. London: Academic Press, 1986:1-23.
12. Russell NJ. Cold adaptation of microorganisms. Phil Trans Roy Soc Lond 1990; 326B:595-611.
13. Ingraham JL. Growth of psychrophilic bacteria. J Bacteriol 1958; 76:75-80.
14. Harder W, Veldkamp H. Physiology of an obligately psychrophilic marine *Pseudomonas* species. J Appl Bacteriol 1968; 31:12-23.
15. Gounot AM. Effects of temperature on the growth of psychrophilic bacteria from glaciers. Can J Microbiol 1976; 22:839-846.
16. Gounot AM. Psychrophilic and psychrotrophic microorganisms. Experientia 1986; 42:1192-1197.
17. Ingraham JL. Factors affecting the lower limits of temperature for growth. In: Gibbons NE, ed. Recent Progress in Microbiology. Toronto: Univ of Toronto Press, 1963:201-212.
18. MacLeod RA, Calcott PH. Cold shock and freezing damage to microbes. In: Gray TRG, Postgate JR, eds. The Survival of Vegetative Microbes. Cambridge: Cambridge Univ Press, 1976:81-109.
19. Farrell J, Rose AH. Temperature effects on microorganisms. In: Rose AH, ed. Thermobiology. London: Academic Press, 1967:147-218.
20. Das HK, Goldstein A. Limited capacity for protein synthesis at zero degrees centigrade in *Escherichia coli*. J Mol Biol 1968; 31:209-226.

21. Broeze RJ, Solomon CJ, Pope DH. Effects of low temperature on *in vivo* and *in vitro* protein synthesis in *Escherichia coli* and *Pseudomonas fluorescens.* J Bacteriol 1978; 134:861-874.
22. Feller G, Narinx E, Arpigny JL, Zekhnini Z, Swings J, Gerday C. Temperature dependence of growth, enzyme secretion and activity of psychrophilic Antarctic bacteria. Appl Microbiol Biotechnol 1994; 41:477-479.
23. Margesin R, Schinner F. Extracellular protease production by psychrotrophic bacteria from glaciers. Int Biodeterioration Biodegradation 1992; 29:177-189.
24. Fairbairn DJ, Law BA. Proteinases of psychrotrophic bacteria: their production, properties, effects and control. J Dairy Res 1986; 53:139-177.
25. Kato N, Nagasawa T, Tani Y, Ogata K. Protease formation by a marine psychrophilic bacterium. Agric Biol Chem 1972; 36:1177-1184.
26. Gügi B, Orange N, Hellio F, Burini JF, Guillou C, Leriche F, Guespin-Michel JF. Effect of growth temperature on several exported enzyme activities in the psychrotrophic bacterium *Pseudomonas fluorescens.* J. Bacteriol 1991; 173:3814-3820.
27. Hoshino T, Ishizaki K, Sakamoto T, Kumeta H, Yumoto I, Matsuyama H, Ohgiya S. Isolation of a *Pseudomonas* species from fish intestine that produces a protease active at low temperature. Lett Appl Microbiol 1997; 25:70-72.
28. Feller G, Thiry M, Arpigny JL, Mergeay M. Gerday C. Lipases from psychrotrophic antarctic bacteria. FEMS Microbiol Lett 1990; 66:239-244.
29. Burini JF, Gügi B, Merieau A, Guespin-Michel JF. Lipase and acidic phosphatase from the psychrotrophic bacterium *Pseudomonas fluorescens*: two enzymes whose synthesis is regulated by the growth temperature. FEMS Microbiol Lett 1994; 122:13-18.
30. Hamamoto T, Horikoshi K. Characterisation of an amylase from a psychrotrophic *Vibrio* isolated from a deep-sea mud sample. FEMS Microbiol Lett 1991; 84:79-84.
31. Feller G, Payan F, Theys F, Qian M, Haser R, Gerday C. Stability and structural analysis of α-amylase from the antarctic psychrophile *Alteromonas haloplanctis* A23. Eur J Biochem 1994; 222:441-447.
32. Okuyama H, Fukunaga N, Sasaki S. Effect of temperature on the activities of the electron transport systems and the coupled phosphorylation in mesophilic, psychrotrophic and psychrophilic bacteria. J Gen Appl Microbiol 1979; 25:299-306.
33. Guillou C, Guespin-Michel JF. Evidence for two domains of growth temperature for the psychrotrophic bacterium *Pseudomonas fluorescens* MF0. Appl Environ Microbiol 1996; 62:3319-3324.
34. Harder W, Veldkamp H. A continuous culture study of an obligately psychrophilic *Pseudomonas* species. Arch Mikrobiol 1967; 59:123-130.
35. Feller G, Narinx E, Arpigny JL, Aittaleb M, Baise E, Genicot S, Gerday C. Enzymes from psychrophilic organisms. FEMS Microbiol Rev 1996; 18:189-202.
36. Orange N. Growth temperature regulates the induction of β-lactamase in *Pseudomonas fluorescens* through modulation of the outer membrane permeation of a β-lactam-inducing antibiotic. Microbiology 1994; 140:3125-3130.
37. Dé E, Orange N, Saint N, Guérillon J, De Mot R, Molle G. Growth temperature dependence of channel size of the major outer-membrane protein (OprF) in psychrotrophic *Pseudomonas fluorescens* strains. Microbiology 1997; 143:1029-1035.
38. Russell NJ. Functions of lipids: structural roles and membrane functions. In: Ratledge C, Wilkinson SG, eds. Microbial Lipids. London: Academic Press, 1989:279-365.
39. Van de Vossenberg JLCM, Driessen AJM, da Costa MS, Konings WN. Homeostasis of the membrane proton permeability in *Bacillus subtilis* grown at different temperatures. J Bacteriol 1998 (in press).
40. Finegold L, Singer MA, Federle TW, Vestal JR. Composition and thermal properties of membrane lipids in cryptoendolithic lichen microbiota from Antarctica. Appl Environ Microbiol 1990; 56:1191-1194.
41. Sinensky M. Homeoviscous adaptation - a homestatic process that regulates the viscosity of membrane lipids in *Escherichia coli.* Proc Natl Acad Sci USA 1974; 71:522-525.

42. Jackson MB, Cronan JE Jr. An estimate of the minimum amount of fluid lipid required for the growth of *Escherichia coli*. Biochim Biophys Acta 1978; 512:472-479.
43. Russell NJ. Mechanisms of thermal adaptation in bacteria: blueprints for survival. Trends Biochem Sci 1984; 9:108-112.
44. Harwood JL, Russell NJ. Lipids in Plants and Microbes. London: George Allen & Unwin, 1984.
45. Morita N, Gotoh M, Okajima H, Hayashi H, Higashi S, Murata N. Both the anaerobic pathway and aerobic desaturation are involved in the synthesis of unsaturated fatty acids in *Vibrio* sp. strain ABE-1. FEBS Lett 1992; 297:9-12.
46. Nichols DS, Nichols PD, McMeekin TA. Polyunsaturated fatty acids in Antarctic bacteria. Antarct Sci 1993; 5:149-160.
47. Nichols DS, Nichols PD, Russell NJ, Davies NW, McMeekin TA. Polyunsaturated fatty acids in the psychrophilic bacterium *Shewanella gelidimarina* ACAM 456^{T}: molecular species analysis of major phospholipids and biosynthesis of eicosapentaenoic acid. Biochim Biophys Acta 1997; 1347:164-176.
48. Nichols DS, Russell NJ. Fatty acid changes in an Antarctic bacterium - changes in primer utilization. Microbiology 1996; 142:747-754.
49. Nichols DS, Brown JL, Nichols PD, McMeekin TA. Production of eicosapentaenoic acid and arachidonic acids by an Antarctic bacterium: response to growth temperature. FEMS Microbiol Lett 1997; 152:349-354.
50. Watanabe K, Yazawa K, Kondo K, Kawaguchi A. Fatty acid synthesis of an eicosapentaenoic acid-producing bacterium: *de novo* synthesis, chain elongation, and desaturation systems. J Biochem 1997; 122:467-473.
51. Keweloh H, Heipieper HJ. *Trans* unsaturated fatty acids in bacteria. Lipids 1996; 31:129-137.
52. Diefenbach R, Keweloh H. Synthesis of *trans* unsaturated fatty acids in *Pseudomonas putida* P8 by direct isomerization of the double bond of lipids. Arch Microbiol 1994; 162:120-125.
53. Lands WEM. Dialogue between membranes and their lipid-metabolizing enzymes. Biochem Soc (UK) Trans 1980; 8:25-27.
54. Aguilar PS, Cronan JE Jr, de Mendoza D. A *Bacillus subtilis* gene induced by cold shock encodes a membrane phospholipid desaturase. J Bacteriol 1998; 180:2194-2200.
55. Fujii DK, Fulco AJ. Biosynthesis of unsaturated fatty acid in bacilli: hyperinduction and modulation of desaturase synthesis. J Biol Chem 1977; 252:3660-3670.
56. Zuber H. Temperature adaptation of lactate dehydrogenase. Structural, functional and genetic aspects. Biophys Chem 1988; 29:171-179.
57. Jaenicke R. Protein structure and function at low temperatures. Phil Trans Roy Soc London 1990; 326B:535-553.
58. Cowman RA, Speck ML. Low temperature as an environmental stress on microbial enzymes. Cryobiology 1969; 5:291-299.
59. Jahns T, Kaltwasser H. Properties of the cold-labile NAD^+-specific glutamate dehydrogenase from *Bacillus cereus* DSM 31. J Gen Microbiol 1993; 139:775-780.
60. Privalov PL. Stability of proteins. Adv Protein Chem 1979; 33:167-241.
61. Marshall CJ. Cold-adapted enzymes. TIBTech 1997; 15:359-364.
62. Malcolm NL. Molecular determinants of obligate psychrophily. Nature 1969; 221:1031-1033.
63. Mitchell P, Yen HC, Mathemeier PF. Properties of lactate dehydrogenase in a psychrophilic marine bacterium. Appl Environ Microbiol 1985; 49:1332-1334.
64. Ochiai T, Fukunaga N, Sasaki S. Two structurally different $NADP^+$-specific isocitrate dehydrogenases in an obligately psychrophilic bacterium, *Vibrio* sp. strain ABE-1. J Gen Appl Microbiol 1984; 30:479-487.
65. Russell NJ, Hamamoto T. Psychrophiles. In: Horikoshi K, Grant WD, eds. Extremophiles: Microbial Life in Extreme Environments. New York: Wiley-Liss, 1997.

66. Aghajari N, Feller G, Gerday C, Haser R. Crystal structures of the psychrophilic α-amylase from *Alteromonas haloplanctis* in its native form and complexed with an inhibitor. Protein Sci 1998; 7:564-572.
67. Russell RJM, Gerike U, Danson MJ, Hough DW, Taylor GL. Structural alterations of the cold-active citrate synthase from an Antarctic bacterium. Structure 1998; 6:351-361.
68. Alvarez M, Zeelen J Ph, Mainfroid V, Rentier-Delrue F, Martial JA, Wyns L, Wierengas RK, Maes D. Triose-phosphate isomerase (TIM) of the psychrophilic bacterium *Vibrio marinus*. J Biol Chem 1998; 273:2199-2206.
69. McCallum K, Butler BJ, Inniss WE. Stress or heat shock protein synthesis and cellular filamentation in psychrophilic and psychrotrophic bacteria. Arch Microbiol 1989; 152:148-153.
70. Gounot AM, Novitsky TJ, Kushner DJ. Effects of temperature on the macromolecular composition and fine structure of psychrophilic *Arthrobacter* species. Can J Microbiol 1977; 23:357-362.
71. Jones LP, Morita LY, Becker RR. Fructose 1,6 biphosphate aldolase from *Vibrio marinus*, a psychrophilic marine bacterium. Z Allg Mikrobiol 1979; 19:97-106.
72. Trimbur DE, Gutshall KR, Prema P, Brenchley JE. Characterization of a psychrotrophic *Arthrobacter* gene and its cold-active β-galactosidase. Appl Environ Microbiol 1994; 60:4544-4552.
73. Margesin R, Schinner F. A comparison of extracellular proteases from three psychrotrophic strains of *Pseudomonas fluorescens*. J Gen Appl Microbiol 1992; 38:209-225.
74. Kobori H, Sullivan C, Shizuya H. Heat-labile alkaline phosphatase from Antarctic bacteria: rapid 5' end-labelling of nucleic acids. Proc Natl Acad Sci USA 1984; 81:6691-6695.
75. Chattopadhyay MK, Uma Devi K, Gopishankar, Shivaji S. Thermolabile alkaline phosphatase from *Sphingobacterium antarcticus*, a psychrotrophic bacterium from Antarctica. Polar Biol 1995; 15:215-219.
76. Reddy GSN, Rajagopalan G. Shivaji S. Thermolabile ribonucleases from antarctic psychrotrophic bacteria: detection of the enzyme in various bacteria and purification from *Pseudomonas fluorescens*. FEMS Microbiol Lett 1994; 122:211-216.
77. Adler E, Knowles J. A thermolabile triosephosphate isomerase from the psychrophile *Vibrio sp.* strain ANT-300. Arch Biochem Biophys 1995; 321:137-139.
78. Neidhardt FC, VanBogelen RA, Vaughn V. The genetics and regulation of heat-shock proteins. Annu Rev Genet 1984; 18:295-329.
79. Yura T, Nagai H, Mori H. Regulation of heat shock response in bacteria. Annu Rev Microbiol 1993; 47:321-350.
80. Bukau B. Regulation of the *Escherichia coli* heat shock response. Mol Microbiol 1993; 9:671-680.
81. McCallum K, Heikkila JJ, Inniss WE.Temperature dependent pattern of heat shock protein synthesis in psychrophilic and psychrotrophic microorgamisms. Can J Microbiol 1986; 32:516-521.
82. Potier P, Drevet P, Gounot AM, Hipkiss A. Temperature-dependent changes in proteolytic activities and protein composition in the psychrotrophic bacterium *Arthrobacter globiformis* SI55. J Gen Microbiol 1990; 136:283-291.
83. Cloutier J, Prévost D, Nadeau P, Antoun H. Heat and cold shock protein synthesis in arctic and temperate strains of *Rhizobia*. Appl Environ Microbiol 1992; 58:2846-2853.
84. Araki T. Changes in rates of synthesis of individual proteins in a psychrophilic bacterium after a shift in temperature. Can J Microbiol 1991; 37:840-847.
85. McCallum K, Inniss WE. Thermotolerance, cell filamentation, and induced protein synthesis in psychrophilic and psychrotrophic bacteria. Arch Microbiol 1990;153:585-590.
86. Araki T. An analysis of the effect of changes in growth temperatures on proteolysis *in vivo* in the psychrophilic bacterium *Vibrio* sp. strain ANT-300. J Gen Microbiol 1992; 138:2075-2082.
87. Hoch JA, Silhavy TJ. Two-component signal transduction. Washington DC: Am Soc Microbiol, 1995.

88. Friedman H, Lu P, Rich A. An *in vivo* block in the initiation of protein synthesis. Cold Spring Harbor Symp Quant Biol 1969; 34:255-260.
89. Oshima A, Fukunaga N, Sasaki S. Effect of temperature on natural mRNA-directed protein synthesis in psychrophilic and mesophilic bacteria. J Fac Sci Hokkaido Univ Ser V (Botany) 1987; 14:1-10.
90. Jones PG, Van Bogelen RA, Neidhardt FC. Induction of proteins in response to low temperature in *Escherichia coli*. J Bacteriol 1987; 169:2092-2095.
91. Graumann P, Marahiel MA. Some like it cold: response of microorganisms to cold shock. Arch Microbiol 1996; 166:293-300.
92. Thieringer HA, Jones PG, Inouye M. Cold shock and adaptation. BioEssays 1998; 20:49-57.
93. Berry ED, Foegeding PM. Cold temperature adaptation and growth of microorganisms. J Food Prot 1997; 12:1583-1594.
94. Levilelt MJ, Kawula TH. Hsc66, an Hsp70 homolog in *Escherichia coli*, is induced by cold shock but not by heat shock. J Bacteriol 1995; 177:4900-4907.
95. Jones PG, Inouye M. The cold-shock response. A hot topic. Mol Microbiol 1994; 11:811-818.
96. Goldstein J, Pollitt S, Inouye M. Major cold shock protein of *Escherichia coli*. Proc Natl Acad Sci USA 1990; 87:283-287.
97. Yamanaka K, Fang L, Inouye M. The CspA family in *Escherichia coli*: multiple gene duplication for stress adaptation. Mol Microbiol 1998, 27:247-255.
98. Schindelin H, Jiang W, Inouye M, Heinemann U. Crystal structure of CspA, the major cold shock protein of *Escherichia coli*. Proc Natl Acad Sci USA 1994; 91:5119-5123.
99. La Teana A, Brandi A, Falconi M, Spurio R, Pon CL, Gualerzi CO. Identification of a cold shock transcriptional enhancer of the *Escherichia coli* gene encoding nucleoid protein H-NA. Proc Natl Acad Sci USA 1991; 88:10907-10911.
100. Brandi A, Pon CL, Gualerzi CO. Interaction of the main cold shock protein CS7.4 (CspA) of *Escherichia coli* with the promoter region of *hns*. Biochimie 1994; 76:1090-1098.
101. Brandi A, Pietroni P, Gualerzi CO, Pon CL. Post-transcriptional regulation of CspA expression in *Escherichia coli*. Mol Microbiol 1996; 19:231-240.
102. Szer W. Cell-free protein synthesis at 0°C; an activating factor from ribosomes of a psychrophilic microorganism. Biochim Biophys Acta 1970; 213:159-170.
103. Goldenberg D, Azar I, Oppenheim AB. Differential mRNA stability of the *cspA* gene in the cold-shock response of *Escherichia coli*. Mol Microbiol 1996; 19:241-248.
104. Goldenberg D, Azar I, Oppenheim AB, Brandi A, Pon CL, Gualerzi CO. Role of *Escherichia coli cspA* promoter sequences and adaptation of translational apparatus in the cold shock response. Mol Gen Genet 1997; 256:282-290.
105. Fang L, Hou Y, Inouye M. Role of the cold-box region in the 5'untranslated region of the *cspA* mRNA in its transient expression at low temperature in *Escherichia coli*. J Bacteriol 1998; 180:90-95.
106. Lottering EA, Streips UN. Induction of cold shock proteins in *Bacillus subtilis*. Curr Microbiol 1995; 30:193-199.
107. Graumann P, Schöder K, Schmid R, Marahiel MA. Cold shock stress-induced proteins in *Bacillus subtilis*. J Bacteriol 1996; 178:4611-4619.
108. Av-Gay Y, Aharonowitz Y, Cohen G. *Streptomyces* contains a 7.0 kDa cold shock like protein. Nucleic Acids Res 1992; 20:5478-5479.
109. McGovern VP, Oliver JD. Induction of cold-responsive proteins in *Vibrio vulnificus*. J Bacteriol 1995; 177:4131-4133.
110. Panoff JM, Thammavongs B, Auffray Y, Boutibonnes P. The cold chock response in *Lactococcus lactis* subsp. *lactis*. Curr Microbiol 1994; 29:213-216.
111. Kim WS, Dunn NW. Identification of a cold shock gene in lactic acid bacteria and the effect of cold shock on cryotolerance. Curr Microbiol 1997; 35:59-63.
112. Mayo B, Derzelle S, Fernandez M, Leonard C, Ferain T, Hols P, Suarez JE, Delcour J. Cloning and characterization of *cspL* and *cspP*, two cold-inducible genes from *Lactobacillus plantarum*. J Bacteriol 1997; 179:3039-3042.

113. Schindelin H, Herrler M, Willimsky G, Marahiel MA, Heinemann U. Overproduction, crystallization and preliminary X-ray diffraction studies of the major cold shock protein from *Bacillus subtilis*. J Bacteriol 1996; 178:4611-4619.
114. Sato N. A cold-regulated cyanobacterial gene cluster encodes RNA-binding protein and ribosomal protein S21. Plant Mol Biol 1994; 24:819-823.
115. Whyte LG, Inniss WE. Cold shock proteins and cold acclimation proteins in a psychrotrophic bacterium. Can J Microbiol 1992; 38:1281-1285.
116. Mayr B, Kaplan T, Lechner S, Scherer S. Identification and purification of a family of dimeric major cold shock protein homologs from the psychrotrophic *Bacillus cereus* WSBC 10201. J Bacteriol 1996; 178:2916-2925.
117. Hebraud M, Dubois E, Potier P, Labadie J. Effect of growth temperatures on the protein levels in a psychrotrophic bacterium *Pseudomonas fragi*. J Bacteriol 1994; 176:4017-4024.
118. Michel V, Lehoux I, Depret G, Anglade P, Labadie J, Hebraud M. The cold shock response of the psychrotrophic bacterium *Pseudomonas fragi* involves four low-molecular-mass nucleic acid-binding proteins. J Bacteriol 1997; 179:7331-7342.
119. Phan-Thanh L, Gormon T. Analysis of heat and cold shock proteins in *Listeria* by two-dimensional electrophoresis. Electrophoresis 1995; 16:444-450.
120. Bayles DO, Annous BA, Wilkinson BJ. Cold stress proteins induced in *Listeria monocytogenes* in response to temperature downshock and growth at low temperatures. Appl Environ Microbiol 1996; 62:1116-1119.
121. Berger F, Morellet N, Menu F, Potier P. Cold shock and cold acclimation proteins in the psychrotrophic bacterium *Arthrobacter globiformis* SI55. J Bacteriol 1996; 178:2999-3007.
122. Berger F, Normand P, Potier P. capA, a *cspA*-like gene that encodes a cold acclimation protein in the psychrotrophic bacterium *Arthrobacter globiformis* SI55. J Bacteriol 1997; 179:5670-5676.
123. Roberts ME, Inniss WE. The synthesis of cold shock proteins and cold acclimation proteins in the psychrophilic bacterium *Aquaspirillum arcticum*. Curr Microbiol 1992; 25:275-278.
124. Julseth CR, Inniss WE. Induction of protein synthesis in response to cold shock in the psychrotrophic yeast *Trichosporon pullulans*. Can J Microbiol 1990; 36:519-524.
125. Ray MK, Sitaramamma T, Ghandi S, Shivaji S. Occurrence and expression of *cspA* a cold shock gene in Antactic psychrotrophic bacteria. FEMS Microbiol Lett 1994; 116:55-60.
126. Aguilar PS, Cronan JE Jr, de Mendoza D. A *Bacillus subtilis* gene induced by cold shock encodes a membrane phospholipid desaturase. J Bacteriol 1998; 180:2194-2200.
127. Panoff JM, Thammavongs B, Gueguen M, Boutibonnes P. Cold stress responses in mesophilic bacteria. Cryobiology 1998; 36:75-83.
128. Van Bogelen RA, Neidhardt FC. Ribosomes as sensors of heat and cold shock in *Escherichia coli*. Proc Natl Acad Sci USA 1990; 87:5589:5593.
129. Jones PG, Cashel M, Glaser G, Neidhart FC. Function of a relaxed-like state following temperature downshifts in *Escherichia coli*. J Bacteriol 1992; 174:3903-3914.
130. Ray MK, Kumar GS, Shivaji S. Phosphorylation of membrane proteins in response to temperature in an Antarctic *Pseudomonas syringae*. Microbiology 1994; 140:3217-3223.
131. Ray MK, Kumar GS, Shivaji S. Phosphorylation of lipopolysaccharides in the Antarctic psychrotroph *Pseudomonas syringae*: a possible role in temperature adaptation. J Bacteriol 1994; 176:4243-4249.
132. Ogata Y, Mizushima T, Kataoka K, Miki T, Sekimizu K. Identification of DNA topoisomerases involved in immediate and transient DNA relaxation induced by heat shock in *Escherichia coli*. Mol Gen Genet 1994; 244:451-455.
133. Mizushima T, Kataoka K, Ogata Y, Inoue RI, Sekimizu K. Increase in negative supercoiling of plasmid DNA in *Escherichia coli* exposed to cold shock. Mol Microbiol 1997; 23:381-386.
134. Rohde JR, Fox JM, Minich SA. Thermoregulation in *Yersinia enterocolitica* is coincident with changes in DNA supercoiling. Mol Microbiol 1994; 12:187-199.
135. Grau R, Gardiol D, Glikin GC, de Mendoza D. DNA supercoiling and thermal regulation of unsaturated fatty acid synthesis in *Bacillus subtilis*. Mol Microbiol 1994; 11:933-941.

136. Smith MG. Survival of *E. coli* and *Salmonella* after chilling and freezing in liquid media. J Food Sci 1995; 60:509-512.
137. Willimsky G, Bang H, Fischer G, Marahiel MA. Characterization of cspB, a *Bacillus subtilis* inducible cold shock gene affecting cell viability at low temperatures. J Bacteriol 1992; 174:6326-6335.
138. Panoff JM, Thammavongs B, Laplace JM, Hartke A, Boutibonnes P, Auffray Y. Cryotolerance and cold adaptation in *Lactococcus lactis* subsp. *lactis* IL1403. Cryobiology 1995; 32:516-520.
139. Thammavongs B, Corroler D, Panoff JM, Auffray Y, Boutibonnes P. Physiological response of *Enterococcus faecalis* JH2-2 to cold shock: growth at low temperatures and freezing/thawing challenge. Lett Appl Microbiol 1996; 23:398-402.
140. Yamasato K, Okuno D, Ohtomo T. Preservation of bacteria by freezing at moderately low temperatures. Cryobiology 1973; 10:453-463.
141. Hubert WA, Ferroni GD, Leduc LG. Temperature-dependent survival of isolates of *Thiobacillus ferrooxidans*. Curr Microbiol 1994; 28:179-183.
142. Duman JG, Olsen TM. Thermal hysteresis protein activity in bacteria, fungi and phylogenetically diverse plants. Cryobiology 1993; 30:322-328.
143. Sun X, Griffith M, Pasternak JJ, Bernard BR. Low temperature growth, freezing survival, and production of antifreeze protein by the plant growth promoting rhizobacterium *Pseudomonas putida* GR12-2. Can J Microbiol 1995; 41:776-784.
144. Hew CL, Yang DSC. Protein interaction with ice. Eur J Biochem 1992; 203:3-42.
145. Wolber PK. Bacterial ice nucleation. Adv Microb Physiol 1993; 34:203-237.
146. Lee MR, Lee RE Jr, Strong-Gunderson JM, Minges SR. Isolation of ice-nucleating active bacteria from the freeze-tolerant frog, *Rana sylvatica*. Cryobiology 1995; 32:358-365.
147. Wynn-Williams DD. Ecological aspects of Antarctic microbiology. Adv Microb Ecol 1990; 11:71-146.
148. Sippola K, Kanerva E, Murata N, Aro E-M. A genetically engineered increase in fatty acid unsaturation in *Synechococcus* sp. PCC 7942 allows exchange of D1 protein forms and sustenance of photosystem II activity at low temperature. Eur J Biochem 1998; 251:641-648.
149. Sakamoto T, Shen G, Higashi S, Murata N, Bryant DA. Alteration of low-temperature susceptibility of the cyanobacterium *Synechococcus* sp. PCC 7002 by genetic manipulation of membrane lipid unsaturation. Arch Microbiol 1998; 169:20-28.

Ecological aspects of cold-adapted plants with a special emphasis on environmental control of cold hardening and dehardening

O. Junttila[1*] and R. Robberecht[2]

[1] Department of Biology, University of Tromsø, N-9037 Tromsø, Norway
[2] Department of Range Resources, College of Forestry, Wildlife and Range Sciences, University of Idaho, Moscow, Idaho 83844-1135, USA

1 Introduction

The distribution of plant species is limited mainly by two environmental factors, the availability of water and temperature. Based on these factors, vegetation on the earth can be divided into a number of major vegetation zones. Biochemical reactions and growth processes in plants are temperature-dependent and have their minimum, optimum and maximum conditions. Temperature conditions are often suboptimal for growth, development, propagation and survival, and low temperatures commonly limit the growth of plants. In large areas of the earth´s land surface, temperatures decrease either periodically in a seasonal pattern, or episodically below 0°C (Fig. 1). Approximately one-third of the land area is frost-free and areas with the minimum temperature below -20°C comprise about 42%.[1,2] In these areas, frost can be one of the most important factors limiting distribution of plants. Of the estimated 250,000 species of vascular plants existing today, approximately 40% grow in tropical forests and the number of species strongly decreases with increasing altitude and latitude. Over evolutionary time, plant species have developed various adaptive strategies and mechanisms for survival in freezing conditions, but still the number of species capable of surviving periodic freezing is very low. However, native plant species are well adapted to survive in their ambient climatic conditions, and are damaged only if temperature and/or precipitation strongly deviates from the long-term mean, either transiently or more permanently in the event of changes in the global climate. For example, maximum hardiness of most northern hardwood tree species is far below the actual minimum temperature, even down to liquid nitrogen (-196°C), and the latitudinal or altitudinal tree line for such species is more related to other climatic factors than to the absolute

* Corresponding author; e-mail: olavi@ibg.uit.no

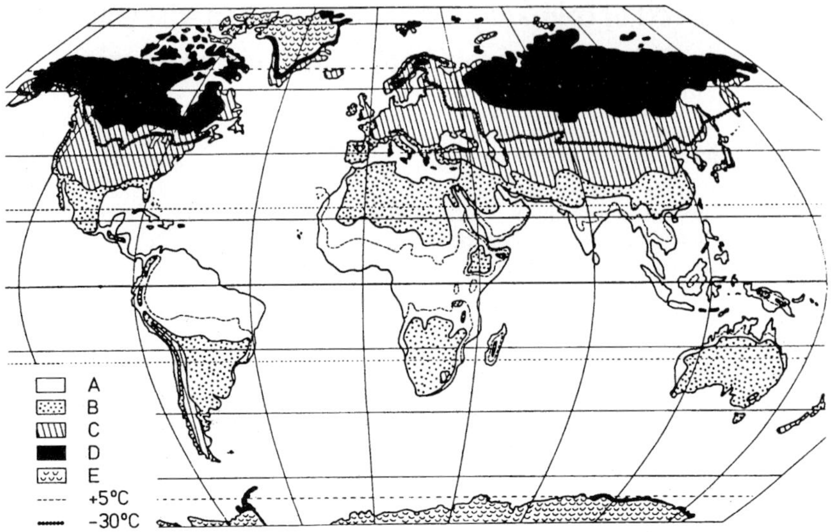

Fig. 1. A map of occurrence of frost on the earth
A frost-free zone; *B* zone with episodic frost down to -10°C; *C* zone with average annual minimum between -10 and -40°C; *D* zone with a average annual minimum below -40°C; *E* polar ice; (---) +5°C lowest temperature isotherm; (····) -30°C average annual minimum isotherm (reproduced with permission from ref. 2)

minimum temperature. For example, Grace[3] concludes that trees do not generally grow in places where the mean temperature of the warmest month is less than 10°C. For cultivated plants in horticulture, agriculture and forestry, low temperature and frost is, however, often a major limiting factor for production, and efforts to increase frost and winter survival through plant breeding and various cultural practicies make an important and intensive area of research.

The purpose of this chapter is to provide an overview on the general and ecological aspects of cold hardiness in plants, with a particular emphasis on environmental control of cold hardening and dehardening. Chilling tolerance will not be covered and responses of the plants to low positive temperatures are mentioned only briefly. Certain physiological aspects of cold hardiness are shortly discussed. More comprehensive reviews on low temperature ecology, including chilling resistance, can be found, e.g., in refs. 1,2,4. Körner and Larcher have reviewed the structural and functional features of plants from high mountain and tundra environments.[5]

2
Adaptation for frost hardiness

Higher plant species vary greatly in their ability to tolerate low temperatures and several classification systems can be found in the literature. Woodward has described five functional groups of broadleaf tree species in North and Central America based on minimum temperatures for survival: >10°C; 10 to 0°C; -1 to -14°C; -15 to -40°C; <-40°C.[6] Figure 2 gives some further examples of hardiness limits for various types of plants, and summarizes the basic mechanisms for frost tolerance. High arctic and high alpine areas, which provide the most extreme growing conditions in respect to low temperature, are characterized not only by cold winters but also by low summer temperatures and short growing seasons. For example, at alpine areas in southern Norway, the growing season, with mean temperatures above 5°C, occurs for only 74 d.[7] The number of vascular plant species growing under such extreme conditions is relatively low and only certain life forms such as small shrubs and perennial herbs are represented. In contrast, the tropical zones are dominated by tree species. The effect of altitude on the number of species of vascular plants is illustrated in Figure 3 with data from southern Norway.[8] The decrease in number of species with increasing altitude is caused by a combination of several factors, including low summer temperatures and long winter seasons. Because of the short and cold summer, most species need two or more seasons to complete their growth cycle from seed to seed. Both alpine and arctic vegetation is dominated by perennial plants:[9] only one annual species (*Koenigia islandica*) is known on Svalbard. In general, arctic species are often able to retain a relatively high rate of respiration at low temperatures and allocate a significant proportion of their photosynthates to below-ground parts.[10] Furthermore, they have a lower threshold temperature for leaf growth and for hydraulic conductance than temperate species.[6] These responses represent unique adaptations to very marginal temperature conditions. On the other hand, the northern tundra ecosystems are also considered to be comparatively recent, and contain species which seem not to be well adaptive to these conditions.[11]

The most cold-hardy species are found in the northern hemisphere. Because of the mild oceanic winters in the southern hemisphere, cold-hardy tree species that could withstand freezing below -30°C seem not to have evolved.[12] For example, the cold-hardy conifers in the southern hemisphere show the same order of resistance as species from warm temperate areas of Japan. The low level of cold resistance of the broadleaf trees in the southern hemisphere coincides with their evergreen habit and tendency to lack specialized resting buds.[12] According to Lipp et al., some selection for freezing tolerance mechanisms has even occurred at high elevations in Hawaii where air temperatures may frequently be below 0°C during winter.[13] However, most of the plant species in these areas do not exhibit characters of typical freezing-tolerant species.[13] On the other hand, Sakai has shown that even tropical willow (*Salix*) species have a genetic potential for extensive cold hardening.[14]

Heritable differences in frost hardiness between various cultivars and between ecotypes of species with a wide geographic range of distribution are well known and described in numerous studies, particularly with plant species that are of interest for forestry, agriculture or horticulture. Such adaptation can be observed along both latitudinal and altitudinal gradients. Generally, the response is a clinal one, as illustrated recently for *Pinus sylvestris*.[15] In populations of *Pinus contorta*, cold hardiness increased with increasing altitude of the population.[16] Frost hardiness is inherited as a complex quantitative character, which is greatly influenced by envi-

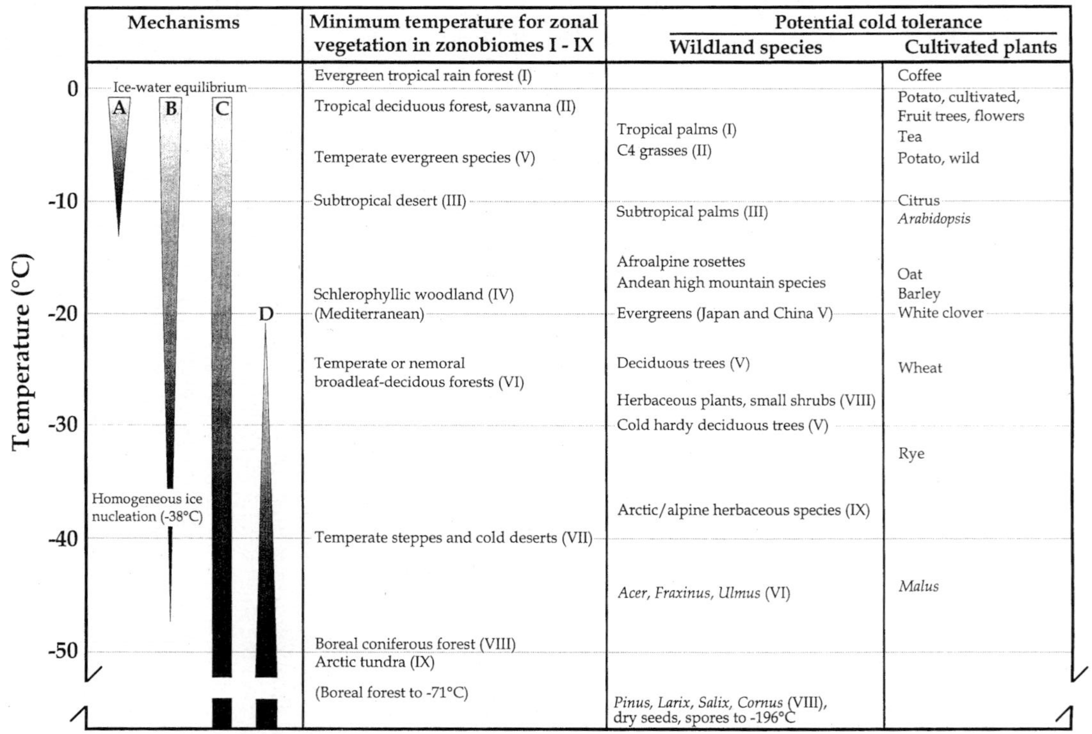

Fig. 2. Cold hardiness in vascular plants
(**1st column**) Temperature ranges for the main mechanisms for cold hardiness: *A* supercooling, avoidance of all freezing; *B* deep supercooling, avoidance of intercellular freezing in xylem ray parenchyma and buds of certain tree species; *C* tolerance of extracellular freezing; *D* intercellular glassformation (vitrification); (**2nd column**) Zonal vegetation based on the nine zonobiomes classified by Walter, and their minimum temperatures (data collected from ref. 76); (**3rd and 4th columns**) Examples of potential cold tolerance (data collected from refs. 2,24,27,49,77–86). For some plant species and categories, their original zonal vegetation is indicated by the Roman numerals

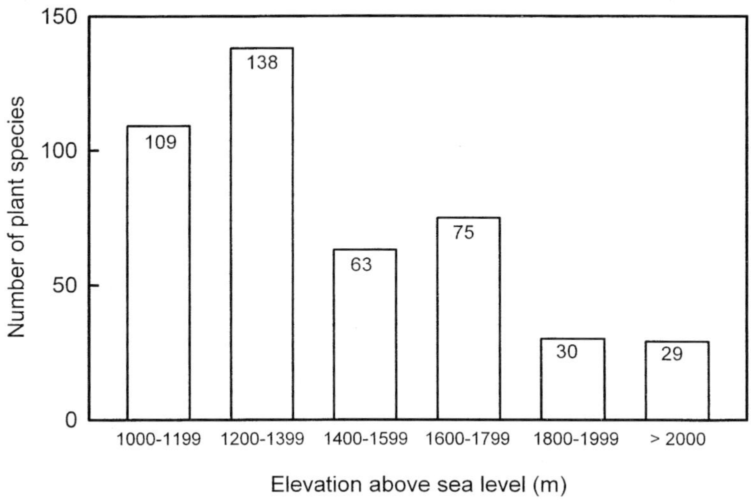

Fig. 3. The elevational distribution of vascular plant species in southern Norway (data from ref. 8)

ronmental conditions.[17] Crosses between contrasting latitudinal ecotypes give F_1-generations with approximately intermediate hardiness compared with the parents (for a review see ref. 18). The genetic background for cold hardiness has been studied in several cultivated plant species. In wheat, for example, the major genes for cold hardiness are in chromosomes 5A and 5D, but four other chromosomes also carry genes controlling frost resistance and winter hardiness.[19] Molecular studies have identified a number of cold-induced genes, which may encode for cold hardiness-related proteins, and rapid development in this area of cold hardiness research is expected. More controversial are the observations that indicate that the climate during sexual reproduction may influence the adaptive properties of *Picea abies* progenies in respect to cold acclimation.[20] Mechanisms for this effect and whether it may have any importance for adaptation to changing climatic conditions are unknown.

3
Mechanisms for frost hardiness

Freezing of water is the most critical single event in respect to frost hardiness of plants. Generally, plant cells will survive subzero temperatures as long as ice nucleation is avoided. Ice nucleation can be either heterogeneous and initiated by external nuclei present in the water, or homogeneous and initiated by aggregates of water molecules (for further discussion see ref. 21). In plant tissue, ice formation is normally initiated extracellularly in the apoplast, intercellular areas and in vascular tissue. Occurrence of intracellular freezing is dependent, among other things, on the freezing rate, the degree of supercooling, and on the properties of plasmalemma. Under natural conditions intracellular ice formation is lethal. Frost survival of plants is based on two main strategies: avoidance or tolerance of freezing.[22] Avoidance can be either avoidance of all ice formation or of intracellular freezing, while tolerance is limited to tolerance of extracellular freezing (Fig. 2). Often these two mechanisms are combined. For example, herbaceous species avoid extreme frost due to snow cover and have developed some tolerance of freezing, although not necessarily enough to tolerate the absolute minimum air temperature at the location. A similar situation can be found for underground plant parts. In very cold-hardy tree species, intracellular ice crystallization can be avoided due to glass formation (vitrification).[23]

The critical role of water is clearly demonstrated in dry seeds. For example, lettuce seeds with a moisture content less than 13% of fresh weight survived immersion in liquid nitrogen. At a moisture content of 15%, cold hardiness dropped to -30°C. At a moisture content between about 15 and 40% the hardiness declined linearly to about -20°C.[24] Thus, freezing stress in dry seeds is effectively avoided through low water content. A similar strategy can be seen in some lichen and moss species, implying that they may change relatively rapidly from a frost-sensitive to a frost-tolerant state – irrespective of the season. Only a few genera of vascular plants contain poikilohydric species.[25] Also, several studies with higher plants that have evolved mechanisms for freezing tolerance show close correlation between the content of water and level of cold hardiness.[22,26,27] Further support comes from a recent study by Yoshida et al., who showed that cold hardening in various cultivars of wheat comprised two stages: the first stage in development of hardiness was related to a reduction in water content, the second stage to the physical state of water without much reduction in water content (Fig. 4).[28]

Generally, succulent plants are not capable of surviving freezing. Cactus species are mostly succulent, grow in warm deserts, and do not acclimate to low temperatures. In a study on cold tolerance of 18 species of cacti, Nobel found that species growing at elevations below 1200 m in a warm desert survived temperatures down to -4 to -7°C.[29] Species that range up to 300 m in southeastern Wyoming were hardy to -17 to -20°C. He also showed that some of the studied species were capable of cold hardening from 2 to 4°C within about a week when exposed to low temperatures. Chlorenchyma tissue of one species (*Carnegiea gigantea*) was found to supercool to -5°C.[29] Some cactus species grow also in Patagonia, Canadian prairies, and on high mountains in South America.[9] On snow free areas in Canadian prairies, they can be exposed to temperatures down to -50°C. The freezing tolerance of a cacti *Opuntia fragilis*, from southern Saskatchewan, increased from -7 to -50°C over a 1-month period in the fall, and was accompanied by a decrease in water content from 5.4 to 2.6 g H_2O g^{-1} dry weight.[30]

Cell membranes, particularly plasmalemma and thylakoid membranes, may be the primary sites for cell injury due to freezing,[31] but mechanisms for cell injury are complex and will not be dealt in this chapter (for further discussion see ref. 32). However, one specific type of damage caused by freezing that should be mentioned here concerns ice formation in the xylem when plants are subjected to subzero temperatures. Freeze-thaw cycles can result in formation of air bubbles in xylem and disturb water transport. Langan et al. have suggested that in evergreen chaparral

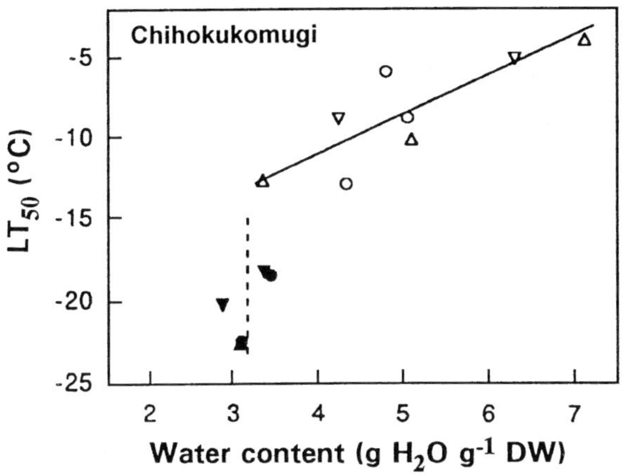

Fig. 4. Relationship between crown tissue water content and frost hardiness in wheat (cv. Chihokukomugi). There was a linear correlation between tissue water content and hardiness when temperature was above 0°C. When temperature dropped below 0°C, increase in frost hardiness was no longer related to water content but to the physical state of tissue water. Field data for 3 years (modified and reproduced with permission from ref. 28)
LT_{50} lethal temperature for 50% of the samples, *DW* dry weight

shrubs in California, a freeze-thaw-induced embolism in the xylem can be a possible factor limiting the growth, survival and distribution of these species.[33] In some species, the freezing-induced embolism occurred at all hydration levels, while in others this embolism occurred when the plants had reduced water potentials due to a drought stress.

Another aspect of frozen xylem tissue is the winter desiccation or "frost drying", a problem particularly for evergreens in mountain zones when the soil is frozen and the leaves are not protected by snow. In deciduous tree species number and structure of the bud scales, e.g. thickness of cuticula, are also important for protection of the bud meristems against desiccation. For example, the buds of *Sorbus aucuparia* are more resistant to desiccation stress than buds of *Betula pubescens*.[34] At timberline areas, winter desiccation and low summer temperature are stronger limiting factors than the minimum temperatures during winter.

4
Survival of episodic frost

Episodic frost events may occur in most vegetation zones, even at high elevations in equatorial latitudes. In these zones frosts result from a negative net radiation balance at night. Episodic frosts during the growing season in temperate, boreal, and arctic zones are not uncommon in spring and fall, and may represent a difficult problem, particularly for cultivated plants. Such frost events often occur relatively rapidly and allow little time for acclimation. For example, in the alpine zones of the equatorial high mountains, night temperatures may sometimes be as low as -12 to -15°C, while the air temperature on sunny days can rise to 20°C.[35] Survival during episodic frosts in such environments must be based primarily on constitutive mechanisms - most commonly on avoidance of ice nucleation due to supercooling, but in some cases on tolerance of extracellular freezing. However, rapid acclimation processes occurring within hours cannot be excluded. In citrus (*Poncirus trifoliate*), e.g. freezing tolerance in leaves and stem was enhanced already after a 6-h exposure to a hardening temperature of 10°C.[36] Also, in leaves of southern beech (*Nothofagus* sp.) increased cold hardiness and proline content were observed within 24-h exposure to low temperature.[37]

Lipp et al.[13] found that four out of five high-elevation Hawaiian plant species appeared to tolerate extracellular ice formation, which occurred at -5 to -9°C. Ice formation was probably initiated in the vascular tissue. Osmotic potential in these plants was lowered during the night due to an increase in levels of low-molecular-weight solutes, which helped to retain water during freezing and provide noncolligative protection to cell membranes. *Dubautia menziesii*, one of the five species studied, did not tolerate tissue freezing. Frost damage in this species has been observed after relatively extreme subzero temperatures. Leaves of a palm species, *Trachycarpus fortunei*, growing in warm temperate regions provides an example of an effective survival due to supercooling; they are able to supercool down to -14°C.[38] In a study on cold resistance in high desert Andean plants (3200-3700 m a.s.l.) it was found that low-statured plants (less than 20 cm tall) showed freezing tolerance (-9 to -20°C), while tall shrubs (27–90 cm) survived through freezing avoidance through supercooling (-5 to -13°C).[39] Certain afroalpine giant rosette plants have a special freezing avoidance mechanism to survive night frost.[40] The central leafbud is protected by an insulating layer of rosette leaves. These leaves have a nyctonastic movement and they form a nightbud that effectively delays cooling of the central, meristematic parts. This movement is related to a exudation and reabsorption of solutes and water by the rosette leaves.[40]

The phenological stage of the plant during an episodic frost can influence the degree and type of frost injury. For example, episodic frost during growing season may cause considerable loss in crop yield, particularly if frosts occur during flowering. Frost survival of flowers is normally based on freezing avoidance. Hummel and Moore[41] have shown that strawberry flowers were strongly damaged (44–95%) already at -1.5°C when ice nucleation was induced at -1.0°C, but survived to -6.0°C (16–52% damage) when frozen without ice nucleation. Similar results were obtained in studies of Norwegian cultivars of strawberry (Dzabic and Junttila, unpubl. data). This study also showed that flower buds were more resistant than open flowers, and that unopened flower buds showed some tolerance of freezing. In species such as *Corylus avellana* that flower during winter, flowers have been stated to survive temperatures down to -15°C (for references see ref. 3). Such flowers may provide interesting objects for further studies.

In Fennoscandia and areas with a similar climate, frost is quite frequent during the summer, and temperatures down to -10°C in the middle of the growing season have been reported in southern Sweden.[42] In these areas, summer frost can be more injurious to forest trees than frost in winter. Growing shoots of Norway spruce

(*Picea abies*) and willow (*Salix* spp.) do not tolerate ice formation, and their survival of summer frost is based on supercooling, while Scots pine (*Pinus sylvestris*) tolerates ice formation to some degree.[42] Generally, pine species seem to be more resistant to summer frost than spruce species.[43]

At high arctic latitudes, frost events during the summer season are not uncommon. The mean summer temperatures are generally low and plants are continually exposed to temperatures that can induce cold acclimation. For example, leaves of an arctic cushion plant, *Saxifraga caespitosa*, survived almost -15°C when grown at 9°C, but only -5°C when grown at 21°C. This survival was based on freezing tolerance.[44] Other studies indicate that high alpine plants survived only -5 to -7°C during the growing season and responded only slightly to a hardening treatment at 0 or -3°C.[45] Körner and Larcher have calculated that high mountain plants in the Alps may possible experience tissue temperatures of -12 to -14°C in summer.[5] Such a situation would, of course, cause serious damage if the cold tolerance is limited to -7°C. Responses of high arctic species to frost during growing season, and their survival mechanisms, are not well studied.

The degree of supercooling of plant tissue is dependent, among other things, on the presence of either extrinsic or intrinsic ice nuclei. Thus, occurrence of ice nucleators is an important factor for avoidance of freezing and, consequently, for survival under episodic frosts. Since the discovery of ice nucleating active bacteria, active even at temperatures of -1.5 to -2°C,[46] biological ice nucleators have also been characterized from lichens, and several species of higher plants.[47] Various aspects of ice nucleation are discussed in more detail by Wisniewski and Fuller.[21] The role of antifreeze proteins for survival under episodic as well as periodic frosts is not known.

5 Survival of periodic frost

Plants that are adapted to climates with a periodic frost season are characterized by seasonal variation in frost hardiness. In woody species, which obtain very high levels of frost hardiness down to the temperature of liquid nitrogen, changes in hardiness are dramatic.[48,49] In these plants, the maximum capacity for cold hardiness is well below recorded minimum temperatures, and successful survival through the winter is mainly dependent on the timing of hardening and dehardening. In contrast, seasonal changes are rather moderate for plants that overwinter under snow cover.[1,50]

Changes in frost hardiness are primarily responses to climatic factors, and they are expressions of the capacity of plants to acclimate, to modify their structure and/or physiology to increase the probability of survival under frost. Naturally, the most important environmental factors causing cold hardening are those that provide cues to the onset of winter: decreasing temperature and photoperiod. The capacity for cold hardening is primarily a feature for above-ground parts, while roots have no or very limited capacity for hardening,[22] as also demonstrated for wheat by Zhou et al.[51] In cold-hardy tree species such as *Pinus sylvestris*, only older roots showed some hardening when exposed to 5°C, but the changes were not at all comparable to those occurring in aerial parts.[52] Similarly, Coleman et al. have shown that cold hardiness levels of roots of subalpine conifers from mild maritime climates (*Abies amabilis*, *Tsuga mertensiana*) or from cold continental climates (*Abies lasiocarpa*, *Pinus contorta*) was not correlated with distribution of these species.[53] Nonacclimated roots were killed by bulk freezing, while hardened roots tolerated bulk freezing to some degree.[53] The inability of roots to harden reflects the lack of strong selection for frost tolerance. Roots generally avoid low temperatures because of the insulating effects of soil and, particularly, snow cover; neither root structure nor physiology are adapted to freezing temperatures. In woody plants, root growth is not affected by photoperiod, and there is no evidence for the existence of dormancy in roots comparable to dormancy in buds. Northern mountain species that grow in snow-free areas and develop very high levels of hardiness in aerial parts might provide interesting objects for further studies on root hardiness. In fact, cold hardiness values of -20 to -30°C have already been reported for roots of conifers and alpine plants.[2,45]

5.1
Environmental regulation of frost hardening in herbaceous species

Cold acclimation in herbaceous species is primarily induced by temperatures slightly above or close to 0°C. For some herbaceous species, e.g. *Brassica*, exposure to slight frost is necessary for maximum rate of hardening.[54] Kacperska distinguishes between hardening caused by nonfreezing and that caused by subfreezing temperatures.[54] Similarly, Yoshida et al. described hardening in wheat crown as a two-stage process, the second stage being induced when temperature dropped below 0°C (Fig. 4).[28] Acclimation is normally a time-consuming process occurring over several days or weeks. However, in *Arabidopsis*, an annual/winter annual herb, significant hardening of leaves can be observed within 1 or 2 d when plants are kept at 1°C in light (Wanner and Junttila, unpubl. data). Perhaps leaf tissue, particularly mature leaf, is more capable of rapid modulations in cold hardiness than stem meristems. Cold hardiness in meristematic tissues involves also regulation of growth activity.

The rate of cold hardening, as well as the final level of frost hardiness, is dependent on both genotype and acclimation temperature. Several *Lolium* cultivars have shown such an interaction of temperature and time on cold hardening.[55] In nature, temperature conditions are seldom constant and may show significant diurnal fluctuation. The physiological effects of fluctuating temperature conditions are not necessarily similar to those evoked by corresponding mean temperatures. Such effects of fluctuating temperatures on cold hardening, and dehardening, have been characterized in *Lolium*.[56] Low temperature enhancement of cold hardening also occurred under fluctuating temperature conditions, irrespective of whether the low temperature exposure (of at least 4–8 h daily in these experiments) occurred during the light or dark period. In perennial grass species and in cereals, the crown tissue generally has the best capacity for cold hardening. In contrast, leaves of strawberry acclimated to tolerate about -20°C while, after a similar cold hardening treatment, the crown tissue died at -6 to -9°C (Dzabic and Junttila, unpubl. data).

Photoperiod generally has a minor effect on cold hardening in herbaceous plants, although there are notable examples of photoperiodic influences. For example, cold hardening in stolons of white clover (*Trifolium repens*) is enhanced by short photoperiods.[57] The rate of elongation growth in stolons is reduced by short days relative to long days, but there is no induction of growth cessation and dormancy. In a less hardy cultivar, AberHerald, some elongation growth occurred even at 6°C, but a hardy genotype from northern Norway showed no measurable elongation growth at this temperature.[27]

Sufficient production of photosynthates is necessary for proper cold hardening, and therefore enhancement of hardening by light is often demonstrated. For example, frost hardiness, as well as an increase in sucrose content, in leaves of *Arabidopsis* was induced at 1°C in light but not in darkness (Wanner and Junttila, unpubl. data). In a northern ecotype of white clover, reduced irradiation during a short day treatment at 18°C, prior to cold hardening for 2 weeks at 6°C, and followed by 2 weeks at 0.5°C, had a negative effect on cold hardening. However, cold hardiness of plants grown under good light conditions for 2 weeks at 18°C, and then for 2 weeks at 6°C, was not affected by differences in irradiance (25–400 μmol m^{-2} s^{-1}) given during a final 2-week treatment at 0.5°C. Even plants kept in darkness during the period at 0.5°C developed hardiness (Junttila, unpubl. data). The role of photosynthesis in cold hardening is indirectly demonstrated by results presented in Figure 5. Increase in frost hardiness, as well as decrease in water content of white clover stolons, was prevented by defoliation at the beginning of the hardening treatment

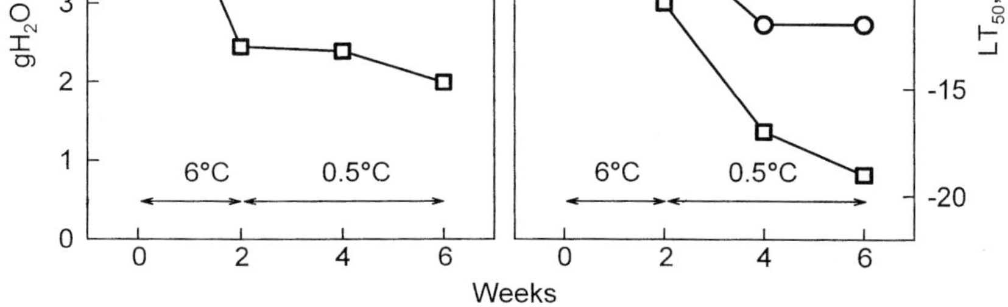

Fig. 5. Effect of defoliation on (**A**) water content and (**B**) cold hardiness in stolons of white clover. Before defoliation, plants were grown for 2 weeks at 18°C in 12-h photoperiod. After defoliation, plants were kept for 2 weeks at 6°C followed by 4 weeks at 0.5°C. Cold hardiness was tested as described[27]

LT_{50} lethal temperature for 50% of the samples, *DW* dry weight

(Junttila, unpubl. data). Defoliation 2 weeks before hardening completely prevented increase in cold hardiness during low-temperature treatment.

In some cases the hardening ability seems to be related to the ability to retain relatively high photosynthetic activity at low temperatures which are needed for development of hardiness. In both wheat and rape, the more hardy winter cultivars were able to maintain higher net assimilation rates and higher *in-situ* CO_2 exchange rates at low temperature than the respective spring cultivars. This allows the overwintering plants to maximize the production of sugars with possible cryoprotective function and to accumulate sufficient carbohydrate storage reserves to support basal metabolism and regrowth in the spring.[58]

5.2 Environmental regulation of frost hardening in woody species

Low temperature is the most universal factor inducing cold hardiness in woody species. Significant cold hardening is obtained at low positive temperatures, but in woody plants the maximum cold hardiness is only obtained after an exposure to freezing temperatures.[48]

Contrary to herbaceous plants, cold hardening in many temperate-zone woody plant species is strongly affected by photoperiod; short-day photoperiods generally enhance the level of cold hardiness.[48] Timing of hardening and dehardening is of primary importance for winter survival of woody plants and photoperiod is the most stable and accurate environmental signal for initiation of hardening processes. In free-growing species, short day-induced hardening is primarily an indirect effect related to the induction of growth cessation and bud set, which is a prerequisite for obtaining the maximum cold hardiness. The importance of apical growth cessation for cold hardening in these species has been shown in several studies (e.g. refs. 18,48) and is illustrated for hybrid aspen in Figure 6. As shown in Figure 6, even an extended exposure to low temperature had a minor effect only on frost hardiness in plants where cessation of apical elongation growth was prevented by too long a photoperiod prior to exposure to low temperature. Recent studies with transgenic hybrid aspen have confirmed that in free-growing species photoperiodic induction of cold hardiness is strictly dependent on apical growth cessation.[59,60] Transgenic plant lines in which overexpression of the oat phytochrome A gene pre-

Table 1. Cessation of apical elongation growth and cold hardiness in hybrid aspen as affected by overexpression of the oat phytochrome A (PhyA OE)
Plants were grown at 18° in 12-h photoperiod for 7 weeks before they were transferred to cold-hardening conditions as indicated (for methods see ref. 60)

Line	Bud set (%)	LT50 (°C)		
		3 weeks at 6°C	3 weeks at 6°C+ 3 weeks at 0.5°C	3 weeks at 6°C+ 6 weeks at 0.5°C
Wild type	100	-15 ± 0.5	-35 ± 1.5	<-50
PhyA OE	0	-4 ± 0.3	-6 ± 0.7	-6 ± 0.3

LT_{50} lethal temperature for 50% of the samples

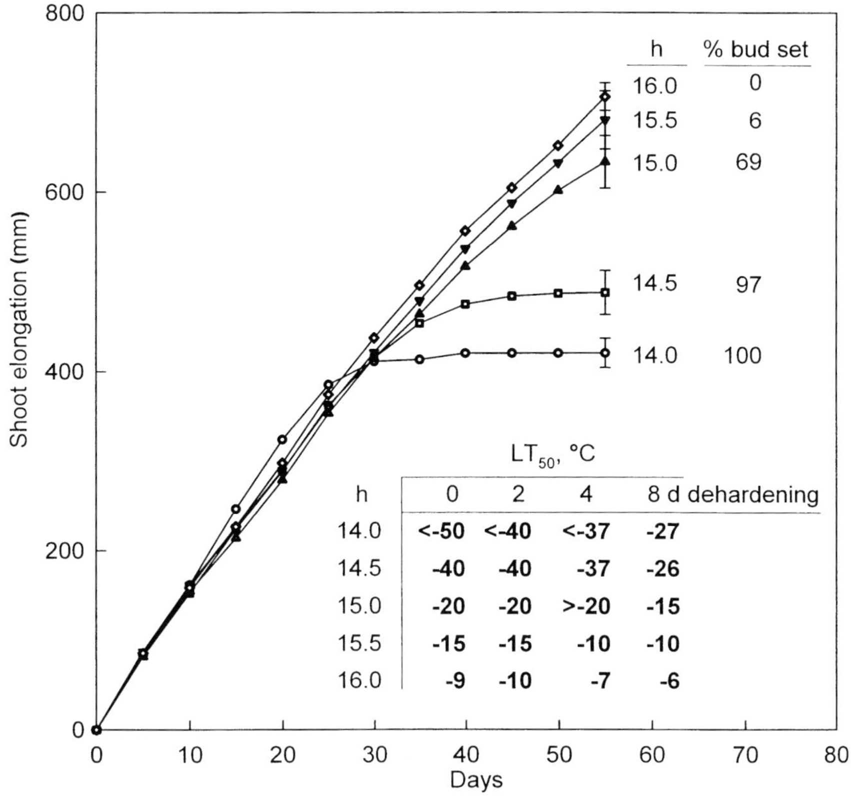

Fig. 6 Effect of photoperiod on cessation of apical growth in hybrid aspen (*Populus tremula* x *P. Tremuloides*). During photoperiodic treatments plants were grown at 18°C. Percentages of bud set at the end of these treatments are given. For cold hardening, plants were then exposed for 3 weeks to 6°C and 7 weeks to 0.5°C in 12-h photoperiod. For dehardening (2–8 d), plants were transferred back to 18°C and 24-h photoperiod. Cold hardiness was measured as described[60] LT_{50} lethal temperature for 50% of the samples

vented short day-induced growth cessation, did not cold acclimate even when exposed to low temperatures for several weeks (Table 1). Thus, in *Populus*, a short photoperiod is not able to initiate cold hardening, unless it causes cessation of elongation growth. Tree species with a same pattern of shoot growth as *Populus* are expected to respond similarly.

Due to the importance of photoperiod as an environmental cue for the cessation of apical growth and cold hardening, temperate zone tree species with a wide distribution range have adapted to the local photoperiod and exhibit a clinal response.[61] The critical photoperiod for cessation of growth and induction of cold acclimation increases with increasing latitude for origin of the population, reaching approximately 22 h at the latitude of 69 to 70°N.[61–63] There is strong evidence

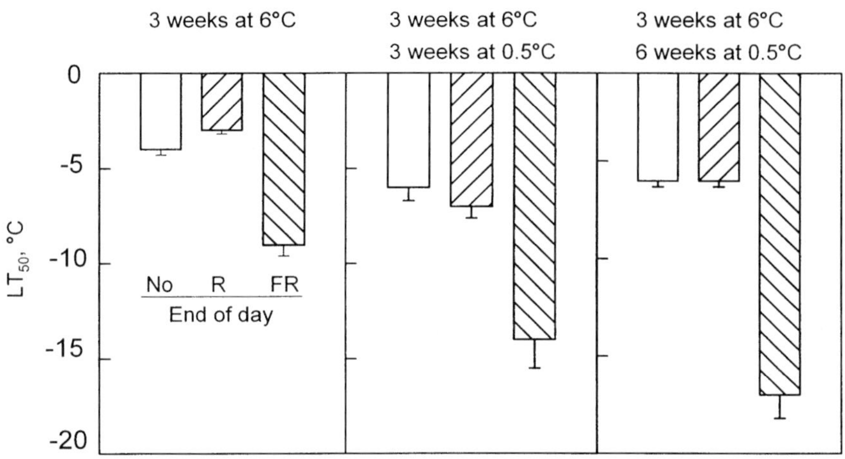

Fig. 7. Cold hardiness in transgenic hybrid aspen, which overexpressed the oat phytochrome A, as affected by end-of-day treatment and length of acclimation period. Results are means of four lines ± standard deviation. End-of-day treatments were given at 18°C, 12-h photoperiod, for 7 weeks. No bud set was induced by any of the treatments. Cold hardiness was measured as described[60] *No* no end-of-day, no shading; *R* 15 min red; *FR* 15 min far red; LT_{50} lethal temperature for 50% of the samples

indicating that plant hormones, particularly gibberellins, are involved in photoperiodic regulation of shoot elongation in trees such as P*opulus*.[59]

Short days can induce limited cold hardiness also in conifer species such as *Chamaecyparis nootkatensis* that do not develop winter buds.[64] However, these species are generally much less hardy than bud-forming tree species. Short days may also have a direct influence on hardening without affecting cessation of growth, as in needles of Scots pine (*Pinus sylvestris*).[65]

Phytochrome, discovered as the photoreceptor involved in red/far red reversible reactions, appears to be the receptor involved in a wide variety of developmental responses of plants to changes in light conditions, including photoperiodism. Five different phytochromes have been identified so far: phytochromes A, B, C, D and E. The roles of the various phytochromes in photoperiodism are still unclear, but both phytochrome A and B have been implicated.[66] Phytochrome-mediated responses other than photoperiodic ones can also affect cold hardening. Results of experiments with both *Cornus* and *Populus* indicate a significant enhancement of hardening by far red end-of-day treatments.[60,67] In *Populus*, end-of-day treatment with far red light significantly enhanced frost tolerance without inducing growth cessation (Fig. 7). Thus, phytochrome may have direct effects on acclimation processes, independent of regulation of shoot growth. Molecular and transgenic approaches open new possibilities to study the environmental regulation mechanisms for cold hardening and dehardening, and we can expect a rapid progress on this field.

In all very cold-hardy species, i.e. those that can survive temperatures below -50°C, and in many less cold-hardy species, the level of cold hardiness is based

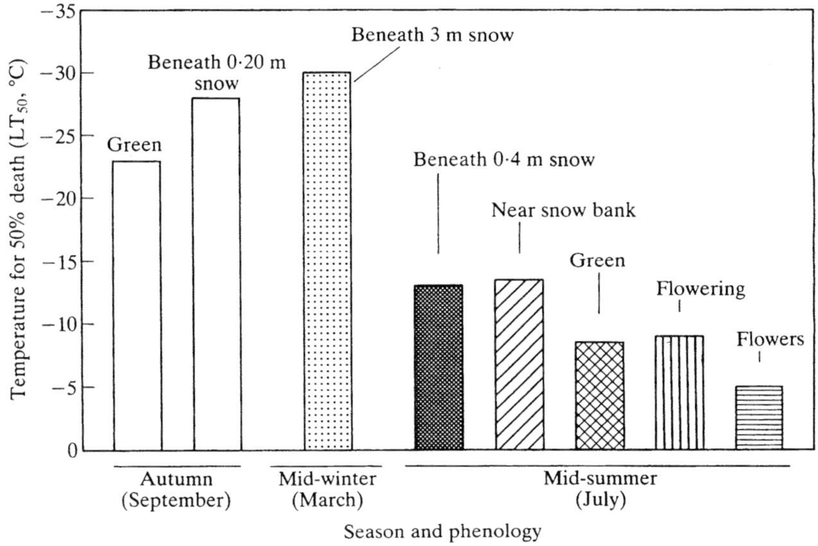

Fig. 8 Cold hardiness in *Silene acaulis*, an arctic-alpine cushion plant, as affected by the time of year and the phenological gradient in midsummer (reproduced with permission from ref. 87)

on tolerance of extracellular freezing. During acclimation, tissues and cells undergo a multitude of physiological and biochemical changes (accumulation of abscisic acid, reduced water content, changes in lipid composition, accumulation of putative cryoprotective compounds, particularly carbohydrates, accumulation of dehydrines, etc.), which increase the capability of the cells to function at low temperatures and to survive freezing stress.[22,68–70] In certain species, xylem ray parenchym cells and buds, particularly flower buds, survive frost by deep supercooling.[21,71,72] The approximate distribution limit for such species is the -40°C isotherm. Although supercooling in these tissues is partly based on anatomical properties and structures, seasonal changes in degree of supercooling are correlated with the level of hardiness. For example, flower buds of persimmon (*Diospyros kaki*) had only a primary exotherm, but the exotherm temperature coincided with hardiness of primordia. Both the bud exotherm and the secondary exotherms in stem tissue showed similar seasonal fluctuation as hardiness values determined in freezing tests.[73] Further examples are provided in ref. 72. Wisniewski and Ashworth have shown that similar changes in cell ultrastructure during cold hardening were observed both in bark (tolerance of freezing) and in xylem ray parenchyma cells (deep supercooling) of peach (*Prunus persica*).[68]

5.3 Environmental regulation of dehardening

Dehardening is primarily a response to increasing temperatures, and the rate of dehardening is positively correlated with the temperature. In northern mountain

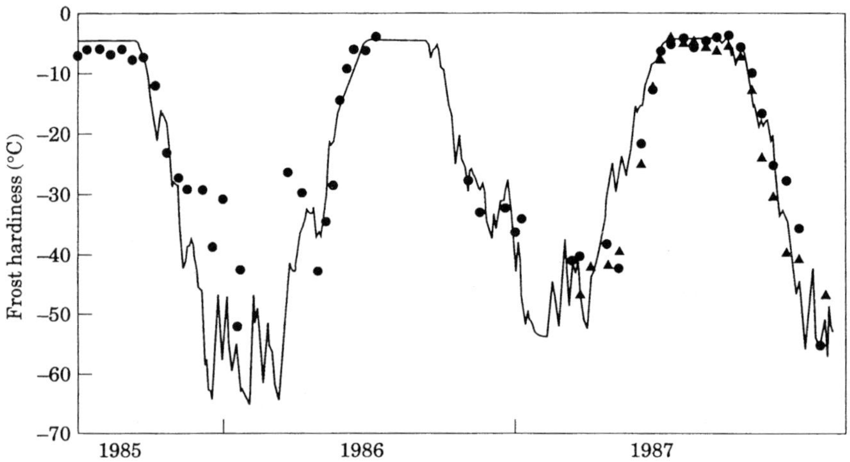

Fig. 9. Model-predicted and measured frost hardiness in *Pinus sylvestris* using impedance method (●) and visual damage scoring (▲); (reproduced with permission from ref. 75)

areas, both the winter and spring conditions can be very variable due to great local differences in snow cover. As illustrated in Figure 8, these conditions affect dehardening of mountain plants. Examples of responses between contrasting cultivars include the effects of a range of temperatures on dehardening in *Lolium*.[56] Studies on white clover show the effects of temperature on the rate of dehardening, and that decrease in hardiness was preceded by a rapid decrease in sucrose content in stolons.[27]

Dehardening can also be relatively slow, as shown by studies on *Populus* (Fig. 6). In woody plants, temperature effects on dehardening, particularly on the rate of dehardening, may vary with phenological stage. This is, at least in part, a reflection of changes in the dormancy level of the tissue. The ability of the tissue to reharden, after a dehardening, may also vary during the winter season. In order to sustain the maximum level of hardiness, woody plant tissue must remain exposed to freezing temperatures. Seedlings of *Pinus contorta* and *P. sylvestris* were not able to retain their cold hardiness when kept at 5.5°C from December to late March, and *P. contorta* lost some of its cold hardiness even when stored at -8.9°C, compared with -13.5°C.[74] However, *Picea abies* maintained its level of cold hardiness under all treatments, and cold hardiness of the needles was strongly correlated with sugar content.[74]

The extensive information on the effects of environmental factors on cold hardiness accumulated from laboratory and field studies provides a basis for development of models to predict cold hardening and dehardening events. Such models may have practical importance as tools to estimate effects of climatic changes on plant cold hardiness In Scots pine, a dynamic model based on temperature, photoperiod and the phase of annual development appeared to give an accurate prediction of frost hardiness during the most frost-sensitive phases in autumn and spring (Fig. 9).[75]

6
Conclusions

Plant species show a remarkable variety of adaptations for growth and survival at low and freezing temperatures. So far, very little is known about the genetic background for cold hardiness, and this remains a significant challenge for further research. Another important aspect of cold hardiness is the development of mechanisms for acclimation, i.e. ability to modify structures and functions as a response to environmental conditions. This capacity for acclimation is a specific feature for plants and makes them able to survive under changing environmental conditions. Consequently, acclimation processes are of primary importance for freezing survival of plants. It is an important goal for future research to aquire more understanding of the ecological, physiological and molecular aspects of the mechanisms for cold hardening, and dehardening. Deeper understanding of the acclimation processes will increase our possibilities to predict responses to changing climatic conditions. Both *in-situ* studies and experimental approaches are needed to achieve these goals. Established model species for molecular studies provide very efficient experimental systems for studies on mechanisms for cold hardiness, and exciting results are to be expected from these studies. However, although the number of plant species which are adapted to frost is rather low, only a small fraction have been studied in any detail. Thus, extension of the cold hardiness research to cover more of this diversity will be a continuous challenge in cold hardiness research.

7
References

1. Larcher W. Physiological Plant Ecology, 3rd ed. Berlin: Springer, 1995.
2. Larcher W, Bauer H. Ecological significance of resistance to low temperature. In: Encyclopedia of Plant Physiology, vol 12A. Berlin: Springer, 1981:403-437.
3. Grace J. Tree lines. Phil Trans R Soc London, 1989; B324:233-245.
4. Sakai A, Larcher W. Frost Survival of Plants. Berlin: Springer, 1987.
5. Körner Ch, Larcher W. Plant life in cold climates. In: Long SP, Woodward FI, eds. Plants and Temperature. Cambridge: The Company of Biologists Ltd, 1988:25-58.
6. Woodward FI, Kelly CK. Plant functional types: towards a definition by environmental constraints. In: Smith TM, Shugart HH, Woodward FI, eds. Plant Functional Types: Their Relevance to Ecosystem Properties and Global Change. International Geosphere-Biosphere Programme Book Series. Cambridge: Cambridge Univ Press, 1997:47-65.
7. Aune B. Klima (Climate). Nasjonalatlas for Norge. Hønefoss: Statens Kartverk, 1993 (in Norwegian, with English summary).
8. Dahl E. Flora and plant sociology in Fennoscandian tundra areas. In: Wielgolaski, FE, ed. Fennoscandian Tundra Ecosystems, Part 1. Plants and microorganisms. Berlin: Springer, 1975:62-67.
9. Billings WD. High mountain ecosystems. Evolution, structure, operation and maintenance. In: Webber PJ, ed. High Altitude Geoecology. Boulder, Colorado: Westview Press, Inc, 1979:97-125.
10. Cooper JE. An ecophysiological investigation of some species of arctic and temperate *Ranunculus* L. with special respect to climate warming. Univ Bradford, UK: PhD Thesis, 1996.

11. Kallio P. Adaptation and evolution at the northern limits of life. Acta Univ Oul F 1. Scripta Acad 1984:131-150.
12. Sakai A, Paton DM, Wardle P. Freezing resistance of trees of the south temperate zone, especially subalpine species of Australasia. Ecology 1981; 62:563-570.
13. Lipp CC, Goldstein G, Meinzer FC, Niemczura W. Freezing tolerance and avoidance in high-elevation Hawaiian plants. Plant Cell Environ 1994; 17:1035-1044.
14. Sakai A. Freezing resistance in willows from different climates. Ecology 1970; 51:458-491.
15. Hurme P, Repo T, Savolainen O, Pääkkönen T. Climatic adaptation of bud set and frost hardiness in Scots pine (*Pinus sylvestris*). Can J For Res 1997; 27:716-723.
16. Rehfeldt GE. Cold acclimation in populations of *Pinus contorta* from the northern Rocky Mountains. Bot Gaz 1980; 141:458-463.
17. Stushnoff C, Junttila O, Kaurin Å. Genetics and breeding for cold hardiness in woody plants. In: Kaurin Å, Junttila O, Nilsen J, eds. Plant Production in the North. Oslo: Norwegian Univ Press, 1985:141-156.
18. Junttila O. Physiological responses to low temperature. Annal Sciences Forestrières 1989; 46 (Suppl.):604-613.
19. Sutka J, Galiba G, Quarrie SA, Veisz O, Snape JW. Cytogenetic studies on frost resistance in wheat (*Triticum aestivum* L.). In: Dörffling K, Brettschneider B, Tantau H, Pithan K, eds. Crop Adaptation to Cool Climates. COST 814 Workshop October 12-14, 1994. Brussels: Europian Commission, 1994:377-388.
20. Johnsen Ø, Skrøppa T, Tigerstedt PMA. Adaptive properties of *Picea abies* progenies are influenced by environmental signals during sexual reproduction. Euphytica 1996; 92:67-71.
21. Wiesniewski M, Fuller M. Ice nucleation and deep supercooling in plants: new insights using infrared thermography. In this book.
22. Levitt J. Responses of Plants to Environmental Stresses, vol. 1. Chilling, Freezing and High Temperature Stresses. New York: Academic Press, 1980.
23. Hirsch AG. Vitrification in plants as a natural form of cryoprotection. Cryobiology 1987; 24:214-228.
24. Junttila O, Stushnoff C. Freezing avoidance by deep supercooling in hydrated lettuce seeds. Nature1977; 269:325-327.
25. Hartung W, Schiller P, Dietz K-J. Physiology of poikilohydric plants. In: Behnke H-D, Esser K, Kadereit JW, Lüttge U, Runge M, eds. Progress in Botany 59. Berlin: Springer, 1997:299-327.
26. Junttila O, Stushnoff C, Gusta LV. Dehardening of flower buds on Saskatoon-berry, *Amelanchier alnifolia*, in relation to temperature, moisture content, and spring bud development. Can J Bot 1983; 61:164-170.
27. Svenning MM, Junttila O, Røsnes K. Frost tolerance and biochemical changes during hardening and dehardening in contrasting genotypes of white clover (*Trifolium repens* L.). Physiol Plant 1997; 101:31-37.
28. Yoshida M, Moriyama M, Shimokawa S, Nakamura Y. Seasonal changes in the physical state of crown water associated with freezing tolerance in winter wheat. Physiol Plant 1997; 99:363-370.
29. Nobel PS. Low-temperature tolerance and cold hardening of cacti. Ecology 1982; 63:1650-1656.
30. Ishikawa M, Gusta LV. Freezing and heat tolerance of *Opuntia* cacti native to the Canadian prairie province. Can J Bot 1996; 74:1890-1895.
31. Steponkus PL. Role of plasma membrane in freezing injury and cold acclimation. Annu Rev Plant Physiol 1984; 35:543-584.
32. Reaney MJT, Gusta LV. Modeling sequential responses of plant cells to freezing and thawing. In this book.
33. Langan SJ, Ewers FW, Davis SD. Xylem dysfunction caused by water stress and freezing in two species of co-occurring chaparral shrubs. Plant Cell Environ1997; 20:425-437.

34. Barclay AM, Crawford RMM. Winter desiccation stress and resting bud viability in relation to high altitude survival in *Sorbus aucuparia* L. Flora 1982; 172:21-34.
35. Beck E, Schulze ED, Senser M, Scheibe R. Equilibrium freezing of leaf water and extracellular ice formation in afroalpine 'giant rosette' plants. Planta 1984; 162:276-282.
36. Tignor ME, Davies FS, Sherman WB. Rapid freeze acclimation of *Poncirus trifoliata* seedlings exposed to 10°C and long days. HortScience 1997; 32:854-857.
37. Meza-Basso L, Guarda P, Rios D, Alberdi M. Changes in free amino acid content and frost resistance in *Nothofagus dombey* leaves. Phytochemistry 1986; 25:1843-1846.
38. Larcher W, Meindl U, Raiser E, Ishikawa, M. Persistent supercooling and silica deposition in cell walls of palm leaves. J Plant Physiol 1991; 139:146-154.
39. Squeo F, Rada F, Garcia C, Ponce M, Rojas A, Azócar A. Cold resistance mechanisms in high desert Andean plants. Oecologia 1996; 105:552-555.
40. Beck E, Senser M, Scheibe R, Steiger HM, Pongratz P. Frost avoidance and freezing tolerance in afroalpine 'giant rosette' plants. Plant Cell Environ 1982; 5:215-222.
41. Hummel RL, Moore PP. Freeze resistance of Pacific Northwest strawberry flowers. J Am Soc Hort Sci 1997; 122:179-182.
42. Christersson L. Frost damage during the growing season. In: Kaurin Å, Junttila O, Nilsen J, eds. Plant Production in the North. Oslo: Norwegian Univ Press, 1985:191-198.
43. Christersson L, von Fircks H, Sihe Y. Damage to conifer seedlings by summer frost and winter drought. In: Li PH, ed. Plant Cold Hardiness. New York: Alan R. Liss Inc, 1987:203-210.
44. Robberecht R, Junttila O. The freezing response of an arctic cushion plant, *Saxifraga caespitosa* L.: Acclimation, freezing tolerance and ice nucleation. Ann Bot 1992; 70:129-135.
45. Sakai A, Otsuga K. Freezing resistance of alpine plants. Ecology 1970; 51:665-671.
46. Lindow SE. The role of bacterial ice nucleation in frost injury to plants. Annu Rev Phytopathol 1983; 21:363-384.
47. Lee RE Jr., Warren GJ, Gusta LV., eds. Biological Ice Nucleation and its Applications. St. Paul, Minnesota: APS Press, 1995.
48. Weiser CJ. Cold resistance and injury in woody plants. Science 1970; 169:1269-1278.
49. Stushnoff C, Junttila O. Seasonal development of cold stress resistance in several plant species at a coastal and a continental location in North Norway. Polar Biol 1986; 5:129-133.
50. Kaurin Å, Junttila O, Hansen J. Seasonal changes in frost hardiness in cloudberry (*Rubus chamaemorus*) in relation to carbohydrate content with special reference to sucrose. Physiol Plant 1981; 52:310-314.
51. Zhou B-L, Arakawa K, Fujikawa S, Yoshida S. Cold-induced alterations in plasma membrane proteins that are spefically related to the development of freezing tolerance in cold-hardy winter wheat. Plant Cell Physiol 1994; 35:175-182.
52. Ryyppö A, Repo T, Vapaavuori E. Development of freezing tolerance in roots and shoots of Scots pine seedlings at non-freezing temperatures. Can J For Res 1998; 28:557-565.
53. Coleman MD, Hinckley TM, McNoughton G, Smit BA. Root cold hardiness and native distribution of subalpine conifers. Can J For Res 1992; 22:932-938.
54. Kacperska A Biochemical and physiological aspects of frost hardening in herbaceous plants. In: Kaurin Å, Junttila O, Nilsen J, eds. Plant Production in the North. Oslo: Norwegian Univ Press, 1985:99-115.
55. Gay AP, Eagles CF. Quantitative analysis of cold hardening and dehardening in *Lolium*. Ann Bot 1991; 67:339-345.
56. Eagles CF, Williams J . Hardening and dehardening of *Lolium perenne* in response to fluctuating temperatures. Ann Bot 1992; 70:333-338.
57. Junttila O, Svenning MM, Solheim B. Effects of temperature and photoperiod on frost resistance of white clover (*Trifolium repens*) ecotypes. Physiol Plant 1990; 79:435-438.
58. Hurry VM, Strand Å, Tobiæson M, Gardeström P, Öquist G. Cold hardening of spring and winter wheat and rape results in differential effects on growth, carbon metabolism, and carbohydrate content. Plant Physiol 1995; 109:697-706.

59. Olsen JE, Junttila O, Nilsen J, Eriksson M, Martinussen I, Olsson O, Sandberg G, Moritz T. Ectopic expression of oat phytochrome A in hybrid aspen changes critical photoperiod for growth and prevents cold acclimation. Plant J 1997; 12:1339-1350.
60. Junttila O, Olsen JE, Nilsen J, Martinussen I, Moritz T, Eriksson M, Olsson O, Sandberg G. Phytochrome A overexpression and cold hardiness in transgenic *Populus*. In: Li PH, Chen TH, eds. Plant Cold Hardiness. New York: Plenum Publishing Corporation, 1998:245-256.
61. Junttila O, Nilsen J. Growth and development of northern forest trees as affected by temperature and light. In: Alden JN, Ødum S, Mastrantonio JL, eds. Forest Tree Development in Cold Climates, Proc Int Symp June 18-24, Laugarvatn, Island, NATO. New York: Plenum Publ Corp, 1993:43-57.
62. Heide OM. Growth and dormancy in Norway spruce ecotypes (*Picea abies*), I. Interaction of photoperiod and temperature. Physiol Plant 1974; 30:1-12.
63. Håbjørg A. Photoperiodic ecotypes in Scandinavian trees and shrubs. Scientific Reports of the Agricultural College of Norway 1978; 57:1-20.
64. Puttonen P, Arnott JT. Influence of photoperiod and temperature on growth, gas exchange, and cold hardiness of yellow cypress stecklings. Can J For Res 1994; 24:1608-1616.
65. Beck E, Hansen J, Heim R, Schäfer C, Vogg G, Leborgne N, Teulieres C, Boudet AM. Cold hardening and dehardening of evergreen trees. In: Sandermann H Jr, Bonnet-Masembert M, eds. Contribution to Forest Tree Physiology, EUROSILVA, Dordan, France, November 7-10, 1994. Paris: INRA, 1995:171-193.
66. Smith H. Physiological and ecological function within the phytochrome family. Annu Rev Plant Physiol Plant Mol Biol 1995; 46:289-315.
67. McKenzie JS, Weiser CJ, Burke MJ. Effects of red and far red light on the initiation of cold acclimation in *Cornus stolonifera* Michx. Plant Physiol 1974; 53:783-789.
68. Wisniewski M, Ashworth EN. A comparison of seasonal ultrastructural changes in stem tissues of peach (*Prunus persica*) that exhibits contrasting mechanisms of cold hardiness. Bot Gaz 1986; 147:407-417.
69. Rutten D, Santarius KA. Cold acclimation of *Ilex aquifolium* under natural conditions with special regard to the photosynthetic apparatus. Physiol Plant 1988; 72:807-815.
70. Chen THH, Burke MJ, Gusta LV. Frezing tolerance in plants: An overview. In: Lee RE Jr, Warren GJ, Gusta LV, eds. Biological Ice Nucleation and its Applications. St. Paul, Minnesota: APS Press, 1995:115-136.
71. Wisniewski M. Deep supercooling in woody plants and the role of cell wall structure. In: Lee RE Jr, Warren GJ, Gusta LV, eds. Biological Ice Nucleation and its Applications. St. Paul, Minnesota: APS Press, 1995:163-182.
72. Quamme HA. Deep supercooling in buds of woody plants. In: Lee RE Jr, Warren GJ, Gusta LV, eds. Biological Ice Nucleation and its Applications. St. Paul, Minnesota: APS Press, 1995:183-200.
73. Kang SK, Motosugi H, Yonemori K, Sugiura A. Exothermic characteristics of dormant buds of persimmon (*Diospyros kaki* Thunb.) in relation to cold hardiness. HortScience 1997; 32:840-843.
74. Ögren E, Nilsson T, Sundblad LG. Relationship between respiratory depletion of sugars and loss of cold hardiness in coniferous seedlings over-wintering at raised temperatures: indications of different sensitivities of spruce and pine. Plant Cell Environ 1997; 20:247-253.
75. Leinonen I. A simulation model for the annual frost hardiness and freeze damage of Scots pine. Ann Bot 1997; 78: 687-693.
76. Walter H. Vegetation of the Earth. Ecological Systems of the Geo-biosphere, 3rd ed. Berlin: Springer, 1994.
77. Anderson JA, Gusta LV, Buchanan DW, Burke MJ. Freezing of water in *Citrus* leaves. J Am Soc Hort Sci 1983; 108:397-400.
78. Beck E, Scheibe R, Hansen J. Mechanisms of freezing avoidance and freezing tolerance in tropical alpine plants. In: Li PH, ed. Plant Cold Hardiness. New York: Alan R. Liss, Inc, 1987:155-168.

79. Burke MJ, Stushnoff C. Frost hardiness: A discussion of possible molecular causes of injury with particular reference to deep supercooling of water. In: Mussell H, Staples R, eds. Stress Physiology in Crop Plants. New York: John Wiley & Sons, Inc, 1979:198-224.
80. Gu Y, Sun ZJ, Bi HC, Huang SZ, Chai JH, Geng XM, He SA. Introduction and selection of hardy bread-leaved evergreen woody plants in central China. In: Li PH, ed. Plant Cold Hardiness. New York: Alan R. Liss, Inc, 1987:363-368.
81. Gusta LV, Tyler NJ, Chen THH. Deep undercooling in woody taxa growing north of the -40°C isotherm. Plant Physiol 1983; 72:122-128.
82. Hömmö L. Resistance of winter cereals to various winter stress factors – inter- and intraspecific variation and the role of cold acclimation. Agricultural Science in Finland, 1994; 3 (Suppl 1):1-32.
83. Kaku S, Iwaya M. Low temperature exotherms in xylems of evergreen and deciduous broad-leaved trees in Japan with reference to freezing resistance and distribution range. In: Li PH, ed. Plant Cold Hardiness. New York: Alan R. Liss, Inc, 1987:227-239.
84. Sakai A. Low temperature exotherms of winter buds of hardy conifers. Plant Cell Physiol 1978; 19:1439-1446.
85. Sakai A, Hakoda N. Cold hardiness of the genus *Camellia.* J Am Soc Hort Sci 1979; 104:53-57.
86. Toivio-Kinnucan MA, Chen THH, Li PH, Stushnoff C. Plasma membrane alterations in callus tissues of tuber-bearing *Solanum* species during cold acclimation. Plant Physiol 1981; 67:478-483.
87. Junttila O, Robberecht R. The influence of season and phenology on freezing tolerance in *Silene acaulis* L., a subarctic and arctic cushion plant of circumpolar distribution. Ann Bot 1993; 71:423-426.

Plant responses to low temperature: signaling pathways involved in plant acclimation

A. Kacperska

Institute of Experimental Plant Biology, Warsaw University, Pawińskiego 5A, PL-02-106 Warsaw, Poland; e-mail: <alka@ibbrain.ibb.waw.pl

1
Introduction: plant resistance to chilling and freezing stresses

1.1
Chilling and freezing injuries

Plants, like other poikilothermic organisms (i.e. organisms which are not able to regulate the temperature of their bodies), may suffer from low temperature-induced injuries if not adapted to this climatic stressor. Depending on the temperature range and plant genetic characteristics, irreversible cell injury and plant death may occur or a complex set of metabolic events is induced, leading to the development of stress tolerance.

In chilling-sensitive plants of tropical or subtropical origin as well as in fruits of some species with chilling-resistant vegetative organs, chilling injuries occur at temperatures between 15 and 0°C, in the absence of ice nucleation in plant tissues. The important elements in the development of such injuries are the degree of cooling, its duration and the speed of temperature changes during cooling and rewarming (for a review see refs. 1,2). The primary event in chilling injury is a change in membrane semipermeable properties, resulting in the increased membrane leakiness. The effect may be due to a change in membrane fluidity (for a review see refs. 3,4) or to a depolarization of the plasma membrane, followed by a rapid calcium influx into cytosol (for a review see ref. 5). In the photosynthetically active tissues, chilling and the excess of radiation may lead to photoinhibition caused by photoinactivation of photosytem II in the thylakoid membranes.[6–8] This is due to a downregulation of the water-splitting mechanisms and a degradation of the D1 reaction centre protein in the photosystem.[9] Depending on the type of a membrane affected by chilling temperature, different secondary, time-dependent injuries can be observed. Imbalance in ion homeostasis and deficits in energy equivalents, ATP and reduced pyridine nucleotides ($NADH^{+}$ and $NADPH^{+}$), seem to be the main causes of numerous metabolic disorders, resulting either in decreased contents of certain metabolic intermediates or, on the contrary, in the accumulation of certain metabolic products (for a review see refs. 2,10). Overproduction of oxygen free radicals is also a secondary response to the stress-evoked lesions in electron transport systems, located in different membranes such as plasma membrane, thylakoids and the inner mitochondrial membranes.[11–13] Severe imbalance in plant water economy, manifested by wilting of above-ground plant organs, is also observed in some annual chilling-sensitive plants, such as cucumber, bean or maize.[14] This is due to alterations in osmotic properties of cells as well as to imbalance between water uptake by plant roots and water losses by leaves. Severe disturbances in cellular metabolism and water economy lead to the degradation of cellular constituents and to cell and, ultimately, plant death.

Chilling-insensitive and chilling-resistant plants, i.e. plants which do not display membrane dysfunctions upon chilling or have an inherent ability to resist or tolerate the secondary effects of chilling,[15] may suffer from injury in consequence of ice formation in their tissues. When environmental temperature decreases a few degrees below the freezing point of water, ice formation takes place in plant tissues, first at locations having a higher (less negative) osmotic or water potential, i.e. in large vessels of the xylem in leaves and stems and in intercellular spaces (for a review see ref. 16). Ice-nucleation active bacteria (species of *Erwinia* and *Pseudomonas*) may provoke ice formation slightly below 0°C (for a review see ref. 17). The effect is due to the arrangement of water molecules by specific surface proteins of these bacteria. Ice, once formed, will spread throughout xylem vessels and into the extracellular spaces of other tissues. Ice crystals do not penetrate intact plasma membranes but they enlarge in the apoplastic space at the expense of water vapor and the surface film of liquid water on the cell walls.[2] This creates a chemical potential gradient between the unfrozen water in the protoplast and that in the apoplast. As a result, water is withdrawn from the protoplasts, which may shrink to two thirds of their volume.[2] The concentration of salt ions and organic acids in the remaining unfrozen portion of cell sap increases accordingly. Water movement and extracellular formation of ice continue until thermodynamic equilibrium between the ice layer and the cell sap is reached. Its position depends on temperature.[2] As a consequence, injurious effects of cell dehydration occur: the cell surface/cell volume ratio is decreased, the structure of plasma membrane and other cellular membranes is altered,[18] enzymes involved in mebrane lipid hydrolysis are activated,[19,20] proteins undergo degradation or conformational changes, the concentration of ions in the cytoplasm highly increases, which in turn leads to the modification of protein properties, etc. (for a review see refs. 10,16). A characteristic feature of the frost-induced changes in the properties of cell components is that their consequences appear during the postfreezing period, after thawing. Thawing conditions may, therefore, influence plant recovery from the freezing stress.[10]

If the speed of cooling is relatively high (>3°C min^{-1}) and thin-walled cells contain much water, or if the water in the cells has been deeply undercooled (for a review see ref. 16), ice is formed intracellularly which causes total destruction of the protoplasmic structure.This is the reason for the death of many chilling-insensitive but freezing-sensitive plants of cool-temperature climate. Also, this is the reason for the exclusion of certain woody perennials, containing deeply supercooled tissues, from the regions where the air temperature may drop below the threshold subcooling temperature.[21]

Plants may survive the periods of episodic or prolonged periods of a stressful temperature if they are endowed with an ability to avoid or to tolerate the stressor.[10] The avoidance strategy of annual plants may rely on the accomplishement of a full life cycle in the period with temperatures favorable for plant growth and reproduction, the survival of the next generation being dependent on the survival of their dormant (metabolically not active) organs, such as seeds or bulbs.[2] The avoidance of frost by withdrawal of growing shoot tips beneath covering leaves or layers of litter, burial in the ground or by closing of leaves over the shoot tips at nights is also observed in certain plants.[2] The avoidance of ice formation upon freezing is achieved by many biannual or perennial plants through freezing point depression or by increased supercooling ability of cell fluids (for a review see ref. 16).

The tolerance of chilling or freezing temperatures by chilling- or frost-tolerant plants relies on those plant features which either prevent the occurrence of the primary, membrane-located injuries or are the elements of mechanisms involved in the protection of cell components against the secondary effects of stress, such as oxidative stress, metabolic disorders or dehydration-dependent changes in the structure and activity of cell constituents. The role of membrane lipid unsaturation degree, certain phospholipids, specific proteins, low molecular compounds, such as certain sugars (e.g. saccharose, raffinose) and amino acids (e.g. proline) and other cellular compounds in plant resistance to low temperature has been well characterized in quite a few monographic books and review articles[2,4,11,16] to which the interested reader is referred. Modifications in gene expression and protein synthesis in cold-affected plants have been also recently overviewed.[22,23]

1.2 Acclimation of plants to low temperatures

The resistance of a plant to low temperature may be increased providing that the plant is exposed to low but not injurious temperatures under a specific light regime (for a review see refs. 2,10,11,16). The process, known as plant cold hardening or acclimation, is defined as a "non-heritable modification of structure and function as a response to a given climatic constraint which minimizes damage and improves the fitness of an individual plant".[16] Chilling-sensitive plants may acquire tolerance to chilling temperatures if exposed to suboptimal growth temperature in a water-saturated atmosphere for a certain time[14] in darkness[12] or under a proper light regime.[24] However, they are not able to acquire resistance to ice nucleation temperatures. On the other hand, in plants classified as chilling-insensitive, according to Raison and Lyons,[15] acclimation in the cold (at temperatures just above 0°C) brings about the development of a certain degree of resistance to freezing temperatures as well. The cold-dependent events which lead to the lowering of the freezing survival temperature are termed hardening to freezing.[10]

The process of frost hardening proceeds in recognizable phases. During the phase of intensive elongation growth most plants can scarcely become hardened at all. Therefore, the termination (in woody plants) or interruption (in herbaceous plants) of growth is the first prerequisite for a successful cold acclimation and frost hardening of chilling-insensitive plant species. This stage is defined as readiness to harden.[2] In woody plants readiness to harden is induced by short photoperiods and is associated with the development of winter dormancy.[25] Dormancy is a temporary, endogenously regulated and environmentally influenced interruption of growth.[2] The transition to winter dormancy is evidenced in the selective supression of gene activity, inhibition of translation, reduction of mitotic activity, suppression of metabolism, changes in enzyme patterns and activities, as well as in the compartmentation and ultrastructure of cells (for a review see ref. 2).

The occurrence of the stage of readiness to harden in herbaceous plants is questioned by Larcher.[2] However, it was shown recently that the interaction between photosynthetic active radiation and low (5–12°C) temperature controls morphogenesis of winter wheat and winter oilseed rape plants and influences the process of frost hardening.[26–28] The beneficial effects of such an interaction for a full expression of frost tolerance upon further decrease of the day/night temperature has been defined as prehardening.[27] Prehardening results in the inhibition of cell elongation growth and in "compact" morphology of the whole plant,[26,27] in adjustments in the photosynthetic capacity, nonphotochemical quenching capacity and tolerance to photoinhibition, as well as in the accumulation of certain cold-induced transcripts.[26]

Changes in the irradiance conditions concomitant with the lowering of the day/night temperature are characteristics of autumn in a temperate climate. Therefore, it seems that the stage of readiness for hardening is achieved by both perennial woody plants and overwintering herbs, the mechanism of light action in this process being different for these two types of plants. In woody perennials, phytochrome-controlled photoperiodic reactions are involved in the transition to winter dormancy and in the induction of acclimation processes.[25] In photosynthetically active tissues of herbaceous plants, a complex interaction of photosynthetic photon flux density and temperature alters photosynthetic activities and affects the pattern of growth and development.[26–28]

There is no doubt that the maximal expression of freezing tolerance in plants depends on a consecutive action of cold (prolonged exposure to temperatures close to 0°C) and moderate frosts (night exposures to temperatures between -3 and -5°C). Such a pattern of frost hardening has been described both for woody perennials[25] and for overwintering herbaceous plants, such as winter oilseed rape,[29] alfalfa,[30] rye,[31] and winter wheat.[11] However, the main input of processes taking place in the cold seems to differ in woody and herbaceous plants. In woody perennials, the development of dormancy, associated with an improvement of tissue freezing tolerance[25] and significant changes in cell components and ultrastructure take place in the cold.[32] In metabolically active tissues of herbs, the first stage of acclimation leads to two distinct effects: (1) improved performance of cells at low temperature and (2) increased resistance to freezing.[33] The adjustment of growth and metabolism to low temperature conditions allows plants to resume growth and accomplish further developmental stages in the cold.[33]

The metabolic consequences of the cold-induced first stage of plant acclimation have been reviewed in other articles which can be referred to for further information.[5,10,16,33] Shifts in metabolic pathways allow for synthesis and incorporation of highly unsaturated, low temperature-stable phospholipids into membranes, accumulation of soluble carbohydrates (saccharose and certain oligosaccharides) and polyols, increased contents of low molecular nitrogen-containing substances (amino acids, polyamines) and water-soluble proteins (for a review see ref. 16). These compounds, interacting with water molecules, decrease water potential of protoplasm and decrease the risk of frost-induced dehydration. Certain sugars and low molecular proteins protect membranes and other structural cell constituents against conformational changes upon desiccation.[34]

In herbs, the other attribute of acclimation in cold is increased ability of plants to avoid water stress. This is achieved both by decreased stomatal aperture[35] and by osmotic adjustment (for a review see ref. 36), which relies on the accumulation of osmotically active substances in the cell sap and protoplasm. Better protection of photosynthetically active tissues against water losses is accomplished also by a decrease in the leaf surface/fresh weight ratio[37] and by cell wall modifications.[38,39]

The information on the nature of hardening effects of transient, moderate frosts is rather limited. Extracellular freezing which results in the dehydration of cell constituents was proposed to induce conformational changes in protein structure in cells of dormancy-developing woody plants.[25] A comparison of seasonal variations in frost tolerance as well as in protein, RNA and phospholipid contents in cells of *Robinia pseudoaccacia*[32] or poplar[40] shows, however, that these processes were induced at the time when night frosts are likely to occur. This may indicate that transient frosts affected the operation of metabolic pathways in dormancy-developing cells. In leaves of overwintering herbaceous plants transient frosts were found to induce an increase in the amount of phospholipids[41] and to promote accumulation of sugars.[36] Reversible alterations in membrane functioning and phospholipid metabolism, associated with modification of phospholipase D activity were also reported.[20] Depending on the degree of the frost-induced membrane alterations and the presence of alcohol (glycerol), phosphatidic acid or phosphatidylglycerol were formed in phospholipase D-catalyzed reactions. Recent reports indicate that such a mode of action of the enzyme allows for its pivotal role in transmembrane signaling and cellular regulation, including membrane trafficking (for a review see ref. 42). Also, freezing and dehydration (but not cold treatment) appeared to be the necessary requirement for the induction of the phosphoinositide signaling system in winter oilseed rape leaf tissue.[43] A frost-dependent expression of certain genes has been also reported.[30,44] The product of one of these genes, CAS 15, is proposed to be a nuclear protein,[30] whereas another one codes for extensin, a cell wall protein. Accumulation of dehydrins (proteins of the *LEA* D11 family, which typically accumulate in plants during the late stages of embryogenesis in response to any environmental constraint which has a dehydration component) was also reported in tissues subjected to extracellular freezing.[45] Recently, the cryoprotective activity of the protein (β-1,3-glucanase) which accumulated in leaves during frost hardening under natural conditions was reported.[46]

Taking into consideration all the discussed data and observations, it is reasonable to propose that the requirement of the consecutive action of cold and subfreezing temperatures for a full expression of frost resistance in chilling-insensitive tissues may rely on the different nature of the cellular events they induce.

The brief review of the effects of chilling and freezing temperatures on plant cells presented above indicates that acclimation of plants to low temperature is a very complex phenomenon. The involvement of a single signaling pathway in the perception of the stimulus can hardly be expected, the more so that the existence of a specific receptor for such a diffuse signal as low temperature can hardly be envisaged. The cold-induced changes in molecular motion of membrane lipids are too small to be detected by the putative sensor of the temperature shift.[47] Also, the perception of this environmental constraint may be different in cells of different function and membrane composition. The involvement of the secondary effects of the low temperature stress, such as cell dehydration and ion imbalance in the operation of the signaling pathways in the low temperature-affected plants, ought to be taken into consideration as well. In the following section a brief review of what is known on the operation of signaling pathways in plants and a survey of plant early responses to the environmental factors which may trigger plant acclimation to low temperature are given.

2 Signaling pathways operating in cold- and frost-affected cells

2.1 Signal transduction pathways in plants

There is no doubt that plant responses to developmental and environmental cues such as light, temperature, water availability, salinity, wounding, hormones or other molecules on the cell surface, occur by stimulus-response coupling.[48] A stimulus is perceived by the cell, a signal is generated and transmitted (signal transduction) and a physiological or biochemical change is induced (the response). The whole process requires the recognition of the stimulus by a receptor and the subsequent use of a chemical and/or effector proteins to transmit a signal that will trigger the appropriate response. Key properties of signal transduction are speed, sensitivity (achieved by amplification), and specificity, all of which are controlled by a network of positively or negatively acting elements.[48]

It is accepted now that there are three elements of signal perception and transduction pathways in plant cells (Fig. 1).

2.1.1 Cell surface receptors

Cell surface receptors, located in plasma membrane, play a prominent role in the initiation of cellular signaling. Until quite lately, the site of perception and molecular candidates for the receptors for environmental and hormonal stimuli have been elusive,[49] with the one exception of light. Aside from the blue light receptor (cryptochrome), the mode of action of different phytochromes in photoreception and signaling of "induction" reactions has been described recently.[50] In the high

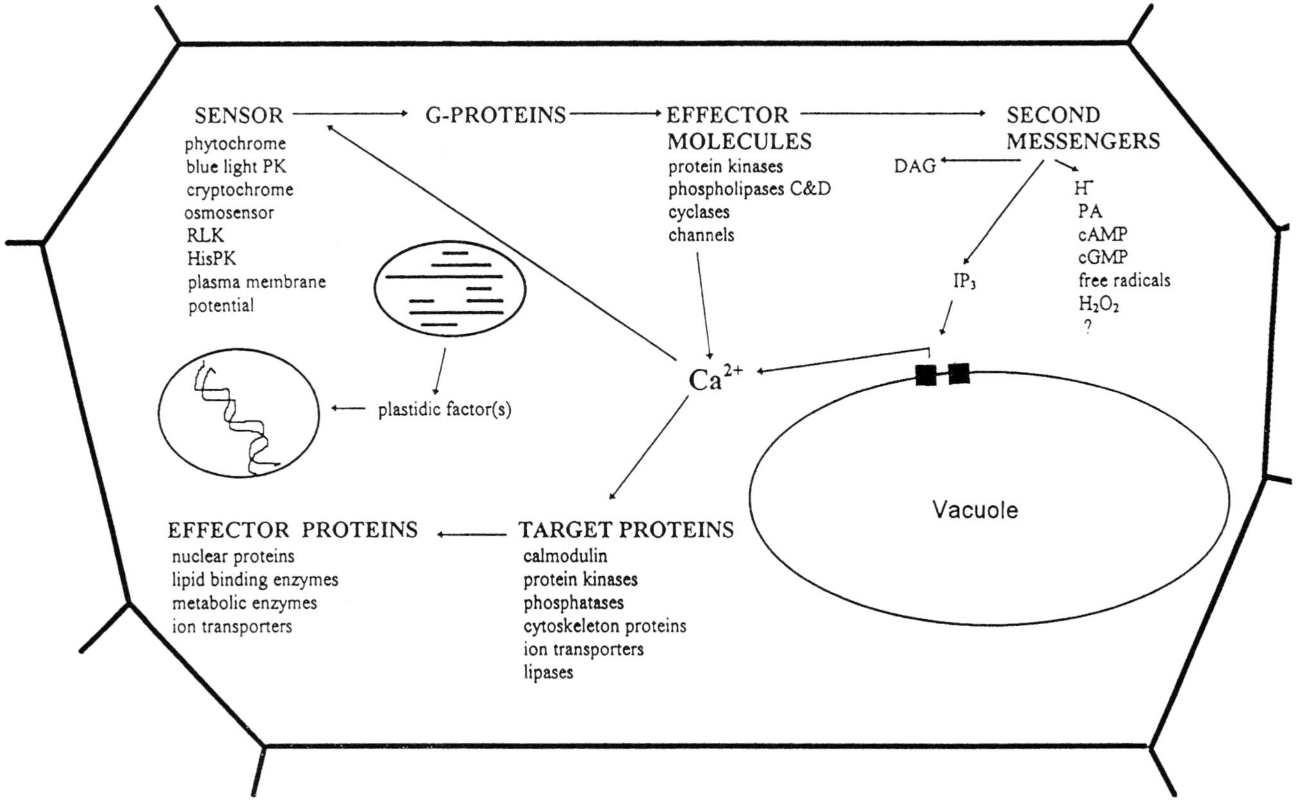

Fig. 1. Schematic presentation of signal transduction pathways in plants
RLK receptor-like protein kinase; *HisPK* membrane-located protein kinase autophosphorylating on a histidine residue; *DAG* diaglycerol; IP_3 inositol-1,4,5-trisphosphate; *PA* phosphatidic acid

irradiance responses of plants the participation of another pigment, provisionally named heliochrome, is proposed.[51] In plants grown in the sunlight this pigment controls the synthesis of phytochrome or interacts with a substance produced by phytochrome. Light receptors other than those located in plasma membrane are also found in higher plants. Chloroplasts are considered to be involved in the sensing of environmental temperature changes through a redox sensing signaling system,[7] monitored by photosystem II (PSII). Regulation of expression of genes involved in scavenging of H_2O_2 in the cytosol by the redox events in the proximity of PSII, probably by the redox status of the plastoquinone pool, was discovered recently.[52]

According to recent developments in the field, one of the earliest responses of a plant cell to a stimulus is a change in plasma membrane polarization which alters functioning of ion channels, including the voltage-dependent calcium channels (for a review see ref. 53). Electrophysiological studies indicated that some stimuli may interact with channels in the plasma membrane directly or through modulation of H^+-ATPAase activity.[49] Transient membrane depolarization was actually elicited by wounding[54] or rapid-cooling stimuli in plant leaves.[5] The generation of a strong positive electrostatic charge in the cell membrane is proposed to be the first event initiated by phytochrome and heliochrome.[51] Action potentials were proposed earlier to constitute multifunctional signals in plants.[54]

There is also growing evidence that plant cells carry cell surface receptors with intrinsic protein kinase activity, known as the receptor-like protein kinases (RLPK, for a review see ref. 55). Two distinct components of the signal transduction pathway induced by the plant hormone, ethylene, showing protein kinase activity, have been identified.[56]

It is thought that membrane-spanning receptors subsequently activate heterotrimeric or small (monomeric) GTP-binding proteins (G proteins). Activated G proteins regulate the activity of effector molecules including phospholipase C (the enzyme catalyzes the hydrolysis of phosphatidylinositol-4,5-bisphosphate to generate two compounds involved in the control of cytosolic calcium level or in regulation of protein kinase C activity, inositol-1,4,5-trisphosphate (IP_3) and diacylglycerol, respectively), nucleotide cyclases, and channels, which in turn regulate second messenger production (for a review see ref. 57).

2.1.2 Second messengers

The stimulus/receptor complex is thought to elicit changes in a cytoplasmic factor that amplify, integrate and propagate the initial signal.[49] Many second messengers operating in different signaling systems have been identified in plant cells (for a review see ref. 57). These include phosphatidic acid, IP_3, cGMP, cAMP, Ca^{2+}, H^+, redox agents such as ascorbic acid or glutathione, hydrogen peroxide, and free radicals. Moreover, diacylglycerol and lysophospholipids, the products of the activation of phospholipases C and D, are likely to remain in the plasma membrane where they could further modify functions of membrane proteins.[57] Cytosolic Ca^{2+} levels have been found to change in response to many environmental and hormonal stimuli (e.g. touch, wind, turbulescence or shaking, hypoosmotic stress, cold and heat shocks, oxidative and anaerobic stresses, salination, and drought) and became the best-characterized second messenger in plants (for a review see ref. 49). Released from intracellular (endoplasmic reticulum and vacuoles) or extracellular (apoplast) stores Ca^{2+} combines with calmodulin, the primary plant calcium receptor, or other calcium-binding proteins. Subsequent cellular processes are then initiated usually via the activation of appropriate protein kinase.[58]

2.1.3
Protein kinases

Posttranslational modification of serine, threonine and tyrosine residues by protein kinases and phosphatases is undoubtedly a major transduction route for many signals. Protein kinases may phosphorylate several substrates whilst the substrates themeselves may be phosphorylated by several kinases, allowing for "cross-talk" between regulatory pathways (for a review see refs. 49,56). Several groups of protein kinases are known: calmodulin-dependent kinases; calcium-dependent kinases (CDPKs), which have a domain related to calmodulin; plant receptor-like protein kinases (RLPKs); mitogen-activated protein kinases (MAPK) involved in the MAPK signal transduction cascade.[56] Potential substrates for CDPKs include cytoskeletal proteins, membrane transporters and H^+-ATPase from plasma membrane.[49] The MAPK cascade is now implicated in the transduction of mechanical signals and signals triggered by such plant hormones as gibberelins, abscisic acid (ABA), ethylene, auxin, and by fungal elicitors, and factors regulating cell proliferation.[57] Calcium is also involved in the regulation of activities of phosphatases and several other enzymes, e.g. phospholipases. Presently, the participation of protein kinases/phosphatases in plant responses to hormonal and environmental stimuli is gaining more and more attention.

2.2
Signaling pathways operating in metabolically active plants during their acclimation to low temperatures

As mentioned previously, the beginning of acclimation of woody perennials to low temperatures is conditioned by the phytochrome-mediated onset of winter dormancy. This aspect of plant development and acclimation is well described in numerous papers and books to which the interested reader is referred.[2] Also, the information on signaling pathways operating in woody plants is rather limited with the exception of the plant hormone-dependent ABA-induced reactions. Therefore, the discussion of possible signaling pathways induced and operating in the low temperature affected plants is limited here to nondormant plants and tissues.

The first step in the control of plant acclimation to low temperature is the recognition of this environmental cue. Therefore, it is necessary to look for those early plant responses, which are registered during the alarm phase of the stress syndrome.[2] This phase is defined as the onset of disturbances, followed by destabilization of the structural and functional conditions required for the normal pursuance of vital activities.[2] Depending on the intensity of the stress factor, and plant genetic properties, the stress reactions initiated during the alarm phase may lead either to acute collapse of cell integrity or to the initiation of processes required for acclimation.

2.3 Plant reactions to the temperature decrease

The examination of the abundant literature data describing different plant responses to low temperature indicates that two categories of plant reactions displayed by plant tissues and cells upon cooling ought to be distinguished: (1) reactions evoked by a slow temperature decrease; (2) reactions evoked by a rapid temperature drop, termed by Minorsky[1] rapid-cooling responses. A comparison of the early events registered during these reactions (Table 1) points out their most characteristic features.

2.3.1 *Slow responses*

In slow responses, the alteration of water status of leaf cells seems to be the primary event in the recognition of the stressful temperature. It is indicated by a small (ca. 0.5 MPa) but meaningful decrease in tissue water potential.[59,61] The water potential decrease is possibly due to imbalance between water uptake by roots and water loss by transpiration. The first and the most sensitive response to water deficit is a decrease in turgor and an inhibition of growth processes, particularly elongation growth of cells. The decrease in turgor and/or change in cell volume may cause alterations in the cell wall-plasma membrane continuum and lead to modifications of cytoplasmic free calcium concentration. The source of calcium and the mechanism involved in its influx to cytosol of the cold-affected cells are, for the time being, not clear. The involvement of calcium channels on plasma membrane, activated by a decrease in membrane fluidity,[94] can hardly be taken into consideration since changes in membrane fluidity in chilling-insensitive tissues are observed at temperatures lower than that allowing for acclimation in cold.[47] Neither the activation of mechanosensitive Ca^{2+} channels[96,97] in the plasmalemma is likely to be involved in calcium influx from external pool (apoplast) in cells showing decreased turgor pressure. Activation of such channels was observed in algae cells responding to hypotonic treatment, leading to an increased turgor pressure.[98] It seems that in cells responding to a slow temperature decrease and showing decreased turgor pressure the intracellular sites for the released Ca^{2+} ought to be considered. The intracellular response system, activated by cytoskeletal links transmitting forces in the walls, represents the attractive candidate for further studies, as suggested previously for *Arabidopsis* roots responding to touch.[99] It seems that the low temperature-induced alterations in free calcium homeostasis ought to be compared for cells responding to a slow or to a sudden temperature decrease.

In photosynthesis-conducting cells, any environmental stress including temperature, irradiance, water availability or nutrient status can potentially affect the balance between energy absorption and energy utilization. Energy imbalance may be sensed through the modulation of the redox state of PSII because the turnover of the PSII reaction centers is a limiting step in the conversion of light energy into ATP and NADPH. The chloroplastic redox sensing/signaling mechanism was proposed to act synergistically with other known signal transduction pathways to elicit the appropriate molecular and physiological responses in plants exposed to low temperature.[7] Interaction of low temperature and light was proved to be involved in time-dependent adjustment of photosyntetic apparatus, modification of plant morphology, specific gene expression and increased freezing tolerance.[26] The relative reduction state of PSII was shown recently to affect thylakoid protein phosphorylation and PSII protein turnover through the redox-dependent activation/deactivation of the membrane-bound protein kinase(s), mostly localized in grana.[100] The question arises what is the primary factor which leads to the alterations of the redox state of PSII: is it low temperature *per se* or a change in water balance of CO_2 assimilating cells? In nonacclimated plants responding to the temperature change during the prehardening period, the excitation pressure may be an important element in the regulation of photosynthesis and plant morphogenesis, as proposed by Gray et al.[26] and by Rapacz.[27,28] In the prehardened plants adjusted to an excess of light, a higher efficiency of photosynthetic electron transport may condition the proper course of the acclimation events occurring in the cold.

It is known that a redox signaling system in cells subjected to different stressors (including chilling and water stress) is involved in the induction of synthesis of nonenzymic antioxidants such as the tripeptide thiol, glutathione, and vitamins C (ascorbate) and E (α-tocopherol), as well as in the increases in antioxidant enzymes such as superoxide dismutase, glutathione peroxidase, glutathione reductase, ascorbate peroxidase and catalase (for a review see ref. 101). Promotion of pathways involved in energy transduction or dissipation such as photorespiration or the KCN-insensitive alternative pathway of electron transport in mitochondria may be also an important element in plant protection against the oxidative stress (for a review see refs. 103,104).

A decrease in energy consumption for growth processes as well as increased availability of reducing equivalents in cold-affected cells allows for the promotion of certain metabolic pathways, including those which are involved in polyunsaturated fatty acid synthesis.[33] The increased production of some metabolites (e.g. certain sugars, amino- and organic acids) allows for the osmotic adjustment of cells which is a prerequisite for the improvement of water balance in plants (for a review see ref. 78). On the other hand, certain sugars (glucose or fructose) are capable of acting as regulatory signals that affect gene expression (for a review see refs. 104,105). The hexokinase-dependent sugar signal transduction pathway was recently thoroughly discussed by Jang and Sheen.[105]

Another effect of water imbalance in cold-affected herbaceous plants is a transient accumulation of ABA, as indicated in Table 1. The role and contribution of stress-induced ABA transients was examined recently by Cowan et al.,[106] who pointed out a complex interaction between apoplastic (tissue) and cellular pools of ABA in stimulus-response-coupling. As a consequence of increased ABA_{tissue} concentration, apoplastic hormone-receptor binding initiates slow responses of a plant

Table 1. Comparison of plant responses to slow and rapid temperature decreases

Plant responses to low temperature Slow responses	Rapid responses
Decrease in water potential of a tissue[59-61] (for a review see ref. 36) Inhibition of cell expansion[37] Decreased leaf surface area[37] Increased root/shoot ratio[60]	Membrane functions impaired: – membrane depolarization[4,81,82] – enhanced efflux of ions,[81] primarily K[83,84] – rapid elevation of cytosolic free Ca concentration[85,86]
Shifts in metabolism	
Increased activity of systems engaged in the prevention of oxidative stress: – increased levels of antioxidants[12,13,62,63] (for a review see ref. 11) – increased activity of oxygen-scavenging enzymes[64] – transient increase in the activity of alternative respiratory electron pathway[65] – transient increase in photorespiration[66] Modifications in energy metabolism: imbalance between energy supply and energy consumption[6,7] – redox potential modified[6] – increased adenylate[67] and anabolic reduction[37] charges Promotion of NADPH- and NADH-dependent metabolic pathways[33] Increased unsaturation of membrane lipids[68] Modification of carbon photosynthetic metabolism – decreased incorporation of ^{14}C to sugars[69] – increased ^{14}C incorporation to C4 metabolites[66] Accumulation of osmolytes such as sugars[65] or proline[70]	Development of oxidative stress: – enhanced superoxide or hydrogen peroxide formation[62,87,88] Induction of lipoxygenase activity[89] Induction of phospholipase D transphosphatidylation activity[20] Induction of phospholipase C activity[43] Activation of cell wall β-glucosidase[90] Activation of phenylalanine ammonia-lyase[91] (promotion of phenylpropanoid synthesis)
Changes in hormone level	
Hormonal balance modified: (for a review see refs. 71–72) – ABA content increase[73–75]	Enhanced synthesis of ethylene[89,92] (for a review see ref. 93) Further accumulation of ABA[75]

Table 1 – continuation

Plant responses to low temperature	
Slow responses	Rapid responses
Modifications in protein metabolism	
Changes in protein turnover: – accumulation of soluble proteins[29,76] – promoted synthesis of certain proteins[22,23,76] – synthesis of antifreeze proteins, similar to pathogenesis-related proteins[77] – induction of LT-polypeptides involved in adjustment of membrane lipid composition or in osmoprotection (for a review see ref. 23)	Transient degradation of a bulk of proteins[91] Accumulation of β-1,3-glucanase showing cryoprotective activity[46]
Modifications in gene expression	
ABA- and/or? dehydration-induced genes (for a review see refs. 22,23,78-80)	Induction of *cas15* gene (in alfafa) by subfreezing[30] Ca^{+2}-dependent induction of cold-acclimation specific genes (*cas*)[94] * Induction of extensin genes[44] Cold-shock induced *Arabidopsis tch* genes[95]

* Rapid transfer from 25 to 4°C.
ABA abscisic acid; *LT* low temperature

to the hormone. Changes in the cellular and tissue concentration of ABA result in modifications of enzyme activities,[107–108] and result in modifications in gene expression (for a review see ref. 109).

The ABA-dependent signaling pathways operating in plant cells were recently discussed by Bray,[78–79] Merlot and Giradudat,[110] and Shinozaki and Yamaguchi-Shinozaki.[111] In these excellent overviews the interested reader may find all the data currently available on the subject. Therefore, the discussion of the problem has not been included in the present chapter. However, it ought to be pointed out that in the ABA-responsive expression gene promoters contain the ABA-responsive element, and that the ABA signal transduction pathway probably comprises a protein kinase/phosphatase cascade interacting with Ca^{2+}. In an ABA-insensitive mutant of *Arabidopsis thaliana* (*aba1*) the ABA1 gene encodes a polypeptide with a protein phosphatase 2C domain at the C-terminus and an ethylene-forming enzyme hand-like calcium-binding domain at the N-terminus.[48] Recently, ABA induction of MAPK activation has been demonstrated in barley aleurone protoplasts and was shown to occur via a tyrosine phosphatase leading to the induction of *rab16* gene expression.[112] Protein farnesyl transferase was found to be a component of the ABA signal transduction pathway.[113] It is proposed to catalyze the modification of a receptor or a component of the signal transduction pathway for membrane localization.

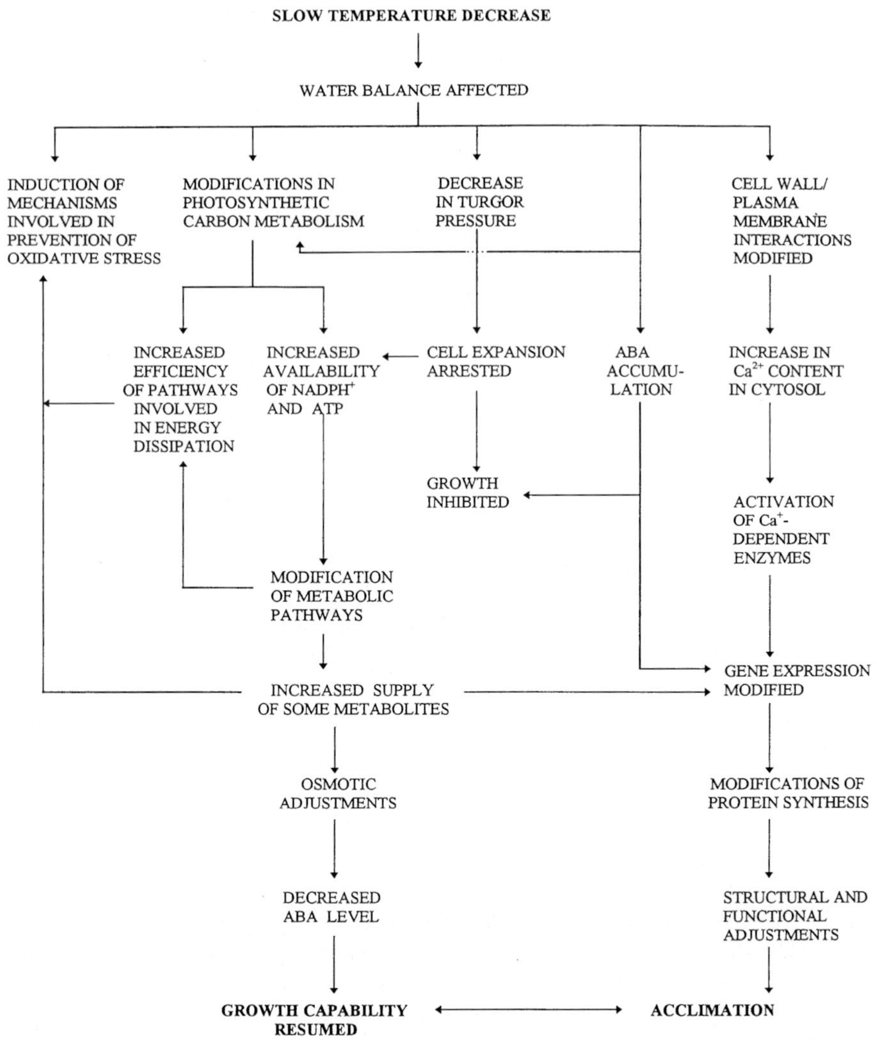

Fig. 2. Proposed sequence of events occurring during plant reaction to a slow temperature decrease; *ABA* abscisic acid

In plant cells responding to low temperature and/or to water stress, the ABA-independent genes operate as well.[22–23,78,79,111] The involvement of the dehydration responsive element (DRE) in the promoter regions of many cold- and drought-inducible genes has been reported.[22,78,79,111] A protein factor that specifically interacts with the DRE sequence and is involved in the regulation of the ABA-independent pathway induced by water deficit, low temperature and salinity has been detected in *A. thaliana* seedlings.[114] The question arises whether the drought-inducible genes operate in plants which are exposed to a mild water stress, brought about by a slow temperature decrease, as discussed above.

The main outputs of plant responses to slow temperature decrease are metabolic adjustments including development of resistance to photoinhibition, modifications in growth pattern of different organs, changes in the root/shoot interrelations, increased capability of cells to maintain positive water balance and to avoid oxidative stress under low temperature conditions. Osmotic adjustment and gene-dependent modifications in cell structure[115] and functions allow the restoration of growth capability and adjustment of metabolism to low temperature conditions, i.e. they permit acclimation of plants to cold, as has been pointed out earlier.[33]

In conclusion, I want to propose that the increased supply of ABA, in concert with the increased PSII excitation pressure, the increased levels of reducing equivalents and transient increase in cytosol Ca^{2+} content, may constitute the signal systems which operate in plants exposed to a slow, moderate temperature decrease (Fig. 2). Cold-induced modifications in activities of different protein kinases and phosphatases which make possible a cross-talk between different signaling pathways are not indicated in the scheme in order not to complicate the whole picture. Proteins that accumulate in the cold-grown tissues seem to play an important role in the maintenance of a proper water regime in the cold acclimated cells. Certain proteins that accumulate in the apoplast may also exhibit antifreeze activity and, which is interesting, they show a similarity to the pathogen-related proteins.[77]

It ought to be pointed out that reactions induced by a slow temperature decrease are in many aspects similar to those described for plants subjected to a moderate water stress. The observation is in accordance with the Bray[78] and Bohnert et al.[80] propositions, indicating the key role of water stress in plant adaptation to environmental stresses.

2.3.2
Reactions induced by a rapid temperature drop

A comparison of the cellular events induced by a rapid temperature decrease or a sudden change in water balance due to extracellular formation of ice (Table 1) indicates that they are different from those produced by a slow change in temperature. The final temperatures reached in the rapid cooling treatment seem to be of minor importance: the sensitivity of an individual plant to such a treatment will be a function of its genetic predisposition as well as its developmental and physiological state.[1,5]

The analysis of the events shown in Table 1 indicates that the plasma membrane transient distortion, associated with its depolarization (inhibition of proton excrection into apoplast) seems to be the primary reaction of cells to rapid cooling, as pointed out by Minorsky.[1] Extracellular ice formation may bring about a similar effect in chilling-insensitive plants.[82] The stress-induced inhibition of the proton-motive force at the plasma membrane[116] results in alkalization of apoplast which acts as an anion trap in which ABA is accumulated (for a review see ref. 106). This may be the reason for the increased level of ABA in leaves collected from plants subjected to the subfreezing treatment.[75]

Another direct consequence of the plasma membrane dysfunction is a change in ion fluxes and a rapid calcium influx into the cytoplasm, possibly due to the opening of the voltage-dependent channels in plasmalemma.[97] Cold activation of Ca^{2+} permeable channels at the plasma membrane, resulting in an inward Ca^{2+} current which depolarizes the plasma mebrane and affects anion channel activity, has been also proposed.[81] It seems that the cold-increased influx of extracellular[45] Ca^{2+} into cytosol in alfalfa cells, which could be mimicked by application of a calcium ionophore or a calcium channel agonist,[117] was also due to a membrane dysfunction in response to a rapid transfer of plants from 25°C to the cold (4°C). The low temperature-induced changes in calcium gradients within a cell affect different cellular functions, such as ion balance, motility, carbohydrate metabolism, mitosis, secretion, cytoskeleton organization, protein (de)phosphorylation, MAPK activation, etc. (for a review see ref. 97).

Alterations in membrane properties result also in osmotic loss of water which is secreted from the temperature-affected cells into intercellular spaces or to a medium. This brings about the dehydration of protoplast, shrinkage of cells, and alterations in cell wall/plasma membrane interactions. All these reactions may affect the operation of different signal transduction pathways.

The important feature of plant responses to a rapid temperature decrease is the activation of phospholipases C[43] and D,[20] which leads to the increased levels of such second messengers as inositol-1,4,5-trisphosphate (IP_3) and diaglycerol. The involvement of IP_3 in the control of calcium influx from intracellular source in the cold-shocked cells was proposed recently.[86] In plants phospholipase D signaling is stimulated by the activation of G protein[118] and its role as a potential signal transduction mechanism should not be neglected now. It is worth mentioning that formation of phosphatidyl glycerol in the phospholipase D-catalyzed transphosphatidylation has been proposed to play important role in cell recovery from freezing.[20]

There is no doubt that the low temperature-induced membrane dysfunctions result also in the development of oxidative stress and in the promotion of the lipoxygenase-mediated peroxidation of lipids, as shown in Table 1. Lipoxygenase-dependent formation of jasmonates[119] can also be expected, although no experimental evidence supporting this possibility has been given so far for cold/frost-affected plants. Increased synthesis of stress (wound) ethylene in cells showing membrane dysfunction was also observed (for a review see ref. 93). It is worth mentioning that ethylene has been shown to be involved in the regulation of defense genes in plants.[125] In addition to local cellular responses, increased synthesis of ethylene may affect a whole tissue or a whole plant and play an important role in the control of systemic responses of plants to the stress factor.

Genetic analysis of the ethylene pathway in *A. thaliana* indicates the possible involvement of a MAPK module. From one of the mutants showing a constitutive triple response in the absence of ethylene, the *etr1* gene was isolated, which encodes a protein with homology to two-component sensor-regulator proteins which are well-known signaling transducers in bacteria. The sensor is a membrane-located kinase that autophosphorylates on a histidine residue upon receiving extracellular signals. The regulator then becomes phosphorylated on an aspartate residue. The

Table 2. Protein phosphorylation increase as result of a sudden temperature decrease or other shocks

Protein	System	Treatment
Protein phosphorylation[94]	*Alfalfa* cell suspension	1–3 h after a rapid transfer from 25 to 4°C
Several proteins, including CAS 15[94]	*Alfalfa* cell suspension	3 h cold stress
MAP kinases (independent of ABA)[120]	Alfalfa roots	A sudden shift to 4°C or severe dehydration
Receptor-like kinase[121]	*Arabidopsis thaliana*	3 h after transfer from 25 to 4°C (induced also by 1 h dehydration or salt treatments)
	Other shock responses	
Protein kinase (45 kDa, Ca^{2+}- and ABA-independent)[122]	Maize primary roots	Water stress (-1.6 MPa, 1 h)
CDP kinases (49 and 52 kDa)[123]	Sorghum roots	Osmotic stress (100 or 200 mOsm)
Protein kinase (46 kDa)[124]	Variety of species	Wounding (1 min after cutting)

CAS 15 cold acclimation specific polipeptide (15 kDa); *MAP* mitogen-activated protein; *ABA* abscisic acid; *CDP* calcium dependent protein

ethylene receptor (ETR1) protein resembles the membrane-located protein (SLN1) which is responsible for signaling the hyperosmotic stress (shock) in yeast. The protein is the upstream (earlier in the sequence) regulator of the MAPK cascade (for a review see ref. 56), involved in yeast adaptation to a hyperosmolar medium.

The available experimental data indicate that rapid responses of plant cells to different stressors result in increased activities of certain protein kinases (Table 2). Therefore, the participation of certain protein kinases in the signaling pathways induced by shock responses may be assumed.

In conclusion, I would like to propose that in plants, affected by a sudden temperature decrease, plasma membrane transient dysfunctions, resulting in cell shrinkage, elevation of apoplastic ABA, rapid changes in calcium gradients within a cell, activation of certain protein kinases as well as an induction of the ethylene-dependent reactions, may constitute a complex signaling system the operation of which leads to the modification of gene expression (Fig. 2). The up- or downregulation of certain genes results in modified protein composition in cells affected by a rapid temperature decrease (Fig. 3).

The role of proteins induced by a rapid temperature decrease is not yet fully understood. However, it is reasonable to speculate that some of them, like dehydrins, show chaperone activities or may act as surfactants capable of inhibiting the coagulation of macromolecules, thereby preserving structural integrity of cell components.[45] The others, like the cell wall-located extensins,[44] may prevent cell wall

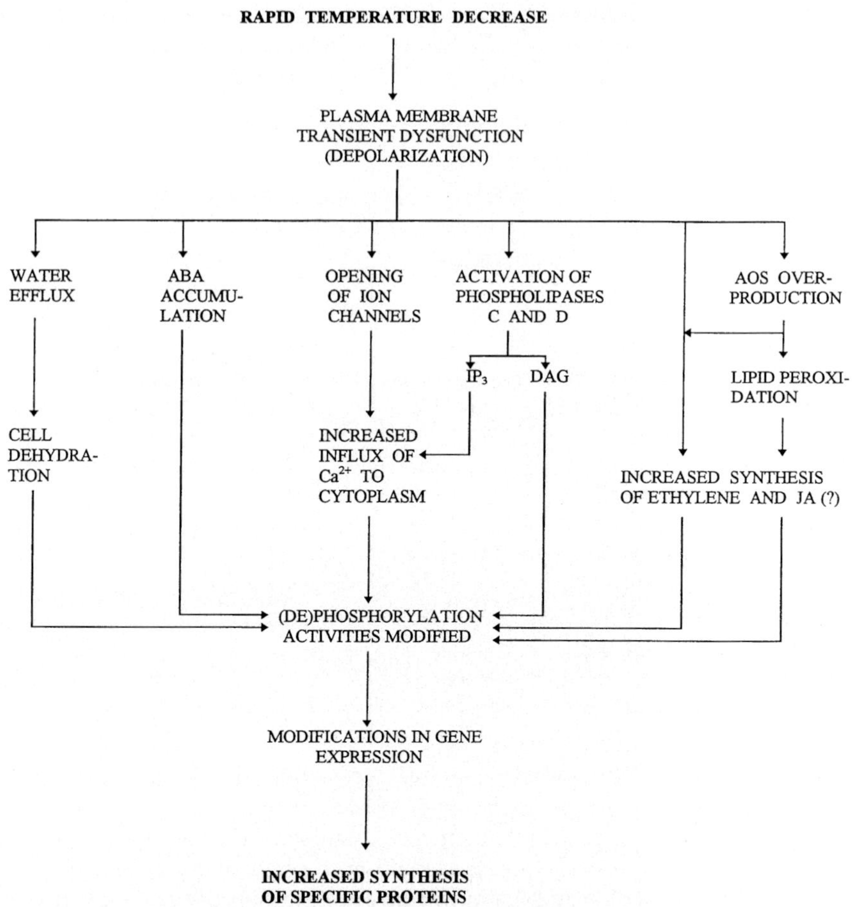

Fig. 3. Proposed sequence of events occurring during plant reaction to a rapid temperature decrease
ABA abscisic acid; IP_3 inositol-1,4,5-trisphosphate; *DAG* diaglycerol; *AOS* active oxygen species; *JA* jasmonic acid

deformations brought about by cell dehydration during extracellular ice formation. Still others, like β-1,3-glucanase, may reduce osmotic stress and vesicle rupture during thawing.[44]

In summary, beneficial effects of the subfreezing treatment, inducing the second stage of frost hardening in chilling-insensitive plants, may rely on the activation of defense systems engaged in the preservation of cellular structures under frost-induced dehydration and thawing.

In many aspects plant responses to a sudden temperature decrease are similar to those observed in mechanically wounded or pathogen-affected tissues. It is of interest that changes in gene expression, induced by a sudden temperature drop, result in the synthesis of proteins which can be ranked as dehydrins, osmotins, and even pathogen-related proteins. It seems, therefore, that shock responses may be considered as alarm reactions, according to the Chessin and Zipp definition,[126] formulated for the pathogen-induced responses.

3 Conclusions and future perspectives

In view of recent research developments, acclimation of plants to low temperature (and to any other stress factor) may be defined as the stress-induced modifications in gene expression allowing for a recasting of plant cell components. Such recasting enables metabolic adjustment and continuation of growth under stress conditions as well as a protection of macromolecules against irreversible deformations induced by a change in the temperature and/or by the low temperature-induced dehydration of a cell. Upon a sudden temperature decrease the induction of a rapid defense system takes place which permits enduring the stress and resuming cellular activities after its retreat.

Aside from the hormone-related perception of stress stimuli, the stress-induced alterations in membrane functions seem to be engaged in signal transduction, especially in the reactions to a rapid temperature change. Modulation of the proton-pumping activity and ion channel gating seems to be the most promising candidates for further studies on the perception of low temperature, and possibly of water and osmotic stresses.

It seems that further progress in the reseach on molecular mechanisms involved in plant acclimation to low temperature and on signal transduction pathways engaged may depend on the fulfilment of the following conditions:

1. Distinction should be made between the effects of chilling or cold treatments (temperatures >0°C) and the effects of slight frosts, sometimes described as the effects of the subfreezing temperature (understood as the temperature just below freezing point of water).
2. It will be desirable to test whether the mechanism under study is put into operation as a result of a slow or a rapid temperature decrease. This is especially important in the studies on plasma membrane properties, ion channel functions and Ca^{2+} influxes to the cytosol.
3. The possible involvement of jasmonic acid[127] in control of plant responses to a rapid temperature decrease (leading to peroxidation of membrane lipids), may be an interesting object for further studies.

Acknowledgements. The work was supported by the grant no. ERBIC15CT960901 from the European Commission. The skilful help of Ms. Danuta Solecka in the preparation of Tables and Figures is gratefully acknowledged.

4 References

1. Minorsky PV. An heuristic hypothesis of chilling injury in plants: a role for calcium as the primary physiological transducer of injury. Plant Cell Environ 1985; 8:75-94.
2. Larcher W. Physiological Plant Ecology, 3rd ed. Berlin, Heidelberg: Springer, 1995:321-327.
3. Lyons JM. Chilling injury in plants. Annu Rev Plant Physiol 1973; 24:445-466.
4. Nishida I, Murata N. Chilling sensitivity in plants and cyanobacteria: the crucial contribution of membrane lipids. Annu Rev Plant Physiol Plant Mol Biol, 1996; 47:541-586.
5. Minorsky PV. Temperature sensing by plants: a review and hypothesis. Plant Cell Environ. 1989; 12:119-135.
6. Huner NPA, Öquist G, Hurry VM, Krol M, Falk S. Griffin M. Photosynthesis, photoinhibition and low temperature acclimation in cold tolerant plants. Photosynth Res 1993; 37:19-39.
7. Huner NPA, Maxwell DP, Gray GR, Savitch LV, Krol M, Ivanov AG, Falk. Sensing environmental temperature change through imbalances between energy supply and energy consumption: Redox state of photosystem II. Physiol Plant 1996; 98:358-364.
8. Anderson JM, Park Y-I, Chow WS. Photoinactivation and photoprotection of photosystem II in nature. Physiol Plant 1997; 214-223.
9. Giardi MT, Masojídek J, Godde D. Effects of abiotic stresses on the turnover of D1 reaction centre II protein. Physiol Plant 1997; 101:635-642.
10. Levitt J. Responses of Plants to Environmental Stress: Chilling, Freezing and High Temperature Stress. New York: Academic Press, 1980:23-290.
11. McKersie BD, Leshem YY. Stress and Stress Coping in Cultivated Plants. Dordrecht, Boston London: Kluwer Academic Publ, 1994:79-128.
12. Prasad TK. Mechanism of chilling-induced oxidative stress injury and tolerance: changes in antioxidant system, oxidation of proteins and lipids and protease activities. Plant J 1996; 10:1017-1026.
13. Prasad TK. Role of catalase in inducing chilling tolerance in pre-emergent maize seedlings. Plant Physiol 1997; 114:1369-1376.
14. Wilson JM. The mechanism of chill- and drought-hardening of *Phaseolus vulgaris* leaves. New Phytol 1976; 76:257-268.
15. Raison JK, Lyons JM. Chilling injury: a plea for uniform terminology. Plant Cell Environ 1986; 9:685-686.
16. Sakai A, Larcher W. Frost Survival of Plants. Berlin, Heidelberg: Springer, 1987:21-138.
17. Lindow SE. The role of bacterial ice nucleation in frost injury in plants. Annu Rev Phytopathol 1983; 21:363-384.
18. Steponkus PL. Role of plasma membrane in freezing injury and cold acclimation. Annu. Rev Plant Physiol 1984; 35:543-548.
19. Yoshida S, Sakai A. Phospholipid degradation in frozen plant cells associated with freezing injury. Plant Physiol 1974; 53:509-511.
20. Sikorska E, Kacperska A. Freezing-induced membrane alterations: Injury or adaptation? In: Li P, Sakai A, eds. Plant Cold Hardiness and Freezing Stress. Mechanisms and Crop Implications, vol. 2. London, New York: Academic Press, 1982:261-272.
21. George MF, Burke MJ, Pellet, HM, Johnson AG. Low temperature exotherm and woody plant distribution. HortSci 1974; 9:519-522.
22. Thomashow MF. Genes induced during cold acclimation in higher plants. In: Steponkus P, ed. Advances in Low Temperature Biology. London, England, Greenwich, Connecticut: JAI Press Ltd, 1993:183-210.
23. Palva ET. Gene expression under low temperature stress. In: Basra AM, ed. Stress-Induced Gene Expression in Plants. Amsterdam: Harwood Academic Publishers, 1994:103-130.

24. Mauro S, Dainese P, Lannoye R, Bassai R. Cold-resistant and cold-sensitive maize lines differ in the phosphorylation of the photosystem II subunit, CP29. Plant Physiol 1997; 115:171-180.
25. Weiser CJ. Cold resistance and injury in woody plants. Science 1970; 169:1269-1278.
26. Gray GR, Chauvin L-P, Sarhan F, Huner NP. Cold Acclimation and freezing tolerance. Plant Physiol. 1997; 114:467-474.
27. Rapacz M. The effect of day and night temperature during early growth of winter oilseed rape seedlings on their morphology and cold acclimation responses. Acta Physiol Plant 1998; 20:68-72.
28. Rapacz M. The after effects of temperature and irradiance during early growth of winter oilseed rape seedlings on the progress of their acclimation under different temperature conditions. Acta Physiol Plant 1998; 20:73-78
29. Kacperska-Palacz A. Mechanisms of cold acclimation in herbaceous plants. In: Li PH, Sakai A, eds. Plant Cold Hardiness and Freezing Stress. New York: Academic Press, 1978:139-152.
30. Monroy AF, Castonguay Y, Laberge S, Sarhan F, Vezina LP, Dhindsa RS. A new cold-induced alfalfa gene is associated with enhanced hardening at subzero temperature. Plant Physiol 1993; 102:873-879.
31. Olien CR. An adaptive response of rye to freezing. Crop Sci 1984; 24:51-54.
32. Siminovitch D, Rheaume B, Pomeroy K, Lepage M. Phospholipid, protein and nucleic acid increases in protoplasm and membrane structures associated with development of extreme freezing resistance in black locust tree cells. Cryobiology 1968; 5:202-225.
33. Kacperska A. Metabolic consequences of low temperature stress in chilling-insensitive plants. In: Li PH, ed. Low Temperature Stress Physiology in Crops. Boca Raton, Fl: CRC Press, 1989:27-40.
34. Santarius KA. The mechanism of cryoprotection of biomembrane systems by carbohydrates. In: Li PH, Sakai A, eds. Plant Cold Hardiness and Freezing Stress, vol. 2. London, New York: Academic Press, 1982:475-486.
35. Kacperska-Palacz A, Egierszdorff S. Effects of cold and CCC treatments on hydration and frost and desiccation hardiness of plant tissues. Bot Gaz 1972; 133:355-360.
36. Kacperska A. Water potential alterations – a prerequisite or a triggering stimulus for the development of freezing tolerance in overwintering herbaceous plants? In: Li PH, Christersson L, eds. Advances in Plant Cold Hardiness. Boca Raton, Ann Arbor, London, Tokyo: CRC Press, 1993:73-91.
37. Maciejewska U, Kacperska A. Changes in the levels of oxidized and reduced pyridine nucleotides during cold acclimation of winter rape plants. Physiol Plant 1987; 69:687-691.
38. Chen PM, Li P, Cunningham WP. Ultrastructure differences in leaf of *Solanum* species in relation to their frost resistance. Bot Gaz 18′977; 138:267-285.
39. Griffith M, Huner NPA, Espelle KE, Kollattukudy PE. Lipid polymers accumulate in the epidermis and mestome sheath cell walls during low temperature development of winter rye leaves. Protoplasma 1985; 125:53-64.
40. Yoshida S. Studies on lipid changes associated with frost hardiness in cortex of in woody plants. Contrib Inst Low Temp Sci 1974, Ser B; 18:1-43.
41. Sikorska E, Kacperska-Palacz A. Phospholipid involvement in frost tolerance. Physiol Plant 1979: 47:144-150.
42. Wang X. Molecular analysis of phospholipase D. Trends Plant Sci 1997; 2:261-266.
43. Smoleñska-Sym G, Kacperska A. Inositol 1,4,5-trisphosphate formation in leaves of winter oilseed rape plants in response to freezing, tissue water potential and abscisic acid. Physiol Plant 1996; 96: 692-698.
44. Kozbial P, Jerzmanowski A, Shirsat A, Kacperska A. Transient freezing regulates expression of extensin-types genes in winter oilseed rape. Physiol Plant 1998; 103:264-270.
45. Close TJ. Dehydrins: Emergence of a biochemical role of a family of plant dehydration proteins. Physiol Plant 1997; 97:795-803.

46. Hincha DK, Meins FM Jr, Schmitt JM. β-1,3-Glucanase is cryoprotective in vitro and is accumulated in leaves during cold acclimation. Plant Physiol 1997; 114:1077-1083.
47. Murata N, Los DA. Membrane fluidity and temperature perception. Plant Physiol 1997; 115:875-879.
48. Bowler C, Chua N-H. Emerging themes of plant signal transduction. Plant Cell 1994; 6:1529-1541.
49. Gilroy S, Trevawas AT. A decade of plant signals. BioEssays 1994; 16:677-682.
50. Furuya M, Schäfer E. Photoreception and signaling of induction reactions by different phytochromes. Trends Plant Sci 1:301-307.
51. Tanada T. The photoreceptors in the high irradiance response of plants. Physiol Plant 1997; 101:451-454.
52. Karpiñski S, Escobar C, Karpinska B, Cressen G, Mullineaux P. Photosynthetic electron transport regulate the expression of cytosolic ascorbate peroxidase genes in *Arabidopsis* during excess light stress. Plant Cell 1997; 9:627-640.
53. Ward JM, Pei Z-M, Schroeder JI. Roles of ion channels in initiation of signal transduction in higher plants. Plant Cell 1995; 7:833-844.
54. Davies E. Action potential as multifunctional signals in plants: unifying hypothesis to explain disparate wound responses. Plant Cell Environ 1987; 10:623-631.
55. Walker JC. Structure and function of the receptor-like protein kinases of higher plants. Plant Mol Biol 1994; 26:1599-1609.
56. Hirt H. Multiple role of MAP kinases in plant signal transduction. Trends Plant Sci 1997; 2:11-15.
57. Trevawas AT, Malhó R. Signal perception and transduction: the origin of the phenotype. Plant Cell 1997; 9:1181-1195.
58. Trevawas T, Knight M. Mechanical signalling, calcium and plant form. Plant Mol Biol 1994; 26:1329-1341.
59. Kacperska A, Kulesza L. Frost resistance of winter rape leaves as related to the changes in water potential and growth capability. Physiol Plant 1987; 71:483-488.
60. Kacperska A, Szaniawski R. Frost resistance and water status of winter rape leaves as affected by different shoot/root temperature. Physiol Plant 1993; 89:775-782.
61. Lang V, Mäntylä E, Welin B, Sundberg B, Palva ET. Alterations in water status, endogenous abscisic content, and expression of *rab18* gene during the development of freezing tolerance in *Arabidopsis thaliana*. Plant Physiol 1994; 104:1341-1349.
62. Prasad TK, Anderson MD, Martin BA, Stewart CR. Evidence for chilling-induced oxidative stress in maize seedlings and a regulatory role for hydrogen peroxide. Plant Cell 1994; 6:65-74.
63. Hodges DM, Andrews CJ, Johnson DA, Hamilton RI. Antioxidant compound responses to chilling stress in differentially sensitive inbred maize lines. Physiol Plant 1996; 98:685-692.
64. Hakam N, Simon J-P. Effect of low temperatures on the activity of oxygen-scavenging enzymes in two populations of the C4 grass *Echinocchloa crus-galli*. Physiol Plant 1996; 97:209-216.
65. Rychter A, Ciseœla E, Kacperska A. Participation of the cyanide-resistant pathway in respiration of winter rape leaves as affected by plant cold acclimation. Physiol Plant 1988; 73:299-304.
66. Sosiñska A, Maleszewski S, Kacperska-Palacz A. Carbon photosynthetic metabolism in leaves of cold-treated rape plants. Z Pflanzenphysiol 1977; 83:285-299.
67. Sobczyk EA, Kacperska-Palacz A. Adenine nucleotide changes during cold acclimation of winter rape plants. Plant Physiol 1978; 62:875-878.
68. Kuiper PJC. Lipids in alfalfa in relation to cold hardiness. Plant Physiol 1970; 45:684-687.
69. Maciejewska U, Maleszewski S, Kacperska-Palacz A. Effect of cold hardening on the carbon photosynthetic metabolism. Bull Acad Pol Sci 1974; 22:513-517.
70. Fahl E, Kunkel S, Dörfling K. Low-temperature induced changes in frost tolerance and proline content in three winter rape varieties under filed and growth chamber conditions. In:

Dörfling K, Bretteschneider B, Tantau H, Pithan K, eds. Crop Development for the Cool and Wet Regions of Europe. Crop Adaptation To Cool Climates. Brussels: COST 814; 1994:99-106.

71. Reaney MJT, Ishikawa M, Robertson AJ, Gusta LV. The induction of cold acclimation: the role of abscisic acid. In: Li PH, ed. Low Temperature Stress Physiology in Crops. Boca Raton, Fl: CRC Press Inc, 1989:3-25.
72. Lee SP, Chen THH. Molecular biology of plant cold hardiness development. In: Li PH, Christersson L, eds. Advances in Plant Cold Hardiness, Boca Raton, Fl: CRC Press, 1993:1-30.
73. Capell B, Dörfling K. Low temperature-induced changes of abscisic acid contents in barley and cucumber leaves in relation to their water status. J Plant Physiol. 1989; 135:571-575.
74. Janowiak F, Dörfling K. Chilling of maize seedlings:changes in water status and abscisic acid content in ten genotypes differing in chilling tolerance. J Plant Physiol 1996; 147:582-588.
75. Smoleñska-Sym G, Gawroñska H, Kacperska A. Modifications of abscisic acid level in winter oilseed rape leaves during acclimation of plants to freezing temperatures. Plant Growth Regulators 1995; 17:61-65.
76. Guy CL. Cold acclimation and freezing stress tolerance: role of protein metabolism. Annu Rev Plant Physiol Plant Mol Biol 1990; 41:187-223.
77. Hon, W-C, Griffith M, Mlynarz A, Kwok YC, Yang DSC. Antifreeze proteins in winter rye are similar to pathogenesis-related proteins. Plant Physiol 1995; 109:879-889.
78. Bray EA. Molecular responses to water deficit. Plant Physiol 1993; 103:1035-1040.
79. Bray EA. Plant responses to water deficit. Trends Plant Sci 1997; 2:48-54.
80. Bohnert HJ, Nelson DE, Jensen RG. Adaptations to environmental stresses. Plant Cell 1995; 7:1099-1111.
81. Lewis BD, Karlin-Neumann G, Davis RW, Spalding EP. Ca^{2+}-activated anion channels and membrane depolarizations induced by blue light and cold in *Arabidopsis* seedlings. Plant Physiol 1997; 114:1327-1334.
82. Piotrowska G, Kacperska A, Filek M. May changes in membranes' electric potential be connected with freezing injuries in winter rape plants? Acta Physiol Plant 1998; 20:62-63 (Suppl).
83. Erlandson G, von Fircks H, Jensén P. $K^+(Rb^+)$ and Ca^{2+} fluxes in young winter wheat plants exposed to sub-zero temperatures. Physiol Plant 1987; 69:258-264.
84. Kasamo K, Shimomura T. Effect of cold shock on potassium efflux from *Nicotiana tabacum* leaf discs induced by abscisic acid and ionophores. Plant Cell Physiol 1991; 31:846-852.
85. Knight MC, Campbell AK, Smith SM, Trevawas AJ. Transgenic plant aequorin reports the effects of touch and cold-shock and elicitors on cytoplasmic calcium. Nature 1991; 352:524-526.
86. Knight H, Trevawas AT, Knight M. Cold calcium signaling in *Arabidopsis* involves two cellular pools and a change in calcium signature after acclimation. Plant Cell 1996; 8:489-503.
87. Jahnke LS, Hull MR, Long SP. Chilling stress and oxygen metabolizing enzymes in *Zea Mays* and *Zea diploperennin*. Plant Cell Environ 1991; 14:97-104.
88. Okuda T, Matsuda Y, Sugawara M, Sagisaka S. Metabolic response to treatment with cold, paraquat or 3-amino-1,2,4-triazole in leaves of winter wheat. Biosci Biotechnol Biochem 1992; 56:1911-1915.
89. Kacperska A, Kubacka-Zêbalska M. Formation of stress ethylene depends on both on ACC synthesis and on the activity free radical-generating system. Physiol Plant 1985; 77: 231-237.
90. Sawicka T, Kacperska A. Soluble and cell wall-associated β-galactosidases from cold-grown winter rape (*Brassica napus* L var. *oleifera* L) leaves. J Plant Physiol 1995; 145:357-362.
91. Solecka D, Kacperska A. Phenylalanine ammonia lyase activity in leaves of winter oilseed rape plants as affeceted by acclimation of plants to low temperature. Plant Physiol Biochem 1995; 33:585-591.

92. Janowiak F, Dörfling K. Chilling-induced changes in the contents of 1-aminocyclopropane-1carboxylic acid (ACC) and its N-malonyl conjugate (MACC) in seedlings of two maize inbrreds differing in chlling tolerance. J Plant Physiol 1995; 147:257-262.
93. Kacperska A. Ethylene synthesis and a role in plant responses to different stressors. In: Kanellis AK, Chang C, Kende H, Grierson D, eds. Biology and Biotechnology of the Plant Hormone Ethylene. Dordrecht, Boston, London: Kluwer Academic Publ, 1997:207-216.
94. Monroy AF, Sarhan F, Dhindsa RS. Cold-induced changes in freezing tolerance, protein phosphorylation, and gene expression. Evidence for a role of calcium. Plant Physiol 1993; 102:227-1235.
95. Polisensky DH, Braam J. Cold-shock regulation of *Arabidopsis TCH* genes and the effects of modulating intracellular calcium levels. Plant Physiol 1996; 111:1271-1279.
96. Ding JP, Pickard BG. Mechanosensory calcium-selective channels in epidermal cells. Plant J 1993; 3:713-720.
97. Bush DS. Calcium regulation in plant cells and its role in signaling. Annu Rev Plant Physiol Plant Mol Biol 1995; 46:95-122.
98. Okazaki Y, Tazawa M. Calcium ion and turgor regulation in plant cells. J Membrane Biol 1990; 114:189-194.
99. Legué V, Blancaflor E, Wymer C, Perbal G, Fantin D, Gilroy S. Cytoplasmic free Ca^{2+} in *Arabidopsis* roots changes in response to touch but not gravity. Plant Physiol 1997; 114:789-800.
100. Gal A, Zer H, Ohad I. Redox-controlled thylakoid protein phosphorylation. News and views. Physiol Plant 1997; 100:869-885.
101. Foyer CH, Lopez-Delgado H, Dat JF, Scott IM. Hydrogen peroxide- and glutathione associated mechanisms of acclimatory stress tolerance and signalling. Physiol Plant 1997; 100:241-254.
102. Ogren W. Photorespiration: pathways, regulation and modification. Annu Rev Plant Physiol 1984; 35:415-442.
103. Purvis AC, Shewfelt RL. Does the alternative pathway ameliorate chilling injury in sensitive plants tissues? Physiol Plant 1993; 88:712-718.
104. Koch KE. Carbohydrate-modulated gene expression in plants. Annu Rev Plant Physiol Plant Mol Biol 1996; 47:509-540.
105. Jang J-C, Sheen J. Sugar sensing in higher plants. Trends Plant Sci 1997; 2:208-214.
106. Cowan AK, Richardson GR, Maurel JCG. Stress-induced abscisic acid transients and stimulus-response-coupling. Physiol Plant 1997; 100:491-499.
107. Sanhla N, Huber W. Effects of abscisic acid on the activities of photosynthetic enzymes and $^{14}CO_2$ fixation products in *Pennisetum typhoides* seedlings. Physiol Plant 1974; 30:291-299.
108. Sankhla N, Huber W. Regulation of balance between C3 and C4 pathway: role of abscisic acid. Z Pflanzenphysiol 1975; 74:267-271.
109. Chandler PM, Robertson M. Gene expression regulated by abscisic acid and its relation to stress tolerance. Annu Rev Plant Physiol Plant Mol Biol 1994; 45:113-141.
110. Merlot S, Giraudat J. Genetic analysis of abscisic acid signal transduction. Plant Physiol 1997; 114:751-757.
111. Shinozaki K, Yamaguchi-Shinozaki K. Gene expression and signal trandsuction in water-stress response. Plant Physiol 1997; 115:327-334
112. Knetsch MLW, Wang M, Snaar-Jagalska E, Heimovaara-Dijkstra S. Abscisic acid induces mitogen-activated protein kinase activation in barley aleurone protoplasts. Plant Cell 1996; 8:1061-1067.
113. Cutler S, Ghassemian M, Bonetta D, Cooney S, McCourt P. A protein farnesyl transferase involved in abscisic acid signal transduction in *Arabidopsis*. Science 1996; 273:1239-1241.
114. Yamaguchi-Shinozaki K, Shinozaki K. A novel *cis*-acting element in *Arabidopsis* gene is involved in responsiveness to drought, low temperature, or high salt stress. Plant Cell 1994; 6:251-254.

115. Ristic Z, Ashworth E.N. Changes in leaf ultrastructure and carbohydrates in *Arabidopsis thaliana* L. (Heyn) cv. Columbia during rapid cold acclimation. Protoplasma 1993; 172:11-123.
116. Arora R, Palta JP. A loss in plasma membrane ATPase activity and its recovery coincides with incipient freeze-thaw injury and post-thaw recovery in onion bulb scale tissue. Plant Physiol 1991; 95:846-852.
117. Monroy AF, Dhindsa RS. Low-temperature signal transduction: induction of cold-acclimation specific genes of alfalfa by calcium at 25°C. Plant Cell 1995; 7:321-331.
118. Munnik T, Arisz A, de Vrije T, Musgrave A. G protein activation stimulates phospholipase D signaling in plants. The Plant Cell 1995; 7:2197-2210.
119. Wasternack C, Parthier B. Jasmonate-signalled plant gene expression. Trend Plant Sci 1997; 2:302-307.
120. Jonak C, Kiegler S, Ligterink W, Barker PJ, Huskisson NS. Stress signaling in plants: a mitogen-activated protein kinase pathway is activated by cold and drought. Proc Natl Acad Sci USA 1996; 93:11274-11279.
121. Hong SW, Jon JH, Kwak JM, Nam HG. Identification of a receptor-like kinase gene rapidly induced by abscisic acid, dehydration, high salt and cold treatments in *Aradopsis thaliana.* Plant Physiol 1997; 113:1203-1212.
122. Conley TR, Sharp RE, Walker JC. Water deficit rapidly stimulates the activity of a protein kinase in the elongation zone of the maize primary roots. Plant Physiol 1997; 113:219-226.
123. Pestenacz A, Erdei L. Calcium-dependent protein kinase in maize and sorghum induced by polyethylene glycol. Physiol Plant 1996; 97:360-364.
124. Usami S, Banno H, Ito Y, Nishihama R, Machida Y. Cutting activates 46-kilodalton protein kinase in plants. Proc Natl Acad Sci USA 1995; 92:8660-8664.
125. Ecker JR, Davis RW. Plant defense genes are regulated by ethylene. Proc Natl Acad Sci USA, 1987; 86:1539-1554.
126. Chessin M, Zipf AE. 1990. Alarm system in higher plants. Bot Rev 1990; 56:193-235.
127. Seo S, Sano H, Ohashi Y. Jasmonic acid in wound signal transduction pathway. Physiol Plant 1997; 101:740-745.

Ice nucleation and deep supercooling in plants: new insights using infrared thermography

M. Wisniewski[1]* and M. Fuller[2]

[1] US Department of Agriculture, Agricultural Research Service, 45 Wiltshire Road, Kearneysville, West Virginia 25430, USA
[2] Department of Agriculture and Food Studies, Seale-Hayne Faculty of Agriculture, Food and Land Use, University of Plymouth, Newton Abbot, Devon TQ12 6NQ, United Kingdom

1 Introduction

In the course of a year, perennial plants in temperate climates are exposed to several types of freezing stress including low temperature extremes, ice encasement, and unseasonable episodes of frost. Many plants can adapt to survive freezing through a process of cold acclimation. This is a complex process with many contributing factors.[1] Cold acclimation is also a dynamic process. Not only does it exhibit a distinct seasonality but the mechanisms that confer cold hardiness in midwinter may differ from those in late winter or early spring. This is further complicated by the discovery that in many perennial woody plants, and some herbaceous plants, tissues within the same plant can respond very differently to exposure to freezing temperatures.

When plants are exposed to subzero temperatures, they are exposed not only to the stress of low temperature but also the potential presence of ice within their tissues. Once ice forms, cells are subjected to both a severe dehydrative stress and a mechanical stress brought about by the effects of cytorrhysis (freeze-induced collapse of cells) and the presence of large ice crystals. Cold-hardy plants have developed several unique mechanisms in order to handle both the mechanical and dehydrative stresses imposed by the presence of ice within their tissues. These adaptations fall into two general categories: freeze tolerance and freeze avoidance.[2] Tissues exhibiting freeze tolerance respond to freezing temperatures by formation of extracellular ice followed by a concomitant loss of cellular water in order to come into chemical equilibrium with the external ice. The loss of water results in the collapse of the entire cell and an increase in the solute concentration of the cell sap leading to a lowering of the freezing point of the cell. In contrast, tissues that avoid stress

* Corresponding author; e-mail: mwisniew@afrs.usda.gov

survive freezing temperatures by supercooling, a process in which cell solutions are kept between the equilibrium freezing point and the homogeneous ice nucleation temperature of water (usually between -1 and -41°C).[3] The array of biochemical responses that plants have developed to adapt to freeze-induced dehydration is described elsewhere (see Chapter by Hiilovaara-Teijo and Palva, in this book). The objective of this chapter is to provide an overview of ice nucleation and deep supercooling in plants and discuss how the use of infrared thermography can be used to better understand the freezing process in plants. Additionally, brief mention will be made of the ability of some very cold-hardy plants to form glassed solutions within their cells. The commonality that unites the processes of ice nucleation, deep supercooling, and glass formation in plants is that each one represents a condition where water is present in a metastable condition.[3] Ice nucleation and supercooling describe mechanisms that plants have evolved to inhibit the phase change of water from a liquid to a solid whereas the process of glass formation describes a mechanism by which cells become osmotically, thermally, and mechanically inert to the presence of external ice even at temperatures as low as -196°C.

2 Ice nucleation

In order for ice to form on or within a plant, ice nucleation must first occur. Although the melting point of ice is 0°C, the freezing temperature of water is not as defined.[4] In fact, although it is not commonly recognized, pure water has a low probability of freezing at temperatures warmer than -40°C.[5] This is because a small ice crystal embryo is necessary in order for ice to form and grow to any substantial size. The probability of forming such an ice crystal embryo in pure water, as well as the half-life of such a crystal, is low until temperatures approaching -40°C. This temperature is referred to as the homogeneous ice nucleation point.[5]

In nature, it is rare for water to exist in a pure state but it rather exists as an ionic or colloidal solution. In such solutions heterogeneous ice nucleation is initiated on the surface of objects or on suspended particles.[4] Heterogeneous ice nucleators are very effective in inducing ice formation and are very abundant. As a consequence, freezing occurs in nature at much warmer temperatures than the homogeneous nucleation temperature.

The role of heterogeneous ice nucleators in inducing ice formation in plants is important because if methods can be developed for regulating ice nucleation, significant advances could be made in limiting frost injury to both freezing-sensitive and cold-adapted plants. A major question concerns the relative importance of extrinsic ice nucleation agents produced by ice-nucleation-active (INA) bacteria (e.g. *Pseudomonas syringae*) and intrinsic nucleation agents synthesized by plants.[6] While all plants can supercool (i.e. have tissues below 0°C without freezing) to some extent,[6–9] the extent of supercooling varies between plant species and is influenced by the presence of ice nucleating agents which may be of plant[9–12] or bacterial1[3–15] origin.

The abundance of ice nuclei on plants can be estimated by freezing of droplets of plant macerates or small portions of plant tissue[6,15] but these procedures are destructive and do not provide information on where ice formation was initiated. Ice formation in intact plants can be readily detected by measuring, with thermocouples, the heat that is released upon the freezing of the water in the plant.[16–19] Nevertheless, even when arrays of temperature-measuring devices are attached to plants, the actual site of ice initiation and the temperature at the site where ice nucleation occurred can only be inferred.[20] This is a significant technical limitation and more details of the freezing process are required in order to accurately predict freezing patterns and determine under what conditions the reduction or interference of extrinsic ice nuclei would provide significant frost protection. Recently, the ability to use infrared video thermography to directly observe ice nucleation (i.e. initial ice formation) and propagation in plants has been demonstrated.[21–25] The use of this technology to study the freezing process is based on the fact that ice formation is an exothermic event and the release of the heat of fusion as water changes phase from a liquid to a solid can be monitored and visualized. The temperature and spatial resolution of the device used in these studies has enabled the researchers to clearly define the initial site of ice nucleation as well as monitor the ice front as it spread into the surrounding tissues. Importantly, these studies have shown that the site of placement of a thermocouple was often the site of initial freezing. This suggests that an erroneous conclusion may have been derived from earlier studies of ice nucleation in plants that relied on thermocouple-generated data.

Using infrared thermography it is possible to determine the role of extrinsic and intrinsic ice nucleating agents in the freezing process, rates of ice propagation, the effect of plant structure on the freezing process, and how the specific pattern of freezing relates to visual patterns of injury. It is also possible to clearly evaluate if the reduction of ice nuclei or inhibiting their activity is a feasible approach to frost protection. An illustration of the use of infrared technology to study the freezing process is illustrated in Figures 1–3.

Figure 1 illustrates that, with infrared thermography, the degree and extent of freezing can be directly associated with the amount of visual injury. In this experiment, a series of different leaves of tomato (*Lycopersicon esculentum*) were removed from an environmental chamber at various periods of time after initiation of freezing. Visual injury (Fig. 1A) was then compared to infrared images of the same leaves just before they were removed from the freezing chamber (Fig. 1B–1F). It appears that it was not the freezing process that was directly responsible for injury but rather the extent of dehydration brought about by the loss of cellular water to extracellular ice. Initial freezing of a droplet of water containing the Ice$^+$ bacterium, *Pseudomonas syringae*, did not result in injury. Only as the exothermic process proceeded and dehydration increased did visual injury begin to appear. Interestingly, visual injury in leaf E indicated the possibility of numerous independent nucleation events; however, the infrared image clearly shows that the entire leaf had frozen and was induced by a single freezing event. Such data are not obtainable using thermocouple-based technology.

Figure 2 shows the influence of plant structure on ice nucleation in leaves of bean (*Phaeseolus vulgaris*). When droplets of water containing Ice$^+$ bacteria are

Fig. 1. Infrared thermography of freezing injury in leaves of tomato (*Lycopersicon esculentum* cv. Roma). In all cases, nucleation of a droplet of water containing Ice$^+$ bacteria (*arrow*) occurred at approximately -1.3°C
(A) Visual injury of tomato leaves after increasing times of exposure to subzero temperatures, letters B-F correspond with infrared images B-F; **(B)** Leaf removed after freezing of surface droplet results in no injury; **(C)** Leaf removed after initial freezing of leaf at approximately -3.2°C also results in little injury; **(D–E)** As the exotherm increases (visualized by warmer leaf temperature) due to more water being removed from cells (hence creating a dehydrative stress) injury increases; **(F)** Dramatic differential between leaf temperature and air temperature brought about by the freezing of water in the leaf tissue which is an exothermic process, this results in extensive injury as shown in **(A)**

placed on either the adaxial surface, having no stomates (Fig. 2A), or the abaxial surface, having stomates (Fig. 2B), the droplets freeze at approximately the same time (Fig. 2C). Freezing of the leaf first occurs, however, on the abaxial surface, presumably from propagation of ice crystals from the frozen droplet through stomatal

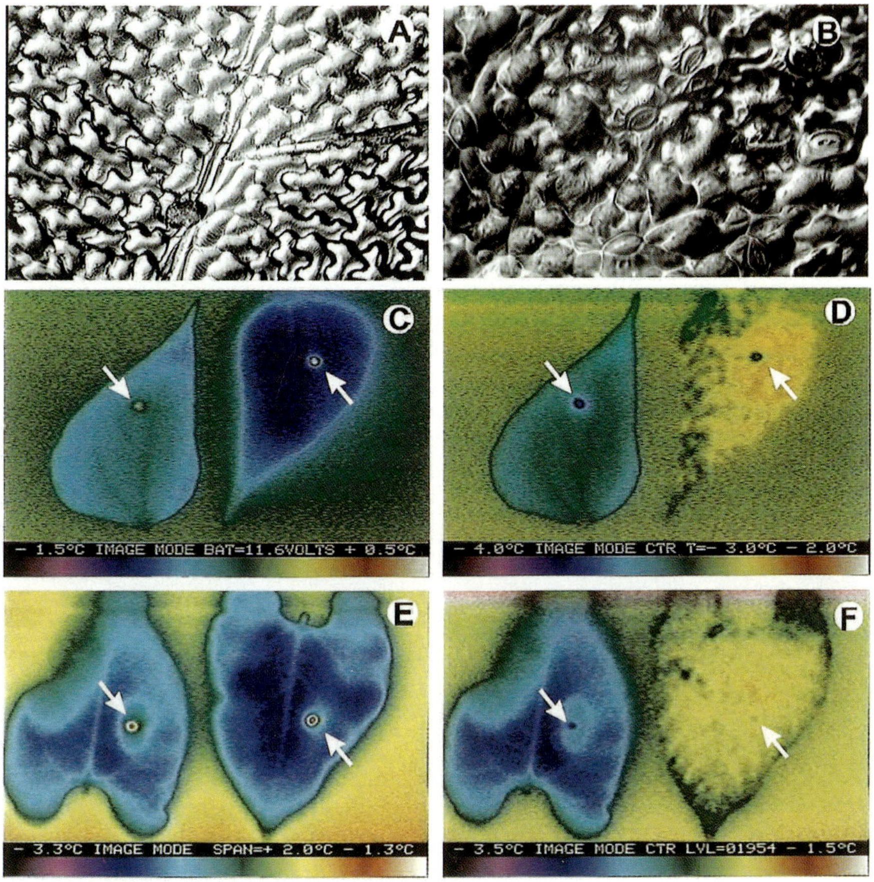

Fig. 2. Impression of the adaxial and abaxial leaf surface of bean (*Phaseolus vulgaris*) and infrared thermography of freezing process
(**A**) Adaxial leaf surface; (**B**) Abaxial leaf surface; (**C**) Initial freezing of droplet of water (*arrows*) containing Ice$^+$ bacteria on adaxial (*left*) and abaxial (*right*) leaf surface of bean; (**D**) Freezing of leaf induced by frozen droplet present on the abaxial leaf surface (*right*), frozen droplet on adaxial surface (*left*) has not yet induced freezing of leaf indicating that the ice crystals on abaxial surface had penetrated the leaf through stomatal openings whereas the cuticle on adaxial surface provided a temporary barrier to the growing ice crystal; (**E**) Initial freezing of droplet of water (*arrow*) containing Ice$^+$ bacteria on abaxial surface of bean leaves, leaf on left had a coating of silicon grease under the droplet; (**F**) Freezing of bean leaf on right, that did not have the barrier of silicon grease, is induced by the frozen droplet; leaf on left, with the silicon grease barrier, is unfrozen and remains unfrozen to as low as -6.0°C, providing further evidence that ice crystals must physically grow into the interior of the leaf to induce the leaf to freeze

openings (Fig. 2D). This freezing of the leaf through the abaxial surface can be prevented by applying a layer of silicon grease between the droplet and the leaf surface (Fig. 2E–2F). This further supports the idea that ice must physically penetrate the leaf surface to induce freezing of the leaf tissue and that the cuticle provides a barrier to the growth of ice crystals.

We have observed an even greater influence of plant structure on the freezing process in leaves of *Rhododendron* sp. (Fig. 3). Freezing of leaves is easily induced through the abaxial leaf surface (Fig. 3C) at relatively warm temperatures but freezing of leaves via the adaxial surface is strongly inhibited, even at temperatures below -4.0°C (Fig. 3D). It is plausible to suggest that the thick cuticle of rhododendron and the leaf curling (which shelters the abaxial surface) at low temperature may represent an adaptation to avoid freezing.

Clearly, the process of ice nucleation and ice propagation in whole plants under field conditions is incompletely understood. This can be partially attributed to the methods by which these processes are monitored. Current methods rely on estimating the number of ice nuclei present in a plant via destructive sampling or the use of thermocouples to monitor the occurrence of freezing in intact plants. These assays, however, cannot determine either the primary site of nucleation or the

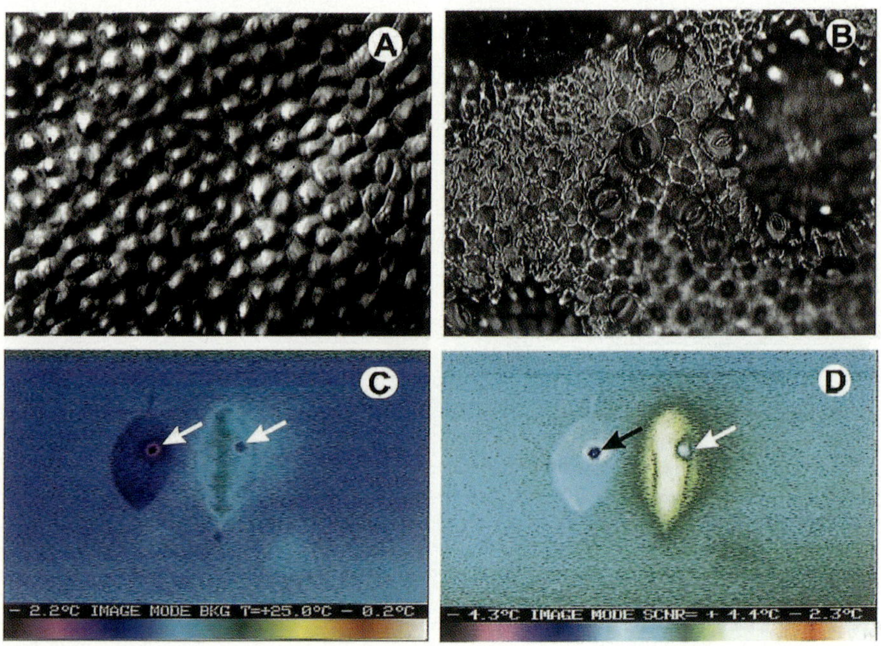

Fig. 3. Impression of the adaxial and abaxial *Rhododendron* sp. leaf surface, and infrared thermography of freezing process
(A) Adaxial surface, no stomates; **(B)** Abaxial surface, with stomates; **(C)** Frozen droplets of water containing Ice^{+} bacteria (*arrows*) on adaxial (*left*) and abaxial (*right*) leaf surface; frozen droplet on abaxial surface has induced ice formation in leaf tissue while leaf with frozen droplet on adaxial surface remains supercooled and unfrozen; **(D)** Leaf with frozen droplet on adaxial leaf surface (*left*) continues to remain supercooled and unfrozen at -3.0°C; leaf on left remained supercooled to -6.2°C before freezing, indicating that the cuticle on adaxial surface presents a substantial barrier to ice propagation

details of subsequent ice propagation through the plant. The use of infrared thermography has revealed new information about the complexity of the freezing process and it is expected that this technology will contribute greatly to the development of frost-protection strategies.

Although discussed in detail by Hiilovaara-Teijo and Palva (in this book), relevant to the discussion of ice nucleation is the recent report of the presence of extracellular antifreeze proteins in cold acclimated plants.[26,27] First reported in fish,[28] antifreeze proteins have subsequently been found to be common in a wide array of insects, bacteria, fungi, and plants.[27,29] Antifreeze proteins are complex proteins that depress the freezing temperature of water in a noncolligative manner (i.e. they exhibit thermal hysteresis) and affect the morphology of ice crystals by binding to the surface and restricting the growth of crystals except in defined planes of growth. Resulting ice crystals are small and needle-like in appearance rather than large and amorphous. In plants, these proteins show homology to chitinases and glucanases, proteins that are important to disease resistance.[30] Although plant antifreeze proteins depress the freezing point by only 0.2–0.4°C, they may inhibit the activity of extrinsic and intrinsic ice nucleating agents and thereby play an important role in freeze avoidance.[28,29]

3
Extraorgan freezing and deep supercooling

The response of dormant buds of cold-hardy woody plants is different than in other portions of the same plant. Freezing is initiated within the bud scales and subtending stem tissue and the shoot apex or flower primordium is isolated from the mechanical damage caused by the presence of large ice crystals.[31,32] The presence of the ice in the bud scales results in a water potential gradient and in many species of woody plants, water then migrates from the shoot or floral apex to the sites of extracellular ice. This response has been termed extraorgan freezing.[33] When buds are killed, mortality is due to dehydrative stress rather than low temperatures or ice.

In other species of woody plants, however, a portion of water remains supercooled within the primordium rather than migrating to the sites of extracellular ice in the bud scales and subtending stem. The supercooled water remains unfrozen to very low subzero temperatures (hence the term deep supercooling) as long as it is isolated from any nucleation event; however, once the primordium does freeze, intracellular ice is formed and the freezing results in bud mortality.[34] Of the many aspects of biological ice nucleation and cold hardiness of plants, deep supercooling is perhaps the most enigmatic. It can be defined as the ability of a population of cells to retain water in a liquid phase at subfreezing temperatures approaching the homogeneous nucleation temperature of water by remaining free from internal, heterogeneous ice nuclei and isolated from the nucleating effect of extracellular ice.

Deep supercooling of flower buds has been documented in grape,[35–37] blueberry,[38] several species of *Prunus* sp.,[39] and many other plants as well.[34] As illustrated in Figure 4, when flower buds of peach are subjected to differential thermal analy-

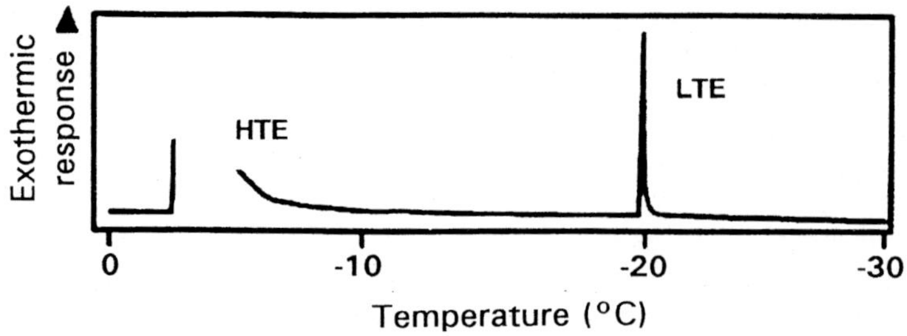

Fig. 4. Differential thermal analysis profile of peach flower bud
A high-temperature exotherm (HTE), representing the freezing of extracellular water, and a low-temperature exotherm (LTE), representing intracellular ice formation in floral tissue, are visible. The LTE represents a lethal event resulting in death of the floral tissue

sis (DTA), two distinct exotherms are visible. The high-temperature exotherm (HTE) is associated with the freezing of water in the bud scales and subtending stem tissue and the low-temperature exotherm (LTE) is associated with the freezing of intracellular, deep-supercooled water contained within the primordial tissues. In species with multiple flowers within a mixed or floral bud, each floret freezes as an independent unit.[39] This is demonstrated by the appearance of multiple LTEs using DTA. Multiple LTEs are present in grape, sweet cherry, sour cherry, azalea, and flowering dogwood.[1,34] Importantly, the LTE is correlated with the degree of cold hardiness of the tissue and is used extensively as a tool to evaluate the level of cold hardiness.[40]

In order for deep supercooling of buds to occur, a barrier to water movement and ice propagation must exist. The factors responsible for regulating deep supercooling of buds, however, are not completely understood. Using a combination of light and scanning-electron microscopy, several authors have demonstrated preferential formation of ice in the bud scales and subtending stem.[32,33,41] These large crystals cause a mechanical disruption of the tissue seen as large voids upon thawing (Fig. 5). In subsequent freezing events, ice again forms in these voids. Ashworth et al.[41] suggested that these voids serve to accommodate ice throughout the dormant season. Recently, Quamme et al.[42] have suggested that the small size of cells associated with procambial strands, as well as tissues at the base of the primordium, along with thick cell walls, and an absence of intercellular spaces, all contribute to forming a barrier to ice propagation. Wisniewski and Davis[43] have also reported the presence of nonesterified pectins in intercellular spaces of cold-acclimated bud tissues and suggested that these pectins may also inhibit the growth of ice into the bud primordium.

The extent of deep supercooling varies on a seasonal basis. Deep supercooling and hence cold hardiness is progressively lost as buds begin to swell and approach budbreak. The rapid loss of supercooling in peach floral buds is associated with maturation of the procambial strands within the primordium leading to vascular continuity.[44] The functional strand of xylem serves as a conduit for the rapid spread

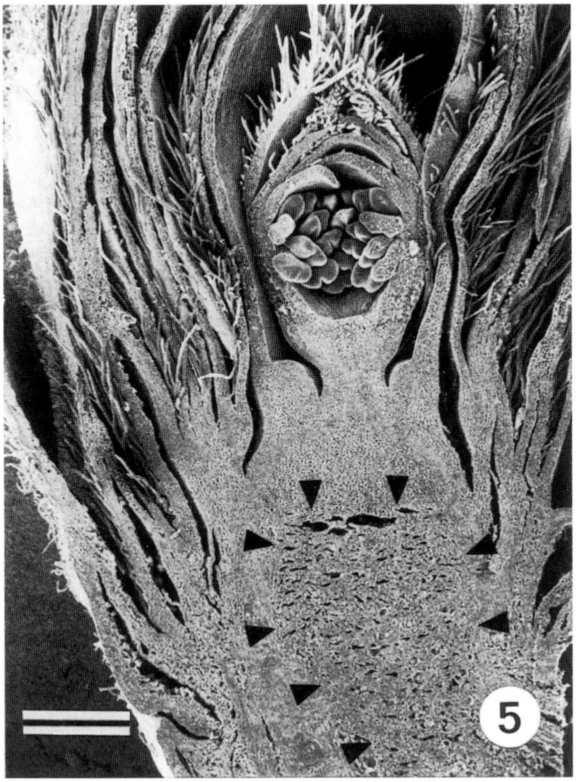

Fig. 5. Scanning electron micrograph of peach (*Prunus persica* cv. Loring) flower bud tissue prepared by freeze fixation and freeze substitution
Arrowheads indicate voids in axis tissue subtending floral buds where ice crystals had formed. Voids are also present in the bud scales. Note absence of voids in the flower itself (reprinted with permission from ref. 41)

of ice into the primordium resulting in the formation of ice within primordial tissues.

Deep supercooling of xylem tissues is also common to many temperate, perennial fruit crops, and temperate native species of woody plants as well.[45] Using DTA (Fig. 6), it is possible to obtain evidence of an LTE at very low temperatures (-25 to -40°C) whose appearance is correlated with the death of xylem parenchyma cells.[46,47] Because of this association with mortality, DTA has been used extensively to evaluate the degree of cold hardiness of stem tissues of many important fruit and landscape species of woody plants. As with deep supercooling of dormant buds, a barrier must exist in xylem tissues that prevents the rapid loss of water to extracellular ice and also prevents the growth of ice crystals into living cells. In this regard, the porosity and/or permeability of the cell wall appears to play an essential role in the regulation of deep supercooling.[45]

In a series of papers,[48–50] Wisniewski and coauthors have demonstrated that the structure and composition of the pit membrane of xylem parenchyma cells appear to play an important role in regulating the extent of deep supercooling of xylem tissues. The pit membrane is a thin portion of the cell wall that allows for the passage of solutes and other materials, as well as plasmodesmatal connections, between cells. It is composed mainly of cellulose and pectic materials and unlike secondary wall material is unlignified, at least during the first 2 years of development. By chemically or enzymatically altering the structure of the pit membrane, the extent of deep supercooling could be reduced or eliminated. In particular, pectins appeared to play a significant role in defining the permeability or porosity of the cell wall.

As noted by Wisniewski,[45] although pectin-mediated regulation of deep supercooling may account for many of the reported observations, many questions must be resolved before a definitive knowledge of the regulation of deep supercooling is obtained. Only as scientific evidence for the control and regulation of deep supercooling develops will we be able to determine whether or not this trait can be manipulated in a manner that will prevent freeze injury in woody plants. In this regard, Burke and Stushnoff[51] have proposed that deep supercooling as a genetic trait be selected against in breeding programs in favor of other mechanisms of cold hardiness that impart a greater degree of freeze tolerance. The inheritance and genetic regulation of deep supercooling, however, remain to be elucidated.

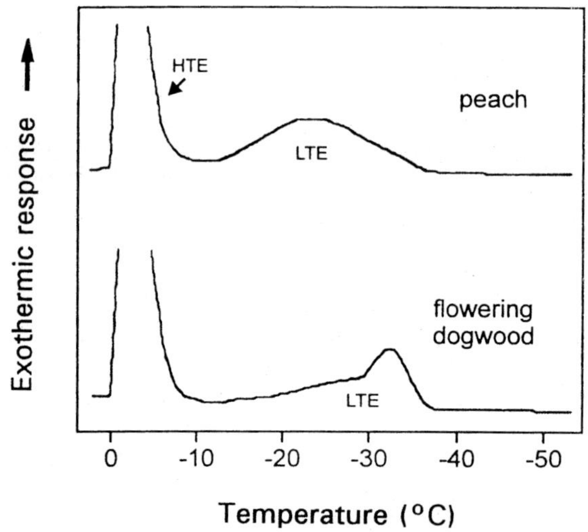

Fig. 6. Differential thermal analysis profiles of xylem tissues of peach (*Prunus persica* cv. Loring) and flowering dogwood (*Cornus florida*)
Both high-temperature (HTE) and low-temperature (LTE) exotherms are visible. The LTEs are correlated with injury to xylem tissues, i.e. as the exotherm progresses greater amounts of xylem injury are visible. Temperatures below the LTEs characteristic for the species are lethal

4 Glass formation

Another adaptation that results in a metastable state of water is the formation of glassed cell solutions.[3] This is believed to occur in extremely hardy plant species that can survive cooling to -196°C in liquid nitrogen. The formation of glasses is a unique metastable condition that cannot be terminated below the glass transition temperature. Aqueous glasses are extremely viscous solution with the viscosity brought about by high solute (sugar) concentrations at a sufficiently low temperature. In poplar (*Populus tremuloides*), glasses are reported to form below -20°C.[52] Glasses in such systems may be supercooled, supersaturated, and under hydrostatic tension. However, they are not subject to ice nucleation, solute crystallization, or water vapor cavitation so long as the solution remains below the melting temperature of the glassed solution. Thus cell sap in such a state is extremely stable and relatively impervious to all the stresses associated with low temperature and the presence of ice. The glassed state is frequently relied upon for the survival of plant tissues during cryopreservation.[53]

5 Conclusions

The thermomechanical response of plants to freezing temperatures is complex. Plants vary greatly in their ability to survive freezing temperatures and the resulting dehydrative and mechanical stresses that occur as a result from the presence of ice. Initially this is represented by the ability to inhibit the formation of ice (ice nucleation) by supercooling. Significant questions exist, however, about the role of intrinsic and extrinsic ice nucleating agents in determining the extent to which any particular plant can supercool. Additionally, little is known about how plant structure can affect ice nucleation and propagation. It is believed that the use of high resolution infrared thermography will reveal significant details about the freezing process that will lead to new approaches in developing frost protection technology.

Dramatic suppression of ice formation in certain tissues of woody plants by deep supercooling was also discussed. It represents a unique adaptation of woody plants to avoid freezing injury but is limited by the physical constraint of the homogeneous nucleation temperature of water. Therefore, deep supercooling is a characteristic of only moderately hardy woody plants. Elucidation of the mechanism that allow tissues to deep supercool will require further research before it can be determined whether or not this trait can be genetically manipulated. In contrast, it has been proposed that the most cold-hardy woody plants have the ability to form glassed solutions. Although a glass is also a metastable condition of water, it is very stable as long as the cell remains below the melting temperature of the glassed solution, and allows tissues to become relatively impervious to the stresses associated with extremely low temperatures.

All the mechanisms described have thermomechanical properties associated with them and present many opportunities for further research. A better understanding of these processes may provide insight into the development of materials adapted for environmental extremes, enhancement of cryopreservation and stabilization of pharmaceuticals, as well as the development of new frost protection technologies applicable to horticulture and agriculture.

6
References

1. Sakai A, Larcher W. Frost Survival of Plants. Berlin Heidelberg: Springer, 1987.
2. Levitt J. Responses of Plants to Environmental Stress, vol. 1. Chilling, Freezing, and High Temperature Stresses, 2nd ed. London: Academic Press, 1980.
3. Chen THH, Burke MJ, Gusta LV. Freezing tolerance in plants: An overview. In: Lee RE Jr, Warren GJ, Gusta LV, eds. Biological Ice Nucleation and its Applications. APS Press, 1995:115-136.
4. Ashworth EN. Formation and spread of ice in plant tissues. Hort Rev 1992; 215-255.
5. Franks F. Biophysics and Biochemistry at Low Temperatures. Cambridge: Cambridge University Press, 1985.
6. Ashworth EN, Kieft TL. Ice nucleation activity associated with plants and fungi. In: Lee RE Jr, Warren GJ, Gusta LV, eds. Biological Ice Nucleation and its Applications. APS Press, 1995:137-162.
7. Burke MJ, Gusta LV, Quamme HA, Weiser CJ, Li PH. Freezing injury in plants. Annu Rev Plant Physiol 1976; 27:507-528.
8. Lindow SE. Control of epiphytic ice-nucleation-active bacteria for management of plant frost injury. In: Lee RE, Jr, Warren GJ, Gusta LV, eds. Biological Ice Nucleation and its Applications. APS Press, 1995:239-256.
9. Lindow SE, Arny DC, Upper CD. Distribution of ice- nucleation-active bacteria on plants in nature. Appl Environ Microbiol 1978; 36:831-838.
10. Fuller MP, White GG, Charman A. The freezing characteristics of cauliflower curd. Ann Appl Biol 1994; 125:179-188.
11. Andrews PK, Sandridge CR, Toyama TK. Deep supercooling of dormant and deacclimating *Vitis* buds. Am J Enol Vitic 1984; 35:175-177.
12. Gross DC, Proebsting EL, Jr, MacCrindle-Zimmerman H. Development, distribution, and characteristics of intrinsic, non-bacterial ice nuclei in *Prunus* wood. Plant Physiol 1988; 88:915-922.
13. Gross DC, Proebsting EL, Jr, Andrews PK. The effects of ice-nucleation-active bacteria on the temperatures of ice nucleation and low temperature susceptibilities of *Prunus* flower buds at various stages of development. J Am Soc Hort Sci 1984; 109:375-380.
14. Hirano SS, Baker LS, Upper CD. Ice nucleation temperature of individual leaves in relation to population sizes of ice nucleation active bacteria and frost injury. Plant Physiol 1985; 77:259-265.
15. Lindow SE. The role of bacterial ice nucleation in frost injury to plants. Annu Rev Phytopath 1983; 21:363-384.
16. Cary JW, Mayland HF. Factors influencing freezing of supercooled water in tender plants. Agron J 1970; 62:715-719.
17. Quamme HA, Stushnoff C, Weiser CJ. The relationship of exotherms to cold injury in apple stem tissues. J Am Soc Hort Sci 1972; 97:608-613.
18. Ashworth EN. Formation and spread of ice in plant tissues. Hort Rev 1992; 13:215-255.
19. Proebsting EL, Jr, Andrews PK, Gross D. Supercooling in young developing fruit and flower buds in deciduous orchards. HortScience 1982; 17:67-68.

20. Ashworth EN, Anderson JA, Davis GA, Lightner GW. Ice formation in *Prunus persica* under field conditions. J Am Soc Hort Sci 1985; 110:322-324.
21. LeGrice P, Fuller MP, Campbell A. An investigation of the potential use of thermal imaging technology in the study of frost damage to sensitive crops. In: Proceedings of the Int Conference on Biological Ice Nucleation. Laramie: Univ Wyoming, 1993:4.
22. Ceccardi TL, Heath RL, Ting IP. Low-temperature exotherm measurement using infrared thermography. HortScience 1995; 30:140-142.
23. Wisniewski M, Lindow SE, Ashworth EN. Observations of ice nucleation and propagation in plants using infrared video thermography. Plant Physiol 1997; 113:327-334.
24. Wisniewski M. the use of infrared video thermography to study freezing in plants. In: Li PH, Chen THH, eds. Plant Cold Hardiness. Plenum Press, 1988:311-316.
25. Fuller MP, Wisniewski M. The use of infrared thermal imaging in the study of ice nucleation and freezing in plants. J Therm Biol 1998; 23:81-89.
26. Griffith M, Ala P, Yang SC, Hanand WC, Moffatt BA. Antifreeze protein produced endogeneously in winter rye leaves. Plant Physiol 1992; 100:593-596.
27. Duman JG, Olsen TM. Thermal hysteresis protein activity in bacteria, fungi, and phylogenetically diverse plants. Cryobiology 1993; 30:322-328.
28. Duman JG, Wu DW, Olsen TM, Urratia M, Tursman D. Thermal-hysteresis proteins. Adv Low Temp Biol 1993; 2:131-182.
29. Griffith M, Ewart KV. Antifreeze proteins and their potential use in frozen foods. Biotechnol Adv 1995; 13:375-402.
30. Hon WC, Griffith M, Chang P, Yang DSC. Extraction and isolation of antifreeze proteins from winter rye leaves. Plant Physiol 1994; 104:971-980.
31. George MF, Burke MJ, Weiser CJ. Supercooling in overwintering azalea flower buds. Plant Physiol 1974; 54:29-35.
32. Quamme HA. Mechanism of supercooling in overwintering peach flower buds. J Am Soc Hort Sci 1978; 103:57-61.
33. Sakai A. Freezing avoidance mechanism of primordial shoots of conifer buds. Plant Cell Physiol 1979; 20:1381-1386.
34. Quamme HA. Deep supercooling in buds of woody plants. In: Lee RE Jr, Warren GJ, Gusta LV, eds. Biological Ice Nucleation and its Applications. APS Press, 1995:183-200.
35. Pierquet P, Stushnoff C, Burke MJ. Low temperature exotherms in stem and bud tissues of *Vitis riparia*. J Am Soc Hort Sci 1977; 102:54-55.
36. Quamme HA. Use of thermal analysis to measure freezing resistance of grape buds. Can J Plant Sci 1986; 66:945-952.
37. Wolf TK, Pool RM. 1987. Factors affecting exotherm detection in the differential thermal analysis of grapevine dormant buds. J Am Soc Hort Sci 1987; 520-525.
38. Biermann J, Stushnoff C, Burke MJ. Differential thermal analysis and freezing injury in cold hardy blueberry flower buds. J Am Soc Hort Sci 1974; 99:315-318.
39. Quamme HA. An exothermic process involved in freezing injury to flower buds of several *Prunus* species. J Am Soc Hort Sci 1974; 99:315-318.
40. Wisniewski M, Arora R. Adaptation and response of fruit trees to freezing temperatures. In: Biggs AR, ed. Cytology, Histology, and Histochemistry of Fruit Tree Diseases. Boca Raton: CRC Press, 1993:299-320.
41. Ashworth EN, Davis GA, Wisniewski M. The formation and distribution of ice within dormant and deacclimated peach flower buds. Plant Cell Environ 1989; 15:521-528.
42. Quamme HA, Su WA, Veto LJ. Anatomical features facilitating supercooling of the flower within the dormant peach flower bud. J Am Soc Hort Sci 1995; 120:814-822.
43. Wisniewski M, Davis G. Immunogold localization of pectins and glycoproteins in tissues of peach with reference to deep supercooling. Trees 1995; 9:253-260.
44. Ashworth EN. Xylem development in *Prunus* flower buds and the relationship to deep supercooling. Plant Physiol 1984; 74:862-865.

45. Wisniewski M. Deep supercooling in woody plants and the role of cell wall structure. In: Lee RE Jr, Warren GJ, Gusta LV, eds. Biological Ice Nucleation and its Applications. APS Press, 1995:163-182.
46. Quamme HA, Stushnoff C, Weiser CJ. The relationship of exotherms to cold injury in apple stem tissues. J Am Soc Hort Sci 1972; 97:608-611.
47. Hong SG, Sucoff E. Lee-Stadelman OY. Effects of freezing deep supercooled water on the viability of ray cells. Bot Gaz 1980; 141:464-469.
48. Wisniewski M, Davis G. Evidence for the involvement of a specific cell wall layer in the regulation of deep supercooling of xylem parenchyma. Plant Physiol 1989; 139:105-116.
49. Wisniewski M, Davis G, Arora R. Effect of macerase, oxalic acid, and EGTA on deep supercooling and pit membrane structure of xylem parenchyma of peach. Plant Physiol 1991; 139:1354-1359.
50. Wisniewski M, Davis G, Schaffer K. Mediation of deep supercooling of peach and dogwood by enzymatic modifications in cell-wall structure. Planta 1991; 184:254-260.
51. Burke MJ, Stushnoff C. Frost hardiness: a discussion of possible molecular causes of injury with particular reference to deep supercooling of water. In: Mussel H, Staples RC, eds. Stress Physiology of Crop Plants. Wiley-Interscience, 1979:197-215.
52. Hirsh AG, Williams RJ, Meryman HT. A novel method of natural cryopreservation: Intracellular glass formation in deeply frozen *Populus*. Plant Physiol 1985; 79:41-56.
53. Steponkus PL, Lang R, Fujikawa S. Cryopreservation of plant tissues by vitrifications. In: Steponkus PL, ed. Advances in Low-Temperature Biology. London: JAI Press, Ltd, 1992:1-61.

Modeling sequential responses of plant cells to freezing and thawing

M. J. T. Reaney[1] and L. V. Gusta[2*]

[1] Crop Utilization Research, Agriculture and Agri-Food Canada, 107 Science Place, Saskatoon Saskatchewan, Canada S7N 0X2
[2] Crop Development Centre, 51 Campus Drive, University of Saskatchewan, Saskatoon Saskatchewan, Canada S7N 5A8

1 Introduction

The geographic distribution of many perennial and biennial temperate plant species is determined largely by the minimum temperature encountered during the winter and the ability of the plants to survive winter temperatures at which the bulk of plant water will freeze.[1] Although many temperate plants achieve winter survival by avoiding ice crystallization, most species that overwinter in a vegetative state are tolerant to the growth of ice crystals in their tissues.[2] Plants respond to low temperature and ice formation through molecular interaction[3] and thermodynamic and kinetic processes.[4,5] Freezing injury is influenced by the rate of freezing, depth of freezing, solute composition, rate of thawing[1] and degree and duration of dehydration.[2] Plants are complex organisms with many internal microscopic structures and it is not yet known how the growth of ice in plant tissues results in injury.

Several theoretical models have been proposed to explain the events that lead to freezing injury in plants,[5] plant cells[5] and plant protoplasts.[6] A model of the events that occur during freezing has been proposed for Puma rye (*Secale cereale* L.) protoplasts subjected to rapid freezing and thawing in both neutral and ionic isotonic solutions.[4] The events outlined for protoplasts may also occur in whole cells, but the freezing tolerance of whole cells often differs from that of their protoplasts.[8–10] A model that describes events occurring during the freezing of plant cells must consider all cell structures, gradients in water and solute potential and kinetic processes.[5] The protoplast model therefore falls short of describing freezing injury to whole cells, but provides information useful for developing a model for a cellular system.

* Corresponding author; e-mail: gusta@duke.usask.ca, reaneym@em.agr.ca

In dilute solutions, ice formation during freezing causes cellular dehydration.[11] The physical consequences of ice formation in colloids, living cells and organisms have been investigated.[11] Sequential models have been developed to explain events occurring in soil,[12] red blood cells[13] and plant protoplasts[4,6] during freezing.

Cryomicroscope observations of freezing tissues reveal that the plasma membrane appears to be in contact with the cell wall during freezing but often detaches from the cell wall during thawing.[14–16] Upon thawing, the collapsed cell wall may return to its original shape before the cell is fully rehydrated.[16] Based on this observation, Iljin proposed that the mechanism of cellular injury was the physical tearing of the plasma membrane due to rapid movements of the cell wall.[16]

During mild freezing and during thawing, solutes accumulate in the extracellular spaces of plant cells.[5] Olien proposed that these solutes prevented ice adhesion to the plasma membrane.[5] Analysis of these solutes by Livingston and Henson[17] showed that they included small and large carbohydrate molecules as well as apoplastic enzymes. They also were able to demonstrate that these enzymes were released through a specific pathway and not from the rupture of cells since the cytoplasmic marker enzyme malate dehydrogenase was not exported during a freeze-thaw cycle.

We have incorporated information regarding the interactions of water and ice with solutions, porous materials and membranes into a comprehensive model that allows the understanding and prediction of phenomena associated with freezing tolerance and freezing behavior. The model was tested on bromegrass (*Bromus inermis* Leyss) cell suspension cultures under defined experimental protocols. The model can be used to predict and understand cell responses to freezing and to design strategic approaches for engineering plants with improved freezing tolerance. Understanding these responses is important in developing conventional breeding or genetic engineering programs for improving winter tolerance.

2 Definition of model systems

2.1 Why do we need comprehensive models?

Steponkus et al.[18] and Uemura and Steponkus[19] suggested freezing injury in winter rye (*Secale cereale* L.) and arabidopsis (*Arabidopsis thaliana* L.) was due to freeze-induced destabilization of the plasma membrane and that the presence of the cell wall did not affect freezing tolerance. In contrast, Tao et al.[8] and Murai and Yoshida[9,10] demonstrated that the freezing tolerance of cells with walls is less than that of protoplasts derived from suspension-cultured potato cells (*Solanum tuberosum* L.) and Jerusalem artichoke slices (*Helianthus tuberosus* L.). Analysis of the freezing protocols used by these researchers and subsequent experimentation demonstrated that the differences in the results obtained by the researchers are most likely due to the experimental protocols. The freezing tolerance of protoplasts and cells (or tissues) was dependent on the volume of interstitial water,[8–10] while field-grown plants do not usually contain significant amounts of interstitial water unless they are exposed to a midwinter thaw or flooding.

Repeated freezing and thawing can also affect the freezing tolerance displayed by an organism.[1] Trunova[20] reported that exposure of cold hardened plants to mild nonlethal freezing conditions resulted in a significant increase in freezing tolerance. This observation has since been verified many times.[17,21–23] Olien[21] and Livingston and Henson[17] associated this phenomenon of enhanced freezing tolerance to the ability of cells to selectively release protective solutes to the apoplast in response to freezing. Though it is clear that these solutes were released at low temperatures, neither Olien[21] nor Livingston and Henson[17] demonstrated how the solutes arrived in the apoplastic space as a result of freezing. If cells and tissues mitigate the impact of freezing by changing the apoplastic environment, then the presence of a large volume of interstitial water will dilute the protective substances exported by the cells during freezing. Models to study freezing injury that include large interstitial volumes of water like protoplasts would not reveal these defensive phenomena.

In the following discussion the events that occur and conditions that exist during freezing, thawing and a second cycle of freezing will be described for three systems: cells with little interstitial water, cells with large volumes of interstitial water, and protoplasts. In this work, a new model is proposed that attempts to unite the previous observations of behavior of freezing and acclimation processes with information regarding the physical processes of freezing of non-living systems. The resulting model describes the effects of cell and tissue behavior on membranes, cell walls and cell solutes during freezing and thawing. The model describes mechanisms of acclimation and cryoprotection and points to potential artefacts in the measurement of freezing tolerance.

2.2 Walled cells with little interstitial water

In most healthy living plant tissues, the amount of interstitial water present is small due to the presence of xylem tension, nevertheless, interstitial water may collect in special areas of the plant defined as sumps.[24] We have been working with a bromegrass (*Bromus inermis* Leyss) cell suspension culture that can be cold hardened from -8°C (tender state) to -45°C (hardy state).[7] The measured freezing tolerance, as we will demonstrate, varies with the experimental protocol. We believe that measurements of freezing tolerance should be performed under experimental protocols that most closely represent those of field-grown plants. Bromegrass cell cultures are readily manipulated to represent a wide variety of field conditions.

2.3 Walled cells with a large volume of interstitial water

In order to maintain viability, cell cultures and tissues on cryostage microscopes require large volumes of interstitial water. It is assumed in most living tissues, that the ratio of apoplastic volume to cytoplasmic volume is less than 0.20 (the apoplastic volume of potato was approximately 3.2%[25]). Cells, tissues or cell clusters with a ratio of apoplast volume to cytoplasm volume of greater than 0.2 are considered to have a large volume of interstitial water.

2.4 Protoplasts

Our modeling of the freezing behavior of protoplasts has made the following assumptions: protoplast solutions were nucleated at their freezing point and did not supercool. The volume of the protoplast cytoplasm is small when compared with the extracellular volume. The protoplasts maintain a spherical shape during dehydration and therefore surface area is related to volume by simple geometry. Surface area loss was achieved by mechanisms such as endocytosis.[6,20] The plasma membrane of the protoplasts is relatively inelastic and surface area changes are achieved by addition or subtraction of material from the plasma membrane.[4,6]

3 Ice growth

Ice growth in non-living systems has been studied in depth and these studies are predictive of the behavior of ice in certain plant tissues. Ice will grow in water, dilute solutions and tubes at velocities determined by simple kinetic and thermodynamic principles (as reviewed by Fletcher[11]). These same principles can be used to predict the behavior of interstitial water and xylem water. The cell wall is a water-conducting matrix with very small pores in which water freezes as in other matrix materials. The size of the pores determines the freezing point of cell wall water, i.e. the freezing point is lower in small pores than in large pores. Although the plasma membrane is a barrier to ice growth, it can be mechanically ruptured by ice.

As with other crystallization phenomena, growing ice crystals exclude solutes and impurities.[11] Thus under mild freezing conditions solutes are concentrated. In a mixture of a matrix and a frozen solution, solutes excluded from the ice during freezing will accumulate in the matrix solution. The matrix structure depresses the freezing point of water and the accumulated solutes depress the freezing point of the matrix even further.[11]

Ice growth has been visualized in tender plants by infrared thermometry[26] (see Chapter by Wisniewski and Fuller, in this book). As predicted, once nucleation occurs ice grows rapidly through the xylem until the whole plant is frozen. At present the infrared video method does not have sufficient resolution to depict freezing at the cellular level. In certain cases plants are able to control the growth of ice in their tissues.[27] Plant tissues contain ice nucleating proteins that may be used by the plant to initiate freezing in specific tissues. Furthermore, plants also contain antifreeze proteins[28] that depress the freezing point of cell wall water and modify the shape of growing ice crystals. Although whole tissue responses will not be described in great detail in this discussion, it is apparent that plants are not passive in their interaction with growing ice.

3.1
Ice growth rates in interstitial spaces and the cell wall

The rate of water loss from the cell wall during freezing depends on the quantity of interstitial water.[7] The difference in the rate of freezing among cells is partially dependant on the amount of interstitial water and the least resistant pathway for either water migration to the site of nucleation or for the pathway of ice growth. The fast rate of water loss during freezing of cells with a larger volume of interstitial water indicates that the resistance to water flux is reduced.

At a low cell wall water potential, the hydraulic resistance of the cell wall increases thereby reducing the rate of ice formation.[29,30] The hydraulic resistance also increases as the path length for water flow increases. In models of soil freezing, ice segregates into large crystals and capillary suction occurs at the ice-water interface.[12] As the temperature falls in a freezing soil, suction is generated in the matrix water. Under mild freezing conditions the growing ice does not enter matrix pores but draws liquid water to the crystal through those pores. A similar process occurs when extracellular ice draws water through pores in the cell wall of plants.

A second path for water movement from cells is through the vapor phase, i.e. water evaporation from cell walls. The vapor phase water potential is determined by the minimum temperature and the presence of ice. The cell wall potential is higher than that of the ice and therefore ice draws water from the cell wall through the vapor phase.[11] Water is released to the vapor phase from the cell wall solution or directly from matrix components such as cell wall fibers.[5] As observed for other matrix materials, the presence of cell wall fibers, solutes and colloids lowers the vapor pressure below that of pure water and thereby increases the resistance to water loss from the cell wall.[5]

Olien and Chao[31] found the electrophoretic mobility of dyes in frozen filter paper was a function of the liquid water content of the filter paper. They postulated that the resistance of the plant cell wall to water loss also increases in response to drying.[29,30] Hydraulic resistance rapidly increases during freezing in a soil due to the formation of ice and the reduction of distance between soil particles.[32] In silt frozen to -0.4°C, the hydraulic resistance increased by four orders of magnitude.[32] The distance between particles in bentonite soils decreased by as much as 63% when frozen to -10°C.[33] This evidence indicates that dehydration of any matrix (e.g. a cell wall) increases its resistance to water and solute flow. The resistance for water movement to the ice crystal in both liquid and gaseous pathways increases as the water potential of the cell wall becomes more negative.

3.2
Behavior of walled cells and protoplasts during interstitial ice growth

3.2.1
The freezing of walled cells

The volume and composition of the interstitial medium of a bromegrass cell culture can be controlled experimentally by equilibrating the cells with a medium and subsequently removing excess water by centrifugation.[7] This bromegrass cell system which can readily be cold-hardened provides an ideal system to determine the effect of both interstitial water and solutes on the freezing behavior of cells.

If the volume of interstitial water is reduced, bromegrass cells frozen to -2°C lose symplastic and apoplastic water through both the liquid and vapor phases (Fig. 1A). The driving force for water movement at -2°C is the difference in chemical potential (-2.7 MPa) between ice and water.[11] The plasma membrane and cell wall provide little resistance to water flow[34] across the cell wall. The cell wall water does not freeze at -2°C due to the porous nature of the cell wall matrix.[11,35] Thus water must travel from the cell wall (liquid state) to the growing ice crystal either through the vapor phase or through the liquid phase in the cell wall.

Bromegrass cells frozen to -2°C in a large volume of interstitial water lose water by a pathway different from cells frozen in a low quantity of interstitial water. Upon nucleation of the cell mass with ice, the ice crystals grow rapidly (cm s^{-1}) through the interstitial water[11] and ultimately contact the cell wall. Resistance to cell water loss is determined by both the hydraulic resistance and thickness of the cell wall.[24,25] Figure 1B illustrates that the cell wall provides resistance in its transverse axis so the path length for water flux shortens if ice is adjacent to the cell. In addition, water travels directly through the liquid phase to the surrounding ice, thereby eliminating the resistance of the vapor phase. Thus the observation that the presence of large volumes of interstitial water increases the freezing rate may be explained by simple physical principles (Fig. 1).

Gusta et al.[36] demonstrated that the rate of freezing of cells was dependant on the ice nucleation temperature. In this study, bromegrass cells with little interstitial water, were nucleated at -2°C and held at -2, -3 or -4°C in a 20 MHz ^{1}H-nuclear magnetic resonance spectrometer (NMR). The samples were allowed to equilibrate for at least 60 min, at which time the temperature was reduced 2°C in less than 5 s and the decrease of the water signal was monitored. The rate of freezing was similar for cells frozen from -2 to -4°C or from -3 to -5°C, but the rate of freezing of cells frozen from -4 to -6°C was higher.[7] Water movement to the ice crystal occurs by the same pathway in cells frozen either from -2 to -4°C or from -3 to -5°C, but in cells cooled from -4 to -6°C a lower resistance path of crystallization was established. Our interpretation is that cell wall water did not freeze until the temperature was lowered below -5°C. The slower rate of freezing during the mild freeze is attributed to the larger resistance required for water to pass through the cell wall than for ice to grow to the cells. If the temperature falls below the freezing point depression of the water in the cell wall, ice grows rapidly through the wall. When cell wall water freezes, the distance that water must migrate to join the ice crystal is greatly reduced and the rate of freezing increases. An alternate explanation is that the cells were injured due to a lower nucleation temperature, however, viability tests did not confirm this.

The freezing temperature of the cell wall is determined by the concentration of solution in the cell wall, the size of cell wall pores and the matrix potential of the cell wall.[11] Tyree and Richter[37] reported that cell wall may be very rigid in plants; however, the cell walls of bromegrass cell suspension cultures are demonstrably compressible.[7] The pore size of the cell wall varies with the extent of compression.

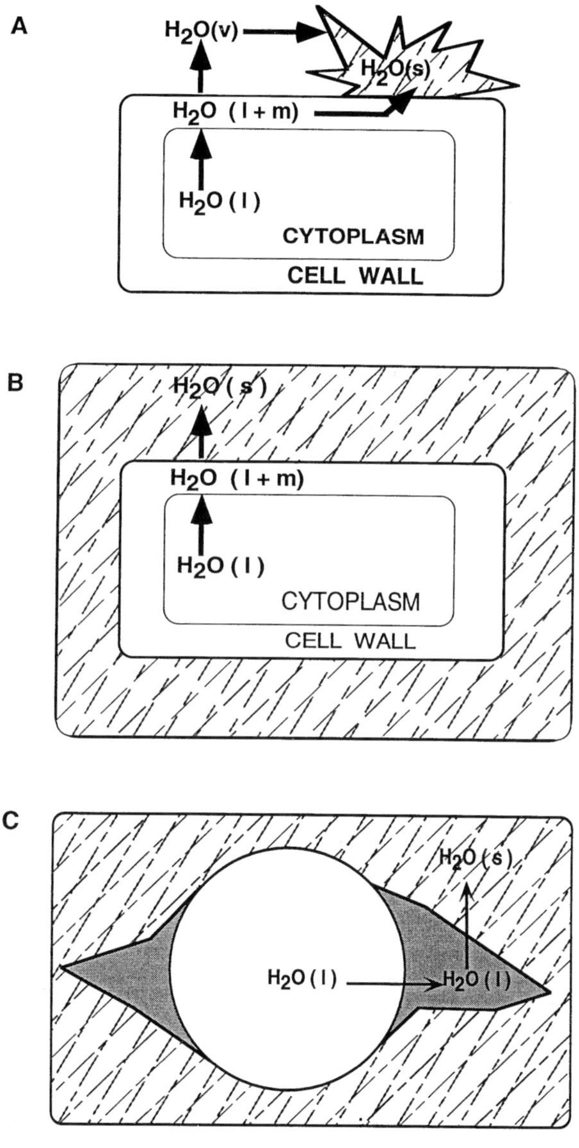

Fig. 1. Predicted paths of water movement in cells and protoplasts of bromegrass when frozen to -2°C. (**A**) Water movement in cells with loosely bound interstitial water removed by centrifugation before freezing; (**B**) Water movement in cells frozen in the presence of a large volume of interstitial water; (**C**) Water movement in protoplasts frozen in the presence of a large volume of interstitial solution

v vapor phase water; *l* liquid phase water; *m* matrix phase water; *s* solid phase (ice)

The extent to which compression may depress the freezing point of the cell wall solution has not been determined.

3.2.2
The freezing of protoplasts in a suspending osmoticum

Relative to whole cells, protoplasts are maintained in a large volume of medium. Thus, protoplasts frozen to -2°C experience an environment completely different from cells with walls. Ice crystals grow rapidly through the supporting osmoticum, leaving solution channels where the osmoticum is concentrated to the same water potential as the surrounding ice (Fig. 1C). In this case, only the hydraulic conductivity of the plasma membrane limits the rate of water loss of the protoplast. Estimates of hydraulic conductivity of protoplast plasma membranes indicate that protoplasts rapidly equilibrate with surrounding osmotic solutions and that equilibration is more rapid than expected for whole cells.[34,36] The major potential consequence of slow equilibration during freezing is the possibility of ice penetrating the plasma membrane of a supercooled protoplast.[1] Although this form of injury may occur under rapid freezing conditions, it probably does not represent a potentially injurious process to tissues nucleated at mild temperatures and then cooled slowly (1°C h^{-1} or less).[1]

4
Events and conditions for cells at a constant nonlethal subzero temperature

The concentration gradient for slightly permeable solutes across the plasma membrane increases with freezing. During freezing, transient pores may form in the plasma membrane that temporarily allow bulk flow of solutes.[38] The direction and rate of bulk flow are determined by the size and direction of pressure gradients across the membrane. If bulk flow occurs, both low and high molecular weight solutes will cross the membrane to the periplastic space. The low molecular weight solutes can enter the cell wall but the high molecular weight solutes will be trapped in the spaces between the cell wall and the plasma membrane.

4.1
The behavior of walled cells

As discussed previously, ice generates a suction force on cell wall water that draws free water to the matrix. Figure 2A depicts the direction of pressure and solute gradients in a frozen cell with little interstitial water. Until the cell wall freezes or the cell wall is completely water saturated, the plasma membrane will tightly adhere to the cell wall and the cells experience cytorhysis[39] (cytorhysis is the dehydration of cells with the cell wall and plasma membrane in contact). Thus the plasma membrane surface area will depend on the surface area of the inner cell wall. Cell wall fibers will compress to some extent during freezing[7] and the plasma membrane will be induced to compress in response.

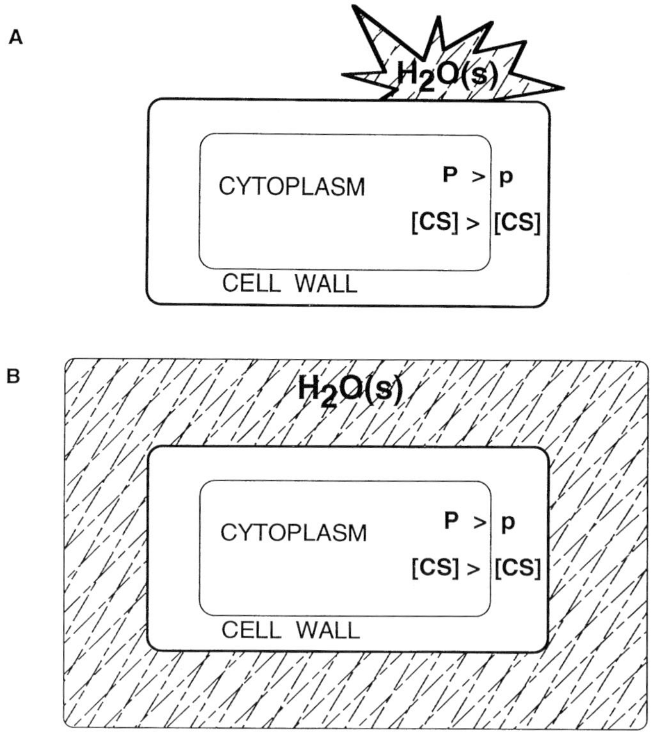

Fig. 2. Predicted gradients of solute concentration and pressure in bromegrass cells. (**A**) Frozen following removal of loosely bound interstitial water by centrifugation; (**B**) Frozen in the presence of a large volume of interstitial water
P,p pressure; *CS* cell solutes; *s* solid phase (ice)

As a result of freezing, the cell wall is compressed both perpendicular and parallel to the surface of the plasma membrane. If the adhering plasma membrane is also compressed the membrane free energy will increase and it is postulated that such forces would induce the loss of membrane material by the release of lipids into the solution phase.[40] Membrane compression may also induce dielectric poration.[40]

Cells frozen in the presence of a large volume of interstitial water experience similar suction forces and solute gradients to those experienced by cells frozen with little interstitial water present. During crystallization ice expansion in confined spaces can generate extreme pressures.[12,41,42] When ice pressure is generated in plants, the fibers of the cell wall and the symplast will be compressed. Figure 2B depicts the direction of pressure and solute gradients in a frozen cell with large amounts of interstitial water. Under these conditions solute can be forced from the cytoplasm to the cell wall and even to pores in interstitial ice.[11] The importance of ice pressure on plant survival is not known, however, antifreeze proteins[28] and arabinoxylan gums[2] that modify ice crystal formation may be a mechanism for ameliorating the injurious effects of ice pressure.

4.2 Protoplasts

While the behavior of the plasma membrane during freezing is not known for walled cells, the behavior of the plasma membrane of protoplasts has been described thoroughly.[4,6,18,19] The thermodynamic model of membrane behavior proposed by Wolfe and Steponkus indicates that plasma membranes lose surface area during freezing until a resting tension is achieved.[4] We believe that the protoplast model will describe the behavior of the tonoplast of both protoplasts and walled cells. Membrane material is deleted in part through the formation of endocytotic vesicles. Similarly, the tonoplast will develop a tension due to deletion of the surface area.[4] If the plasma membrane or tonoplast develops pores large enough to allow bulk flow of solutes, the tension in the membrane will generate a weak driving force from the inside of the membrane to the outside. This driving force will be based on the resting tension in the protoplast which is approximately 10 Pa.[4] This is roughly five orders of magnitude less than the suction generated by ice or ice pressure observed in walled cells.[7]

5 Events and conditions occurring during thawing

The rate of thawing affects plant survival,[1] i.e. fast thawing is more injurious than slow thawing. In cells thawed slowly, the cell wall is under suction until sufficient free water is released from the ice to hydrate the cell wall and the cytoplasm.[7] Solutes released during freezing and thawing would result in the presence of a lower osmotic potential in the apoplastic space (Fig. 3). Under these conditions plasmolysis and separation of the plasma membrane and cell wall is possible. Plasmolysis during thawing would result in a decrease in membrane surface area which is exactly the opposite of the behavior of thawing protoplasts which invariably increase in surface area during thawing.

If plasmolysis does not occur during thawing, tension in the plasma membrane at the cell wall interface will determine if there is any change in the surface area of the plasma membrane. The cell membrane is highly inelastic.[4] For example, the plasma membrane of rye protoplasts will begin to rupture if stretched 2% in surface area.[4,6] Elastic expansion of more than 4% is invariably fatal and greater increases in surface area require the addition of new membrane material. In cells that do not plasmolyze during thawing it is not known if a freeze-thaw event has any effect on the surface area of the membrane. Freezing and thawing (dehydration and rehydration) of whole cells may not generate membrane tension if the cell wall stabilizes the plasma membrane during rehydration. Therefore injury to the plasma membrane experienced by protoplasts during hydration or thawing is avoided by walled cells.

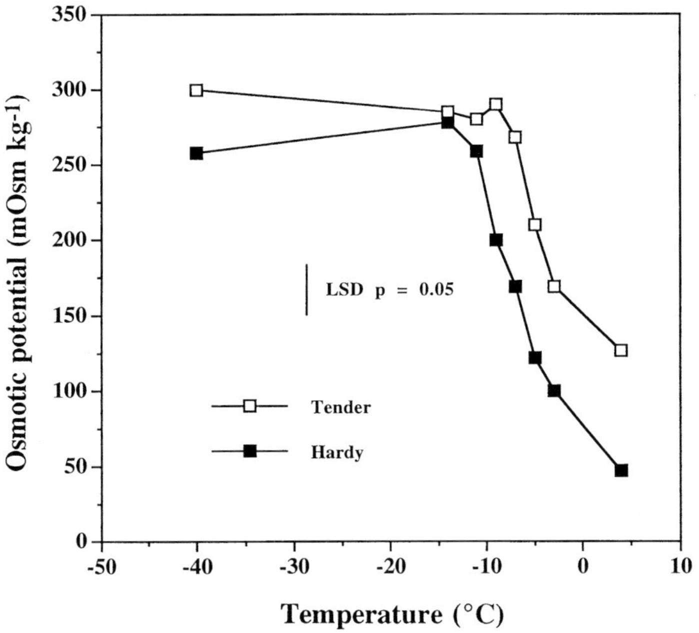

Fig. 3. The effect of freezing conditions on the osmotic potential of walls of tender and cold hardy bromegrass cells

The interaction of the cell wall and plasma membrane prevents the loss of plasma membrane surface area but the tonoplast membrane is predicted to behave in a manner similar to a protoplast. Thus phenomena associated with the behavior of protoplasts during freezing and thawing, specifically, large changes in surface area, lysis and poration, will also occur in the tonoplast. Nonlytic poration injury could occur through mixing of components of cellular compartments during and after thawing.

5.1 The behavior of walled cells

During thawing of tissues with little interstitial water, ruptured cells and uninjured (but porated) cells may leak solutes into the apoplastic space to a concentration that can induce plasmolysis. As mentioned above, plasmolysis could actually cause a decrease in membrane surface area and prevent injury induced by expansion induced tension in the plasma membrane. To determine if solute leakage could significantly alter the cell wall osmotic potential after a freeze–thaw cycle, tender and hardy bromegrass cells were washed with distilled water, centrifuged (at 1840 *g*) to remove interstitial water, and frozen to the temperatures indicated in Figure 3. The cells were then thawed overnight at 4°C. The osmotic potential of the frozen bromegrass cells was measured by placing thawed cells directly into the sample chamber of a vapor pressure osmometer. In tender and hardy bromegrass suspension cultures the cell osmotic potential was elevated in response to a mild freezing treatment (Fig. 3). The increase in osmotic potential of cells of tender bromegrass cells could be partly attributed to cell injury. In contrast, the osmotic potential of cells of hardy cells is apparently a noninjurious response to freezing as determined from viability tests.[7]

To determine the volume of solution retained after freezing and thawing, tender and hardy bromegrass cells were frozen to a series of test temperatures from -4 to -40°C. Then the cells were held at 0°C while being centrifuged at 1840 *g* in an apparatus that allowed the liquids released to pass through a fiber support but retained the cells and solids. The water retained by the cells was measured immediately after thawing. The lower the freezing test temperature, the lower the cellular water content upon thawing (Fig. 4). Tender cells, that were totally viable to -7°C, had a relative water content of 0.59 after being frozen to -3°C compared to 1 for nonfrozen tender cells. A further decline in water content of cells exposed to lower temperatures and thawed was not significant in tender cells, even for cells that were killed by freezing at -11°C and below. In contrast to tender cells, the relative water content of hardy cells was unaffected by freezing to -3 or -5°C. However, at lower temperatures the relative water content declined from 1 to approximately 0.50 in cells frozen to -9°C or lower temperatures. In hardy cells, water loss occurred at freeze temperatures that produced no injury detectable by viability tests. Solute leakage, as measured by conductivity of the extracellular solution, is frequently used as a

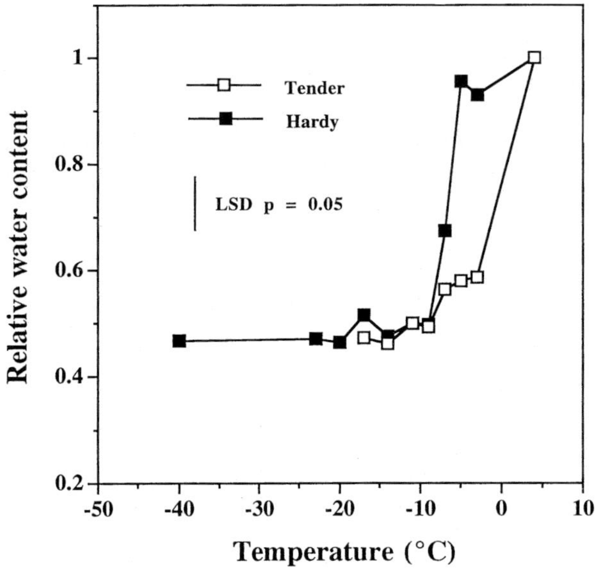

Fig. 4. Relative water content of hardy and tender bromegrass cells following controlled freezing and thawing

measure of cell viability but our findings indicate that solute leakage is part of the plants normal defensive mechanism against freezing.

When large extracellular volumes are present, the cytoplasm of thawed cells will continue to leak solutes if tension was generated in the plasma membrane. The plasma membrane surface area is restricted by the cell wall, therefore, it will not change surface area substantially during thawing. Continuous loss of cytoplasmic solutes will cause the tonoplast to continuously undergo increases in surface area well after thawing is complete. This may explain why vacuoles lyse as described by Murai and Yoshida.[10]

5.2 Protoplasts

Protoplasts are suspended in osmotic solutions to prevent lysis. Since the osmotic media have a much greater volume than the protoplast itself, it is impossible for the protoplast to regulate its environment. Following a freezing event, protoplasts regain surface area during thawing as described previously.[6] Tension will remain in the membrane either until enough solutes have diffused from the cytoplasm or sufficient membrane material is added to the plasma membrane.[4] Due to their geometry and their greater loss of surface area during contraction protoplasts experience a larger gain in surface area during thawing than whole cells.

6 Events and conditions occurring during a second freeze

In temperate climates, it is normal for the temperature to oscillate diurnally. Diurnal fluctuations around the freezing point of water may cause tissues to cycle between freezing and thawing prior to conditions which lead to continuous freezing. Diurnal freeze-thaw cycling may be an important event in preparing tissues to tolerate winter freezing conditions. Hardy bromegrass cells, frozen to noninjurious temperatures and thawed in an NMR tube, froze more rapidly during an identical second freezing treatment.[7] The cells appeared dry before the first freezing event, but appeared water-soaked upon thawing. This condition is consistent with observation of freezing and thawing treatments of whole plants.[1] The second freezing event was more rapid as determined by NMR analysis due to the presence of interstitial water. Cells killed by freezing rapidly lose solutes to the apoplastic water following thawing. Furthermore, the osmotic potential of the walls of brome grass cells frozen to nonlethal temperatures was similar to the osmotic potential of lethally frozen cells, suggesting temperature induced movement of solutes. Both observations agree with the finding that solutes are found at high concentrations in the apoplast in hardy rye crowns subjected to noninjurious freezing to -3°C.[6,21]

7
Acclimation

In order for plants to survive the formation of ice in their tissues a period of cold acclimation that involves multiple changes in the physiology of the plant tissues is required.[2] The accumulation of antifreeze proteins[28] and ice nucleation active proteins[27] allows the plant to direct the formation of ice to specific tissues that are tolerant of freezing. The cell wall is important to the freezing survival of plants due to its ability to prevent certain injurious phenomena. The cell has the ability to regulate the composition of solutes in the cell wall and the periplasmic space. Specific proteins and carbohydrates are targeted for release on freezing.[17] The tendency of the cell to export fructans and fructan exohydrolase is an important adaptation that can protect the plasma membrane and tonoplast from high tensions caused by a freeze thaw cycle.

Vadnais and McKersie recently transformed alfalfa (*Medicago sativa* L.) to express invertase in the cytosol, the apoplast and the vacuole (Vadnais and Mckersie, pers. communication).[43] Plants expressing invertase in the cytosol and the apoplast had similar freezing tolerance to untransformed controls. However, targeting invertase to the vacuole increased shoot carbohydrates, reduced root and shoot growth and reduced freezing tolerance. Our model predicts the following: during the freeze-thaw cycle sucrose would be transported into the vacuole where it would be hydrolyzed to glucose and fructose by invertase. The hydrolysis would increase the osmotic strength of the vacuole solution and the vacuole membrane would expand and remain under tension after thawing. This tension could allow more sucrose to enter the vacuole and eventually result in its lysis.

8
Cryopreservation

The conceptual model elaborated above can be used to predict the mechanism of action of cryoprotectants. Only their thermodynamic and kinetic properties are considered here. When a cell equilibrates with interstitial ice, permeating solutes, e.g. glycerol and dimethyl sulfoxide (DMSO), accumulate in the cell wall and cytoplasm and therefore add to the total osmotic potential of the apoplast. When a cell rehydrates after thawing, the plasma membrane will come under tension during re-expansion as described previously. Permeating solutes concentrated in the cell interior will diffuse to the extracellular space, relaxing tension on the plasma membrane. Nonpermeating solutes may also be cryoprotective if they continuously induce cell wall compression and pull the plasma membrane firmly against the cell wall. This action would prevent rapid surface area changes during freezing and thawing.

9
Conclusions and discussion

The environment of the plasma membrane of protoplasts is artificial, as it is selected by the experimenter. Protoplasts must be maintained in an osmoticum, the nature of which will determine most protoplast responses during freezing and thawing. With this understanding, it is apparent that the experimenter literally selects the mechanism of freezing injury and the temperature at which that injury occurs when selecting the solute constituent of the isotonic solution.

Walled cells frozen slowly lose less surface area than protoplasts during freezing. During freezing and thawing, whole cells experience much greater pressure gradients than protoplasts. Finally, mechanisms of cryoprotection of walled cells may not be possible in protoplasts. Care must be taken in working with walled cells to avoid the problems outlined for protoplasts. When freezing tissues and cells, freezing and thawing rate, as well as apoplastic solution volume, are selected by the experimenter. These factors may also select both the killing temperature and the mechanism of injury.

The conceptual model described here has allowed the identification of new injurious processes, potential mechanisms of cryoprotection and major shortcomings in previous models of freezing injury in cells and protoplasts. More research is required to make the model numerical and to address other aspects of the model, such as the mechanism of membrane removal under high pressures. McGrath[40] identified molecular desorption, endocytosis and blebbing as possible mechanisms and Gordon-Kamm and Steponkus[43] observed endocytotic vesicles and lipid bodies in frozen plant tissues. However, the amount of material removed from the plasma membrane by coated vesicles[44] and molecular desorption is unknown. When the data are more complete, determination of the mechanisms of cryoprotection, acclimation, freezing injury and freezing tolerance of cells will be possible.

10
References

1. Levitt J. Responses of Plants to Environmental Stresses, vol I. Chilling, Freezing and High Temperature Stresses. New York: Academic Press, 1980.
2. Olein CR, Smith MN. Protective systems that have evolved in plants. In: Olein CR, Smith M, eds. Analysis and Improvement of Plant Cold Hardiness. Boca Raton: CRC Press, Inc, 1981:61-87.
3. Crowe JH, Crowe LM, Mouradian R. Stabilization of biological membranes at low water activities. Cryobiology 1983; 20:346-356.
4. Wolfe J, Steponkus PL. Mechanical properties of the plasma membrane of isolated plant protoplasts. Plant Physiol 1983; 71:276-285.
5. Olein CR, Smith MN. Analysis of midwinter freezing stress. In: Olein CR, Smith M, eds. Analysis and Improvement of Plant Cold Hardiness. Boca Raton: CRC Press, Inc, 1981:35-59.
6. Steponkus PL. Role of the plasma membrane in freezing injury and cold acclimation. Annu Rev Plant Physiol 1984; 35:543-584.

7. Reaney MJT. The measurement of physiological parameters of cell suspension cultures: relative water content, cell osmotic potential, non-osmotic volume, turgor and cell wall modulus of elasticity. PhD Thesis. Saskatoon: University of Saskatchewan 1989:135-162.
8. Tao D, Li PH, Carter JV. Role of cell wall in freezing tolerance of cultured potato cells and their protoplasts. Physiol Plant 1983; 58:527-532.
9. Murai M, Yoshida S. Evidence for the cell wall involvement in temporal changes in freezing tolerance of jerusalem artichoke (*Helianthus tuberosus* L.). Plant Cell Physiol 1998; 39:97-105.
10. Murai M, Yoshida S. Vacuolar membrane lesions induced by a freeze-thaw cycle in protoplasts isolated from deacclimated tubers of jerusalem artichoke (*Helianthus tuberosus* L.). Plant Cell Physiol 1998; 39:87-96.
11. Fletcher NH. The Chemical Physics of Ice. Cambridge: Cambridge University Press, 1970.
12. Miller RD, Baker JH, Kolaian JH. Particle size, overburden pressure, pore water pressure and freezing temperature of ice lenses in soil. In: Transactions of the 7th Int Congress of Soil Science, vol. 1. Madison, Wisconsin, 1960:122-129.
13. Meryman HT. Modified model for the mechanism of freezing injury in erythrocytes. Nature 1968; 218:333-336.
14. Asahina E. The freezing process of plant cells. Contributions from The Institute of Low Temperature Science, Hokkaido University, Series B 1956, 10:83-126.
15. Siminovitch D, Scarth GW. A study of the mechanism of frost injury to plants. Can J Bot 1983, 16:467-480.
16 Iljin WS. Über den Kältetod der Pflanzen und seine Ursachen. Protoplasma 1933, 20:105-124.
17. Livingston DP, Henson CA. Apoplastic sugars, fructans, fructan exohydrolase, and invertase in winter oat: responses to second-phase cold hardening. Plant Physiol 1998; 116:403-408.
18. Steponkus PL, Uemura M, Webb MS. A contrast of the cryostability of the plasma membrane of winter rye and spring oat. In: Steponkus PL, ed. Advances in Low Temperature Biology, vol 2. London: JAI Press Ltd, 1993:211-312.
19. Uemura M, Joseph RA, Steponkus PL. Cold acclimation of *Arabidopsis thaliana*. Effect on plasma membrane lipid composition and freeze-induced lesions. Plant Physiol 1995, 109:15-30.
20. Trunova TL. Light and temperature systems in the hardening of winter wheat and the significance of oligosaccharides for frost resistance. Fiziol Rast 1965, 12:70-77.
21. Olein CR. An adaptive response of rye to freezing. Crop Sci 1984; 24:51-54.
22. Livingston DP III. The second phase of freezing tolerance and fructan isomer change in winter cereal crowns. Crop Sci 1996; 36:1568-1573.
23. Siminovitch D. Common and disparate elements in the process of adaptation of herbaceous and woody plants to freezing - A perspective. Cryobiology 1981; 18:166-185.
24. Canny MJ. Apoplastic water and solute movement: new rules for an old space. Annu Rev Plant Physiol Plant Mol Biol 1995; 46:215-36.
25. Michael W, Schultz A, Meshcheryakov AB, Ehwald R. Apoplasmic and protoplasmic water transport through the parenchyma of the potato storage organ. Plant Physiol 1997; 115:1089-1099.
26. Wisniewski IR, Lindow SE, Ashworth EN. Observations of ice nucleation and propagation in plants using infrared video thermography. Plant Physiol 1997; 113:327-334.
27. Brush RA, Griffith M, Mlynarz A. Characterization and quantification of intrinsic ice nucleators in winter rye (*Secale cereale* L.) leaves. Plant Physiol 1994; 104:725-735.
28. Antikainen M, Griffith M, Zhang J, Hon W-C, Yang DSC, Pihakaski-Maunsbach K. Immunolocalization of antifreeze proteins in winter rye leaves, crowns, and roots by tissue printing. Plant Physiol 1996; 110:845-857.
29. Fisher RA. Resistance to water loss in the mesophyll of leek (*Allium porrum* L.). J Exper Bot 1968; 19:135-145.

30. Jarvis PG, Slayter RO. The role of the mesophyll cell wall in leaf transpiration. Planta 1970; 90:303-322.
31. Olein CR, Chao S. Liquid water content of cell walls in frozen tissues evaluated by electrophoresis of indicators. Crop Sci 1973; 13:674-676.
32. Williams PJ, Burt TP. Measurement of hydraulic conductivity of frozen soils. Can Geotech J 1974; 11:647-650.
33. Anderson DM, Hoekstra P. Migration of interlamellar water during freezing and thawing of Wyoming bentonite. Soil Sci Soc Am Proc 1965; 29:498-504.
34. Stout DG. Plant plasma membrane water permeability and slow freezing injury. Plant Cell Environ 1979; 2:273-275.
35. Carpita N, Sabularse D, Montezinos D, Delmer DP. Determination of pore size of cell walls of living plant cells. Science 1979; 205:1144-1147.
36. Gusta LV, Rajashekar C, Chen PM, Burke M. Freeze induced membrane permeability changes in winter rye. Cryoletters 1982; 3:27-34.
37. Tyree MT, Richter H. Alternate methods of analyzing water potential isotherms: some cautions and clarifications. II. Curvilinearity in water potential isotherms. Can J Bot 1982; 60:911-916.
38. Zimmerman U. Electric field mediated fusion and electrical phenomena. Biochim Biophys Acta 1982; 694:227-277.
39. Pearce RS. Extracellular ice and cell shape in frost-stressed cereal leaves: a low-temperature scanning-electron-microscopy study. Planta 1988; 175:313-324.
40. McGrath JJ. Thermodynamic modeling of membrane damage. In: Morris GJ, Clark A, eds. Effects of Low Temperature on Biological Membranes. London: Academic Press, 1981:35-77.
41. Hoekstra P. Conductance of frozen bentonite suspensions. Soil Sci Soc Am Proc 1966; 29:519-522.
42. Penner E. The mechanism of frost heaving in soils. Highway Research Bd Bull 225. National Research Council Washington Publ 685. Washington: Nat Acad Sci 1959:1-22.
43. Gordon-Kamm WJ, Steponkus PL. The behavior of the plasma membrane following osmotic contraction of isolated protoplasts: implications in freezing injury. Protoplasma 1984; 123:83-94.
44. Tanchak MA, Griffing LR, Mersey BG, Fowke LC. Endocytosis of canonized ferritin by coated vesicles of soybean protoplasts. Planta 1984; 162:481-486.

Cold adaptations in terrestrial invertebrates

L. Sømme

Department of Biology, University of Oslo, POB 1050 Blindern, N-0316 Oslo, Norway;
e-mail: l.s.somme@bio.uio.no

1
Insects and other terrestrial arthropods

1.1
Introduction

With his improvement of the thermometer, Réaumur[1] was the first biologist to record the environmental temperatures of insect. In his experiments, he observed that some insects can withstand freezing temperatures, while others cannot. Early work on insect cold hardiness has been reviewed by Bachmetjev[2] and Payne,[3] and a modern approach to the field summarized by Salt.[4,5]

The strategies of overwintering, or how terrestrial arthropods are able to survive low temperatures, has became a wide field. During later years, different aspects of the cold hardiness of terrestrial arthropods have been discussed in a number of review articles,[6–18] as well as in recent books.[19,20,21]

1.2
The concept of cold hardiness

Insects and other terrestrial arthropods have different strategies of overwintering. Some species, often in their egg stage, overwinter on twigs and branches of trees, where they are exposed to very low ambient temperatures. Others overwinter in the ground well protected from low air temperatures by plant litter and a snow cover.

1.2.1
Categories of cold hardiness

With regard to their cold hardiness, terrestrial arthropods usually fall into one of two categories.[21] Freeze-tolerant species can survive the actual formation of ice within their tissues, but there is always a lower temperature limit to their tolerance. All mites and springtails, as well as most insects, are freeze-susceptible and depend on their ability to avoid ice formation by supercooling. For this reason the latter category is also often termed freeze-avoiding. The lower limit of supercooling during a gradual decrease in temperature (e.g. 1°C min^{-1}) is called the supercooling point (SCP; Sect. 1.4). In all species, whether freeze-susceptible or freeze-tolerant, there is a lower temperature limit, and the duration of exposure is important for survival.

According to Salt,[5] cold hardiness should be divided into more than two categories and a similar system was later suggested by Bale.[22] The diagram in Figure 1 is partly based on a chart by Salt[5] but highly modified due to increased knowledge during the past 30 years. Terrestrial arthropods that are not adapted to low temperatures may be chill-intolerant and will suffer from lethal chill injuries even at temperatures above zero. Prolonged chilling may result in irreversible metabolic imbalances.[11] Other species are chill-tolerant or become chill-tolerant during a hardening process, which enables them to survive temperatures both above and below the freezing point (FP) of their body fluids. Consequently, both freeze-susceptible and freeze-tolerant species are also adapted to avoid the risks of chilling. In some insects, rapid chill hardening may be achieved within a few hours.[23,24]

1.2.2 Preexposure history

The preexposure history, or acclimatization, is important for the degree of supercooling in both freeze-susceptible and freeze-tolerant species, and for the lower lethal temperature in freeze-tolerant species. As explained in later sections, the supercooling capacity is increased by removal of ice nucleating agents (INAs) and the accumulation of substances like polyols, sugars and thermal hysteresis proteins (THPs). In freeze-tolerant species, the lower lethal temperature is also affected by cryoprotectant substances (Fig. 1) and by the presence of INAs. In addition to

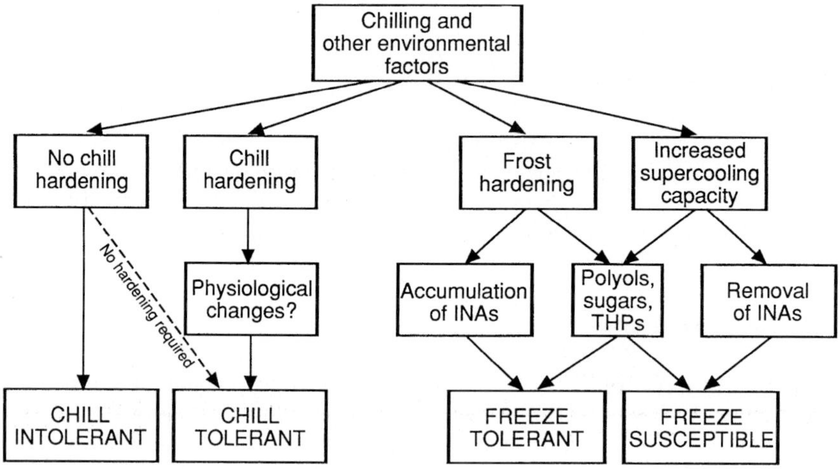

Fig. 1. Chart of processes above and below the freezing point of body liquids resulting in different categories of cold hardiness in terrestrial arthropods
INAs ice nucleating agents; *THPs* thermal hysteresis proteins

decreasing temperatures, physiological changes in both freeze-tolerant and freeze-susceptible species are triggered by environmental factors such as photoperiod and desiccation.

Although generalizations cannot be made,[25] a relation between diapause and cold hardiness has been demonstrated in some insects. In the heteropteran bug *Pyrrhocoris apterus*, the SCPs are lowered by short photoperiod and further by low temperatures, provided the bugs are in diapause.[26,27] Extirpation of the retrocerebral complex from nondiapausing *P. apterus* enhances supercooling, suggesting that cold hardiness may partly be regulated by endocrinological processes.

1.2.3 Prolonged exposures

Due to all these factors, there is a striking difference between tropical and summer-acclimated temperate insects, that cannot even tolerate short periods at moderately low temperatures, and winter-adapted specimens that survive several months at temperatures far below zero. In species that live at the limit of their temperature tolerance, great variation in winter mortality may be expected. Experimental data on insect survival under different time and temperature regimes are relatively scarce,[28] and studies on the effect of prolonged exposures at low temperatures should be given more attention in models of population dynamics.

The presence of a species in an area of severe climatic conditions, however, is actually the best evidence of its ability to survive prolonged exposures at low temperatures. As an example, Lozina-Lozinskii[29] reported that during the winter of 1941/1942 air temperatures in the Soviet Tatar Republic remained below -40°C for long periods, below -45°C for weeks, and even dropped to -55°C for 1 d. These conditions were endured by supercooled larvae of the bark beetle *Scolytes multistriatus* and frozen caterpillars of the cabbage moth *Mamestra* sp. hibernating above the snow cover.

Studies on cold hardiness should also be seen in relation to other aspects of life history.[18,30] In addition to winter cold, any given species must survive a variety of environmental challenges such as site selection, dormancy, energy storage and adaptations for feeding, mating, avoidance of predators, etc. Winter mortality in itself can vary widely among species, population and sites. In short, according to Danks,[30] a more holistic view is needed considering cold hardiness in the context of the entire life cycles and the tradeoffs required by selection. Although these comments are very appropriate, it will be outside the scope of the present chapter to discuss all such possible relations.

1.3 Winter mortality in pest insects

Data on the survival of insect of economic importance are of particular interest to predict the intensity of outbreaks the following spring. Low winter minimum temperatures, extended periods of frost and periods of thawing in the middle of the winter are critical for hibernating insects.[6]

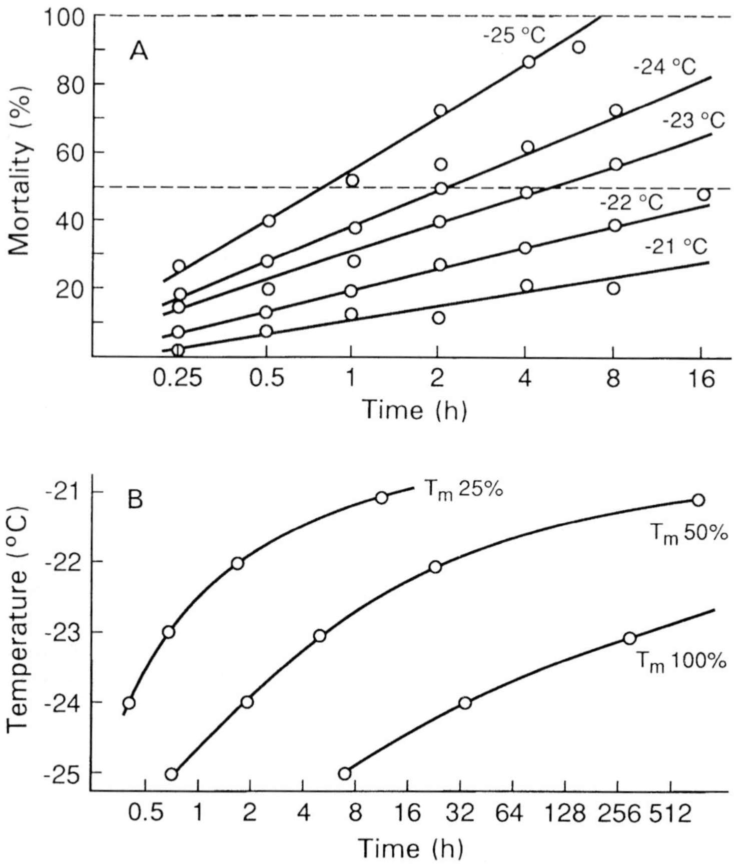

Fig. 2. The effect of long exposure to low temperatures in diapausing pupae of the white cabbage butterfly *Pieris brassicae*
(**A**) Rates of mortality at five constant temperatures above the mean supercooling point; (**B**) Duration of exposure causing 25, 50 or 100% mortality at these experimental temperatures, T_m temperature of mortality (reprinted with permission from ref. 6)

Methods to predict winter mortality have often been based on the ability of pest insects and mites to supercool. In a study of 24 species of insects and mites from fruit orchards in Nova Scotia,[31] some species were found to be sufficiently cold-hardy to survive any winter conditions in the area. The mortality of other species vary with the winter and the location. As an example, the European red spider mite *Panonychus ulmi*, a pest in fruit orchards, has mean SCP of -31°C and suffers high mortality rates during the coldest winters.

By exposing insects to constant temperatures of 1–5°C above their mean SCPs, Merivee[6] showed that the slope of the regression lines decreases with increasing temperatures (Fig. 2A). For this reason the relationship between temperatures and log time required to give a certain rate of mortality flattens out. As an example, by extrapolating the data for pupae of the white cabbage butterfly *Pieris brassicae*, it appears that 100% mortality will not occur for several months at temperatures above -22°C (Fig. 2B) and 50% mortality will not occur above -19 to -20°C. In fact, among 700 diapausing pupae of *P. brassicae* maintained at different freezing temperatures above the critical level, only 6% died after 5 months.[32] Merivee[6] concluded that insect mortality depends upon temperatures that approach the SCP only. To predict survival rates, mortality data for various pest insects must be compared with recorded winter temperatures, and in particular with the hourly sum below the critical temperature range for each species.

It should be kept in mind, however, that winter mortality is also affected by other factors, such as moisture, desiccation and predation.[18,30,33] Monitoring of the natural population size over the winter, correlated with data on climatic factors like temperature, rainfall, surface wetness and snow cover is a better approach to forecasting. If field and laboratory studies in ecology and physiology of pest insects are integrated, research on cold hardiness can make effective contribution to pest management.[33]

Critical factors for pupae of the bertha armyworm *Mamestra configurata*, a pest on canola and rape, overwintering in the soil of the Canadian prairies, are depth of snow cover and the duration of periods with temperatures below -10°C.[34] Even layers of snow as thin as 5 cm can protect from low temperatures at the depth of the soil at which the pupae overwinter. The model is useful for smaller fields where soil temperatures are recorded but since small changes in snow depth greatly affect the soil temperature, pupal survival for large areas is difficult to predict.

More examples on the implication of cold hardiness in pest management are given in a review of the literature by Bale.[33] In Britain, overwintering adult aphids are killed within few minutes at temperatures of -5 to -15°C, although they supercool to -25°C under experimental conditions.[35,36] In this case the SCP is a poor indication of their cold hardiness. Presumably, the adult aphids have to overwinter in microhabitats were severe freezing temperatures are not encountered, but the egg stage is more cold hardy. Relatively accurate methods have been developed to forecast the intensity of the outbreak of the peach-potato aphid *Myzus persicae* and cereal aphids *Sitobion* spp. in Britain based on preceding winter temperatures.[33]

In comparing pupae of the root cabbage fly *Delia radicum* from localities in Canada, Estonia and Britain, Turnock et al.[37] concluded that SCPs are inadequate to describe differences in cold hardiness within the populations. All populations had relatively similar SCPs in the range of -22 to -25°C. By calculating the "lower limit of cold injury zone" from experimental data on long time survival at subzero temperatures, they found that the population from Ascot, Britain, was killed by short exposures at -19.6°C, while those from St. Jean and Winnipeg survived temperatures close to their SCPs. In all populations the estimated "upper limit of cold injury zone" was close to -12.5°C, suggesting that these values are not related to winter weather in the areas of the different populations.

Although relatively few detailed studies are available, it is reasonable to assume that the ability of the overwintering stage to survive prolonged periods of chilling changes with the season. As an example, in the apple blossom weevil *Anthonomus pomorum* the ability to survive exposures at subzero temperatures and the ability to supercool change simultaneously.[38] In the Czech Republic, adults of this species hide in their shelters for estivo-hibernation from July, having completed their pre-dormancy feeding on apple leaves. Following cessation of feeding, the supercooling capacity increases (Fig. 3) and remains at a low level throughout the winter and until feeding is resumed in the spring. Parallel to the changes in supercooling capacity, the survival rates at -7, -15 and -18°C increase in July, but are greatly reduced in April. While 1-d exposure at -15 and -18°C was tolerated at an early stage of estivo-hibernation, survival rates after 5-d exposure at these temperatures were highest in the middle of the winter. The results suggest that the ability to sur-

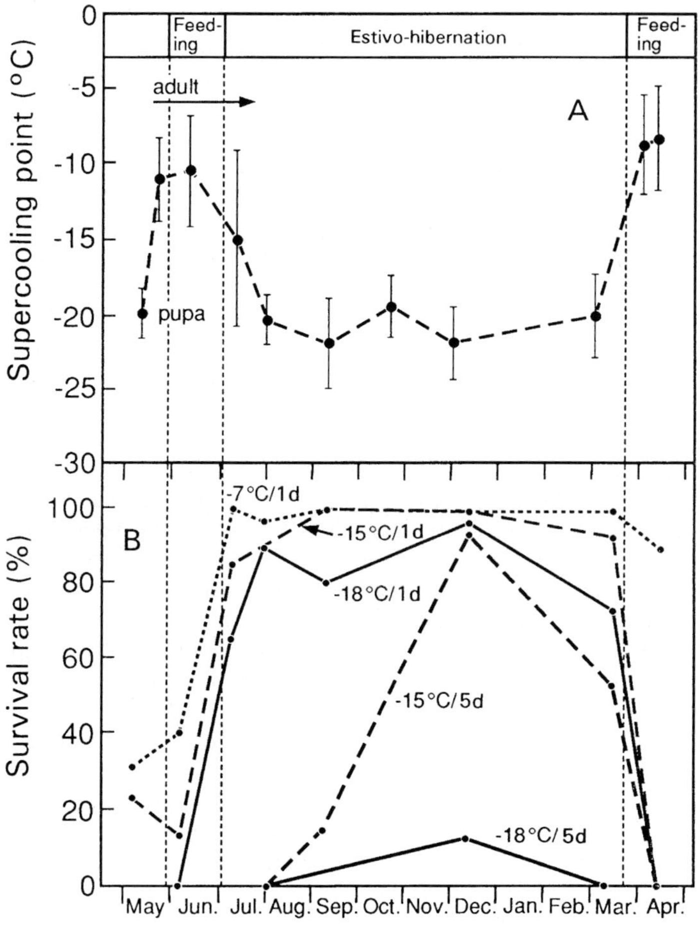

Fig. 3. Seasonal changes in the cold hardiness of the apple blossom weevil *Anthonomus pomorum* (**A**) Supercooling points; (**B**) Survival rates at -7, -15 or -18°C for 1 or 5 d (reprinted with permission from ref. 38)

vive prolonged chilling develops as a gradual process, reaching its highest level at the coldest part of the winter.

1.4
Supercooling in water and arthropods

Contrary to what would be expected, pure water will typically not freeze at 0°C. In a small container, water is easily supercooled several degrees below its melting point (MP). If a small crystal of ice is added to supercooled water, freezing is immediately initiated and the temperature rises to the MP of 0°C. Small volumes of highly purified water, e.g. droplets suspended in mineral oil, supercool to temperature close to -40°C.[39] This is the temperature of homogeneous nucleation. Theoretically, ice-like clusters of water molecules are formed and increase in size as the temperature is lowered below the MP. When they reach a critical size, the clusters act as small embryo ice crystals, nucleating the supercooled water. In nature, contact with mineral particles or with protein molecules in biological systems cause heterogeneous nucleation by enhancing the arrangement of water molecules into the structure of ice.

For freeze-susceptible insects and other terrestrial arthropods, the avoidance of freezing is essential. This is achieved by their greater or lesser ability to supercool. In the same way as small volumes of highly purified water will supercool far below the MP, winter-adapted insects may be considered as liquid containers in which the conditions for supercooling are exceptionally favorable.[7]

The limit of supercooling in insects is traditionally measured as the SCP, which represents the lower limit of survival in winter-adapted freeze-susceptible species. The SCP is the temperature at which spontaneous freezing occurs during gradual cooling, e.g. at a rate of 1°C min^{-1} in the laboratory. When a supercooled insect freezes, the released heat of fusion causes the temperature to rise (Fig. 4), and a rebound appears on the freezing curve. Just as heat is required to melt ice, heat is released when ice is formed. When the thermocouple is not penetrating the insect, as in Figure 4, the rebound is arbitrary, depending on the contact between the thermocouple and the insect itself. Theoretically, the temperature inside the insect will rise almost to the MP of its body fluid.[40] Most insects will tolerate freezing at a temperature corresponding to their hemolymph MP, and will survive when removed at B in Figure 4. Subsequently, more ice is formed as the temperature is again lowered, and freeze-susceptible species will be killed at C (Fig. 4) or lower temperatures.

As reviewed by Sømme,[7] insect eggs often have low SCPs during the winter, commonly in the range of -30 to -40°C. A number of them overwinter in exposed situations such as on twigs and branches of trees. Overwintering freeze-avoiding larvae present a wide range of SCP values from -16 to -60°C, depending on their microhabitat. Most pupae have mean SCPs in the range of -20 to -30°C. Values are frequently higher in adult insects, although several species have mean SCPs of -25 to -35°C. Mean SCPs in the range of -20 to -35°C have been recorded in alpine and polar Collembola.[10,21] Mites have generally low SCPs, while spiders, although freeze-susceptible, usually have poor supercooling capacity,[7] suggesting that they depend on more protected sites for overwintering.

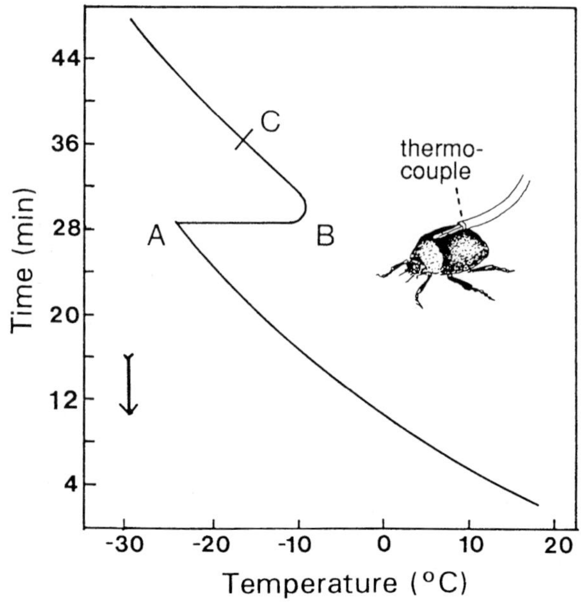

Fig. 4. Measurement of supercooling point of an insect in contact with a copper-constantan thermocouple connected to a recording potentiometer
When spontaneous freezing occurs **(A)**, a rapid increase in temperature is recorded **(B)**. Most insects will survive if removed to nonfreezing temperatures at B, but freeze susceptible species are killed when the temperature again decreases, e.g. to **(C)** in the diagram, or lower. The *arrow* indicates the direction of movement of the paper of the recording potentiometer (reprinted with permission from ref. 21)

1.5
Freeze-tolerant species

Most animals are killed by freezing, but in some insects protective physiological mechanisms have evolved, enabling freeze-tolerant species to survive. Freezing may take place within the cells or in the hemolymph that surrounds the cells (Fig. 5). Intracellular freezing is lethal to most organisms,[9,13,41] since metabolic processes are disturbed[11] and osmotic swelling may damage the cytoskeleton and cell membranes. The nature of these injuries has rarely been demonstrated in insects, but Tanno[42,43] showed that in the Japanese poplar sawfly *Trichiocampus populi* intracellular freezing of different groups of fat body cells prevents the development from prepupae to pupae or the subsequent metamorphosis from prepupae to adult.

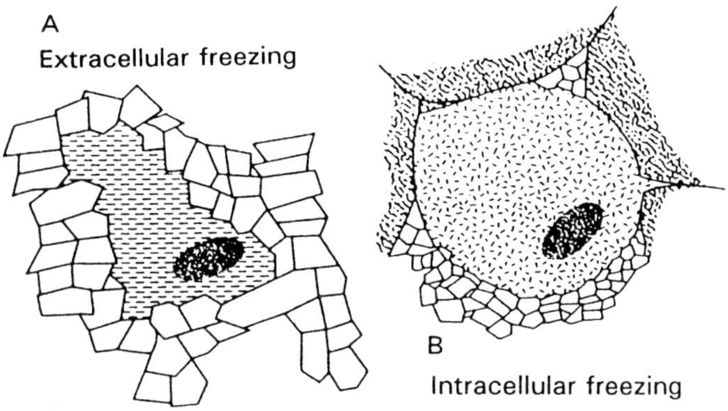

Fig. 5. Extracellular and intracellular freezing of fat body cells from the poplar sawfly *Trichiocampus populi*
(**A**) Cells surrounded by ice crystals, cell content unfrozen; (**B**) Cell frozen intracellularly, small ice crystals are also seen in the extracellular liquid (reprinted with permission from ref. 21)

1.5.1
Supercooling and freeze tolerance

The formation of extracellular ice is also injurious to most terrestrial arthropods, but the harmful effect is more or less controlled in freeze-tolerant insects. Some selected examples of freeze-tolerant insects are presented in Table 1. Freeze-tolerant insects are found in different orders, locations and habitats. The lower lethal temperature, which represents the lowest temperature that these insects are able to survive in the frozen state, is different for different species. Some species like the arctic woolybear caterpillar *Gynaephora groenlandica* may be cooled to -70°C or lower,[44] while several others have lower lethal temperatures in the range of -30 to -40°C.[21] Several freeze-tolerant insects have been reported from arctic areas, in particular from North America.[45,46] In some species, like the alpine weeta *Haemideina maori* from New Zealand,[47] the lower lethal temperature is higher than -10°C, which is apparently in agreement with their ecological requirements.

From Table 1 it also appears that several other freeze-tolerant species have SCPs in the range of -4 to -9°C. Examples are found among adult Coleoptera[6,45,54] and among larvae of Lepidoptera,[6,44,45,48] Diptera[4,46,49] and Hymenoptera.[42,50,51] Some freeze-tolerant species, however, have great capacity of supercooling. They are able to survive the initiation of freezing at very low temperatures, indicating that different physiological and biochemical mechanisms of survival may be found in freeze-tolerant insects. In this respect there is a striking difference between larvae of the two *Pytho* spp. from the Rocky Mountains.[46]

Table 1. Some selected examples of freeze tolerance in insects from different orders, locations and habitats

Species	Stage	Location	Habitat	Lowest mean SCP (°C)	Lower lethal temp. (°C)	Ref.
Orthoptera						
Hemideina maori	Ad	New Zealand	Under rocks	-4	-9	47
Neuroptera						
Hemerobius simulans	Ad	Alaska	Bark of aspen	-36	-50	45
Coeloptera						
Pterostichus brevicornis	Ad	Alaska	Stubs; timber	-11	-70	45
Upis ceramboides	Ad	Alaska	Under bark	-6	-60	45
Phosphuga atrata	Ad	Extonia	Bark, stubs	-6	-22	6
Pelophila borealis	Ad	Norway	Soil	-21	-5	54
Pytho americanus	Larva	Rocky Mtns	Under bark	-5	-44	46
Pytho deplenatus	Larva	Rocky Mtns	Under bark	-54	-55	46
Lepidoptera						
Nymphalis antiopa	Ad	Alaska	Hibernacula	-20	-34	45
Papilio machaon	Pupae	Hokkaido	Lab. strain	-	<-50	48
Arctica caja	Larva	Estonia	?	-9	-78	6
Gynaephora groenlandica	Larva	Ellesmere Island	Litter	-7	-70	44
Diptera						
Belgica antarctica	Larva	Antarctica	Soil	-6	-15	49
Xylophagus sp.	Larva	NW Territory	Under bark	-6	-43	46
Eurosta solidaginis	Larva	N. America	In galls	-9	<-50	4
Hymenoptera						
Trichiocampus populi	Pupae	Hokkaido	Litter	-9	<-70	42
Bracon cephi	Larva	Canada	Parasite	-47	<-47	51
Amauromenatus sp.	Larva	NW Territory	Willow	-7	-40	50

SCP supercooling point; *temp* temperature

1.5.2
Ice nucleating agents

As originally demonstrated in the tenebrionid beetles *Eleodes* and *Coelocnemis* from mountain ranges in southern California,[52] the high SCPs of freeze-tolerant insects are caused by INAs in the hemolymph. These substances are inactivated by heat, indicating a proteinaceous nature, and INAs from the cranefly *Tipula trivittata* has been described as a 800-kDa lipoprotein.[14]

The function of hemolymph INAs is apparently to inhibit lethal intracellular ice formation.[9,13,14] When an insect is highly supercooled, the risk of intracellular freezing is greatly increased, but due to hemolymph INAs, extracellular freezing takes place at higher temperatures. Due to the presence of hemolymph INAs, freeze-tol-

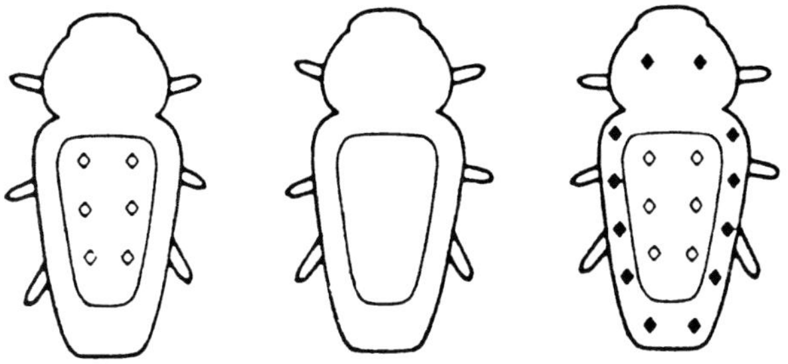

Fig. 6. The distribution of ice nucleating agents (INAs) in body fluids of freeze-susceptible (avoiding) and freeze-tolerant species
During the summer, INAs are present in the gut or other internal compartments (*open symbols*) of both categories. In winter, freeze-susceptible insects are free of INAs, while many freeze-tolerant species have INAs in their hemolymph (*closed symbols*); (reprinted with permission from ref. 53)

erant insects differ from freeze-susceptible ones,[9,53] in which INAs in both gut and hemolymph are eliminated in preparation for overwintering (Fig. 6).

The freeze tolerance of insects is limited both in time and by temperature. An example is the alpine carabid beetle *Pelophila borealis*[54] that is killed by 10 min at -10°C, while some specimens survive more than 40 d at -5°C (Table 2). At -3°C, which is closer to the ambient temperature of this species' overwintering site, most of the beetles will live for several months.

The degree of freeze tolerance in most species shows great seasonal fluctuations. The chrysomelid beetle *Phyllodecta laticollis* from the alpine region in Norway, overwintering below loose bark on willows, has SCPs around -6°C all the year around.[55] The lower lethal temperature, however, decreases from -10°C in September to -40°C in February, following closely the ambient temperatures at the site of collection.

Table 2. Mortality rates in the alpine carabid beetle *Pelophila borealis* at different temperatures and for different time intervals in the frozen state[54]

Temperature (°C)	Time interval	Number of specimens (n)	Mortality (%)
-10	1 h	24	100
-5	8 d	13	2
	16 d	24	46
	32 d	12	75
	42 d	19	95
-3	32 d	12	0
	64 d	8	25

1.6
Freeze-susceptible species

1.6.1
Removal of INAs

In freeze-susceptible insects, removal of INAs are essential to avoid lethal effects of freezing (Fig. 6). The cessation for feeding and clearing of the gut at the onset of winter increase the supercooling capacity in many species.[7] Components causing nucleation in the gut are probably mineral dust and food particles.

Feeding experiments with the antarctic springtail *Cryptopygus antarcticus* serve as an example on the effect of the gut content on supercooling.[10,56] As shown in Figure 7, almost all specimens collected in the austral summer had supercooling points above -15°C, and with a mean of -7.2°C. When starved for 6 d at 5°C in the laboratory, about 55% of them lowered their SCPs below -15°C. The mean of this low group was -24.8°C. Following starvation, the Collembola were fed a homogenate of moss turf or a diet of purified unicellular green algae. In the first case, probably due to mineral grains within the turf, the SCP of most specimens increased considerably. In the group fed green algae, the ratio of low group SCPs actually increased, suggesting that this kind of food is free of nucleators. During the short summers of the antarctic continent, springtails may be feeding during daytime and be exposed to freezing temperatures during the night.[21] In this case a nucleator-free diet is essential to avoid freezing.

1.6.2
Ice nucleating bacteria and their application

Ice nucleating bacteria are known from plants, where they promote freezing injury by nucleating water on the plant surface, causing freezing of the underlying tissues.[16,57] These rod-shaped, asporogenous, Gram-negative bacteria are extremely potent nucleators, with the capacity to cause ice nucleation at temperatures as high as -2°C. They are widespread in nature, and may explain the reduced supercooling capacity of many feeding insects better than the presence of mineral dust particles in the gut. Recently, Olsen and Duman[58] demonstrated that the SCP of the larvae of the beetle *Dendroides canadensis* is lowered partly due to gut evacuation and partly because ice nucleating bacteria are masked by increased concentrations of certain proteins in the gut fluid.

Due to their high nucleating capacity, ice nucleating bacteria have great potential as biological control agents. Ingestion of ice nucleating bacteria raises the SCPs of feeding insects considerably.[16,57] The mean SCP of diapausing adults of the Colorado potato beetle *Leptinotarsa decemlineata* rose from -7 to -3°C after exposure to various concentrations of the ice nucleating bacteria *Pseudomonas syringae* in the soil.

The results of laboratory studies suggest that ice nucleating bacteria may be used for the control of stored product insects, all of which are freeze-susceptible. According to Lee et al.[57] the SCPs of several species were raised by between 6 and 11°C

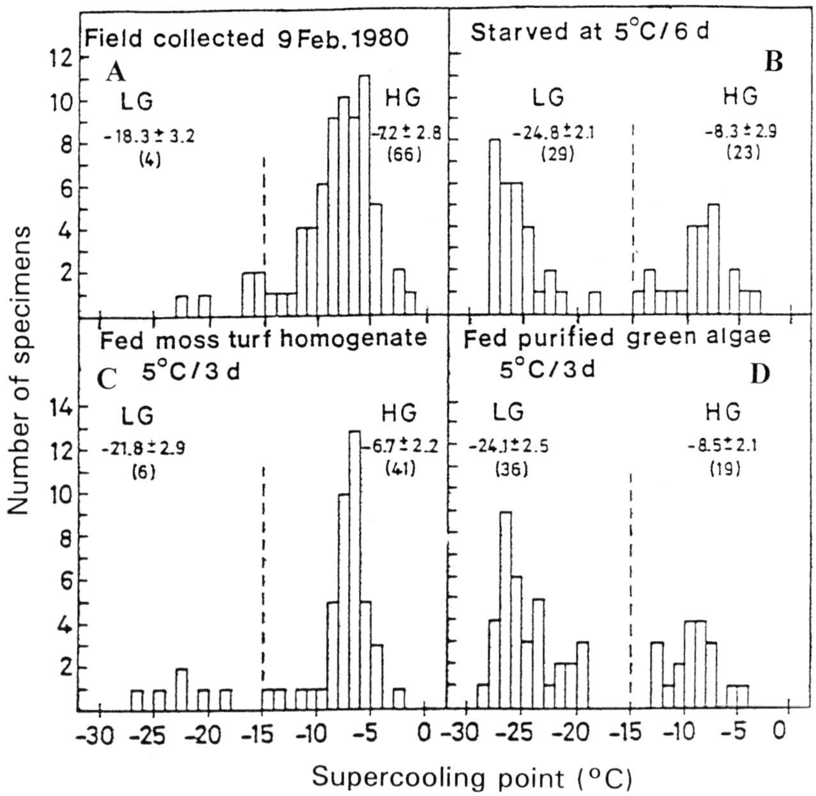

Fig. 7. Supercooling point distribution histogram for the antarctic Collembola *Cryptopygus antarcticus*; number of specimens plotted against their individual supercooling points within 1°C intervals. (**A**) Field-collected specimens; (**B**) Specimens starved for 6 d at 5°C; (**C**) Specimens fed a diet of homogenated moss turf; (**D**) Specimens fed a diet of purified green algae (a mixture of *Stichococcus* and *Monodus*)
LG low group supercooling points; *HG* high group supercooling points; mean ± SD and number of specimens (in parentheses) are given for each LG and HG (reprinted with permission from ref. 56)

when preparations of *P. syringae* were added to their substrate. Apparently, the chill tolerance of some species at -5°C was also greatly reduced. For the practical application of these results, it may be a problem to lower the temperature of grain and flower storage sufficiently.

1.7 Low molecular weight cryoprotectant substances

A striking characteristic of overwintering insects is their capacity to accumulate high concentrations of polyols and sugars.[9,15,21] Due to their effect on supercooling in freeze-susceptible species and on survival in freeze-tolerant species, these substances are often referred to as cryoprotectants.

The most widely distributed cryoprotectant in freeze-avoiding terrestrial arthropods is glycerol.[7,13] As a nonionizing solute, one molal concentration of glycerol depresses the MP of water by 1.86°C. In insects, however, it has been shown repeatedly that glycerol depresses the SCP about twice as much as the MP. The reason for this is that water molecules are bound to the surface of glycerol molecules, and in this way fewer water molecules are available for homogeneous ice nucleation. In overwintering insects, the hemolymph frequently contains 1-3 Mol of glycerol, and concentrations of 5 Mol have been reported in some species.[51]

In addition to glycerol, several other polyols, e.g. sorbitol, mannitol and myoinositol, have been found to act as cryoprotectants in freeze-susceptible insects and other arthropods. Among the sugars, glucose and trehalose may act as cryoprotectants. Polyols and sugars are often found in combinations, and Ring[41] suggested that the toxic effect of high concentrations of a single substance is avoided in such multicomponent systems. Ethylene glycol, as found in concentrations of 2.7 Mol in the bark beetle *Ips typographus*,[59] is toxic to humans.

Seasonal change in the capacity of supercooling is typical for freeze-susceptible insects, and this is related to parallel changes in the concentration of glycerol and

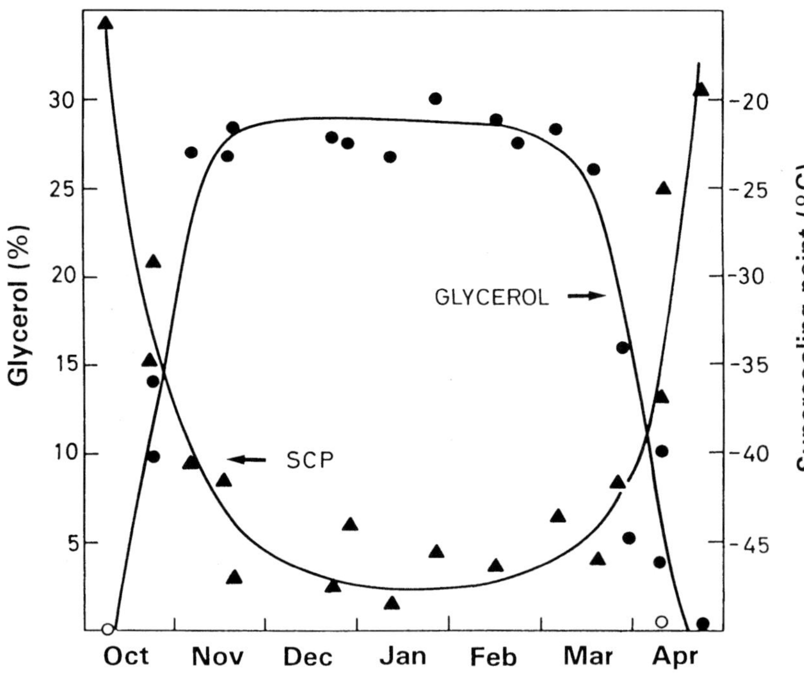

Fig. 8. Seasonal changes in glycerol content (% of fresh weight; *circles*) and supercooling points (SCPs; *triangles*) in larvae of the pine shoot tortricid *Petrova recinella* (reprinted with permission from ref. 21)

other cryoprotectant substances; for example, the seasonal changes observed in larvae of the pine shoot tortricid *Petrova resinella* from Estonia[60] (Fig. 8). Low midwinter SCPs are concurrent with high concentrations of glycerol, while specimens without glycerol in autumn and spring have high SCPs.

In freeze-tolerant insects, increasing concentrations of glycerol and other polyols apparently depress the lower lethal temperature.[9] The effect of glycerol in freeze-tolerant insects is to reduce the amount of ice formed in the tissues. Since water molecules are bound to glycerol, the percentage of frozen body water at any given temperature is less than in the absence of this cryoprotectant.

In beetles, and perhaps in other insects as well, it appears that similar concentrations of polyols have greater effect on the cold hardiness of freeze-tolerant than of freeze-susceptible species.[9] When the lower lethal temperatures of the chrysomelid beetle *Phyllodecta laticollis* or of the pythid beetle *Pytho depressus* is plotted against the concentrations of solutes in their hemolymph, a steep regression line appears. In comparison, similar concentrations of solutes have less effect on the SCPs of freeze-susceptible beetles. Consequently, as suggested by Zachariassen,[9] freeze tolerance is probably the most successful strategy of cold hardiness in areas with extremely low temperatures. In fact, several species of insects from the interior of Alaska and arctic Canada are tolerant to freezing.

1.8 Thermal hysteresis proteins

The discovery of the so-called antifreeze proteins or glycoproteins in insects has opened for a better understanding of the nature of cold hardiness. These substances were originally described from antarctic fishes,[61] in which they prevent freezing in species swimming in ice-filled waters by lowering the FP of the blood to ca. -2°C. Later, antifreeze proteins were reported from a number of insects,[14] where they appear to increase cold hardiness in several ways.

1.8.1 The function of thermal hysteresis proteins

Antifreeze proteins have the unusual characteristic that they depress the FP of the hemolymph, but not its MP.[14] The difference between the MP and FP is termed thermal hysteresis. For this reason, antifreeze proteins are usually called thermal hysteresis proteins (THPs). Apparently, they adhere to the surface of seedling ice crystals, inhibiting further growth and in this way they are expected to stabilize the supercooled state in insects over the entire supercooled range.[9] Furthermore, THPs may prevent or delay the penetration of ice from the outside through the cuticle in overwintering insects, as shown by Gehrken[62] in the beetles *Ips acuminatus* and *Bolitophagus reticulatus*, and may increase survival rates during prolonged exposure at low temperatures in freeze-susceptible species. THPs have also been found in a few freeze-tolerant insects,[14] and survival in frozen cells from the gut of the centipede *Lithobius forficatus* is greatly increased by THPs.[63] The effect of THPs in

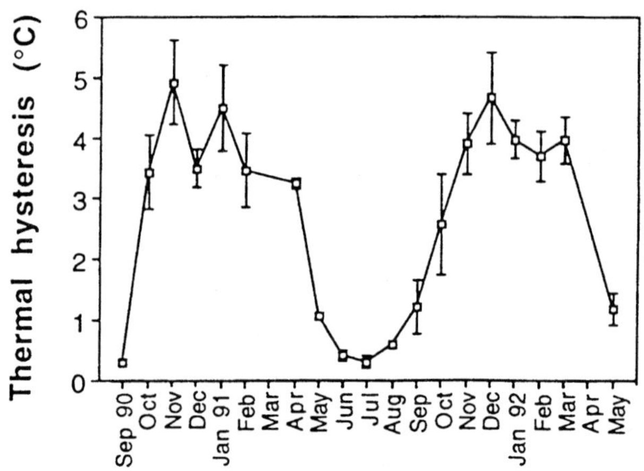

Fig. 9. Seasonal changes in thermal hysteresis (mean ± SE) in larvae of the pyrochroid beetle *Dendroides canadensis* through 2 years (reprinted with permission from ref. 64)

freeze-tolerant organisms may be to prevent a harmful effect of ice recrystallization during thawing.

In larvae of the pyrochroid beetle *Dendroides canadensis* thermal hysteresis reaches a level of 5°C in midwinter (Fig. 9), but declines rapidly during the spring.[64] The low FP will in themselves prevent freezing at temperatures down to -6°C, but seasonal lowering of the SCP down to -20°C is also associated with increased THP concentrations. The THPs of *D. canadensis* are able to depress the nucleating activity of proteins and lipoproteins in hemolymph of the larvae, thus lowering its SCP. Extension of the hemolymph supercooling capacity is only functional if other fluid-filled compartments undergo similar seasonal changes. As mentioned previously (Sect. 1.6.2), the THPs of *D. canadensis* mask the activity of ice nucleating bacteria in the gut.[58]

The THPs of insects are comprised of several components. In addition, activating proteins have been isolated, increasing the thermal hysteresis activity severalfold in *D. canadensis*[14] and in larvae of the meal worm *Tenebrio molitor*.[65] A 12.86-kDa THP isolated from *T. molitor* was found to be packed into protein granules of the fat body cells.[65] Beside protecting from intracellular freezing of the large fat body cells, they represent a store of THP available for secretion into the hemolymph. Horwath et al.[65] have evidence to suggest that THPs are produced and stored in the cells during the summer and autumn, and released into the hemolymph when triggered by stimulating photoperiods. The low level of THPs in fat body cells of winter-acclimated larva are in agreement with this hypothesis.

Tracking of the 12.86-kDa THP in *T. molitor* has revealed a complex cellular regulation of its biosynthesis, storage and secretion, factors which may be controlled by environmental and endocrine triggers.[65] This opens for further studies on the precise regulation of THPs and their activators in generating an efficient cold-hardening system.

1.8.2
Multifactorial systems

In the bark beetle *Ips acuminatus* the mean SCP is lowered by 2–3°C in the presence of THPs,[66] but further analysis of the cold hardiness of this beetle reveals a multifactorial system (Fig. 10). Lowering of the SCPs in *I. acuminatus* is initiated by the evacuation of gut content and a decrease in water content,[67] accompanied by accumulation of the disaccharid trehalose. Another drop in SCPs from -17 to -20°C is closely correlated with an increased level of thermal hysteresis, while further lower-

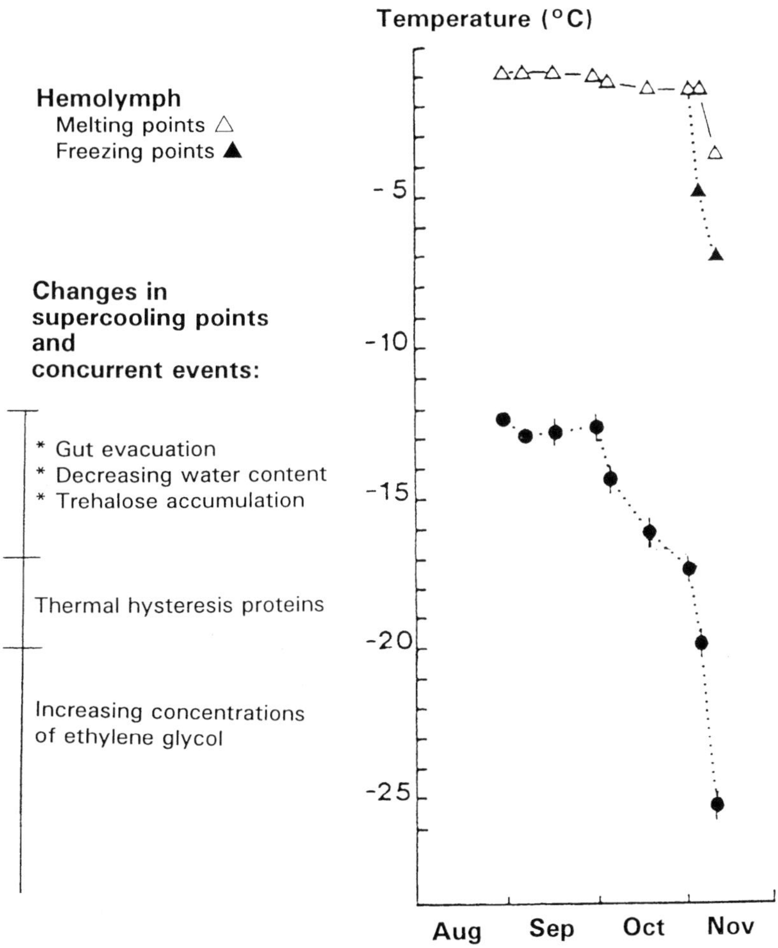

Fig. 10. A multifactorial process of cold hardening in the bark beetle *Ips acuminatus*
Hemolymph melting points (*open triangles*) and freezing points (*closed triangles*) are shown in the upper part of the diagram; supercooling points (mean ± SE; *closed circles*) in the lower part (reprinted with permission from ref. 62)

ing of the SCPs is caused by increasing contents of ethylene glycol, which reaches concentrations of 2.7 Mol.

The cold hardiness of the springtail *Entomobrya nivalis* from the Swiss Prealps is also based on more factors.[68] While ribitol and other polyols are synthesized during the time the Collembola are inactive in their hibernacula under loose bark of spruce, THPs additionally protect against freezing during the active transition phases in spring and autumn. In *E. nivalis* THPs, the synthesis of THPs is triggered in the autumn by short daylength and decreasing temperatures.

1.8.3
Applications of THPs

The unusual characteristics of THPs from fish and insects give several possibilities for their applications. Added to ice cream and other frozen food they may reduce the problem of recrystallization. By transplanting genes for THP, it may be possible to produce plant varieties with enhanced cold hardiness. In the past, it has often been suggested that studies on freeze tolerance in insects may give results of value for the cryopreservation of biological materials. If THPs are found to protect against freezing,[63] this is an aspect that merits further investigations.

1.9
Desiccation in terrestrial arthropods

1.9.1
Overlapping adaptations

During hibernation, insects are not only exposed to cold but may also experience serious stress of desiccation. Due to its higher vapor pressure, water in supercooled insects will be lost to the frozen surroundings, resulting in different degrees of dehydration. In polar regions, nonavailability of moisture is more limiting than low temperatures to biological activity.[69] Ring and Danks[70] have argued that cryoprotection and resistance to desiccation are overlapping adaptations. Solute production, regulation of water content and selection of overwintering sites are important to survive low temperatures, but the same factors may also protect against dehydration or enhance resistance to desiccation. Increased concentrations of polyols and other low molecular weight substances lower the water vapor pressure of a supercooled organism, reducing its water loss in a frozen environment. In contrast, frozen insects lose little water.[46,71] As the temperature is lowered, more and more ice is formed but the water vapor pressure of the unfrozen fraction of their body fluids is in equilibrium with the ambient temperature.

In some insects, water content is drastically reduced during the winter, combined with high concentrations of cryoprotectants. In larvae of the beetle *Pytho deplanatus*[46] the water content decreased from 69% of fresh weight in nonacclimated specimens to 30% of fresh weight in fully cold-acclimated ones. This dehydration is accompanied by an increase in concentration of glycerol to 13% and of sugars to 6% of body fresh weight. Similarly, in larvae of the goldenrod moth *Epi-*

blema scudderiana,[72] the water content decreased from 57 to 25% of fresh weight, while glycerol concentration rose to 19%.

More data on the relation of desiccation and polyols are wanted, but in some cases it has been shown that the production of such substances is triggered by water loss. The bark beetle *Ips acuminatus* does not accumulate ethylene glycol until the water content has fallen below 52% of fresh weight,[59] and the antarctic mite *Alaskozetes antarcticus* is able to synthesize glycerol at low relative humidity.[10]

1.9.2
Dehydration in Collembola

All Collembola studied up to present are freeze-susceptible[10,21] and depend on supercooling to survive low temperatures. In this context it is surprising that the arctic springtail *Onychiurus arcticus* normally has mean SCPs of -6 to -7°C.[73,74] At Spitsbergen, *O. arcticus* overwinters in bird cliffs where the temperature in their habitat falls below -20°C.[75]

When specimens of *O. arcticus* were dehydrated over ice at temperatures from -3 to -19.5°C (Fig. 11), the water content decreased substantially.[74] At the lowest temperature, the springtails lost 92% of their total water content, but more than 80% of them survived when rehydrated. Specimens that froze spontaneously dur-

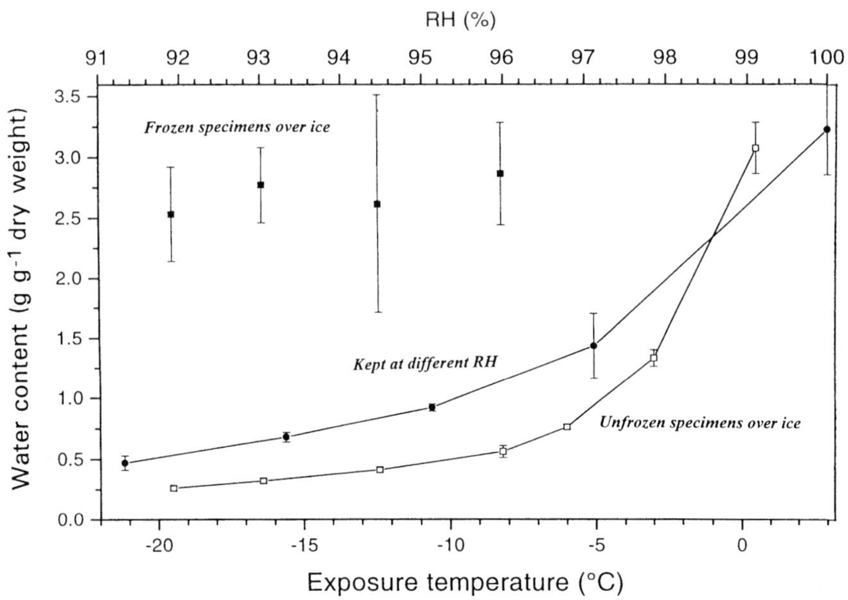

Fig. 11. The relationship between water content (mean ± SD) of unfrozen specimens of *Onychiurus arcticus* kept over ice for 14 d at temperatures down to -19.5°C, and of specimens kept at 0.5°C and relative humidity (RH) from 100 to 91.3%. Specimens frozen spontaneously during exposure over ice maintained higher levels of water content (reprinted with permission from ref. 74)

ing the experiment lost a small part of their water only, and did not survive after thawing. High water losses and high survival rates were also found in specimens kept at 0.5°C and different relative humidities. The high survival rates of dehydrated Collembola may be related to the accumulation of trehalose during desiccation.[73]

At a water content of 0.6 g g^{-1} dry weight the mean SCP was depressed to -22°C.[74] At even lower water contents, freezing could not be detected at all, apparently since no freezable water is left.

While Collembola from hot deserts and other dry environments may be subjected to desiccation at high temperatures, as reviewed by Sømme,[21] dehydration as a protection against low temperatures is only known from *O. arcticus*. The physiological mechanisms of anhydrobiosis in hot desert springtails are probably quite different from the dehydration mechanism observed in *O. arcticus*, since the arctic species is quite susceptible to desiccation at higher temperatures.[76]

1.10 Anaerobiosis

In alpine and polar areas, many terrestrial arthropods are enclosed by ice in their overwintering habitats.[21] On mountain ridges, most of the snow is blown away, and the surface is turned into a compact layer of ice. Insects, mites and springtails are enclosed in tiny air spaces within ice and soil,[77] where an oxygen deficiency is likely to occur. As shown in Figure 12, some of these species are able to survive long periods without oxygen, and during anaerobiosis lactate is formed as a metabolic end product. Anaerobiosis is apparently an important part in the winter strategies of many terrestrial arthropods, but more studies in this field are wanted.

2 Cold adaptation in nematodes

2.1 Freeze avoidance and freeze tolerance

As in terrestrial arthropods, two strategies of low temperature survival are found in nematodes.[78,79] Some species are killed by freezing and avoid ice formation by supercooling, while others survive the formation of ice within their tissues. The distinction between freeze tolerance and freeze susceptibility in nematodes, however, is often less clear than in arthropods. As an example, a combination of freeze tolerance and freeze avoidance has been demonstrated in three species of *Eudorylaimus* inhabiting mosses (*Andreaea*) in the maritime Antarctic.[78] In the laboratory, *Endorylaimus* spp. supercooled to temperatures in the range of -5 to -25°C, of which the lower values would be sufficient to withstand the extreme temperatures of their natural habitats. Specimens with low SCPs were killed by freezing, while individuals with SCP above -15°C showed increasing freeze tolerance with increasing temperatures.

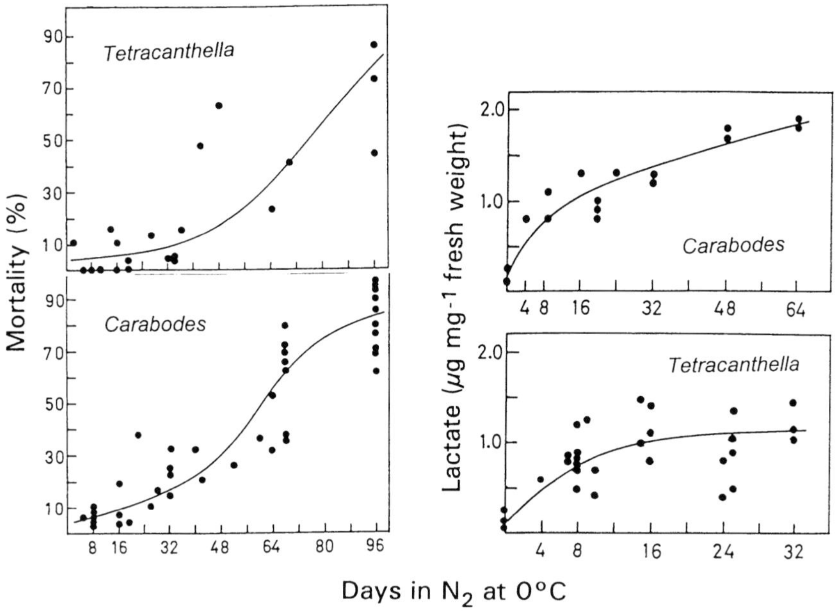

Fig. 12. Mortality and accumulation of lactate in the alpine Collembola *Tetracanthella wahlgreni* and the oribatid mite *Carabodes labyrinthicus* during anoxia (storage in N_2) at 0°C
Mortality: each dot represents a sample of 40–50 specimens; *lactate*: each dot represents 1–2 specimens (reprinted with permission from ref. 77)

Nematodes are more difficult to extract and handle than arthropods,[79] and in studies on cold hardiness the results are greatly influenced by the experimental procedure. Thus, the antarctic nematode *Panagrolaimus davidi* displays a freeze-avoiding strategy when cooled in air or liquid paraffin. When cooled in water, the nematodes freeze by exogenous ice formation and 5–20% of them survive freezing at temperatures from -10 to -40°C. Recently, Wharton and Ferns[80] showed that *P. davidi* actually survive intracellular freezing, and will subsequently grow and reproduce.

In principle, nematodes are aquatic organisms and depend on moist conditions for growth and development. In nature, freeze avoidance may not be a reliable strategy, since most species will freeze on contact with ice that is formed in their surroundings. There are exceptions, however, and the free-living third instar larva of the sheep parasite *Trichostrongylus colubriformis* is protected against inoculative freezing by an external sheath.[79] Similarly, the second stage larva of the potato cyst nematode *Globodera rostochiensis* is protected from ice nucleation within the egg shell.

2.2 Anhydrobiosis

Some nematodes enter a state of anhydrobiosis following the loss of a large proportion of their body water.[79,81] The free-living mycophagous nematode *Aphelencus avenae* enters anhydrobiosis following slow dehydration. Worms with water contents below 21% of fresh weight may survive short exposures at -196°C, while worms with 38% water content are killed. The free-living antarctic nematode *Dytilenchus* sp. survived exposures to -40 and -80°C for 48 h in the anhydrobiotic state.[82]

3 Adaptations in tardigrades

3.1 Anhydrobiosis

As in nematodes, many other invertebrates are able to tolerate an almost complete loss of body water.[12,81,83] According to Crowe et al.,[81] the physiological principle by which damage from extreme dehydration is avoided, may be universal for such different organisms as bacteria, yeast cells (*Saccharomyces cerevisiae*), nematodes, tardigrades and brine shrimps (*Artemia*). Their conclusion is based on the presence of trehalose in anhydrobiotic organisms. The effect of trehalose is to replace water and thus stabilize membrane phospholipids and different protein molecules during dehydration.

The earliest observation on anhydrobiosis in tardigrades was described by Spallanzani more than 200 years ago. As cited by Keilin,[83] Spallanzani's experiments included rotifers, tardigrades and nematodes found in the dry sand of roof gutters. "There are animalcula", Spallanzani wrote, "which notwithstanding they inhabit infusions, are so much distinguished and privileged by nature, so to enjoy the advantage of real resurrection after death."

Westh and Ramløv[84] kept *Adorybiotus coronifer* in dry moss for 8 months, and older literature, as reviewed by Keilin,[83] contains numerous accounts on the duration of anhydrobiosis in tardigrades. The record of survival is supposed to be 120 years, but data like that may be difficult to verify.

3.2 Freeze tolerance

Tardigrades in anhydrobiosis are extremely resistant to physiological stress[12,83,85,86] and may tolerate the freezing temperatures of liquid gases for several months. The freeze tolerance of tardigrades, however, is not restricted to the dehydrated state. According to Wright et al.[86] hydrated specimens of *Adorybiotus coronifer* survived exposure to -196°C for 15 min, but longer periods were not tested.

Although the tolerance of tardigrades to extremely low temperatures has fascinated many biologists, little is known about tolerance to temperature more representative of their natural environments. Tardigrades from Greenland, however, survive very low winter temperatures.[87] The limnic *Amphibolus nebulosus* overwinters as frozen cysts and the moss-inhabiting *Adorybiotus coronifer* overwinters both in a frozen and anhydrobiotic state. At the Inland Ice of Greenland, *Hypsibius klebesbergi* and *Diphascon recamieri* live in cryoconite holes at the surface,[88] and this habitat is solidly frozen for most of the year. Tardigrades are also found in crevices 70 m below the glacier surface.[88]

Following gradual cooling in the laboratory, *Macrobiotus harmsworthi* from the antarctic continent survived freezing at a series of temperatures down to -80°C.[89] Antarctic tardigrades are also able to survive extremely long periods at freezing temperatures. Dehydrated specimens of *Echiniscus jenningsi*, *Macrobiotus furciger* and *Diphascon chilense* from nunataks in Dronning Maud Land survived for 8 years in a freezer at -22°C.[85] In hydrated specimens, high survival rates were found after 590 d at -22°C and 150 d at -80°C. At -180°C, mortality of dehydrated tardigrades did not increase in 14 d, while all hydrated specimens were killed in 1–2 d.

3.3
Comparison of tolerance to freezing and dehydration

While resistance to cold and desiccation in arthropods may be considered as overlapping adaptations (Sect. 1.9.1), Crowe et al.[90] pointed out that the specific effect of trehalose during anhydrobiosis is in contrast with the low specificity of solute requirements for the stabilization of proteins during freezing. When cells or proteins in solution are frozen, a fraction of the water remains bound or unfrozen. Many solutes, such as polyols, sugars and amino acids, may act as cryoprotectants and inhibit destructive changes during freezing and thawing. In contrast to freezing, nonfreezable water can be removed during dehydration. In this respect, freezing and dehydration are fundamentally different stress factors.[81,90]

4
Protective dehydration mechanism in oligochaetes

4.1
Earthworm cocoons

According to Holmstrup and Zachariassen,[91] lumbricid earthworms do not survive freezing, and those that survive the winter have presumably migrated to deeper frost-free soil layers. Earthworm cocoons with their embryos, however, cannot escape winter frost, and a gradual water loss takes place in a frozen environment.[92] The water loss is due to the difference in water vapor pressure between ice and the body fluids, but the deficit decreases as the concentration of solutes increases. At -3°C, the cocoons of *Dendrobaena octaedra* lost ca. 60% of their fresh weight within a few days. At this point, the water vapor pressure of the unfrozen part of their body fluids is in equilibrium with the surrounding temperature, and the body fluids cannot freeze. As a response to dehydration, the embryo accumulates sorbitol which may prevent deleterious effects of the extensive water losses.

Cocoons of several species survived for 110 d in the dehydrated state at -3 and -8°C. This "protective dehydration mechanism", where dehydration is caused by ice outside the organism, resembles freeze tolerance in insects where extracellular ice formation induces a protective dehydration of the cells.[91]

4.2
Enchytraeids

Compared with temperate regions, the diversity of the fauna in arctic areas is limited, and most of the larger soil invertebrates are missing.[21] Enchytraeids are an important group in the Arctic, and densities of up to 30,000 specimens m^{-2} have been recorded from moist soil on Spitsbergen.[93]

During the winter, adult and juvenile enchytraeids survive for several months in the frozen arctic soil. In spite of this, specimens of eight species from Svalbard were rapidly killed by freezing at -3 to -5°C.[94] The supercooling capacity of these species was also low, with mean SCP in the range of -5 to -8°C. When subjected to slow dehydration in a frozen environment, however, high survival rates were recorded following slow thawing. All specimens of *Mesenchytraeus flavus* recovered from a loss of 56% of their water content at -3°C, and all specimens of *Enchytraeus kinkaidi* from a loss of 73% at -6°C. The results clearly demonstrate that the dehydration as protection against freezing is an important adaptation of arctic enchytraeids.

5
Conclusions

Insects and other terrestrial arthropods overwinter in a variety of habitats and are found in some of the coldest polar and temperate environments. As may be expected from their enormous diversity, many different systems of cold hardiness have evolved. Some insects survive extracellular ice formation, while most insects, as well as mites and springtails, are killed by freezing and depend on supercooling to survive. Both the degree of supercooling and the degree of freeze tolerance are increased by the accumulation of low molecular weight cryoprotectant substances, e.g. glycerol. Thermal hysteresis proteins stabilize the supercooled state in insects and may protect during freezing. Ice nucleating agents are important to freeze-tolerant insects, while ice nucleating agents and ice nucleating bacteria must be eliminated or masked in freeze-avoiding species. More studies on prolonged exposure to low temperatures and survival in the frozen state are needed and may lead to results of practical application.

Supercooled insects and other terrestrial arthropods are at risk of desiccation in a frozen environment. Low molecular weight substances may protect against dehydration as well as cold. Earthworm cocoons and enchytraeids are protected against freezing due to dehydration at the low water vapor pressure in frozen surroundings. A similar mechanism has recently been found in an arctic Collembola. Following dehydration, tardigrades and nematodes enter a state of anhydrobiosis in which they are very resistant to low temperatures. Some species also survive freezing in their hydrated state. Although anhydrobiosis in invertebrates has fascinated biologists for more than two centuries, many questions about its nature remain to be answered.

6 References

1. Réaumur RAF. Mémoirs pour servir à l'Histoire des Insectes, Tome 2. Paris: d'Imprimerie Royal, 1736.
2. Bachmetjew P. Experimentelle entomologische Studien vom physikalisch-chemischen Standpunkt aus, Band 1. Leipzig: Wilhelm Engelmann, 1901.
3. Payne MM. Freezing and survival in insects at low temperatures. Q Rev Biol 1926; 1:270-282.
4. Salt RW. Principles of insect cold hardiness. Annu Rev Entomol 1961; 6:55-74.
5. Salt RW. The survival of insects at low temperatures. Symp Soc Exp Biol 1969; 23:331-350.
6. Merivee E. Cold hardiness in insects. Tallin: Acad Sci Estonian SSR, 1978.
7. Sømme L. Supercooling and winter survival in terrestrial arthropods. Comp Biochem Physiol 1982; 73A:519-543.
8. Baust JG, Rojas RR. Insect cold hardiness: facts and fancy. J Insect Physiol 1985; 31:755-759.
9. Zachariassen KE. Physiology of cold tolerance in insects. Physiol Rev 1985; 65:799-832.
10. Cannon RJC, Block W. Cold tolerance of microarthropods. Biol Rev 1988; 63:23-77.
11. Storey KB, Storey JM. Freeze tolerance in animals. Physiol Rev 1988; 68:27-84.
12. Ushatinskaya RS. Latent life and anabiosis. Moscow: Hayka, 1990 (in Russian).
13. Duman JG, Wu DW, Xu L, Tursman D, Olsen TM. Adaptations of insects to subzero temperatures. Q Rev Biol 1991; 66:387-410.
14. Duman JG, Xu L, Neven LG, Tursman D, Wu DW. Haemolymph proteins involved in insects sub zero-temperature tolerance: ice nucleators and antifreeze proteins. In: Lee RE Jr, Denlinger DL, eds. Insects at Low Temperature. New York: Chapman and Hall, 1991:94-127.
15. Lee RE Jr. Principles of insect low temperature tolerance. In: Lee RE Jr, Denlinger DL, eds. Insects at Low Temperature. New York: Chapman and Hall, 1991:46-17.
16. Lee RE Jr, Costanzo JP, Mugnano JA. Regulation of supercooling and ice nucleation in insects. Eur J Entomol 1996; 93:405-418.
17. Block W. Cold or drought – the lesser of two evils for terrestrial arthropods? Eur J Entomol 1996; 93:325-339.
18. Danks HV. Winter habitats and ecological adaptations for winter survival. In: Lee RE Jr, Denlinger DL, eds. Insects at Low Temperature. New York: Chapman and Hall, 1991:231-259.
19. Lee RE Jr, Denlinger DL, eds. Insects at Low Temperatures. New York: Chapman and Hall, 1991.
20. Leather SR, Walters KF, Bale JS. The Ecology of Insects Overwintering. Cambridge: Cambridge University Press, 1993.
21. Sømme L. Invertebrates in Hot and Cold Arid Environments. Berlin: Springer, 1995.
22. Bale JS. Classes of insect cold hardiness. Funct Ecol 1993; 7:751-753
23. Lee RE Jr, Chen CP, Denlinger DL. A rapid cold hardening process in insects. Science 1987; 238:1415-1417.
24. Coulson SJ, Bale JS. Characterisation and limitations of the rapid cold-hardening response in the house fly *Musca domestica* (Diptera: Muscidae). J Insect Physiol 1990; 36:207-211.
25. Denlinger DL. Relationship between cold hardiness and diapause. In: Lee RE Jr, Denlinger DL, eds. Insects at Low Temperature. New York: Chapman and Hall, 1991:174-198.
26. Hodkova M, Hodek I. Control of diapause and supercooling by the retrocerebral complex in *Pyrrhocoris apterus*. Entomol Exp Appl 1994; 70:237-245.

27. Hodkova M, Hodek I. Temperature regulation of supercooling and gut nucleation in relation to diapause of *Pyrrhocoris apterus* (L.) (Heteroptera). Cryobiol 1997; 34:70-79.
28. Sømme L. The effect of prolonged exposures at low temperatures in insects. Cryo-Lett 1996; 17:341-346.
29. Lozina-Lozinskii LK. Studies in Cryobiology. New York: John Wiley, 1974.
30. Danks HV. The wider integration of studies on insect cold-hardiness. Eur J Entomol 1996; 93:383-403.
31. MacPhee AW. Cold-hardiness, habitat and winter survival of some orchard arthropods in Nova Scotia. Can Entomol 1964; 96:617-625.
32. Hansen T, Merivee E. Cold-hardiness of the cabbage butterflies *Pieris brassicae* L. and *Pieris rapae* L. Eesti NSV Ted Akad Toim Biol 1970; 20:298-303
33. Bale JS. Implications of cold hardiness in pest management. In: Lee RE Jr, Denlinger DL, eds. Insects at Low Temperature. New York: Chapman and Hall, 1991:461-498.
34. Lamb RJ, Turnock WJ, Hayhoe HN. Winter survival and outbreaks of bertha armyworm, *Mamestra configurata* (Lepidoptera: Noctuidae), on canola. Can Entomol 1985; 117:727-736.
35. Pullin AS, Bale JS. Cause and effect of prefreeze mortality in aphids. Cryo-Lett 1988; 9:101-113.
36. Bale JS, Harrington R, Clough MS: Low temperature mortality of the peach-potato aphid *Myzus persicae*. Ecol Entomol 1988; 13:121-129.
37. Turnock WJ, Boivin G, Ring RA. Inter-population differences in the cold hardiness of *Delia radicum* (L.). Can Entomol 1998; 130:119-129.
38. Kostál V, Simek P. Biochemistry and physiology of aestivo-hibernation in the adult apple blossom weevil, *Anthonomus pomorum* (Coleoptera: Curculionidae). J Insect Physiol 1996; 42:727-733.
39. MacKenzie AP. Non-equilibrium freezing behaviour of aqueous systems. Philos Trans R Soc London 1977; B278:167-189.
40. Zachariassen KE. The water relation of overwintering insects. In: Lee RE Jr, Denlinger DL, eds. Insects at Low Temperature. New York: Chapman and Hall, 1991:47-63.
41. Ring RA. Insects and their cells. In: Ashwood-Smith MJ, Farrant J, eds. Low Temperature Preservation in Medicine and Biology. London: Pitman Medical, 1980:187-217.
42. Tanno K. Freezing injury in fat body cells of the poplar sawfly. In: Asahina E, ed. Cellular Injury and Resistance in Freezing Organisms. Sapporo: Institute of Low Temperature Science, 1967:245-257.
43. Tanno K. Frost resistance in the poplar sawfly, *Trichiocampus populi* Okamoto. IV. Intracellular freezing in fat-cell and injury occurring upon metamorphosis. Low Temp Sci Ser B 1968; 26:71-78.
44. Kukal O. Behavioral and physiological adaptations to cold in a freeze-tolerant arctic insect. In: Lee RE Jr, Denlinger DL, eds. Insects at Low Temperature. New York: Chapman and Hall, 1991:276-300.
45. Miller K. Cold-hardiness strategies of some adult and immature insects overwintering in interior Alaska. Comp Biochem Physiol 1982; 73A:595-604.
46. Ring RA. Freezing tolerant insects with low supercooling points. Comp Biochem Physiol 1982; 73A:605-612.
47. Ramløv H, Bedford J, Leader J. Freezing tolerance of the New Zealand alpine weta, *Hemideina maori* Hutton (Orthoptera; Stenopelmatidae). J Term Biol 1992; 17:51-54.
48. Asahina E. Frost resistance in insects. Adv Insect Physiol 1969; 6:1-49.
49. Baust JG, Edwards J. Mechanisms of freeze tolerance in the Antarctic midge (*Belgica antarctica*). Physiol Entomol 1979; 4:1-5.
50. Humble LM, Ring RA. Inoculative freezing of a larval parasitoid within its host. Cryo-Lett 1985; 6:59-66.
51. Salt RW. Role of glycerol in the cold-hardening of *Bracon cephi* (Gahan). Can J Zool 1959; 37:59-69.

52. Zachariassen KE, Hammel HT. Nucleating agents in the haemolymph of insects tolerant to freezing. Nature (London) 1976; 262:285-287.
53. Zachariassen KE. The role of polyols and nucleating agents in cold-hardy beetles. J Comp Physiol 1980; 140:227-234.
54. Sømme L. The overwintering of *Pelophila borealis* Payk. III. Freezing tolerance. Norw J Entomol 1974; 21:131-134.
55. Van der Laak S. Physiological adaptations to low temperature in freezing-tolerant *Phyllodecta laticollis* beetles. Comp Biochem Physiol 1982; 73A:613-620.
56. Sømme L, Block W. Cold hardiness of Collembola at Signy Island, maritime Antarctic. Oikos 1982; 38:168-176.
57. Lee RE Jr, Lee MR, Strong-Gunderson JM. Biological control of insect pests using ice-nucleating microorganisms. In: Lee RE Jr, Warren GJ, Gusta LV, eds. Biological Ice Nucleation and its Applications. St. Paul: APS Press, 1995:257-269.
58. Olsen TM, Duman JG. Maintenance of the supercooled state in the gut fluid of overwintering pyrochroid beetle larvae, *Dendroides canadensis*: role of ice nucleators and antifreeze proteins. J Comp Physiol B 1997a; 167:114-122.
59. Gehrken U. Winter survival of an adult bark beetle *Ips acuminatus* Gyll. J Insect Physiol 1984; 30:421-429.
60. Hansen T. Variation on glycerol content in relation to cold-hardiness in the larvae of *Petrova resinella* L. (Lepidoptera, Tortricidae). Eesti NSV Tead Akad Toim Biol 1973; 22:105-112.
61. DeVries AL. Biological antifreeze agents in coldwater fishes. Comp Biochem Physiol 1982; 73A:627-640.
62. Gehrken U. Inoculative freezing and thermal hysteresis in the adult beetles *Ips acuminatus* and *Rhagium inquisitor*. J Insect Physiol 1992; 38:519-524.
63. Tursman D, Duman JG. Cryoprotective effects of thermal hysteresis protein on survivorship of frozen gut cells from the freeze tolerant centipede *Lithobius forficatus*. J Exp Zool 1995; 272:249-257.
64. Olsen TM, Duman JG. Maintenance of the supercooled state in overwintering pyrochroid beetle larvae, *Dendroides canadensis*: role of hemolymph ice nucleators and antifreeze proteins. J Comp Physiol B 1997; 167:105-113.
65. Horwath KL, Easton CM, Poggioli GJ Jr, Myers K, Schnorr, IL. Tracking the profile of a specific antifreeze protein and its contribution to the thermal hysteresis activity in cold hardy insects. Eur J Entomol 1996; 93:419-433.
66. Gehrken U. Supercooling and thermal hysteresis in the adult bark beetle, *Ips acuminatus*. J Insect Physiol 1989; 35:421-429.
67. Gehrken U. Correlative influence of gut appearance, water content and thermal hysteresis on whole body supercooling point of adult bark beetles, *Ips acuminatus*. Gyll. Comp Biochem Physiol 1995; 112A:207-214.
68. Meier P, Zettel J. Cold hardiness in *Entomobrya nivalis* (Collembola, Entomobryidae): annual cycle of polyols and antifreeze proteins, and antifreeze triggering by temperature and photoperiods. J Comp Physiol B 1997; 167:297-304.
69. Kennedy AD. Water as a limiting factor in the Antarctic environment: a biogeographical synthesis. Alp Arctic Res 1993; 25: 308-315.
70. Ring RA Danks HV. Desiccation and cryoprotection: overlapping adaptations. Cryo-Lett 1994; 15:181-190.
71. Lundheim R, Zachariassen KE. Water balance of overwintering beetles in relation to strategies for cold hardening. J Comp Physiol B 1993; 183:1-4.
72. Rickards J, Kelleher MJ, Storey KB. Strategies of freeze avoidance in larvae of the goldenrod gall moth *Epiblema scudderiana*: winter profiles of a natural population. J Insect Physiol 1987; 33:443-450.
73. Worland MR, Grubor-Lajsic G, Montiel PO. Partial desiccation induced by the sub-zero temperatures as a component of the survival strategy of the Arctic collembolan *Onychiurus arcticus* (Tullberg). J Insect Physiol 1998; 44:211-219.

74. Holmstrup M, Sømme L. Dehydration and cold hardiness in the Arctic Collembola *Onychiurus arcticus*. J Comp Physiol B 1998; 168:197-203.
75. Coulson SJ, Hodkinson ID, Strathdee AT, Block W, Webb NR, Bale SJ, Worland MR. Thermal environments of arctic soil organisms during winter. Arctic Alp Res 1995; 27:364-370.
76. Hodkinson ID, Coulson SJ, Webb NR, Block W. Can high arctic soil microarthropods survive elevated summer temperatures? Funct Ecol 1996; 10:314-321.
77. Sømme L, Conradi-Larsen EM. Anaerobiosis in overwintering collembolans and oribatid mites from windswept mountain ridges. Oikos 1977; 29:127-132.
78. Pickup J. Strategies of cold hardiness in three species of free-living predatory Antarctic dorylaimid nematodes. J Comp Physiol B 1990; 160:167-173.
79. Wharton DA. Cold tolerance strategies in nematodes. Biol Rev 1995; 70:237-245.
80. Wharton DA, Ferns DJ. Survival of intracellular freezing by the antarctic nematode *Panagrolaimus davidi*. J Exp Biol 1995; 198:1381-1387.
81. Crowe JH, Hoekstra FA, Crowe LM. Anhydrobiosis. Annu Rev Physiol 1992; 54:579-599.
82. Pickup J, Rothery P. Water loss and anhydrobiotic survival in nematodes from Antarctic fellfields. Oikos 1991; 61:370-388.
83. Keilin D. The problem of anabiosis or latent life: history and current concepts. Proc R Soc Lond B 1959; 150:149-191.
84. Westh P, Ramløv H. Trehalose accumulation in the tardigrade *Adorybiotus coronifer* during anhydrobiosis. J Exp Zool 1991; 258:303-311.
85. Sømme L. Anhydrobiosis and cold hardiness in tardigrades. Eur J Entomol 1995; 93:349-357.
86. Wright JC, Westh P, Ramløv H. Cryptobiosis in Tardigrada. Biol Rev 1992; 67:1-29.
87. Westh P, Kristensen RM. Ice formation in the freeze tolerant eutardigrades *Adorybiotus coronifer* and *Amphibolus nebulosus* studied by differential scanning calometry. Polar Biol 1992; 12:693-699.
88. Grøngaard A. The Invertebrate Fauna of the Inland Ice of Greenland. Prize Essay. Copenhagen: University of Copenhagen, 1995.
89. Sugawara H, Tanno K, Ohyama Y, Fukuda H. Freezing-tolerance of *Macrobiotus harmsworthi* (Tardigrada) and *Plectus antarcticus* (Nematoda) in the Antarctic Region. Antarct Rec 1990; 34:292-302.
90. Crowe JH, Carpenter JF, Crowe LM, Anchordoguy TJ. Are freezing and dehydration similar stress vectors? A comparison of modes of interaction of stabilising solutes with biomolecules. Cryobiol 1990; 27:219-231.
91. Holmstrup M, Zachariassen KE. Physiology of cold hardiness in earthworms. Comp Biochem Physiol 1996; 115A:91-101.
92. Holmstrup M. Cold hardiness strategy in cocoons of the lumbricid earthworm *Dendrobaena octaedra* (Savigny). Comp Biochem Physiol 1992; 102A:49-54.
93. Birkemoe T, Dózsa-Farkas K. New records of Enchytraeidae (Oligochaeta) from Spitsbergen, Svalbard. Fauna Norv Ser A 1994; 15:35-44.
94. Sømme L, Birkemoe T. Cold tolerance and dehydration in Enchytraeidae from Svalbard. J Comp Physiol 1997; 167:264-269.

Nival invertebrate animals in the East Alps: a faunistic overview

K. Thaler

Institute of Zoology and Limnology, University of Innsbruck, Technikerstrasse 25,
A-6020 Innsbruck, Austria; e-mail: konrad.thaler@uibk.ac.at

1 Introduction

Despite the early interest of Heer[1] in animal life of the nival zone of the Alps, only a few groups have been surveyed and studied more intensively. There are also only a few regional studies. As can be seen from the present chapter, main problems of adaptation, ecophysiology, ecology and evolutionary biology of the nival fauna were hardly investigated at all.[2–4] The early history of explorations of the nival zone in the Alps has been summarized in Bäbler,[5] for recent overviews see Mani[6] and Franz.[7] The main contribution of Heer[1] was followed by a survey on the high alpine fauna of the Tyrolean Alps by Heller[8,9] and by investigations in the Swiss Alps.[5,10] Since then, studies on nival fauna and nival biota have been promoted by a few authors only and by their coworkers, i.e. in Tyrol by Steinböck,[11–13] Schmölzer[14] and Janetschek,[15–19] by Franz[20,21] in the Tauern range, and by Marcuzzi[22] in the Dolomite mountains. Specific contributions will be referred to with the respective taxa. Of course, progress of general knowledge about ecology and zoogeography is strongly connected to the state of taxonomic knowledge. Lack of precise taxonomic information in many groups hinders proper analysis and evaluation of fieldwork also in the nival zone. For general information on vegetation, ecology, ecophysiology and research strategies see Reisigl and Pitschmann,[23] Janetschek,[3] Moser et al.,[24] Franz[25] and Larcher.[26]

2 The nival fauna

2.1 Plathelminthes

At first, terrestrial microturbellaria were thought to be absent in the high alpine zone.[27] Leaving out a casual observation by Reisinger,[28] the occurrence of this group at nival sites was first recorded by An der Lan[29] in the Ötztal Alps at about 3200 m and 3035 m a.s.l. Specimens were extracted from soil samples from *Saxifraga oppositifolia* and other subnival plants (not specified). At least three species were recognized and one was formally described: *Acrochordonoposthia ramolia* An der Lan (Typhloplanidae).

2.2 Nemathelminthes

As usual in extreme environments, also in the nival zone bdelloid rotifers and nematodes are present, in primary soils, in soil from cushion plants and in mosses.[18] Species lists from sites above 3000 m a.s.l. mainly come from the West Alps and from the past;[30,31] records of nematoda from Austria are from about 2500 m a.s.l.[32–34] At a subnival site in the Tauern range in 2540 m about 40 species of nematodes were found.[33] Numbers apparently decrease with increase in altitude, for sites above 3000 m the presence of a few species only has been suggested,[30] belonging to the genera *Dorylaimus*, *Mononchus*, *Plectus*, *Teratocephalus*, *Tylenchus*. These nemathelminthes show anabiosis, high resistance to low temperature and desiccation, anemochorous dispersal and are probably widely distributed. Included are both euryoecious and euryzonal species, such as *Plectus cirratus* Bastian, which recently was recorded in the Zillertal Alps at 3100 m,[19] and others eventually preferring cold environments, such as *Teratocephalus* spp.[33]

2.3 Gastropoda

Snails rarely enter the subnival zone in the East Alps. In community studies five Vitrinidae were recorded at the extreme sites, at 2870 m in the Ötztal Alps,[35] at 2770 m and 2812 m in East Tyrol,[36,37] at 2700 and 2970 m in the main ridge and at the southern slope of the Zillertal Alps,[19,38] and at 2800 m in the Lechtal Alps.[39] Nevertheless, casual records from 3000–3100 m belong to other species,[40,41] *Pyramidula rupestris* (Draparnaud) (Pyramidulidae) and *Arianta arbustorum* (L.) (Helicidae). In the latter species postembryonic development to maturity lasts 5 years in a high alpine population at 2600 m a.s.l., as compared to only 2 years in a montane population at 1220 m.[42]

2.4 Annelida

From 60 species of earthworms now known in Austria,[43] 5–6 were recorded from alpine grassland in the Tauern range,[20,44] from the Zillertal[38] and Ötztal Alps[45,46] and from the Sesvenna mountains.[47] All these species disperse efficiently. In the Alps endemic species occur along the southern periphery and in lower regions only. The highest record is for *Dendrobaena octaedra* (Savigny), in sites with sedge mat fragments and subnival vegetation at about 2850 and 2950 m in the southern Zillertal Alps.[38] Highest records of Enchytraeidae are at about 2800 m in the Northern Calcareous Alps and in nival cushion plants at 3400 m in the Central Alps.[48] Species identification of enchytraeids has not been possible for most recent workers.[17,19,46]

A list from the Grossglockner region includes about 30 species from a range of altitude from 1145–2571 m.[49] Glacier-dwelling "ice worms"[50] do not occur in the Alps.

2.5 Tardigrada

In the West Alps tardigrades were recorded early at sites above 4000 m a.s.l.[30,51] In the Austrian Alps nival tardigrades have been almost neglected despite an "*Echiniscus*" record from the summit of the Großglockner in 3790 m by Ehrenberg,[52] some more recent records by Marcus[53] from North Tyrol, Wildspitze 2800 m, and by Mihelcic (for a bibliography see Kofler[54]). Due to anabiosis and anemochorous dispersal, apparently most nival tardigrades are widespread and present also in lowland environments. An exception is *Hypsibius klebelsbergi* Mihelcic[55] of the Ötztal Alps, which is confined to aquatic microcavernes at the glacier surface ("cryoconite holes").[56] Nevertheless, this species is present also in Nepal Himalaya, at Mingbo glacier 5600 m,[57] according to the synonymy proposed by Kraus[58] and by Dastych.[59]

2.6 Arachnida

2.6.1 Araneae

The nival spider fauna of the East Alps is rather diverse: a survey covering 58 mountains from 2900–3500 m a.s.l. yielded 48 species,[60] mainly Linyphiidae (19 Erigoninae, 15 Linyphiinae), but also Gnaphosidae (5), Lycosidae (3), Thomisidae (2), Salticidae (2), Philodromidae (1), Agelenidae (1). Few species occur mainly in the nival zone, all belonging to Linyphiidae. There are two main communities.[61] At extreme sites, in crevices of scree and boulders without phanerogams only two species persist in the Central Alps, *Diplocephalus rostratus* Schenkel, and *Lepthyphantes armatus* Kulczynski. Others are confined to more favorable sites with cushion plants and mosses: *Erigone tirolensis* L. Koch, *Walckenaeria clavicornis* (Emerton), *Hilaira montigena* (L. Koch). Some further spiders exhibit a euryzonal distribution and exist in open habitats across a range in altitude of about 3000 m up to the nival sites: *Pelecopsis parallela* (Wider), *Walckenaeria vigilax* (Blackwall) (Erigoninae), *Drassodes cupreus* (Blackwall), *Haplodrassus signifer* (C.L. Koch) (Gnaphosidae), *Thanatus striatus* C.L. Koch (Philodromidae). Most species exist also in the alpine zone. As casual guests from the agricultural landscape, four euryoecious aeronautic Linyphiidae turn up regularly, even in the nival zone: *Erigone atra* Blackwall, *E. dentipalpis* (Wider), *Oedothorax fuscus* (Blackwall), *Meioneta rurestris* (C.L. Koch). Species lists from nival habitats within the region are provided by Janetschek,[19] Christandl-Peskoller and Janetschek,[38] Puntscher[62] and Thaler.[41,60,61,63]

Like other nival organisms, also spiders include a strong proportion of arctic-alpine species (9) and of species endemic within the Alps (19) and within European mountains (7).[60] This is valid also for single sites. The summit plateau of Mt. Festkogel (Ötztal Alps) presents sedgemat fragments at about 3030 m, inhabited by a diverse spider fauna: 20 species, 9 of them endemic, 9 arctic-alpine (personal observations). Most nival species occur rather continuously along the Central Alps. Two breaks in species number at the western and at the eastern end of the Ötztal/Stubai Alps, respectively,[60,63] probably indicate recolonization from the West.

Some evidence suggests local speciation in isolated mountain groups in the calcareous Alps.[60,64] This might explain the presence of two endemic scree-dwelling Linyphiidae with small ranges in the Dolomites, *Lepthyphantes merretti* Millidge, *L. brunneri* Thaler. Other species occur discontinuously in mountain ranges of the West Palaearctic and even Holarctic region: *Xysticus bonneti* Denis (Thomisidae), *Chalcoscirtus alpicola* (L. Koch), together with some others which up to now have not been found beyond the limit of 3000 m.[63]

Life cycles have not yet been investigated in nival spiders. Apparently only two from five types of annual cycle distinguished by Schaefer[65] are performed at such elevation (personal observations). Most Linyphiidae and *Cryphoeca* (Agelenidae) are diplochronous and mature in late summer, adults reproducing after hibernation. Most non-Linyphiidae apparently are stenochronous, reproducing in early summer, adults being rather short-lived and dying before winter. Postembryonic development of some Lycosidae studied by Czermak[66] is prolonged in alpine grassland for 1 year and lasts for 3 years, as compared to populations at the timberline in 1900 m.

2.6.2 *Pseudoscorpiones, Opiliones, Acari*

In nival sites one species of false scorpions (Neobisiidae) and three scree dwelling harvestmen (Phalangiidae) regularly occur, all endemic in the Alps. *Neobisium jugorum* (L. Koch) has been found up to 3400 m and *Mitopus glacialis* (Heer) up to 3500 m in the Ötztal Alps.[13,67–69] Extreme records for *Dicranopalpus gasteinensis* Doleschall and *Megabunus armatus* (Kulczynski) are at about 3000 m. *Mitopus morio* (F.) which occurs abundantly in alpine grassland locally may also extend into subnival sedge mat fragments at about 3000 m, as in the Samnaun mountains.[63] Only two species are rather evenly distributed in the region considered. *M. armatus* is confined to the Southeastern Calcareous mountains;[70] *N. jugorum* ranges from the West Alps to the eastern limit of the Stubai Alps at Brenner pass. There are no detailed studies on the way of life of these species. *M. glacialis* has been observed preying upon glacier collembola.[11] The palps, which are armed with strong tubercles, apparently form a trapping device against springtail escape, analogous to setal traps in spiders[71] and in Carabidae.[72] Course and duration of postembryonic development are also unknown in the nival species. Females of *N. jugorum* were found brooding in brood nests in mid July at 2400 m.[73] *M. morio* is univoltine, overwintering in the egg stage and maturing in late summer. Its capacity to conduct the same type of life cycle across the whole range of its vertical distribution from lowlands to 3000 m is most remarkable.

Mites are a dominant group in the nival zone as in polar regions. Unfortunately precise ecofaunistic information is not yet available despite several regional studies

(Table 1). It is not even clear whether species exist with center of distribution within the nival zone. At least no such species was found among Oribatida from the Ötztal Alps.[78] Certainly, faunal elements of zoogeographic interest are included: species endemic to the Alps or to the alpine system, arctic-alpine and euryzonal species (see notes in Janetschek[19] and Schmölzer[76]). *Mesoteneriffia steinboecki* Irk (Teneriffiidae) apparently is a relict species with relatives on other continents.[17,80,81] Like springtails, some species show a refugial-cavatic distribution, being present in underground situations at the periphery of the Alps and at surface sites in the Central Alps.[17,19]

2.7 Crustacea, Isopoda

In the East Alps woodlice are absent widely in alpine grassland. Only *Trachelipus ratzeburgii* (Brandt) (Porcellionidae) can be found constantly up to the dwarf shrub heath.[82] In the southern Zillertal Alps this species was also found in grassland at 2550 m and once even in sedgemat fragments at 2950 m.[38]

2.8 "Myriapoda"

Few representatives of three groups of Myriapoda were still found in the sedge mat fragments of the subnival zone at about 3000 m, all of them rather euryzonal, occurring over a wide range of altitude. Sparse records include edaphic Symphyla (*Scutigerella* sp.) in the Ötztal Alps,[68,83] predaceous epigeic Chilopoda (only *Lithobius* spp.),[5,38,41,68] and saprovorous Diplopoda in the Central Alps, orders Chordeumatida, Julida. In the Ötztal Alps there are four species already mentioned by Verhoeff[84] from Ramolhaus in 3000 m. In adjacent mountain groups some further species might be expected in subnival sites[63,68,85,86] due to high endemism and vicariance patterns in this group. Highest records of Pauropoda are at lower elevation, from alpine grassland at about 2200 m.[87] Millipedes of exceptional interest are *Ommatoiulus sabulosus* (L.) (Julidae) due to its range of vertical distribution from almost sea level to 3000 m, and *Trimerophorella rhaeticum* (Rothenbühler) (Neoatractosomatidae) for its relict distribution within the Central Alps,[17,88–90] despite its vertical range from about 1100–3000 m. For this species Meyer[91] gave evidence for

Table 1. Species numbers of mites in the nival zone of the East Alps

Reference	Region	Species total	Nival species
Willmann[74]	Hohe Tauern	114	?
Janetschek[19]	Zillertal Alps	61	23
Mihelcic[75]	Zillertal Alps/South	?	16 (Parasitiformes excepted)
Schmölzer[76]	Zillertal Alps/West	210	?
Irk[77]	Ötztal Alps	79	20
Schatz[78]	Ötztal Alps	81	30 (Oribatida only)
Schweizer[79]	Sesvenna Alps	509	41

prolongation of the life cycle to 4 years in the high alpine region. It also shows a high resistance to water loss, probably tolerating desiccating conditions on windswept ridges.[92]

2.9
Insecta

2.9.1
"Apterygota"

Highest records of Protura and of Diplura in East Alps come from alpine grassland at about 2500–2600 m, for *Acerentulus alpinus* Gisin (Acerentomidae) in Lower Engadine[93,94] and for *Campodea plusiochaeta* Silvestri (Campodeidae) at Innsbruck, Nordkette.[95] Springtails (Collembola) occur abundantly, bristletails (Archaeognatha) at least regularly in nival habitats. In both groups serious taxonomic and faunistic shortcomings still exist.

A thorough study of high alpine collembolans will certainly add substantially to present knowledge,[96] despite an early overview by Handschin.[97] From nival habitats in Zillertal Alps Janetschek[19] recorded 25 species, Lienhard[98] identified 42 species in alpine grassland at 2500 m in the Swiss National Park; main families being Hypogastruridae, Onychiuridae, Isotomidae, Entomobryidae, and Bourletiellidae. Some of the nival species were also found by Haybach[99,100] in the Tauern range in 2960–2990 m and by Topp[101] in the Ankogel mountains. Also in springtails nival, euryzonal and arctic-alpine species are present;[97] of course, their numbers and identities should be reconsidered. As in mites, there are also some refugial-cavatic taxa.[17,102] The "Gletscherfloh" *Isotoma saltans* (Nicolet) has attracted main interest as a permanent inhabitant of the glacier surface[103,104] (see also Chapter by Eisenbeis and Meyer, in this book). Adaptations to the alpine environment in an alpine Isotomidae sp. have been described by Zettel and Zettel.[105]

Bristletails (Machilidae) exist regularly in nival boulders and scree in the East Alps. Specific names must be traced in taxonomic literature:[106] from the Central Alps of North Tyrol *M. engiadina* Wygodzinsky (about 800–2770 m), *M. fuscistylis* Riezler (1500–3400 m), and *M. rubrofusca* Janetschek (1000–2800 m) are known. Apparently these also show a rather broad range of distribution in altitude. Unfortunately, types of distribution and eventual vicariance have not yet been fully investigated in this group.

2.9.2
Saltatoria, Dermaptera, Paraneoptera

Within the region considered, only some thrips and plant lice occur regularly at nival sites. Some grasshoppers and one earwig species still occur in alpine grassland. Franz and Weiss[107] point to a decrease in altitudinal distribution of more than 100 m for *Anechura bipunctata* (F.) (Forficulidae) in the Glockner mountains between 1938 and 1978, depending on weather conditions. Highest records for grasshoppers refer to *Bohemanella frigida* (Boheman) (Catantopidae) and *Aeropus sibiricus* (L.) (Acrididae) as in Switzerland.[108] *B. frigida* was recorded from the southern Zillertal Alps at 2700 m,[38] *A. sibiricus* in the Tauern range up to the upper border of alpine grassland at 2700 m,[20] records for both in Switzerland even up to 2900 m. In the Sesvenna mountains there is a highly isolated occurrence of a further species, *Aeropedellus variegatus* (Fischer-Waldheim) (Acrididae), which is an oreo-tundral holarctic element, occurring discontinuously from the Alps to Alaska. According to Nadig,[109] *A. variegatus* exists in this area from 2250 m onwards up to subnival sites at 3100 m, which are free of snow for only 2–3 months. Of course, it is univoltine, overwintering in the egg stage.

Only 14 out of a total of 55 free-living Psocoptera in Lower Engadine exist also in the upper subalpine zone from about 1900–2300 m, there is no alpine species at all.[110] In the Ötztal Alps Pelikan[111] found two euryzonal graminicolous Thripidae still in subnival sedge mats at about 3000 m (*Aptinothrips rufus* (Haliday)) and even in cushion plants and grass tufts at 3420 m (*A. stylifer* Trybom). Hemiptera rarely occur within the alpine zone.[20,45,112] There are a few grassland records of euryzonal Heteroptera up to about 2500 m: *Saldula saltatoria* (L.) (Saldidae), *Trapezonotus desertus* Seidenstücker (Lygaeidae), *Eurydema rotundicolle* (Dohrn) (Pentatomidae), and *Canthophorus impressus* (Horvath) (Cydnidae); see studies of Heiss in North Tyrol.[113–115] Euryzonal planthoppers even occur in sedge mat fragments and probably feed on Gramineae. High records in the Zillertal Alps are at 2900 m, *Sothanus thenii* (Löw), and at 2970 m, *Aphrodes assimilis* (Signoret) (Cicadellidae),[38] and in the Ötztal Alps at 3150 m (*Dicranotropis divergens* Kirschbaum, Delphacidae).[41,116] In Switzerland two further species were found at the same elevation. These are all univoltine and overwinter either as an egg or as larvae.[117] Despite discoveries of Janetschek[17,19,38,45] and casual descriptions,[118] nival aphids and coccids are almost unknown. *Nasonovia brevipes* (C. Börner) (Aphididae) was discovered in the Zillertal Alps at 3100 m from grasses. Its lifecycle is not yet known and probably will show some adaptation, as in arctic aphids.[119,120] *Heliococcus nivearum austriacus* Balachowsky (Pseudococcidae) was extracted from cushion plants (*Saxifraga* spp.), lichens and mosses in the Ötztal Alps at 3410 m. Of course, the aphids which are passively transported by wind in considerable densities from lowland to the alpine zone support alpine and nival ecosystems in an important way as an allochthonous food input.[121]

2.9.3
Endopterygota

The orders of Endopterygota which are rich in species also occur within the nival zone: Coleoptera (beetles), Hymenoptera, Diptera (flies) and Lepidoptera (moths and butterflies). From the Pala massif in the southern Dolomite mountains furthermore one female *Boreus* sp. (Mecoptera) was recorded from 3190 m.[68] As snowflies are brachypterous, this female probably represents an undescribed high-alpine species.

The species total of beetles known to Kahlen[122] from the small Austrian province North Tyrol is 4145! This large and well-known insect order is hardly present in the nival zone, albeit in Switzerland 46 species have been found beyond the limit of

3000 m, highest records being from 3250 m.[123] In the Zillertal Alps two species only were present in nival sites,[19] in the Ötztal mountains 11 species were recorded in a pitfall survey in sedge mat fragments from 2800–3100 m:[24] families Carabidae (4), Staphylinidae (4), Byrrhidae (*Byrrhus fasciatus* Forster), Scarabaeidae (*Aphodius mixtus* Villa), and Curculionidae (1); for further records see refs. 41,68,125. According to this evidence, there is no eunival species of beetle in the region, albeit some influents. As a carnivorous pioneer, *Nebria germari* Heer (Carabidae) invades also bare nival scree, the highest record being in 3420 m in the Ötztal Alps and in 3480 m in the Zillertal Alps. *Dichotrachelus stierlini* Gredler, a muscivorous weevil (Curculionidae), was found at 3350 m, further phytophagous families appear in sedge mats at about 3000 m: *Chrysomela marginata* L. (Chrysomelidae), and *Selatosomus rugosus* (Germar) (Elateridae). Apparently the difference in life zone preference between beetles and true nival arthropods might have influenced the contrary opinions of Holdhaus[126] and Janetschek[17] on the survival value of nunataks in the Central Alps for the nival fauna in glacial periods.

Also in Hymenoptera species numbers decrease drastically with altitude (see inventories of Schedl[127,128] in the Ötztal Alps). Three sawflies there still occur at the upper limit of alpine grassland: *Pristiphora bensoni* Lindquist, *P. preadalbanensis* (Cameron), and *Tenthredo arcuata* Förster (Tenthredinidae). High records for Apoidea are *Hylaeus nivalis* (Moravitz) (Colletidae) from *Linaria alpina* at 2800 m, *Osmia inermis* (Zetterstedt) (Megachilidae) at 2800 m, and *Alpinobombus alpinus* (L.) at 3000 m (Apidae). The parasitic wasps present also in nival communities have not yet been thoroughly investigated, main hosts probably being Lepidoptera.

Diptera are apparently the pterygote order which is most rich in species in the high alpine and in the nival zone. This may be inferred already from the survey of Müller[129] about flower-visiting insects above the timberline: Diptera (210 species), Lepidoptera (148), Hymenoptera (88), and Coleoptera (33). Flies are also abundant. Müller's percentages for insect flower visits above the timberline and in lowland regions (values in brackets) are: flies 44 (30), butterflies and moths 31 (9), sawflies, bees and wasps 18 (44), beetles 7 (15), others <1 (2). In the epoch-making study of Bezzi[130] for the whole Alps 58 eunival species are listed, mostly from the altitude 2500–3100 m, and further 51 species occur across a wider range, but are able to develop in high regions. As stated by Bezzi, in high alpine and nival communities Anthomyiidae and Muscidae are preponderant. Studies with "quantitative methods" (e.g. pitfalls, Malaise traps, ground photoeclectors) in grassland sites in the Swiss National Park at 2540 m[131] and in the Ötztal Alps from 2000–2630 m corroborated these statements.[132–134] Pont[135] collected 70 species of Muscidae above the treeline around Obergurgl, Ötztal Alps, from sites between 2100 and 3020 m. These include the same faunistic types as other arthropods. Of the ten species recorded that live only above the treeline, four are endemic to the Alps, two others endemic to European mountains, the remaining four having an arctic-alpine distribution. Despite obvious abundance the lifehistories of most species are largely unknown. Two further families frequent in nival pioneer communities are Sciaridae and Chironomidae (see observations of Janetschek[19] in the Zillertal Alps). These also include endemic species, some with reduced wings: *Bryophaenocladius thaleri* Willassen from Pala mountains in about 3100 m,[136] *Lycoriella ventrosa* (Lengersdorf) in the Central Alps and Dolomite mountains in 2500–3450 m.[137-138] Another brachypterous midge, *Dactylolabis pechlaneri* Mendl (Limoniidae),[68,139] exists from 2600–3100 m in the Austrian Alps and in the Dolomite mountains. The true status of further families represented by casual records[68] only is not yet clear: Tipulidae, Trichoceridae, Mycetophilidae, and Bibionidae. These may include euryzonal species, but also wind-blown specimens.

Lepidoptera is less represented in North Tyrol than beetles; Huemer and Tarmann[140] mentioned 2603 species for the region. Nevertheless, the nival fauna of butterflies and moths is fairly diverse. In the only survey available, Vorbrodt[141] enumerates for Switzerland 32 species in 12 families from sites above 3000 m. From nival sites on the main ridge of the Zillertal Alps 11 species were recorded,[19] autochthonous species belonging to the families Geometridae, Noctuidae, Oecophoridae, Pyralidae, and Yponomeutidae. In a more general report by Burmann[142] about ten species are discussed, including further families such as Arctiidae (Lithosiinae), Bucculatricidae, Nepticulidae, Scythridiidae, and Tortricidae. Additionally, migratory species occur; for a casual observation of Sphingiidae at Timmelsjoch at 2500 m see Schedl.[143] Nival Lepidoptera species are basically phytophagous and feed upon mosses, lichens and cushion plants. They certainly exhibit prolonged postembryonic development.[45,142] The life cycle of *Epichnopterix ardua* Mann, a grassland Psychidae in the Tauern range at 2000–2500 m, is 2 years, in its lowland congeners 1 year only.[144] Both faunal types, endemic and arctic-alpine species, are well represented in this group. Butterflies and moths therefore represent important objects for studies on zoogeography, ecology and evolutionary biology of high alpine biota.

3
Conclusions and discussion

Information on the nival fauna of the East Alps is still scanty, especially for the mesofauna groups. As these predominate in nival sites, a better knowledge of habitat preferences, species composition and distribution area of these groups will probably substantially improve our knowledge of the historic zoogeography of the High Alps. Of course, the species involved are only a small fraction of the total diversity present in mid-Europe. Apparently some stenotopic eunival species exist, together with others occurring also in the alpine zone at extreme sites with open vegetation, in ravines, and in bare scree at slopes exposed to north. Some euryzonal species occur even across a range of altitude of 3000 m! The main faunistic groups present in the epigeic macrofauna include species endemic to the Alps, species occurring also in the boreal tundra and in the high Arctic and others with a highly discontinuous range in high mountain biota. Not all species involved are distributed continuously within the region. Faunistic differences betwen eastern and western parts of the Central Alps, but also between the Central, Northern and Southern Calcareous Alps, probably correspond to differences in ecology and history of nival biota and to local speciation phenomena, but also to different reimmigration ways and migratory intensity of the respective species. Therefore, the distribution patterns of the nival fauna should be studied also on a large scale. Evolutionary relationships have not yet been investigated with molecular methods. Also the main problems of adaptation, ecophysiology and ecology of the nival fauna still remain unresolved.

4 References

1. Heer O. Ueber die obersten Grenzen des thierischen und pflanzlichen Lebens in den Schweizer-Alpen. Neujahrsblatt Naturf Ges Zürich 1845; 47:1-19, Fig. 1-16.
2. Handschin E. Ziele und Probleme der zoologischen Erforschung der Hochalpen. Revue Suisse Zool 1925; 32:65-71.
3. Janetschek H. Aktuelle Probleme der Hochgebirgsentomologie. Veröff Univ Innsbruck 92, Alpin Biol Stud 1974; 6:1-23.
4. Meyer E, Thaler K. Animal diversity at high altitudes in the Austrian Central Alps. In: Chapin FS, Körner C, eds. Arctic and Alpine Biodiversity. Berlin Heidelberg: Springer, 1995:97-108.
5. Bäbler E. Die wirbellose, terrestrische Fauna der nivalen Region. Revue Suisse Zool 1910; 18:761-916, Pl. 6.
6. Mani MS. Ecology and Biogeography of High Altitude Insects. The Hague: Junk, 1968.
7. Franz H. Ökologie der Hochgebirge. Stuttgart: Ulmer, 1979.
8. Heller C. Über die Verbreitung der Thierwelt im Tiroler Hochgebirge. Sitzb Akad Wiss Wien (I) 1881; 83:103-175.
9. Heller C, Dalla Torre C. Über die Verbreitung der Thierwelt im Tiroler Hochgebirge. II. Abt. Sitzb Akad Wiss Wien (I) 1882; 86:8-53.
10. Handschin E. Beiträge zur Kenntnis der wirbellosen terrestrischen Nivalfauna der schweizerischen Hochgebirge. Liestal: Lüdin & Co, 1919.
11. Steinböck O. Zur Lebensweise einiger Tiere des Ewigschneegebietes. Z Morph Ökol Tiere 1931; 20:707-718.
12. Steinböck O. Die Tierwelt des Ewigschneegebietes. Z Dt Österr Alpenverein 1931; 1931:29-46.
13. Steinböck O. Die Nunatak-Fauna der Venter Berge. In: Dt Alpenverein, Zweig Mark Brandenburg, ed. Das Venter Tal. München: Bruckmann, 1939:64-73.
14. Schmölzer K. Die Kleintierwelt der Nunatakker als Zeugen einer Eiszeitüberdauerung. Ein Beitrag zum Problem der Prä- und Interglazialrelikte auf alpinen Nunatakkern. Mitt Zool Mus Berlin 1962; 38:171-400.
15. Janetschek H. Nunataktiere – Ein aktuelles Problem der zoologischen Heimatforschung. Jahrb Vorarlberg Landesmuseumverein 1955; 98:75-80.
16. Janetschek H. Tierleben auf den höchsten Alpengipfeln. Der Schlern (Bozen) 1955; 29:170-174.
17. Janetschek H. Das Problem der inneralpinen Eiszeitüberdauerung durch Tiere (Ein Beitrag zur Geschichte der Nivalfauna). Österr Zool Z 1956; 6:421-506.
18. Janetschek H. Die Mikrophytenstufe (Ein neuer Begriff und ein Programm). Anz Math Naturw Kl Österr Akad Wiss 1964; 1964:185-191.
19. Janetschek H. Über Wirbellosen-Faunationen in Hochlagen der Zillertaler Alpen. Ber Nat Med Verein Innsbruck 1993; 80:121-165.
20. Franz H. Die Landtierwelt der mittleren Hohen Tauern. Ein Beitrag zur tiergeographischen und -soziologischen Erforschung der Alpen. Denkschr Akad Wiss Wien Math Naturw Kl 1943; 107:1-552, Taf. 1-14, Karte 1-11.
21. Franz H. Die Tiergesellschaften hochalpiner Lagen. Biol Gen 1950; 18:1-29.

22. Marcuzzi G. Fauna delle Dolomiti. Mem Ist Veneto Sc Lett Arti, Cl Sc Mat Nat 1956; 31:1-595.
23. Reisigl H, Pitschmann H. Obere Grenzen von Flora und Vegetation in der Nivalstufe der zentralen Ötztaler Alpen (Tirol). Vegetatio 1958; 8:93-129.
24. Moser W, Brzoska W, Zachhuber K, Larcher W. Ergebnisse des IBP-Projekts "Hoher Nebelkogel 3184m". Sitzb Österr Akad Wiss Math Naturw Kl I 1977; 186:387-419.
25. Franz H. Die Biogeographie und Ökologie der Alpen im Lichte ihrer jüngsten geologischen Geschichte. Sitzb Österr Akad Wiss Math Naturw Kl I 1991; 198:197-216.
26. Larcher W. Hochgebirge: An den Grenzen des Wachstums. In: Morawetz W, ed. Ökologische Grundwerte in Österreich. Wien: Österr Akad Wiss, 1994:304-343.
27. Steinböck O. Zur Ökologie der alpinen Turbellarien. Z Morph Ökol Tiere 1926; 5:424-446.
28. Reisinger E. Kärntens Hochgebirgsturbellarien. Carinthia II 1955; 65:112-151.
29. An der Lan H. Die ersten terricolen Turbellarien aus dem Ewigschneegebiet. Schlern Schr (Innsbruck) 1958; 188:161-166, Taf. 8.
30. Menzel R. Ueber die mikroskopische Landfauna der schweizerischen Hochalpen (mit spezieller Berücksichtigung des Rhätikon). Arch Naturg 1914; 80A:1-98, Fig. 1-4.
31. Heinis F. Über die Mikrofauna alpiner Polster- und Rosettenpflanzen. Festschrift Zschokke (Basel) 1920; 6:1-30.
32. Franz H. Untersuchungen über die Kleintierwelt ostalpiner Böden. I. Die freilebenden Erdnematoden. Zool Jahrb Syst 1942; 75:365-546, Taf 2.
33. Gerber K. Die Nematodenfauna alpiner Böden im Glocknergebiet (Hohe Tauern, Österreich). Veröff Österr MaB Hochgebirgsprogramm Hohe Tauern 1981; 4:79-90.
34. Gerber K. Einige Nematoden aus alpinen Böden in den Hohen Tauern (Österreich). Linzer Biol Beitr 1991; 23:661-700.
35. Sperling P. Zur Landschneckenfauna von Obergurgl (Ötztal, Nordtirol). Mitt Dt Malak Ges 1972; 2:378-393.
36. Kofler A. Zur Faunistik, Ökologie und Cönotik Osttiroler Landschnecken. Arch Moll 1965; 94:183-243.
37. Kofler A. Die natürlichen Landschneckencönosen im Großvenedigergebiet Osttirols. Ber Nat Med Verein Innsbruck 1967; 55:125-138.
38. Christandl-Peskoller H, Janetschek H. Zur Faunistik und Zoozönotik der südlichen Zillertaler Hochalpen. Mit besonderer Berücksichtigung der Makrofauna. Veröff Univ Innsbruck 101, Alpin Biol Stud 1976; 7:1-134.
39. Starmühlner F. Beiträge zur Kenntnis der Molluskenfauna des Arlberges. Österr Zool Z 1954; 4:587-632.
40. Frank C. Malakologisches aus dem Ostalpenraum. Linzer Biol Beitr 1992; 24:383-662.
41. Thaler K. Weitere Funde nivaler Spinnen (Aranei) in Nordtirol und Beifänge. Ber Nat Med Verein Innsbruck 1992; 79:153-159.
42. Baur B. Seasonal changes in clutch size, egg size and mode of oviposition in *Arianta arbustorum* L. (Gastropoda) from alpine populations. Zool Anz 1990; 225:253-265.
43. Zicsi A. Die Regenwürmer Österreichs (Oligochaeta: Lumbricidae) mit Bestimmungstabellen der Arten. Verh Zool Bot Ges Österreich 1994; 131:37-74.
44. Zicsi A. Regenwürmer des Großglocknergebietes. Veröff Österr MaB Hochgebirgsprogramm Hohe Tauern 1981; 4:91-94.
45. Janetschek H. Tierische Successionen auf hochalpinem Neuland. Schlern Schriften (Innsbruck) 1949; 67:1-215, Taf. 1-7.
46. Kübelböck G, Meyer E. Ökologische Untersuchungen an Wirbellosen des zentralalpinen Hochgebirges (Obergurgl, Tirol). 6. Abundanz und Biomasse der Oligochaeta (Lumbricidae, Enchytraeidae). Veröff Univ Innsbruck 130, Alpin Biol Stud 1981; 15:1-53.
47. Cuendet G. Les peuplements lombriciens des pelouses alpines du Munt La Schera (Parc national suisse). Revue Suisse Zool 1984; 91:217-228.
48. Schmidegg E. Die Enchytraeiden des Hochgebirges der Nordtiroler Kalkalpen. Ber Nat Med Verein Innsbruck 1938; 46:1-45.

49. Nurminen M. Enchytraeidae (Oligochaeta) from the Grossglockner region of the Austrian Alps. Ann Zool Fennici 1977; 14:224-227.
50. Tynen HJ. The geographical distribution of ice worms (Oligochaeta: Enchytraeidae). Canad J Zool 1970; 48:1363-1367.
51. Heinis F. Tardigraden der Schweiz. Zool Anz 1908; 32:633-638.
52. Ehrenberg CG. Bericht über die mikroskopischen Organismen auf den höchsten Gipfeln der europäischen Central-Alpen, und über das kleinste Leben der bayerischen Kalkalpen. In: Schlagintweit A, Schlagintweit H, eds. Neue Untersuchungen über die physicalische Geographie und die Geologie der Alpen. Leipig: Weigel, 1854:233-267.
53. Marcus E. Beiträge zur Tardigradensystematik. Zool Jahrb Syst 1930; 59:363-386.
54. Kofler A. Biographie und Bibliographie des Acarinologen und Tardigradologen Pfarrer Dr. Franz Mihelcic (1898–1977). Ber Nat Med Verein Innsbruck 1978; 65:213-224.
55. Mihelcic F 1959. Zwei neue Tardigraden aus der Gattung *Hypsibius* Thulin aus Osttirol (Österreich). Zool Anz 1959; 163:254-261.
56. Steinböck O. Über die Fauna der Kryokonitlöcher alpiner Gletscher. Der Schlern (Bozen) 1957; 31:65-70.
57. Janetschek H. Als Zoologe am Dach der Welt. Ber Nat Med Verein Innsbruck Suppl 1990; 6:1-120.
58. Kraus H. *Hypsibius* (*Hypsibius*) *klebelsbergi* Mihelcic 1959 (Tardigrada) aus dem Kryokonit des Rotmoosferners (Ötztaler Alpen). Innsbruck: PhD Thesis, 1977.
59. Dastych H. Redescription of the cryoconital tardigrade *Hypsibius klebelsbergi* Mihelcic, 1959, with notes on the microslide collection of the late Dr. F. Mihelcic (Tardigrada). Veröff Mus Ferdinandeum (Innsbruck) 1993; 73:5-12.
60. Thaler K. Arealformen in der nivalen Spinnenfauna der Ostalpen (Arachnida, Aranei). Zool Anz 1988; 220:233-244.
61. Thaler K. Neue Arachniden-Funde in der nivalen Stufe der Zentralalpen Nordtirols (Österreich) (Aranei, Opiliones, Pseudoscorpiones). Ber Nat Med Verein Innsbruck 1981; 68:99-105.
62. Puntscher S. Ökologische Untersuchungen an Wirbellosen des zentralalpinen Hochgebirges (Obergurgl, Tirol) 5. Verteilung und Jahresrhythmik von Spinnen. Veröff Univ Innsbruck 129, Alpin Biol Stud 1980; 14:1-106.
63. Thaler K, Knoflach B. Funde hochalpiner Spinnen in Tirol 1992-1996 und Beifänge (Araneae .. Coleoptera). Ber Nat Med Verein Innsbruck 1997; 84:159-170.
64. Thaler K. *Lepthyphantes severus* n.sp., eine Reliktart der Nördlichen Kalkalpen westlich des Inn (Österreich) (Arachnida: Aranei, Linyphiidae). Zool Anz 1990; 224:257-262.
65. Schaefer M. Experimentelle Untersuchungen zum Jahreszyklus und zur Überwinterung von Spinnen (Araneida). Zool Jb Syst 1976; 103:127-289.
66. Czermak B. Autökologie und Populationsdynamik hochalpiner Araneen unter besonderer Berücksichtigung von Verteilung, Individuendichte und Biomasse in Grasheidebiotopen. Veröff Österr MaB Hochgebirgsprogramm Hohe Tauern 1981; 4:101-151.
67. Stipperger H. Biologie und Verbreitung der Opilioniden Nordtirols. Arb Zool Inst Univ Innsbruck 1928; 3:17-79.
68. Thaler K. Streufunde nivaler Arthropoden in den mittleren Ostalpen. Ber Nat Med Verein Innsbruck 1989; 76:99-106.
69. Schmarda T. Beiträge zur Kenntnis der Pseudoskorpione von Tirol und Vorarlberg: Faunistik; taxonomische Charakterisierung, Aktivitätsdynamik. Innsbruck: Msc (Diplomarbeit), 1995.
70. Martens J. Spinnentiere, Arachnida: Weberknechte, Opiliones. Tierwelt Deutschlands 1978; 64:1-464.
71. Schütt K. *Drapetisca socialis* (Araneae: Linyphiidae): Web reduction - ethological and morphological adaptations. Europ J Entomol 1995; 92:553-563.
72. Bauer T. Beetles which use a setal trap to hunt springtails: The hunting strategy and apparatus of *Leistus* (Coleoptera, Carabidae). Pedobiologia 1985; 38:275-287.

73. Janetschek H. Zur Brutbiologie von *Neobisium jugorum* (L. Koch) (Arachnoidea, Pseudoscorpiones). Ann Naturhist Mus Wien 1948; 56:309-316.
74. Willmann C. Die hochalpine Milbenfauna der mittleren Hohen Tauern insbesondere des Großglockner-Gebietes (Acari). Bonner Zool Beitr 1951; 2:141-176.
75. Mihelcic F. Ein Beitrag zur Kenntnis der Milben der Südseite der Zillertaler Alpen. Veröff Mus Ferdinandeum (Innsbruck) 1971; 51:141-154.
76. Schmölzer K. Die hochalpinen Landmilben der östlichen Brennerberge (Acarina terrestria). Veröff Mus Ferdinandeum (Innsbruck) 1993; 73:47-67.
77. Irk V. Die terricolen Acari der Ötztaler und Stubaier Hochalpen. Veröff Mus Ferdinandeum (Innsbruck) 1941; 19:145-189.
78. Schatz H. Ökologische Untersuchungen an Wirbellosen des zentralalpinen Hochgebirges (Obergurgl, Tirol) 2. Phänologie und Zönotik von Oribatiden (Acari). Veröff Univ Innsbruck 117, Alpin Biol Stud 1979; 10:15-120.
79. Schweizer J. Die Landmilben des Schweizerischen Nationalparkes 4. Teil: Ihr Lebensraum, ihre Vergesellschaftung unter sich und ihre Lebensweise. Ergeb Wiss Unters Schweiz Nationalpark 1957; NF 6 (37):9-107.
80. Irk V. Drei neue Milbenarten aus dem Tiroler Hochgebirge. Zool Anz 1939; 128:217-223.
81. Schmölzer K. *Mesoteneriffia steinböcki* Irk (Acari, Trombidiformes), ein Relikttier der Zentralalpengipfel. Tiroler Heimatblätter 1952; 27:22-23.
82. Schmölzer K. Die Landasseln der Umgebung von Innsbruck und ihre tiergeographischen Beziehungen zu den Nachbarländern. Veröff Mus Ferdinandeum (Innsbruck) 1952; 30:61-76.
83. Friedel H. Ökologische und physiologische Untersuchungen an *Scutigerella immaculata* (Newp.). Z Morph Ökol Tiere 1928; 10:738-797, Taf. 16.
84. Verhoeff KW. Ueber Diplopoden 1 (21): Formen aus Tirol, Italien und Cypern. Arch Naturg 1902; 68A:175-198, Taf. 9.
85. Bigler W. Die Diplopodenfauna des schweizerischen Nationalparks. Ergebn Wiss Unters Schweiz Nationalpark 1929; 5:1-8, 1-87.
86. Kurnik I. Über Diplopoden aus Barberfallen im Bereich der Großglockner-Hochalpenstraße (Kärnten, Österreich). Carinthia II 1985; 175/95:271-283.
87. Janetschek H. Die Tierwelt. In: Ilg K, ed. Landes- und Volkskunde. Geschichte, Wirtschaft und Kunst Vorarlbergs. Innsbruck: Wagner, 1961:173-240, Taf. 11-12.
88. Verhoeff KW. Diplopoden der Germania zoogeographica im Lichte der Eiszeiten. Zoogeographica 1938; 3:494-547.
89. Pedroli-Christen A. Révision de quelques espèces de Diplopodes décrites par Faes, Rothenbühler et Verhoeff. I. *Trimerophorella*, *Janetschekella* (Chordeumatida) et *Polydesmus alticola* (Polydesmida). Revue Suisse Zool 1993; 100:961-969.
90. Pedroli-Christen A. Faunistique des Mille-Pattes de Suisse (Diplopoda). Documenta Faunistica Helvetiae 1993; 14:1-167, Annexe 1+2.
91. Meyer E. Altitude-related changes of life histories of Chordeumatida in the Central Alps (Tyrol, Austria). In: Minelli S, ed. Proceedings 7th Int Congress of Myriapodology. Leiden Köln: Brill, 1990:311-322.
92. Meyer E, Eisenbeis G. Water relations in millipedes from some alpine habitat types (Central Alps, Tyrol) (Diplopoda). Bijdr Dierkunde 1985; 55:131-142.
93. Gisin H. Protoures de la Suisse. Revue Suisse Zool 1945; 52:513-534.
94. Janetschek H. Protura (Beintastler). In: Helmcke A, Starck D, Wermuth H, eds. Handbuch der Zoologie IV (2. Auflage) 2/3. Berlin: De Gruyter & Co, 1970.
95. Janetschek H. Über Felsenspringer aus den Alpen und den Pyrenäen (Insecta: Microcoryphia). Ber Nat Med Verein Innsbruck 1970; 58:277-296.
96. Christian E. U.-Kl. Collembola (Springschwänze). Cat Faunae Austriae 1987; 12a.
97. Handschin E. Ueber die Collembolenfauna der Nivalstufe. Revue Suisse Zool 1919; 27:65-98, Taf. 1.

98. Lienhard C. Zur Kenntnis der Collembolen eines alpinen *Caricetum firmae* im Schweizerischen Nationalpark. Pedobiologia 1980; 20:369-386.
99. Haybach G. Zur Collembolenfauna der Pasterzenumrahmung im Glocknergebiet (Hohe Tauern). Verh Zool Bot Ges Wien 1972; 110/111:7-36.
100. Haybach G. Zur Collembolenfauna verschiedener Gebirgsstandorte in Österreich. Verh Zool Bot Ges Wien 1992; 129:159-214.
101. Topp W. Biozönotische Untersuchungen in einem Kar der östlichen Hohen Tauern. Carinthia II 1975; 165/85:275-284.
102. Christian E. Springschwänze (Collembolen) aus Vorarlberg. Jahrb Vorarlberg Landesmuseumverein (Bregenz) 1985; 1984:83-89.
103. Kopeszki H. Zur Biologie zweier hochalpiner Collembolen – *Isotomurus palliceps* (Uzel, 1891) und *Isotoma saltans* (Nicolet, 1841). Zool Jb Syst 1988; 115:405-439.
104. Schaller F. *Isotoma saltans* und *Cryptopygus antarcticus*, Lebenskünstler unter Extrembedingungen (Collembola: Isotomidae). Entomol Gener 1992; 17:161-167.
105. Zettel J, Zettel U. Adaptations to the alpine environment in *Isotomurus alticola* (Collembola: Isotomide) in the Swiss Alps. Acta Oecologica 1994; 15:93-104.
106. Janetschek H. Über mitteleuropäische Felsenspringer (Ins., Thysanura). Österr Zool Z 1954; 5:281-328.
107. Franz H, Weiss E. Zur Kenntnis der Ökologie der alpinen Verbreitungsgrenze heliophiler Arthropoden. Veröff Österr MaB-Hochgebirgsprogramm Hohe Tauern 1981; 4:237-246.
108. Thorens P, Nadig A. Atlas de distribution des Orthopteres de Suisse. Documenta Faunistica Helvetiae 1997; 16:1-236.
109. Nadig A. Die Verbreitung der Heuschrecken (Orthoptera: Saltatoria) auf einem Diagonalprofil durch die Alpen (Inntal-Maloja-Bregaglia-Lago di Como-Furche). Jahresbericht Naturf Ges Graubünden 1991; NF 106 (2):1-380.
110. Lienhard C. Die Psocopteren des Schweizerischen Nationalparks und seiner Umgebung (Insecta: Psocoptera). Ergebn Wiss Unters Schweiz Nationalpark 1977; 14:415-551.
111. Pelikan J. Vertical distribution of alpine Thysanoptera. Folia Entomol Hung 1996; 57 (Suppl.):121-125.
112. Heiss E, Josifov M. Vergleichende Untersuchung über Artenspektrum, Zoogeographie und Ökologie der Heteropteren-Fauna in Hochgebirgen Österreichs und Bulgariens. Ber Nat Med Verein Innsbruck 1990; 77:123-161.
113. Heiss E. Zur Heteropterenfauna Nordtirols (Insecta: Heteroptera) II. Aradoidea + Saldoidea. Ber Nat Med Verein Innsbruck 1972; 59:73-92.
114. Heiss E. Zur Heteropterenfauna Nordtirols (Insecta Heteroptera) III: Lygaeoidea. Veröff Mus Ferdinandeum (Innsbruck) 1973; 53:125-158.
115. Heiss E. Zur Heteropterenfauna Nordtirols (Insecta: Heteroptera) VI: Pentatomoidea. Veröff Mus Ferdinandeum (Innsbruck) 1977; 57:53-77.
116. Leising S. Über Zikaden des zentralalpinen Hochgebirges (Obergurgl, Tirol). Veröff Univ Innsbruck 107, Alpin Biol Stud 1977; 9:1-69.
117. Günthart H. Zikaden (Hom. Auchenorrhyncha) aus der alpinen Höhenstufe der Schweizer Zentralalpen. Mitt Schweiz Entomol Ges 1984; 57:129-130.
118. Balachowsky AS. Sur un *Heliococcus* Sulc (Coccoidea-Pseudococcini) nouveau vivant à 3600 m alt. dans les Alpes du Dauphiné et du Tyrol. Mitt Schweiz Entomol Ges 1953; 26:236-240.
119. Strathdee AT, Bale JS, Block WC, Webb NR, Hodkinson ID, Coulson SJ. Extreme adaptive life-cycle in a high arctic aphid, *Acyrthosiphon svalbardicum*. Ecol Entomol 1993; 18:254-258.
120. Strathdee AT, Bale JS. Life-cycle and morph production in the arctic aphid *Acyrthosiphon brevicorne*. Polar Biology 1996; 16:293-300.
121. Dethier M. Etude des communautés d'arthropodes d'une pelouse alpine au Parc national suisse. Mitt Schweiz Entomol Ges 1984; 57:317-334.
122. Kahlen M. Nachtrag zur Käferfauna Tirols. Veröff Mus Ferdinandeum (Innsbruck) 1987; Beilageband 3.

123. Besuchet C. Coleopteres des Alpes suisses atteignant ou depassant l'altitude de 3000 m. Bull Romand Entomol 1983; 1:167-176.
124. De Zordo I. Ökologische Untersuchungen an Wirbellosen des zentralalpinen Hochgebirges (Obergurgl, Tirol) 3. Lebenszyklen und Zönotik von Coleopteren. Veröff Univ Innsbruck 118, Alpin Biol Stud 1979; 11:1-131.
125. Thaler K. Fragmenta Faunistica Tirolensia – 6 (Arachnida .. Carabidae). Ber Nat Med Verein Innsbruck 1984; 71:97-118.
126. Holdhaus K. Die Spuren der Eiszeit in der Tierwelt Europas. Abh Zool Bot Ges Wien 1954; 18:1-493, Taf. 1-52.
127. Schedl W. Untersuchungen an Pflanzenwespen (Hymenoptera: Symphyta) in der subalpinen bis alpinen Stufe der zentralen Ötztaler Alpen (Tirol, Österreich). Veröff Univ Innsbruck 103, Alpin Biol Stud 1976; 8:1-85.
128. Schedl W. Über aculeate Hautflügler der zentralen Ötztaler Alpen (Tirol, Österreich) (Insecta: Hymenoptera). Ber Nat Med Verein Innsbruck 1982; 69:95-117.
129. Müller H. Alpenblumen, ihre Befruchtung durch Insekten und ihre Anpassungen an dieselben. Leipzig: Engelmann, 1881.
130. Bezzi M. Studi sulla ditterofauna nivale delle Alpi italiane. Mem Soc It Sc Nat Mus Civ St Nat Milano 1918; 9:1-164.
131. Dethier M, Haenni JP, Matthey W. Recherches sur les Diptères du Caricetum firmae au Parc National Suisse. Bull Soc Neuchateloise Sc Nat 1983; 106:29-54.
132. Meyer E. Ökologische Untersuchungen an Wirbellosen des zentralalpinen Hochgebirges (Obergurgl, Tirol) 4. Aktivitätsdichte, Abundanz und Biomasse der Makrofauna. Veröff Univ Innsbruck 125, Alpin Biol Stud 1980; 13:1-53.
133. Stockner J. Ökologische Untersuchungen an Wirbellosen des zentralalpinen Hochgebirges (Obergurgl, Tirol) 7. Flugaktivität und Flugrhythmik von Insekten oberhalb der Waldgrenze. Veröff Univ Innsbruck 134, Alpin Biol Stud 1982; 16:1-102.
134. Troger H, Janetschek H, Meyer E, Schatz W. Schlüpfabundanz von Insekten (Diptera, Coleoptera, Hymenoptera) im zentralalpinen Hochgebirge (Tirol: Ötztal). Entomol Gener 1994; 18:241-260.
135. Pont AC. Muscidae from above the treeline in the upper Ötz valley (Tyrol, Austria) (Insecta, Diptera). Ber Nat Med Verein Innsbruck 1995; 82:311-318.
136. Willassen E. A nival *Bryophaenocladius* Thienemann, 1934, with reduced wings (Insecta: Diptera: Chironomidae). Ann Naturhist Mus Wien 1996; 98B:507-512.
137. Mohrig W, Thaler K. Drei weitere flügelreduzierte Trauermücken (Diptera, Sciaridae) aus Österreich. Mitt Schweiz Entomol Ges 1982; 55:307-312.
138. Thaler K, Knoflach B, Meyer E. Fragmenta Faunistica Tirolensia – 10 (Arachnida .. Sciaridae). Ber Nat Med Verein Innsbruck 1993; 80:311-325.
139. Mendl H. Limoniiden und Cylindrotomiden aus Nordtirol (Österreich) (Insecta: Diptera, Limoniidae, Cylindrotomidae). Ber Nat Med Verein Innsbruck 1976; 63:269-285.
140. Huemer P, Tarmann G. Die Schmetterlinge Österreichs (Lepidoptera). Veröff Mus Ferdinandeum (Innsbruck) 1993; Beilageband 5.
141. Vorbrodt C. Schmetterlinge der Schneestufe schweizerischer Hochgebirge. Int Entomol Z 1921; 15:145-148, 1922; 15:161-164, 177-179, 201-206, 1922; 16:9-11, 29-31, 45-47, 61-63, 81-83, 97-99, 113-115, 129-131, 145-147, 161-162, 1923; 16:177-178.
142. Burmann K. Aus dem Schmetterlingsleben in den höchsten Vegetationsstufen unserer Alpen. Jahrb Österr Alpenverein 1961; 86:154-167.
143. Schedl W. Das Timmelsjoch (Ötztaler Alpen, Tirol) als Möglichkeit zur Überquerung der Alpen für Wanderfalter. Z Arbeitsgemeinschaft Österr Entomol 1970; 22:147-149.
144. Meyer E. Struktur und Dynamik einer Population von *Epichnopterix ardua* Mann (Lep.: Psychidae) in einem Krummseggenrasen der Hohen Tauern (Kärnten, Österreich). Zool Jb Syst 1983; 110:165-177.

High alpine streams: cold habitats for insect larvae

L. Füreder

University of Innsbruck, Institute of Zoology and Limnology, Technikerstrasse 25,
A-6020 Innsbruck, Austria; e-mail: leopold.fuereder@uibk.ac.at

1 Introduction

Streams are inherently important components of alpine landscapes because they are vital water and nutrient conduents that connect high elevation snow, ice, and meadows to natural and human resources at lower elevations. Greater attention has been recently given to alpine streams in response to increased recognition of the important role that these headwater streams play in major river systems throughout the world.

Alpine streams can show considerable differences in hydrogeomorphological, chemical and biological conditions, which depend greatly on the prevailing water sources. Temperature appears to be the key variable affecting the structure of flora and fauna in these systems. This chapter addresses the insects that inhabit headwater streams above the treeline, and their adaptations to the cold temperatures that contribute to their successful survival and reproduction in these freshwater ecosystems.

2 The alpine life zone

The alpine life zone is located above the upper altitudinal and latitudinal limits of the continuous forest (the timberline) and below the permanent snowline. The altitudinal boundaries of the alpine zone vary with latitude. In the tropics, one has to go very high (above 4000 m) to cross the treeline; however, the alpine zone in the polar regions can be near sea level.[1,2] The alpine life zone occupies approximately 5.7×10^6 km^2 worldwide, which is an area equal to 60% of Europe. Alpine vegetation covers 70% (4×10^6 km^2) of this area, with remaining lands consisting of mountain deserts, truly barren terrain, or glaciers (Fig. 1). The global latitudinal distribution of the alpine life zone is strongly asymmetric; the majority (82%) of the total alpine areas are found in the northern hemisphere, with 23% in arctic-subantarctic alpine areas, 32% in cool-temperate alpine, and 29% in warm-temperate alpine systems. Only 16% of total alpine land area is found in subtropical and tropical regions.

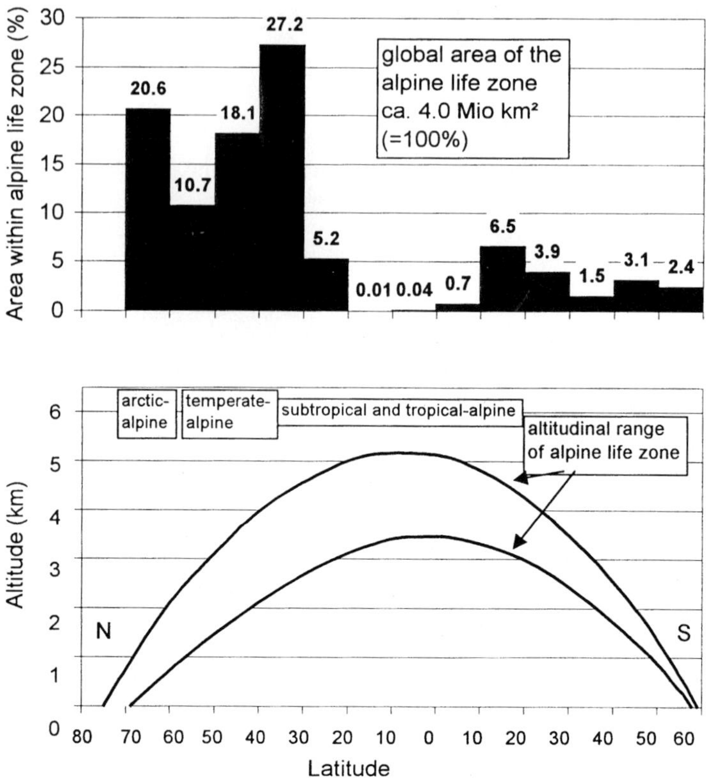

Fig. 1. An estimation of the latitudinal distribution of the alpine life zone
The **upper diagram** illustrates the relative contribution of each 10° latitudinal range to the global area covered by alpine vegetation, the **lower diagram** the mean lower and upper altitudes of the alpine life zone (after ref. 2)

Aquatic and terrestrial ecosystems at high elevations and high latitudes experience extremely harsh conditions such as daily and annual extremes in wind and low temperature, extended periods of snow cover, low humidity, and precipitation. Incident solar radiation during the relatively short snow-free season can be intense in high alpine areas of the temperate and tropical zone, but is low in polar regions. The combination of physiologically demanding conditions, a short growing season, and limited energy and nutrient resources make survival, development, and reproduction difficult. Year-round low temperature appears to be the single most important factor limiting biotic communities; it requires special adaptations for survival in winter and limits development and reproduction in summer. This has been the subject of numerous studies of terrestrial organisms.

Above the timberline, the vegetation changes with increasing altitude and latitude from low shrubs and meadows to cushion plants, mosses and lichens.[1,3,4] Like the flora, the fauna also becomes increasingly restricted as climate changes above the treeline. Species richness of arthropods declines drastically,[4,5] species composition is modified,[4,6] and total biomass of arthropods and other invertebrates is reduced.[7] In response to the decrease in plant and invertebrate production, densities of small mammals are lower than at lower elevations.[1] Also, mammals and birds have developed some adaptations to the environmental conditions in alpine regions (e.g. the generations of phytophagous mammals correspond to the short vegetation period; in rats, the number of offspring declines with increasing altitude).

Aquatic invertebrates generally face less severe conditions because water acts as an effective thermal buffer; snow or ice cover insulates. Together with current in streams it helps to retard ice formation. However, subsurface anchor and frazil ice may form in streams during extremely cold weather.[8] The margins of water bodies may also freeze into the substrate, depending on hydrology, autumn precipitation,[9,10] and insulation of snow and ice cover.[11] Subzero temperatures at the sediment/ice interface are reported from high latitudes, like -13 to -23°C[11] or -4°C.[12]

3
Alpine stream types

From their headwaters to the sea, streams and rivers are not simple, homogeneous habitats; rather, ecological conditions and the associated aquatic communities change as one progresses downstream. Climate (i.e. regional climate and also local weather conditions), magnitude and seasonality of rain fall, watershed geology, and riparian vegetation influence hydrology, temperature gradients, substrate composition, channel geomorphology, and chemistry define the local character of running waters. In alpine areas, however, stream character is most influenced by local water sources and the origin of major tributaries. This has led to the designation of three stream types that can be generally distinguished based on their origin and/or glacial influence: glacier-fed, spring-fed, and rainfall-/snowmelt-dominated streams.[13–22]

An additional factor that affects the character of alpine and arctic rivers is the absence of deciduous riparian vegetation that grows to sufficient height to shade small headwater streams. Alpine vegetation usually consists of a mixture of woody shrubs and grasses, which varies in composition depending on latitude and altitude. Because of the relative absence of plant cover, the amount of leaf litter entering streams is quantitatively small in arctic and alpine rivers. However, allochthonous material, even if small in quantity, still may play an important role in arctic and alpine streams.[1,23] Although solar radiation can be intense in alpine streams, low levels of autochthonous production, related to the long periods of snow cover, low temperatures, and intense early summer peak caused by snowmelt, might be common. Especially in glacier-fed streams, autochthonous production is low because seasonally very low temperatures, fluctuation in discharge accompanied with unstable substrate, and high concentrations of suspended solids provide for highly unfavorable conditions.

4
Zonation in alpine streams

Although lotic systems consist of a complex mosaic of changing habitats and aquatic communities, longitudinal zonation patterns have long been recognized in most stream and river systems.[24–26] These early studies were the basis of major theoretical contributions (e.g. Illies and Botosaneanu[27]) that led to the development of various stream and river typologies especially in Europe (e.g. Braukmann,[28] Moog and Wimmer[29]). Alpine streams were first assigned to the specific krenal and rhithral sections by Illies,[30] Illies and Botosaneanu[27] and Steffan.[31] Initially, the inhabitants of glacier brooks were assigned either to source or mountain-brook communities, although the Austrians An der Lan[32] and Steinböck[14,15] pointed out that glacial brooks differed in both their physiographical factors and their animal communities. Bretschko[33] identified specific differences with data from several glacial streams in the Tyrolean Alps. Finally, Steffan[34] considered glacier-fed streams to be sufficiently different to place them in a distinct freshwater habitat, which he called kryal (the corresponding community was called kryon). He further subdivided the kryal into eukryal (the glacier as a habitat), metakryal, and hypokryal (the two glacier-fed stream sections below the glacier) according to longitudinal faunal changes he observed (Fig. 2).

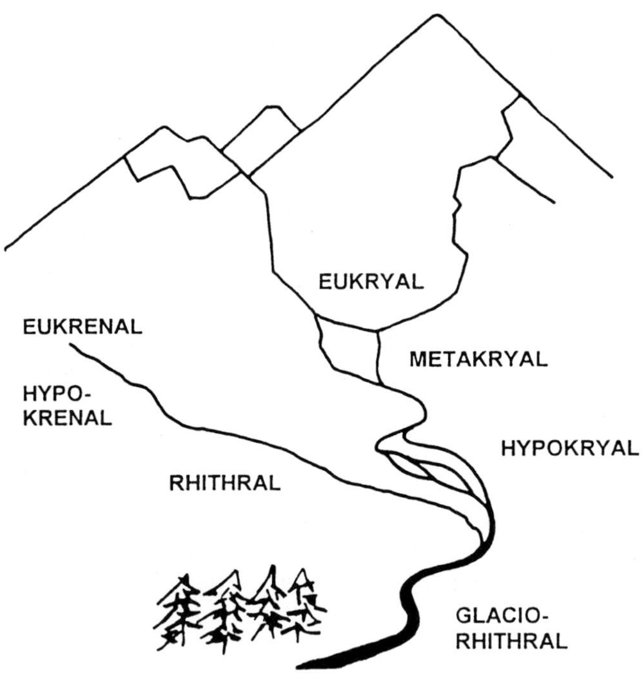

Fig. 2. Alpine stream types above the treeline and the permanent snowline (modified after refs. 30,34)

4.1
Glacier-fed rivers

At both high altitudes and high latitudes, rivers may be fed by meltwater from glaciers and permanent snowfields. The environmental conditions depend on the distance from the glacial mouth, the season, and also the availability of water from nonglacial sources. In general, year-round low temperatures, seasonally dependent, large diel fluctuations in discharge, suspended solids (glacial flour), and low channel stability characterize glacial streams and represent harsh environmental conditions for the benthic fauna. Several studies have described the faunal composition of glacier-fed streams (see reviews by Ward[19] and Milner and Petts[35]).

4.1.1
Eukryal

The eukryal can be characterized as a temporary aquatic habitat on or within the glacier. With increasing solar radiation and temperature in summer, meltwater and liquid snow layers develop and form a freshwater habitat for microbes, plants, and aquatic and terrestrial animals. Temperatures are continuously cold, with very little diel or seasonal variability (Fig. 3). Primary food resources consist of dust (allochthonous particles or airborne detritus) and algae (various species of cyanophytes and green algae) as well as fungi and bacteria associated with algae and detritus. Primary consumers can be species of protozoa, rotifers, tardigrades, and collembolans.[36–39] Despite the harsh temperatures, a highly restricted insect fauna has been reported from crevices in Tyrolean, Himalayan and Patagonean glaciers. This included larvae and imagines of different chironomid species (genus *Diamesa)*,[14,40] and even perlid and capniid stoneflies.[41]

4.1.2
Metakryal

Metakryal stream segments are adjacent to the glacier.[16,19] Water leaving the glacier mouth can be a combination of glacial-melt water, and water originating from nearby rainfall, snowmelt and groundwater sources.[13] Although the relative proportions depend on the specific conditions of the site (e.g. availability of rain, snowmelt, groundwater) and season, flow in summer is largely dominated by glacial ablation. Accordingly, continuously cold temperatures (barely above 0°C) with very little diel and seasonal fluctuations (1–2°C) prevail (Fig. 3).

Metakryal sections are described as detritus based (aerial origin) with very low primary production from a limited biofilm. Algal assemblages consist of a few diatom species at low densities,[42] some cyanophytes, and a filamentous, holarctic chrysophyte (*Hydrurus foetidus*) that is often the predominant source of nutrients and substrate for invertebrates near the glacier.[19] Metakryal segments are considered as one of the most extreme ecosystems for aquatic insects. Representatives of the chironomid genus *Diamesa* are often the dominant (sometimes the only) insect larvae of the metakryal. Other animals like plecopteran, ephemeropteran and other chironomid larvae can also occur as the relative importance of different water sources varies seasonally.[33]

4.1.3 Hypokryal

Hypokryal segments receive water from the metakryal with slightly greater influence from nonglacial sources. Diel and seasonal variation in temperature is greater than in the metakryal, but maximum temperature does not exceed 4°C (Fig. 3).[19,23] During maximum glacier ablation, high sediment loads, large diel fluctuations in discharge, and related increases and decreases in temperature contribute to environmental conditions that are still unfavorable for flora and fauna. Together with the species found, other taxa occur, like other Diptera larvae (e.g. Orthocladiinae and Simuliidae). The Diamesinae, however, may remain the dominant insect group.

4.1.4 Glacio-rhithral

Below the hypokryal, a glacio-rhithral zone continues that may experience runoff patterns generated by snowmelt and icemelt for a considerable distance, providing still for unstable substrate, high sediment loads, and lower summer temperatures. Temporal variations in discharge and temperature reflect the relative proportion of glacial influence in the rivers. As a result of high channel instability, recent and former glacial activities braided flood plains have often developed with complex channel patterns and diversified environmental conditions. Therefore, spatial and temporal differences within a short stream or river section might be very distinct. For example, Tockner et al.[43] suggest that increased spatial heterogeneity might ameliorate the negative effects of diel summer fluctuations, and habitats with the highest environmental stability (e.g. groundwater channels) exhibit highest species diversity.

4.2 Streams fed by springs, snowmelt, and rainfall

Many alpine streams are influenced by springs, snowmelt, and rainfall rather than by glaciers and permanent snowfields. Groundwater-dominated, nonglacial first-order streams (eukrenal-hypokrenal) show more physicochemical constancy. Water temperature is related to annual mean air temperature but might increase slightly with increasing solar radiation. Near the source, temperatures of rheocrenes typically vary only by 1–2°C, whereas the annual maximum for alpine rhithral habitats can approach 10°C (Fig. 3).[19] Because of more stable flow and substrate, clear water, winter-warm conditions, and the absence of ice action, eukrenal areas support of mosses, an abundance populations of diatoms,[44] and mats of the chrysophyte *Hydrurus foetidus*, which is also abundant in krenal, kryal and rhithral segments.[19,44] In recent investigations in alpine springs, Crema et al.[45] found consider-

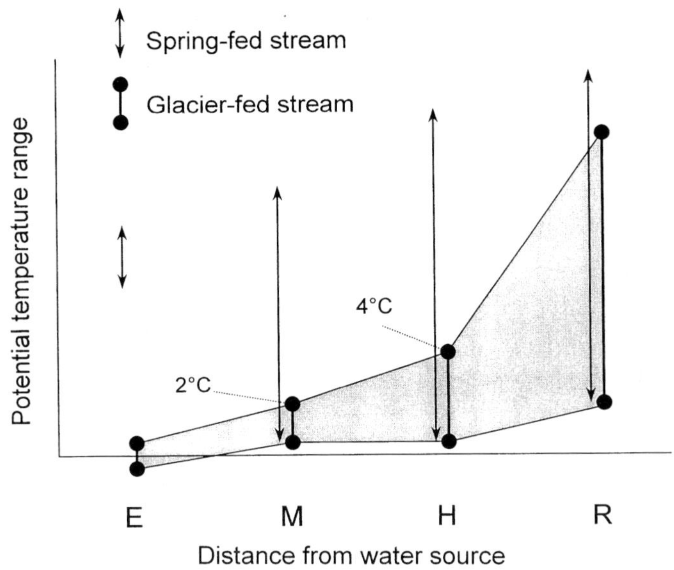

Fig. 3. Potential temperature ranges in a glacier-fed and a spring-fed stream, as two typical stream types in the alpine life zone
E epikryal, epikrenal; *M* metakryal, metakrenal; *H* hypokryal, hypokrenal; *R* rithral

able variation in the invertebrate fauna, with respect of both density as well as occurring taxa. Chironomid larvae were usually the dominant group, however, a diverse aquatic community including Ephemeroptera, Plecoptera, Trichoptera, Acari, Ostracoda, and Copepoda can develop.

4.2.1
Hypokrenal and rhithral

Further downstream (i.e. in the hypokrenal and rhithral zones) temperature gradients expand (Fig. 3), and the influence of seasonal differences in environmental conditions (rainfall, snowmelt, solar radiation, air temperature) become more distinct. The magnitude of the temperature difference experienced along a spring stream depends on distance from the source, surrounding landscape, orientation, and spring discharge. Increased spring discharge lowers the daily fluctuation downstream and extends the stabilizing influence of the spring.[46] Steep canyons reducing the effect of sunlight on water temperature might further extend spring influences,[47] whereas orientation that increases the effect of sunlight on water temperature presumably reduces the extent of spring influences.

In a review on springbrook studies, Mc Cabe[48] found longitudinal changes in the invertebrate communities in most studies examined. Both downstream increases and decreases were recorded. Severe reductions in faunal diversity, such as commonly observed in lowland springbrooks, are not always apparent in alpine springs,[19] however they have been recorded.[45] Alpine rheocrenes typically exhibit species richness values that are intermediate between those of kryal and rhithral zones. Zoobenthos densities generally exceed those in kryal and krenal habitats.[19]

5 Adaptations to cold temperatures in aquatic insects

The zonation pattern in alpine streams reflects the adaptation of the aquatic fauna. In general, there is a change in the species and a reduction of species numbers (different species and less numbers at higher altitudes and latitudes). Therefore many aquatic insects have not been able to adapt to the most extreme alpine conditions, but for those that endure these conditions, we have been able to identify some adaptations in the aquatic and terrestrial life stages. The zonation is, in part, a direct expression of the aquatic insects' ability to tolerate or adapt to the various thermal regimes in alpine streams. Still very little is known about either ecological or physiological adaptations to overwintering in aquatic invertebrates, especially those found in streams.[8,49]

In freshwater ecosystems, temperature is regarded as a key factor that contributes to the spatial and temporal distribution of benthic macroinvertebrates.[18,50,51] Temperature is believed to be the main cause for the longitudinal zonation of aquatic insects by affecting growth, metabolism, reproduction and emergence.[19,30] It influences bioenergetics directly through its effects on behavioral and physiological features (e.g. feeding rates, assimilation efficiency, egestion metabolism) and indirectly through the food resources available to the aquatic insects (effect on primary production). Temperature often interacts with food quality to define primary life history characteristics of aquatic insects.[52,53]

Most of the life stages of aquatic insects have had to adapt to the cold of alpine environments, and these adaptations can be morphological, behavioral, and involve the timing of the life cycle.[54] These adaptations are not usually tied just to low temperature but reflect a summation of the severe conditions in the alpine environment.

Generally, insects and other arthropods respond in different ways to low temperatures.[55] When terrestrial insects from temperate and tropical areas are exposed to temperatures between 0 and 10°C, they are immobilized and enter into a state of chill coma (which may be reversible if the exposure is not too long). Other insects are lethally injured by cold shock even though their body fluids do not freeze. In anticipation and/or response to these conditions, many terrestrial insects survive by entering a dormant (diapause) or quiescent stage,[56,57] or they maintain minimal activity.

Cold hardiness or cold tolerance refers to the capacity of an organism to survive exposure to low temperature. A variety of factors are known to influence this capacity, including the stage of development, genetic potential, season, duration of exposure, nutritional status, and the thermal history.[58] Cold adaptations have been documented for many taxa and by many investigators (for a review see Lee and Denlinger[58]), however was described mainly for terrestrial insects. In recent publications, Sømme[54] reviewed adaptations of terrestrial arthropods to the alpine environment and Strathdee and Bale[59] covered the issues of diversity, distribution, and adaptation of Arctic insects.

For most terrestrial insects and arthropods of the temperate zone, cold hardiness and diapause are both essential components of winter survival.[60] In contrast, few aquatic insects (temperate, tropical, or alpine) exhibit winter diapause or quiescence. Rather, most tend to be active and feeding (and presumably growing and developing) during much of the winter.[61–67] Some species remain active at temperatures as low as 0.1 or 0.2°C.[68,69]

Some species, particularly those inhabiting high-alpine and polar regions, are able to function at very low temperatures. For example, a number of polar and glacial species are mobile at temperatures below 0°C.[36–39,55] Koshima[40,41] reported different species of chironomid midges (genus *Diamesa*), a wingless stonefly, and springtails (genus *Isotoma*) were active even below -10°C in Himalayan and Patagonian glaciers.

5.1 Adaptations concerning the life cycle

During their life cycle most aquatic insects in alpine regions have to overcome conditions both in the terrestrial and in the aquatic environment, respectively (Fig. 4). The terrestrial stages (composed by mostly fragile flying forms of aquatic insects) are exposed to the most threatening conditions, such as heavy winds, extreme cold, and rapid temperature fluctuations. Warmer and less severe conditions on land are often limited to only 2 or 3 months per year. Aquatic environments tend to present fewer environmental extremes.[8,70] Consequently, the aquatic stage (eggs, larvae and pupae) of aquatic insects generally experience subzero temperatures less often than terrestrial insects or terrestrial life stages do in terrestrial environments; but they may also experience fewer days with temperature above 10°C.

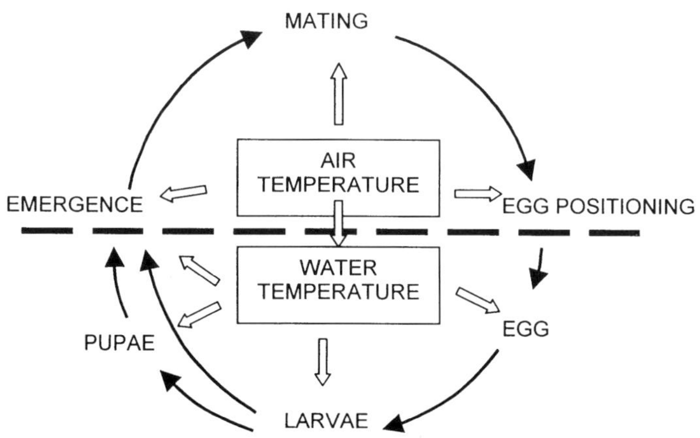

Fig. 4. Potential effects of air and water temperature on the life cycle of aquatic insects

5.1.1
Larval strategies

5.1.1.1
Growth and development

Insects in alpine streams must be able to grow and develop at temperatures that are lower than the minimum of most temperate aquatic insects (which usually are between 4 and 8°C[71,72]). Obviously, some *Diamesa* species are able to grow and successfully complete an entire life cycle at 0–2°C.[35] Even so, there is evidence that some life cycles are extended (due to low temperature and/or limited food availability) from 1 year to 2 or 4 years. Most aquatic and terrestrial arthropods have one generation per year in seasonal temperate areas, while tropical species can have two or more generations per year.[73] With increasing altitude and latitude, some species of terrestrial and aquatic insects need more than 1 year to complete their life cycle.[4,74,75] For example, Saether[76] reported a 2-year life cycle for a *Diamesa* species at the stations near a glacier. Growth and development in the same species was completed within 1 year at lower altitudes. Plecoptera from Norway had a 2-year life cycle,[77] and *Pachyleuctra benlocchi* from the Pyrenees required 4 years to complete its life cycle.[78] For mayfly species, a latitudinal trend in their life cycle as a response to the thermal regime of their habitat is distinct; populations are univoltine in the southern and middle part of their temperature range but semivoltine in more northern areas. Butler[79,80] reported multiyear life cycles (2–7 years) in all studied chironomid midges in an arctic Alaskan tundra pond.

As has been observed in terrestrial insects,[81] aquatic insects might mature at a smaller body size in response to alpine conditions. A smaller body may find shelter more easily in small microhabitats, and is advantageous because it requires less time and food for development.

Growth and development time can also be restricted to times with warmer temperature. Brittain[82] and Lehmkuhl[83] assumed that nymphal growth and development for two *Baetis* species occurred principally during the short ice-free season, because temperatures during the period of ice cover are too low for development.

5.1.1.2
Cold temperature avoidance

Avoiding freezing by moving away from the freezing front and remaining through the winter in a proportion of the stream that will not freeze are the strategies used by most aquatic insects in interior Alaskan streams. In freezing-chamber experiments Irons et al.[10] found nearly all active aquatic insects move away from advancing ice. Stream areas that do not freeze because of groundwater upwelling supported much higher densities of macroinvertebrates than nearby streams and rivers, which freeze completely. Productivity of aquatic macroinvertebrates in the Arctic may be limited by the availability of nonfreezing sites for overwintering.

Seasonal movements to habitats unlikely to freeze have also been reported from a Swedish lake[84] and from rivers.[85,86] The hyporheic zone that can serve as a refuge from a variety of environmental disturbances such as warm summer temperatures or floods[63,87] can also act as a refuge from an advancing freezing front.[10] Therefore vertical movement to deeper sediments provides a refuge from winter mortality for the majority of aquatic insect taxa in streams at high latitudes.

5.1.2
Terrestrial stage of aquatic insects

Specific modifications in adult insects that appear advantageous in the environmental conditions include increased melanism for thermoregulation, reduced adult life span, and reduced swarming and other flight activities. Increased pigmentation with increasing altitude and latitude has been observed in several insect groups,[4,81] enabling them to raise their body temperatures by heat absorption. The importance of melanism for thermoregulation is well documented for butterflies, blowflies, grasshoppers and caterpillars.[54] This is also true for some aquatic insects, (e.g. mosquitoes, chironomids, and other Diptera).[74,76,88] Their body color can be mostly black, very dark brown, or reddish. Increased pigmentation may also protect against the possible harmful effects of the intense ultraviolet radiation.[4,89,90]

Adults can limit exposure of the critical reproductive phase to the harsh alpine climate reducing adult flight and mating activities. Reduction of adult exposure reaches an extreme in the alpine Norwegian species *Prosimulium ursinum*, which does not emerge from the pupae,[91] but sheds its eggs into the stream and when the pupal exuvia disintegrates the following spring. Some species have parthenogenetic populations, further reducing exposure to the harsh environment associated with mating. Rather than swarming in the air, the adults aggregate and mate on the ground. Correlated with this modification of mating behavior, the eyes of the males are often reduced and their claspers enlarged, apparently to secure good contact during copulation. Similar morphological specializations are also found in arctic and alpine chironomids.[75] Saether[76] recognized reduction of antennal segments, antennal ratio, and male antennal plume, as well as enlargement of the hypopygium and legs as adaptations related to copulation on the ground by high altitude chironomids.

Downes[74] reported that adult females did not feed in eight of nine Canadian arctic blackfly; eggs are supplied from nutrients collected by the aquatic larval stages and start their development before the adult emerges from the pupal case.

Reduced swarming and other flight activities, which might be accompanied by wing reduction, appears to be another adaptation to cold temperature. The former prevents accidental displacement, while wing reduction allows more energy and nutrients to be dedicated to growth and reproduction.

Reduction of wings (brachyptery) or complete loss of wings (aptery) occurs in a number of insects.[92] Not only in social and parasitic species but also in insects from extreme environments, such as small islands, polar regions and mountains. Many arctic species have lost the ability to fly,[74,75] and the percentage of alpine species with wing atrophy increases with altitude.[4] The loss of flight would decrease the danger of being blown away to unsuitable surroundings and from potential mates by frequent strong winds in arctic and alpine areas.[81] Other reasons would be that low temperatures are often below the physiological threshold of flight, or that flightless Diptera often occupy habitats in litter and dense vegetation[93] and near the ground, which would enhance the selection for reduced wings.

6 Conclusions

The presence of relatively few species and absence of many common "temperate" taxa suggest the extreme conditions in alpine areas exclude many species. This is true for terrestrial ecosystems as well as for streams. While the stream fauna is greatly reduced, a common alpine assemblage has been repeatedly observed. The harshest alpine conditions (e.g. in glacier-fed streams) eliminate all but a few species. Less extreme locations (i.e. downstream and/or near a spring source) support a more diverse community. We see some evidence of specialization such as growth and development at low temperatures and modification of adult morphology and behavior. The combination of downstream variation in environmental conditions and various degrees of specialization and environmental tolerance result in a downstream zonation structure in aquatic communities that has been observed in alpine streams throughout the world.

Acknowledgements. I would like to thank Dr. John K. Jackson for his comments on earlier versions of this chapter. This work was supported, in part, by the Nationalpark Hohe Tauern.

7 References

1. Franz H. Ökologie der Hochgebirge. Stuttgart: Ulmer, 1979.
2. Körner C. Alpine plant diversity: A global survey and functional interpretation. In: Chapin FS, Körner C, eds. Arctic and Alpine Biodiversity: Pattern, Causes and Ecosystem Consequences. Berlin, Heidelberg: Springer, 1995:45-62.
3. Bliss LC, Heal OW, Moore JJ. Tundra ecosystems: Comparative analysis. Cambridge: Cambridge University Press, 1981.
4. Mani MS. Ecology and Biogeography of High Altitude Insects. The Hague: W. Junk, 1968.
5. Janetschek H, DeZordo I, Meyer E, Troger H, Schatz H. Altitude- and time-related changes in arthropod faunation (Central High Alps: Obergurgl-area, Tyrol). Proceedings XV Int Congress Entomol. Washington DC, 19-27 August 1976, 1977:185-207.
6. Mani MS. Fundamentals of High Altitude Biology, 2nd ed. New Delhi: Oxford and IBH Publ Co, 1990.
7. Franz H. Zusammenfassung der bodenzoologischen Ergebnisse. In: Franz H., ed. Veröffentlichungen des Österreichischen MaB-Hochgebirgsprogrammes Hohe Tauern 4. Innsbruck: Wagner, 1981:247-254.
8. Oswood MW, Miller LK, Irons JG. Overwintering of freshwater benthic macroinvertebrates. In: Lee RE, Denlinger DL, eds. Insects at Low Temperature. New York, London: Chapman and Hall, 1991:360-374.
9. Clifford HF. Limnological features of a northern brown-water stream, with special reference to the life histories of aquatic insects. Am Midl Nat 1969; 82:578-597.
10. Irons JG, Miller LK, Oswood MW. Ecological adaptations of aquatic macroinvertebrates to overwintering in interior Alaska subarctic streams. Can J Zool 1993; 71:98-108.

11. Andrews D, Rigler FH. The effect of an Arctic winter on benthic invertebrates in the littoral zone of Char lake, Northwest Territories. Can J Zool 1985; 63:2825-2834.
12. Olsson TI. Overwintering of benthic macroinvertebrates in ice and frozen sediment in a North Swedish stream. Holarctic Ecol 1981; 4:161-166.
13. An der Lan H. Hydrographische und hydrobiologische Beobachtungen im Liesenser Gletscherbachgebiet. I. Hydrographischer Teil. Veröff Mus Ferdinandeum (Innsbruck) 1936; Jg 1935:32-51.
14. Steinböck O. Die Tierwelt der Gletscherwässer. Jb Dt Österr Alpenverein 1934; 65:263-275.
15. Steinböck O. Arbeiten über die Limnologie der Hochgebirgsgewässer. Int Rev Ges Hydrobiol 1938; 37:467-509.
16. Steffan AW. Qualitative Unterschiede in Energiefluß, Nahrungskreislauf und Produktivität von Fließgewässer-Ökosystemen. Verh Ges Ökol, Saarbrücken, 1973:181-191.
17. Kownacka M, Kownacki A. Vertical distribution of zoocenosis in the streams of the Tatra, Caucasus and Balkan Mts. Verh Int Ver Limnol 1972; 18:742-750.
18. Ward JV. Aquatic Insect Ecology, vol. 1: Biology and Habitat. New York: Wiley, 1992.
19. Ward JV. Ecology of alpine streams. Freshwat Biol 1994; 32:277-294.
20. Gislason GM, Adalsteinsson H. Animal communities in Icelandic rivers in relation to catchment basins. Preliminary results from a study in Iceland. Nordic Association for Hydrology, Nordic Hydrological Conference, Akureyri, Iceland, 1996:1-15.
21. Oswood MW, Irons JG, Milner AM. River and stream ecosystems of Alaska. In: Cushing CE, Cummins KW, Minshall GW, eds. River and Stream Ecosystems. Amsterdam: Elsevier, 1995:9-32.
22. Petersen RC, Gislason GM, Vought LB-M. Rivers of the Nordic countries. In: Cushing CE, Cummins KW, Minshall GW, eds. River and Stream Ecosystems. Amsterdam: Elsevier, 1995:295-341.
23. Steffan AW. Qualitative Unterschiede in Energiefluß, Nahrungskreislauf und Produktivität von Fließgewässer-Ökosystemen. Verh Ges Ökol, Saarbrücken, 1973:181-191.
24. Steinmann P. Die Tierwelt der Gebirgsbäche, eine faunistisch-biologische Studie. Ann Biol Lacustre 1907, 2:30-164.
25. Dodds GS, Hisaw FL. Ecological studies on aquatic insects IV. Altitudinal range and zonation of mayflies, stoneflies and caddisflies. Ecology 1925; 6:380-390.
26. Brehm V, Ruttner F. Die Biozönosen der Lunzer Gewässer. Int Rev Ges Hydrobiol 1926; 16:330-392.
27. Illies J, Botosaneanu L. Problèmes et méthodes de la classification et de la zonation écologique des eaux courantes, considerée surtout du point de vue faunistique. Mitt Int Ver Limnol 1963; 12:1-57.
28. Braukmann U. Zoozönologische und saprobiologische Beiträge zu einer allgemeinen regionalen Bachtypologie. Arch Hydrobiol / Beihefte Ergebn Limnol 1987; 26:1-355.
29. Moog O, Wimmer R. Grundlagen zur typologischen Charakteristik österreichischer Fließgewässer. Wasser und Abwasser 1990; 34:55-211.
30. Illies J. Versuch einer allgemeinen biozönotischen Gliederung der Fließgewässer. Int Rev Ges Hydrobiol 1961; 46:205-213.
31. Steffan W. Zur Statik und Dynamik im Ökosystem der Fließgewässer und zu den Möglichkeiten ihrer Klassifizierung. In: Tüxen R, ed. Biosoziologie. Den Haag: Bericht über das Internationale Symposion in Stolzenau/Weser, 1965:65-110.
32. An der Lan H. Hydrographische und hydrologische Beobachtungen im Liesenser Gletscherbachgebiet. Veröff Mus Ferdinandeum (Innsbruck) 1936; 15:29-51.
33. Bretschko G. Zur Hydrobiologie zentralalpiner Gletscherabflüsse. Verh Dt Zool Ges 1969; Jg 1968:741-750.
34. Steffan AW. Chironomid (Diptera) biocoenoses in Scandinavian glacier brooks. Can Entomol 1971; 103:477-486.
35. Milner AM, Petts GE. Glacial rivers: physical habitat and ecology. Freshwat Biol 1994; 32:295-307.

36. Steinböck O. Der Gletscherfloh. Z Dt Alpenver 1939; 70:138-147.
37. An der Lan H. Tiere im Ewigschneegebiet. Die Umschau in Wissenschaft und Technik 1963; 13:49-52
38. Schaller F. Beobachtungen am Gletscherfloh *Isotoma saltans* (Nicolet, 1841). Zool Anz Suppl Verh Dtsch Zool Ges Wien 1963; 26(1962):679-682.
39. Eisenbeis G, Meyer E. Some ultrastructural features of glacier Collembola, *Isotoma* "sp.G" and *Isotomurus palliceps* (Uzel, 1891) from the Tyrolean Central Alps. In: Dallai R, ed. 2nd Int Seminar on Apterygota, University of Siena, Italy, 1986:257-272.
40. Kohshima S. A novel cold-tolerant insect found in a Himalayan glacier. Nature 1984; 310:225-227.
41. Kohshima S. Patagonian glaciers as insect habitats. In: Nakajima C, ed. Glaciological Studies in Patagonia Northern Icefield. Data Center for Glacier Research, Japanese Society of Snow and Ice, Nogoya University, 1985:94-99.
42. Kawecka B. Sessile algae in European mountain streams. I. The ecological characteristics of communities. Acta Hydrobiologii 1980; 22:371-420.
43. Tockner K, Malard F, Burgherr P, Robinson CT, Uehlinger U, Zah R, Ward JV. Physico-chemical characterization of channel types in a glacial floodplain ecosystem (Val Roseg, Switzerland). Arch Hydrobiol 1997; 140:433-463.
44. Kawecka B. Vertical distribution of algae communities in Maljovica stream (Rila-Bulgaria). Pol Arch Hydrobiol 1974; 21:211-228.
45. Crema S, Ferrarese U, Golo D, Modena P, Sambugar B, Gerecke R. Ricerche sulla fauna betonica e interstiziale di ambienti sorgentizi in area alpina e prealpina. A Research on Benthic and Interstitial Fauna in Alpine and Pre-Alpine Springs. Trento: Centro di Ecologia Alpina, Report no 8, 1996.
46. Smith K, Lavis ME. Environmental influences on the temperature of a small upland stream. Oikos 1975; 26:228-236.
47. Cushing CE. The ecology of cold desert spring-streams. Arch Hydrobiol 1996; 135:499-522.
48. Mc Cabe DJ. Biological communities in springbrooks. In: Botosaneanu L, ed. Studies in crenobiology. The Biology of Springs and Springbrooks. Leiden: Backhuys Publ, 1998:221-228.
49. Moore MV, Lee RE. Surviving the big chill: overwintering strategies of aquatic and terrestrial insects. Am Entomol Summer 1991:111-118.
50. Vannote RL, Sweeney BE. Geographic analysis of thermal equilibria: a conceptual model for evaluating the effect of natural and modified thermal regimes on aquatic insect communities. Am Naturalist 1980; 115:667-695.
51. Ward JV, Stanford JA. Thermal responses in the evolutionary ecology of aquatic insects. Ann Rev Entomol 1982; 27:97-117.
52. Sweeney BW. Factors influencing life history patterns of aquatic insects. In: Resh VH, Rosenberg D, eds. Ecology of Aquatic Insects. New York: Praeger Scientific, 1984:56-100.
53. Sweeney BW, Vannote RL. Growth and production of a stream stonefly: Influences of diet and temperature. Ecology 1986; 67:1396-1410.
54. Sømme L. Adaptations to the alpine environment in insects and other terrestrial arthropods. In: Wielgolaski FE, ed. Ecosystems of the World 3. Polar and alpine tundra. Amsterdam: Elsevier, 1997:11-25.
55. Lee RE. Principles of insect low temperature tolerance. In: Lee RE, Denlinger DL. Insects at Low Temperature. New York: Chapman and Hall, 1991:17-46.
56. Tauber MJ, Tauber CA, Masaki S. Seasonal Adaptations of Insects. New York: Oxford Univ Press, 1986.
57. Danks HV. Insect Dormancy: An Ecological Perspective. Biological Survey of Canada, Ottawa, Ontario, 1987.
58. Lee RE, Denlinger DL. Insects at Low Temperature. New York, London: Chapman and Hall, 1991.

59. Strathdee AT, Bale JS. Life on the edge: Insect ecology in Arctic environments. Annu Rev Entomol 1998; 43:85-106.
60. Denlinger DL. Relationship between cold hardiness and diapause. In: Lee RE, Denlinger DL, eds. Insects at Low Temperature. New York, London: Chapman and Hall, 1991:174-198.
61. Harper PP, Hynes HBN. Life histories of Capniidae and Taeniopterygidae (Plecoptera) in southern Ontario. Arch Hydrobiol Suppl 1972; 40:274-314.
62. Lewis T, Bennett GF. The blackflies (Diptera: Simuliidae) of insular Newfoundland. II. Seasonal succession and abundance in a complex of small streams on the Avalon Peninsula. Can J Zool 1974; 52:1107-1113.
63. Williams DD, Hynes HBN. The occurrence of benthos deep in the substratum of a stream. Freshw Biol 1974; 4:233-256.
64. Corkum LD. The nymphal development of *Paraleptophlebia adoptiva* (McDunnough) and *Paraleptophlebia mollis* (Eaton) (Ephemeroptera: Leptophlebiidae) and the possible influence of temperature. Can J Zool 1978; 56:1842-1846.
65. Colbo MH. Distribution of winter-developing Simuliidae (Diptera) in Newfoundland. Can J Zool 1979; 57:2143-2152.
66. Brittain JE. Mayfly strategies in a Norwegian subalpine lake. In: Flannigan JF, Marshall KE, eds. Advances in Ephemeroptera Biology. New York: Plenum Press, 1980:179-186.
67. Bengtsson BE. The growth of some Ephemeropteran nymphs during winter in a north Swedish river. Aquatic Insects 1981; 3:199-208.
68. Brittain JE, Nagell B. Overwintering at low oxygen concentrations in the mayfly *Leptophlebia vespertina*. Oikos 1981; 36:45-50.
69. Solem JO. Identification of the Norwegian larvae of the genus *Potamophylax* Wallengren, 1891 (Trichoptera, Limnephilidae), with data on life histories, habitats ad food in the Kongsvoll area, Dovrefjell mountains, central Norway. Fauna Norv 1983; 30:69-76.
70. Frisbie MP, Lee RE. Inoculative freezing and the problem of winter survival for freshwater macroinvertebrates. J N Am Benthol Soc 1997; 16:635-650.
71. Gibberson DJ, Rosenberg DM. Effects of temperature, food quantity, and nymphal rearing density on life-history traits of a northern population of *Hexagenia* (Ephemeroptera: Ephemeridae). J N Am Benthol Soc 1992; 11:181-193.
72. Newbold JD, Sweeney BW, Vannote RL. A model for seasonal synchrony in stream mayflies. J N Am Benthol Soc 1994; 13:3-18.
73. Jackson JK, Sweeney BW. Egg and larval development times for 35 species of tropical stream insects from Costa Rica. J N Am Benthol Soc 1995; 14:115-130.
74. Downes JA. Adaptations of insects in the Arctic. Annu Rev Entomol 1965; 10:257-274.
75. Danks HV. Arctic Arthropods. Entomol Soc Canada, Ottawa, 1981.
76. Saether OA. Chironomids of the Finse area, Norway, with special reference to their distribution in a glacier brook. Arch Hydrobiol 1968; 64:426-483.
77. Brittain JE. Semivoltinism in mountain populations of *Nemurella pictetii* (Plecoptera). Oikos 1978; 30:1-6.
78. Lavandier P. Cycle vital de *Pachyleuctra benlocchi* (Navas) dans un ruisseau d'altitude des Pyrenees centrales (Plecoptera). Ann Limnol 1975; 12:1-5.
79. Butler MG. Emergence phenologies of some arctic Alaskan Chironomidae. In: Murray DA, ed. Chironomidae. Ecology, Systematics, Cytology and Physiology. Oxford, New York: Pergamon Press, 1980:307-314.
80. Butler MG. A 7-year life cycle for two *Chironomus* species in arctic Alaskan tundra ponds (Diptera: Chironomidae). Can J Zool 1982; 60:58-70.
81. Sømme L, Block W. Adaptations to alpine and polar environments in insects and other terrestrial arthropods. In: Lee RE, Denlinger DL. Insects at Low Temperature. New York: Chapman and Hall, 1991:318-359.
82. Brittain JE. Studies on the lentic Ephemeroptera and Plecoptera of southern Norway. Norsk ent Tidsskr 1974; 21:135-153.

83. Lehmkuhl DM. A new species of *Baetis* (Ephemeroptera) from ponds in the Canadian arctic, with biological notes. Can Entomol 1973; 105:343-346.
84. Danell K. Overwintering of invertebrates in a shallow northern Swedish lake. Int Rev Ges Hydrobiol 1981; 66:837-845.
85. Tozer W, Resh VH, Solem JO. Bionomics and adult behavior of a lentic caddisfly, *Nectopsyche albida* (Walker). Am Midl Nat 1981; 106:133-144.
86. Olsson TI. Seasonal variation in the lateral distribution of mayfly nymphs in a boreal river. Holarct Ecol 1983; 6:333-339.
87. Williams DD. The hyporheic zone as a habitat for aquatic insects and associated arthropods. In: Resh VH, Rosenberg DM, eds. The Ecology of Aquatic Insects. New York: Praeger Publ, 1984:430-455.
88. Wirth WW, Gressitt JL. Diptera: Chironomidae (midges). Antarct Res Ser 1967; 10:197-203.
89. Hebert PDN, Emery CJ. The adaptive significance of cuticular pigmentation in Daphnia. Funct Ecol 1990; 4:703-710.
90. Hessen DO. Competitive trade-off strategies in Arctic *Daphnia* linked to melanism and UV-B stress. Polar Biol 1996; 16:573-579.
91. Carlsson G. Studies on Scandinavian black flies. Opusc Entomol Suppl 1962; 21:1-280.
92. Roff DA. The evolution of flightlessness in insects. Ecol Monographs 1990; 60:389-421.
93. Hackmann W. On reduction and loss of wings in Diptera. Notulae Entomologicae 1964; 44:73-93.

Ecophysiological and morphological features of glacier-dwelling Collembola

G. Eisenbeis[1] and E. Meyer[2*]

[1] University of Mainz, Institute of Zoology, D-55099 Mainz, Germany
[2] University of Innsbruck, Institute of Zoology, Technikerstrasse 25, A-6020 Innsbruck, Austria

1 Introduction

Collembola predominantly live in soil, and on the soil surface and its neighboring strata. Some species also dwell in extreme habitats in the arctic and antarctic regions and in the high alpine zones.[1–8] Collembola constitute 14–22% of all arctic insect species,[9,10] and in tundra and northern forest regions with snow and frost prevailing for the best part of the year they constitute an essential part of the winter-active fauna.[11–15] Snow Collembola are reported to occur in masses also in the temperate climate zones of central Europe with species like *Hypogastrura socialis*, *Isotomurus palustris* und *Isotoma hiemalis* frequently among them.[16,17] According to Zettel,[18] the surface activity of *I. hiemalis* probably depends on barometric pressure changes.

Generally, activity on ice and snow is mainly based on physiological adaptation mechanisms enabling the springtails to be active at temperatures around the freezing point and to survive periods of the severest frost without freezing damage.[19] Collembola have not yet been reported to morphologically adapt in the sense of cold adaptation. Therefore we have to assume that the hydrophobic cuticular micrograins characteristic of all Collembola are helpful in coping with life in the cold, as they minimize the contact with the cold substrates and counteract in inoculative freezing during spells of severe frost.

Among all Collembola, the glacier-dwelling ones are special. Their preferred winter habitat is the snow layer resting immediately on the ice, also referred to as floating snow layer which they rarely abandon.[20–26] In the case of melting water, for example, penetrating their immediate habitat, they migrate to higher snow and firn layers, the surfaces of which they often populate in masses in early summer (black and red glacier snow, respectively). On glaciers free of snow the glacier-dwelling

* Corresponding author; e-mail: erwin.meyer@uibk.ac.at

Collembola mainly frequent the interstice of the ice layer close to the surface down to a depth of approximately 30 cm. There they live aggregated in nests and appear only occasionally on the surface of the ice and underneath boulders and rock debris at temperatures around 0°C. By now several species of Collembola are known to share the glacier habitat and its marginal areas. Whereas the central zones are preferred by the true ice-dwelling species, the marginal zones are dominated by firn- and debris-dwelling ones. The best known species is *Isotoma saltans*.[20–23,25,26]

2
Survival strategies of Collembola living in permanent cold

Terrestrial arthropods adapted to cold climates have developed two major strategies of cold adaptation, i.e. freeze tolerance and freeze avoidance.[19,29,30] The first of these allows survival by freezing and cooling down to extreme temperatures as low as -100°C,[19] the latter avoids freezing by adopting the so-called hypothermia behavior, referred to as supercooling. In the course of this, freezing is prevented down to temperatures from approximately -25 to -45°C. In the case of the temperature dropping even further below the maximum supercooling point (SCP), the free body fluid freezes and death occurs immediately. Important in this context are the speed of cooling as well as the opportunity of acclimatizing to temperature conditions as long as they are moderate. Collembola are able to endure temperatures as low as approximately -30°C, in extreme cases even around -45°C without freezing,[5,19] but this does not mean they are still active at these temperatures. They remain in a chill coma, awaking at temperatures around 0°C, when they reactivate their metabolism and become mobile. In experiments with *I. saltans*, Steinböck[20] noticed that specimens capable of extreme cold adaptation fall into chill coma only at temperatures as low as -16°C, whereas less adapted ones already become inactive at -6°C. This property distinguishes the glacial Collembola clearly from other cold-adapted Collembola species already becoming inactive around 0°C. Even the best-investigated antarctic Collembolan, *Cryptopygus antarcticus*, only achieves a maximum chill coma temperature of approximately -6.5°C.[2]

The supercooling ability of Collembola usually sets in with the beginning of the cold period in autumn and stops again in spring. In this process, emptying of the gut is generally induced, in this way limiting the probability of inoculative freezing due to nucleators. At the same time, the content of antifreeze agents like polyols (glycerol, mannitol, sorbitol), sugar alcohols (erythriol), sugars (glucose, mannose, trehalose) and amino acids in the hemolymph is increased, lowering the freezing point.[5,19,29,31] These substances increase the osmotic value of the hemolymph and bind water by their colligative properties, which results in a drop in the freezing point. A disadvantage of this is, however, that, mainly because of the accumulation of glycerol, the viscosity of the hemolymph is increased, which has negative effects on the endogenous flow of materials. Therefore generally no glycerol is found in the hemolymph of Isotomidae.[32] The predominant antifreeze agents are glucose und trehalose, both serving at the same time as readily available energy storage.

As a further possibility of lowering the freezing point, Collembola demonstrate the phenomenon of thermohysteresis.[33] In fact, Zettel[34] demonstrated with subalpine and alpine species that during the winter the freezing point was lower than the melting point. This difference (>2°C) was most distinct in *Isotoma* specimens collected from the glacier. This phenomenon could not be observed with lowland species like *Orchesella villosa*. For this, temperature depression thermal hysteresis proteins (THPs) or antifreeze proteins (AFPs) are believed to be responsible, inhibiting further crystal growth by masking ice crystals. This kind of molecular reaction is called noncollegative freeze inhibition. The result of this is improved protection against inoculative freezing. Furthermore, a recrystallization after a short warming period can be avoided. These proteins occur in comparatively low concentrations, which leaves the viscosity of the hemolymph unaffected. Concerning metabolism, Convey[7] regards the ability to produce glycerol as a physiologically primitive phenomenon, whereas the production of THPs is to be considered as derived. The supercooling ability is usually lost abruptly at the beginning of spring. Specimens transferred to the laboratory and reared under warm conditions reduced their content of cryoprotective agents within 24 h.

Experiments with Collembola reared under cold conditions as well as investigations of field populations showed that two supercooling levels exist.[29,35,36] This frequently results in a bimodal distribution of a so-called high group (individuals with comparatively high SCPs) and a low group (individuals with significantly lower SCPs). The specimens of the high group are distinguished mainly for having their guts filled. This seems to indicate that the gut content may be considered to be the source for endogenous freezing, and that improvement of the cold adaptation depends on the removal of the gut nucleators. On the other hand, it may be regarded as an advantage that a population always has some members active and ready to feed, so that it can react rapidly to an improvement in climatic conditions and is better able to utilize the available resources. This is confirmed by Aitchison,[12] who observed that winter-active Collembola can feed selectively at temperatures around -5°C.

Various authors consider the reduction of water content as a further improvement in cold adaptation, yet investigations have not always delivered definite results.[7,37,38] Species with an improved dry resistance usually also demonstrate improved supercooling behavior. Acari improve their cold resistance by reducing their water content in winter to a higher degree than Collembola, and generally have higher glycerol contents available. A further form of adaptation is noticed with Collembola demonstrating extremely high cold stenothermal properties with a comparatively high metabolism at temperatures around the freezing point of the water. As early as 1963 Schaller and Zinkler[39] showed that the glacier springtail *I. saltans* has a comparatively high respiration rate at temperatures close to zero (Q_{10} 1.8). Bale et al.[8] reported that between 0 and 10°C the antarctic Collembolan *Onychiurus antarcticus* shows an even higher rate (Q_{10} 7.0).[40,41] This not only enables the animals to be active at low temperatures but also to react rapidly to the slightest temperature improvement with increased metbolism. Convey[7] describes an increased standard metabolism of antarctic arthropods as compared to species in temperate regions. The fact that particularly low enzymatic activation energies were observed with antarctic mites and insects points to an involvement of metabolic capabilities with cold-stenothermal arthropods.[7,42,43] Finally, Zinkler and Schaller[44] measured increased contents of unsaturated fatty acids occurring with the glacier Collembolan *I. saltans*, which is believed to improve the fluidity especially in the membranes.

In the following list, the main properties contributing to cold adaptation in Collembola are summarized:

- ability to molt at low temperatures;
- high metabolism at temperatures around 0°C;
- increased membrane fluidity due to increased uptake of unsaturated lipids (possibly also low enzymatic activation energies;
- supercooling ability (approximately -25 to -40°C);
- thermal hysteresis by lowering the freezing point against the melting point with thermal hysteresis proteins (THPs or AFPs), avoidance of recrystallization at rising temperatures, and avoidance of increased viscosity in hemolymph with antifreeze agents;
- emptying of the gut to avoid inoculative freezing;
- heterogenous populations concerning cold adaptation by a bimodal distribution of the SCPs: a high group with generally filled guts and high SCPs and a low group with empty guts and low SCPs; this offers more possibilities to react to temperatures changes;
- lowering of water content during cold periods.

3
Glacier-dwelling Collembola found in the Central Alps in Austria

Below, we focus on Collembola found on glaciers and permanent firn areas, one of which is the well-known *I. saltans*, also referred to as glacier flea.[20–22,25–28] By now, several species of Collembola are known to dwell on and in the margins of glaciers. During field work carried out in 1983 and 1985, we collected on the Rotmoos- and Gaisbergferner close to Obergurgl (Ötztal Alps) at least two ice-dwelling springtails, *Isotoma* 'sp. G.', a deep black Isotomidae [a scanning electron microscopic (SEM) description is given in Eisenbeis and Meyer[27] and Eisenbeis and Wichard[45]] and *Onychiurus* 'sp. G', a salmon-red Onychiuridae. A further species (*Isotomurus palliceps*), the firn flea, was common to the marginal firn fields, and another (*Entomobrya* sp.) was living under rock debris lying on the ice. The term 'sp. G' (species glacier) goes back to Zettel,[34] who, when investigating the glacier Collembola of the genus *Isotoma* on the Oberaar glacier in Switzerland, could not clearly identify them as *I. saltans*. So far the taxonomic position of the *Isotoma* species found in the Ötztal Alps is not clear either. We found them to dwell in the same microhabitat as described by Zettel, i.e. in permanent contact with ice. Therefore no significant differences to *I. saltans* are to be expected. Kopeszki[26] gave a detailed description of the biology of *I. saltans* collected on the Mittelberg glacier (Ötztal Alps), and of *I. palliceps* found in the same region. Both differ significantly concerning their habitat and their annual cycle. Kopeszki classified the firnflea as a summer species, where-

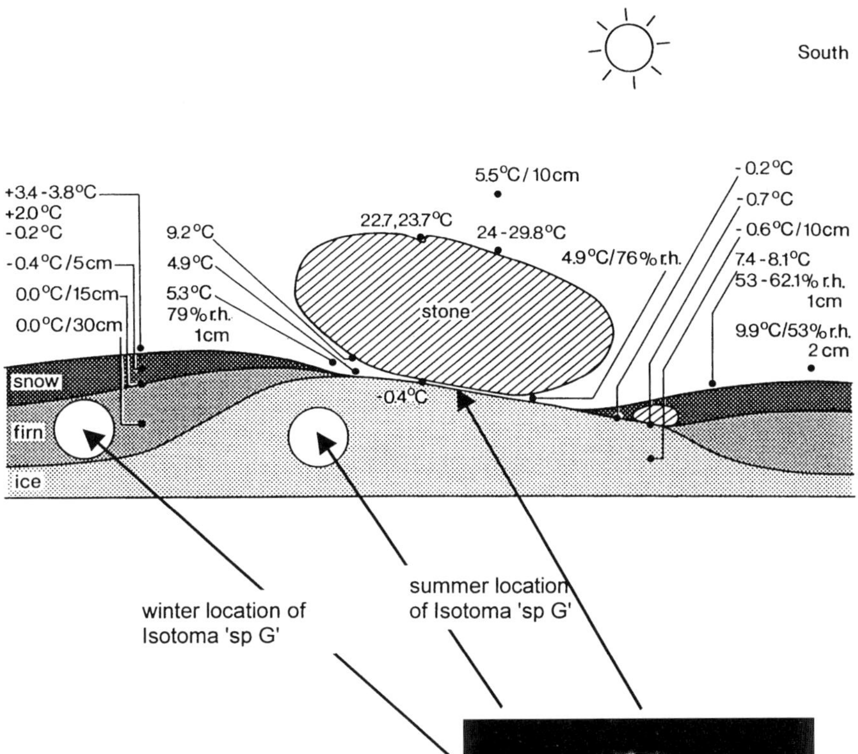

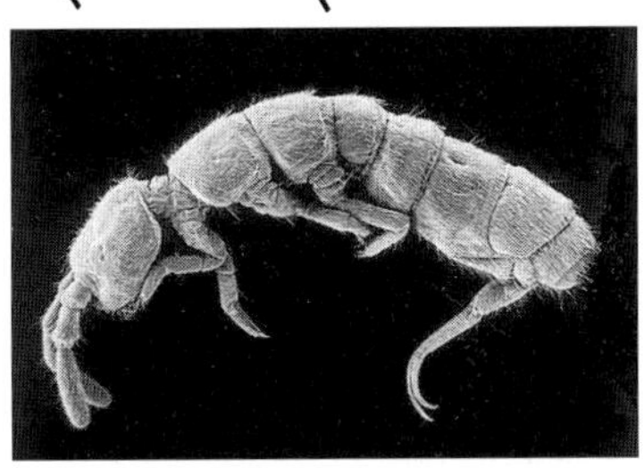

Fig. 1. Schematic view of microhabitats of *Isotoma* 'sp. G' at the Rotmoos glacier (2500 m a.s.l., Ötztal Alps). Temperature and relative humidity were measured during the early afternoon of a clear summer's day (8th August 1983)
SEM picture: *Isotoma* 'sp.G.' in lateral view

as he referred to the glacier flea as a winter species. The main period of activity and reproduction of the glacier flea occurs during the winter months, when the animals prefer to remain underneath the snow cover in the so-called floating snow layer resting immediately on the ice. There they feed and reproduce. In spring and early summer penetrating melt water forces the animals to move up to the snow surface and a mass occurrence of glacier fleas can frequently be observed. During the summer months, when glaciers are free of snow, the glacier fleas retreat into the interstice of the top layer of the ice down to a depth of about 30 cm. However, they can frequently be found under rocks in close contact with the ice.[34] In contrast, the firn flea *I. palliceps* is found in the marginal areas of the glacier on the firn surface preferably under rocks. Yet according to Kopeszki,[26] *I. palliceps* favors areas free of snow and moss cushions, where it feeds and reproduces.

3.1
Microclimatic conditions

Figure 1 indicates the temperature and humidity conditions in the area immediately surrounding a rock boulder resting on the ice surface during the early afternoon of a clear summer's day. From 12.00 to 15.00 p.m. the radiation conditions were 120,000 lux at full exposure with an albedo radiation ranging from 72,000 to 96,000 lux at face level. In the shade of the rock boulder (north side) the radiation was still 17,000 lux. The temperature was measured with several thermocouples (NiCr-Ni, sensitivity 0.1°C), the humidity with a piercing probe (Rotronic, sensitivity 0.1% relative humidity). Two m above the ice surface the temperature of the air was approximately 10–11°C, the relative humidity was approximately 50%. At full exposure maximum temperatures up to 30°C were measured on the dark rock surface. In small hollows partly in the shade only values slightly above 20°C were observed. Close to the snow surface temperatures ranged from 3 to 8°C, depending on the wind speed. Considerable temperature gradients were registered underneath the rock. Along the free gap 5.3°C prevailed, on the lower side of the rock 9.2°C and in the narrow gap between ice and rock -0.4°C were measured. The surrounding snow and ice down to 10 cm also showed temperatures between 0 and -1°C. The relative humidity was between 50 and 60% 1–2 cm above the snow surface. This corresponds to a saturation deficiency of the air of 0.40–0.57 kPa (3.0–4.3 mm Hg), in the gap between ice and rock it dropped to 0.19 kPa (1.4 mm Hg). Temperature recordings under cloudy sky revealed temperatures between 0.3 to 0.5°C on the surface and 0.1–0.3°C within interstices of firn in the marginal areas of the glacier.

3.2
Water loss rates of glacier Collembola at different temperatures

For measuring transpiration rates a continuously recording ultra-microbalance equipped (Sartorius, sensitivity 1 or 0.1 µg) with a built-in cooling system was used.[46–49] Weighing was carried out with small wire cages. Saturated salt solutions[50] were used to adjust a defined surrounding humidity. After weighing the animals, they were exposed to various temperature and humidity levels, with their weight losses being constantly recorded. After the experiment the dry weight was determined to establish the water content of the animal. These data were used as basis for the calculation of the water loss rates.

Table 1 summarizes the data achieved for live weight, water content, dry weight, and the percentage of the water content. W_0 corresponds to the live weight of the animal after removal from the rearing boxes. It is evident that the water content was found to be significantly lower ($P<0.001$) with the two glacier species *Isotoma* 'sp. G' und *Onychiurus* 'sp. G' (65%) in comparison to those of *I. palliceps* und *Entomobrya* sp. (80%). Evaluation of the transpiration experiments showed that the water loss occurred at a constant rate as long as the animals had a sufficient supply of hemolymph available. As other studies[46] (and Eisenbeis, unpubl. data) indicated, glacier Collembola tolerate a loss of water up to 30%. As indicated in Figure 2 the highest water loss with all tested humidities occurred with *Onychiurus* 'sp. G', show-

Table 1. Live body mass (w_o), water mass (m_o), dry mass (dm) and percentage water content (m_o/w_ox100) of four glacier-related Collembola species from the Ötztal Alps (Austria) in 1983

	Isotoma 'sp. G' (n=25)	*Onychiurus* 'sp. G' (n=13)	*Isotomurus palliceps* (n=12)	*Entomobrya* sp. (n=9)
w_o (mg)				
mean ± SE	0.105±0.005	0.093±0.009	0.336±0.015	0.424±0.041
min	0.055	0.043	0.226	0.305
max	0.180	0.151	0.409	0.655
m_o (mg)				
mean ± SE	0.068±0.003	0.060±0.006	0.266±0.012	0.340±0.031
min	0.036	0.026	0.179	0.245
max	0.114	0.100	0.330	0.502
dm (mg)				
mean ± SE	0.037±0.002	0.033±0.003	0.069±0.003	0.084±0.010
min	0.019	0.017	0.047	0.060
max	0.066	0.055	0.079	0.153
m_o/w_ox100 (%)				
mean ± SE	64.7±0.9	63.9±1.7	79.1±0.4	80.4±0.7
min	55.7	53.3	76.4	76.6
max	76.0	73.8	80.8	83.9

SE standard error

ing a maximum value of 375% h^{-1}, at 0% relative humidity and 22°C. Under these conditions the animals die after about 5 min after a water loss of 30%, and they are almost completely dehydrated within approximately 15–20 min. *Isotoma* 'sp. G' shows with slightly lower rates a similar reaction. In comparison to this, the rates of the moraine species *Entomobrya* sp. are lower by one tenth. It is considerably better adapted to dry conditions than the two glacier species and survives at 0% relative humidity and 22°C for more than 1 h.

From Figure 2 a significant effect of the temperatures can also be seen. At 3°C the water loss rates are greatly reduced at all humidities, with the two glacier species, however, losing their water rapidly in dry air. At a water loss rate of 60% h^{-1} *Onychiurus* 'sp. G' can survive for at least more than 30 min. *Isotoma* 'sp. G' loses its water at a rate of 48% h^{-1} at 0% relative humidity and 3°C. In comparison to the humidity classification suggested by Eisenbeis[51] and Eisenbeis and Wichard,[45] the water loss rates demonstrated here place the two glacier species within the group of extremely hygric arthropods, whereas *Entomobrya* sp. belongs only to the hygric ones.

If we compare the transpiration rates (Fig. 3) of the two Isotomidae, the water loss rates of *I. saltans* are three times higher than those of *I. palliceps*. This agrees well with the differing biology that Kopeszki[26] described for both species. *I. palliceps* is able to tolerate the low temperatures of the firn areas, but during the summer it prefers the snow-free marginal zones of glaciers with their significantly high-

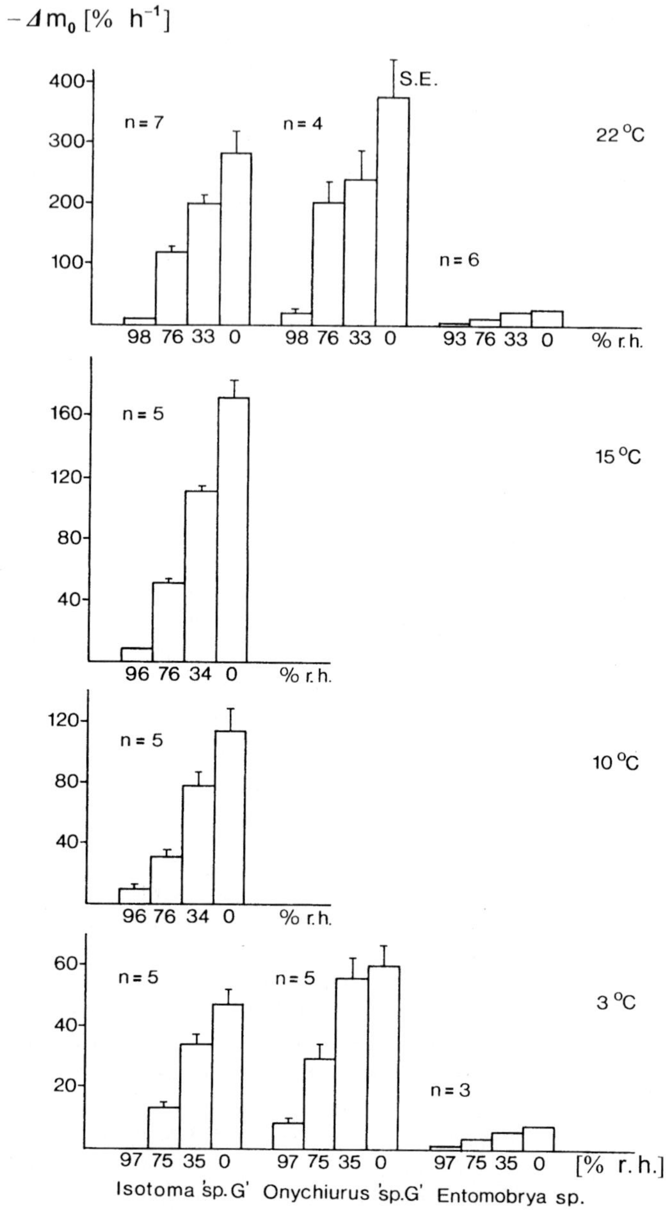

Fig. 2. Comparative water loss rates as percentage of original water (m_0) at certain relative humidities (r. h.) and temperatures with three glacier-related Collembola species *Isotoma* 'sp. G.', *Onychiurus* 'sp. G.' and *Entomobrya* sp.

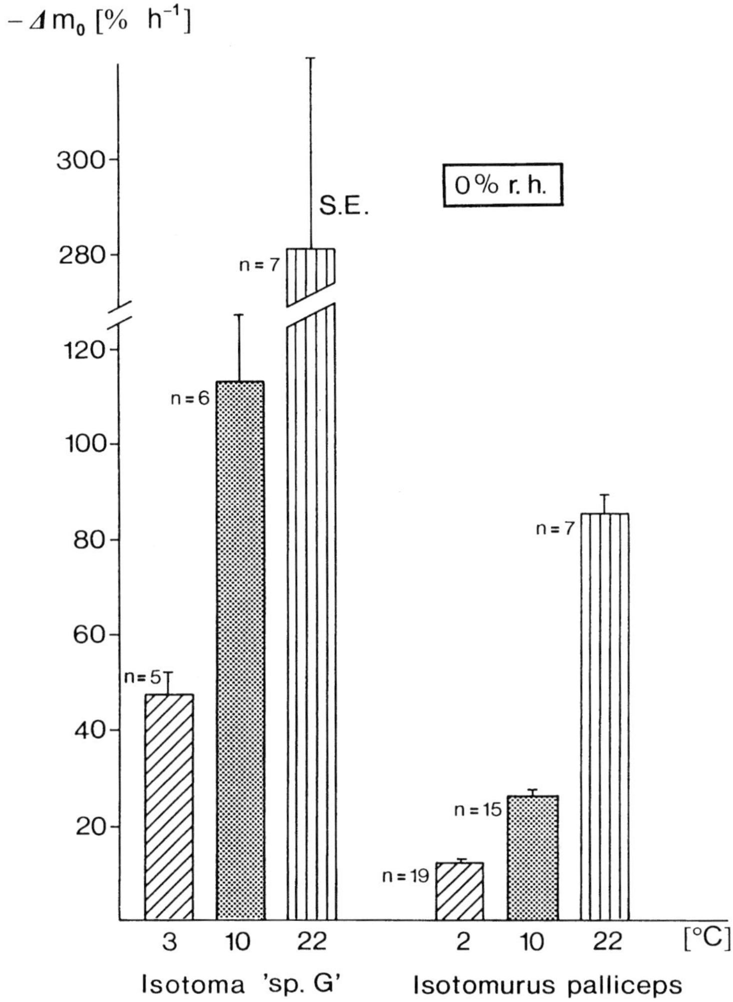

Fig. 3. Comparative water loss rates as percentage of original water (m_0) at 0% relative humidity and three different temperatures with the two Isotomidae *Isotoma* 'sp. G' and *Isotomurus palliceps*

er temperatures. This indicates an improved tolerance for living in drier air. Contrary to this, the two glacier species depend on a high humidity constantly surrounding them and are not able to abandon their moist microhabitat. This corresponds to earlier observations,[20–22,26] that glacier Collembola remain only for a short time at the snow surface. Besides the extreme radiation, most probably the dry air also forces them to rapidly seek protection in the interstice of the snow layer.

Figure 4 shows the changes of the cuticular permeability of *Onychiurus* 'sp. G' und the moraine species *Entomobrya* sp. in relation to ambient relative humidity at

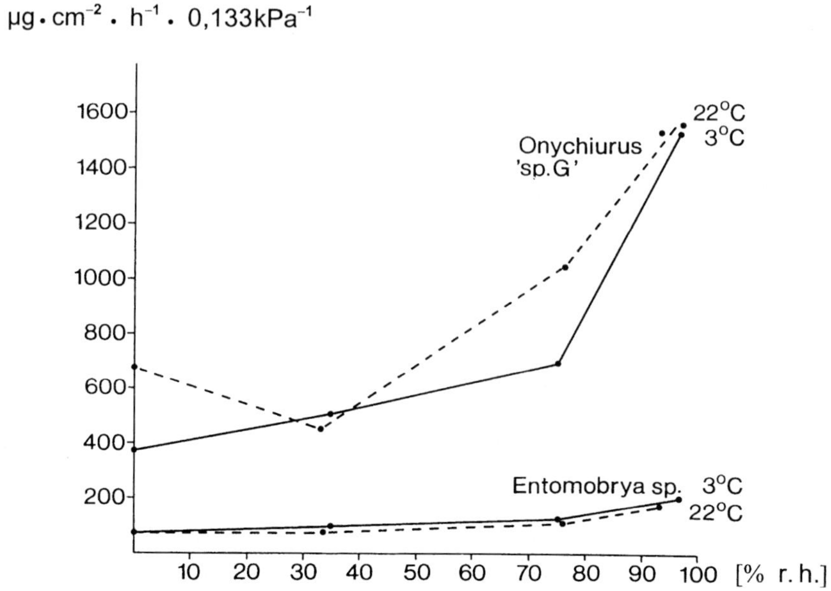

Fig. 4. Changes of the cuticular permeability for water with two glacier-related species *Onychiurus* 'sp. G.' and *Entomobrya* sp. in relation to different relative humidities (r. h.) at 3 and 22°C

3 and 22°C. With *Entomobrya* sp. humidity and temperature have little influence, and the permeability reaches not more than 200 µg cm^{-2} h^{-1} 0.133 kPa^{-1}. With *Onychiurus* 'sp. G' the permeability (400–1600 µg cm^{-2} h^{-1} 0.133 kPa^{-1}) increases when the surrounding humidity is high, whereas the influence of the temperature seems to be comparatively insignificant. With *Isotoma* 'sp. G' values between 200 and 300 µg cm^{-2} h^{-1} 0.133 kPa^{-1} were calculated at 3°C, at 22°C the values increased to 400 and 1000 µg cm^{-2} h^{-1} 0.133 kPa^{-1}. From this we conclude that *Isotoma* 'sp. G' reacts to changes of both humidity and temperature.

3.3
Absorption rate of water via the ventral tube of glacier Collembola

Before the absortion experiments the animals were subjected to the above described dehydration using the recording ultramicrobalance. Then they were placed on a cooled moist filter paper. The ventral tube vesicles of *I. palliceps* proved to be larger and more distinctly everted than those of *Isotoma* 'sp. G'. *I. palliceps* everted its vesicles immediately after contact with the moist filter paper and kept sucking continuously with only brief interruptions. Contrary to this, *Isotoma* 'sp. G' demonstrated a restless behavior, running around on the filter paper and only occasionally touching it with its vesicles. Therefore its presence on the filter paper lasted accordingly longer. After the absorption phase the animals were weighed again, and the amount of water taken in was calculated. Generally, the absorption rates of

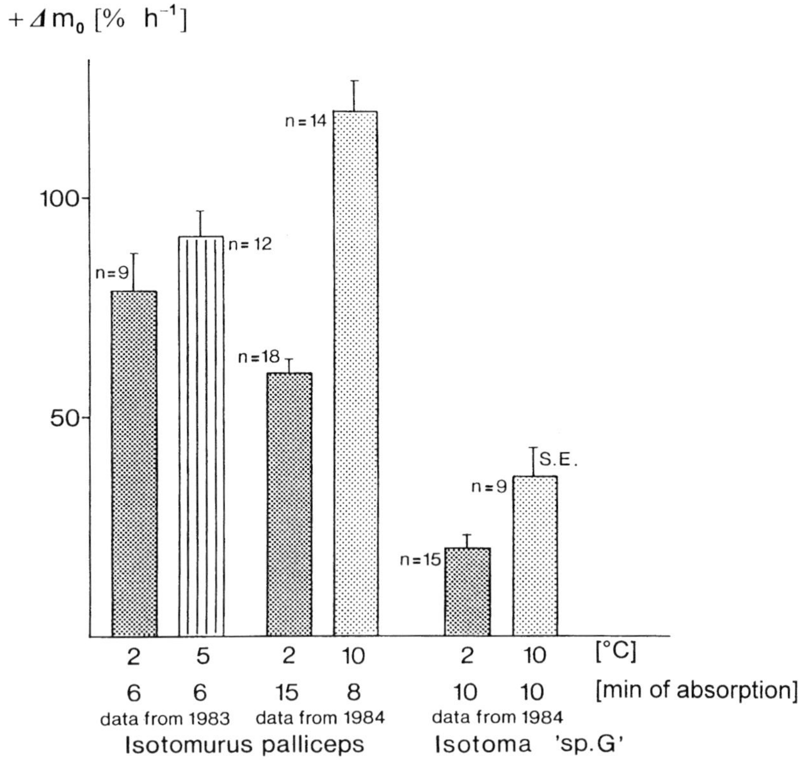

Fig. 5. Comparative water absorption rates as percentage of original water content (m_0) of the two Isotomidae *Isotoma* 'sp. G.' and *Isotomurus palliceps* after exposing them on moist filter paper at different temperatures

I. palliceps were significantly higher ($P<0.001$) than those of *Isotoma* 'sp. G' (Fig. 5). At 10°C the rate was more than three times higher. Comparing the transpiration (-11.8% h^{-1}) and the absorption rates (60-80% h^{-1}) of *I. palliceps* at 2°C and in dry air the ability of the animals to rapidly rehydrate via their ventral tubes becomes apparent. With *Isotoma* 'sp. G' comparable conditions result in a negative balance (-47.5 or +20.1% h^{-1}). However, taking the microclimatic conditons in the habitat of *Isotoma* 'sp. G' into account, hardly more than 20% water loss h^{-1} are to be expected. This allows the assumption that both species are able to compensate their cuticular water loss efficiently with the help of their ventral tubes.

3.4 Ultrastructural features of glacier Collembola

3.4.1 *Structures of the body surface*

Various typical structures of the body surface of glacier Collembola have already been described by Eisenbeis and Meyer[27] and Eisenbeis and Wichard.[45] SEM pictures of the cuticle of the glacier Collembola exhibit the same hexagonal microgranular pattern typical of all Collembola (Fig. 10). A striking feature of the third segment of the antenna is the postantennal organ III, which consists of two shortened and thickened sensory bristles inserted in a sensory pit (Fig. 10). When observing the springtails in their rearing boxes, *Isotoma* 'sp. G.' are noted to constantly explore their surroundings with the tips of their antennae by alternatively touching the substrate in rapid oscillations. A further sensory organ important for the Collembola is the postantennal organ with a delicately perforated epicuticle (Fig. 11), located closely in front of the eyes (Fig. 8). Both sensory organs are chemoreceptors, the postantennal organ usually being identified as the hygroreceptor. Furthermore numerous tactile chaetae cover the body surface (Figs. 6–11), which are pennate with *I. palliceps*. It also has longer ones (macrochaetae) protruding from the body surface, particularly along the back region. The antennae with their four segments are somewhat longer than the head (Fig. 6). The eyes of *Isotoma* 'sp. G' consist of six ommatidia in an S-shaped arrangement. With *I. palliceps* all eight ommatidia are present and arranged in pairs.[27,45] The delicately segemented furca carries flexible distal branches (dentes) with serrate terminal segments (mucrones). All these structures are also exhibited by other Collembola species. They do not constitute any specific form of adaptation to glacial conditions.

3.4.2 *Remarks on histological features*

The body color of *Isotoma* 'sp. G' ranging from deep black to bluish black when exposed to light derives from an intensive pigmentation of the epidermis, completely screening the tissue underneath (Fig. 12). In the areas of muscle insertion long rows of pigment granules are arranged along epidermal tonofibrils. In contrast, the epithelium of the ventral tube vesicles contains only few pigment granules (Fig. 13). The intensive epidermal pigmentation certainly functions as protection against the high radiation on the glacier surface. Especially during periods of bright sunshine in spring and early summer, the springtails appear in masses on the snow surface, only to retreat soon after. Euedaphic Collembola, e.g. the white *Onychiurus* species, would certainly suffer damage due to radiation; but already the salmon-red species *Onychiurus* 'sp. G' that we found on the Rotmoos and Gaisberg glacier in the Ötztal Alps in close contact with ice under rocks seems to be protected against radiation. K. Thaler (Innsbruck) could document this by photographing a mass occurrence of these red Collembola on the Kuchen glacier (Verwall, Central Alps)

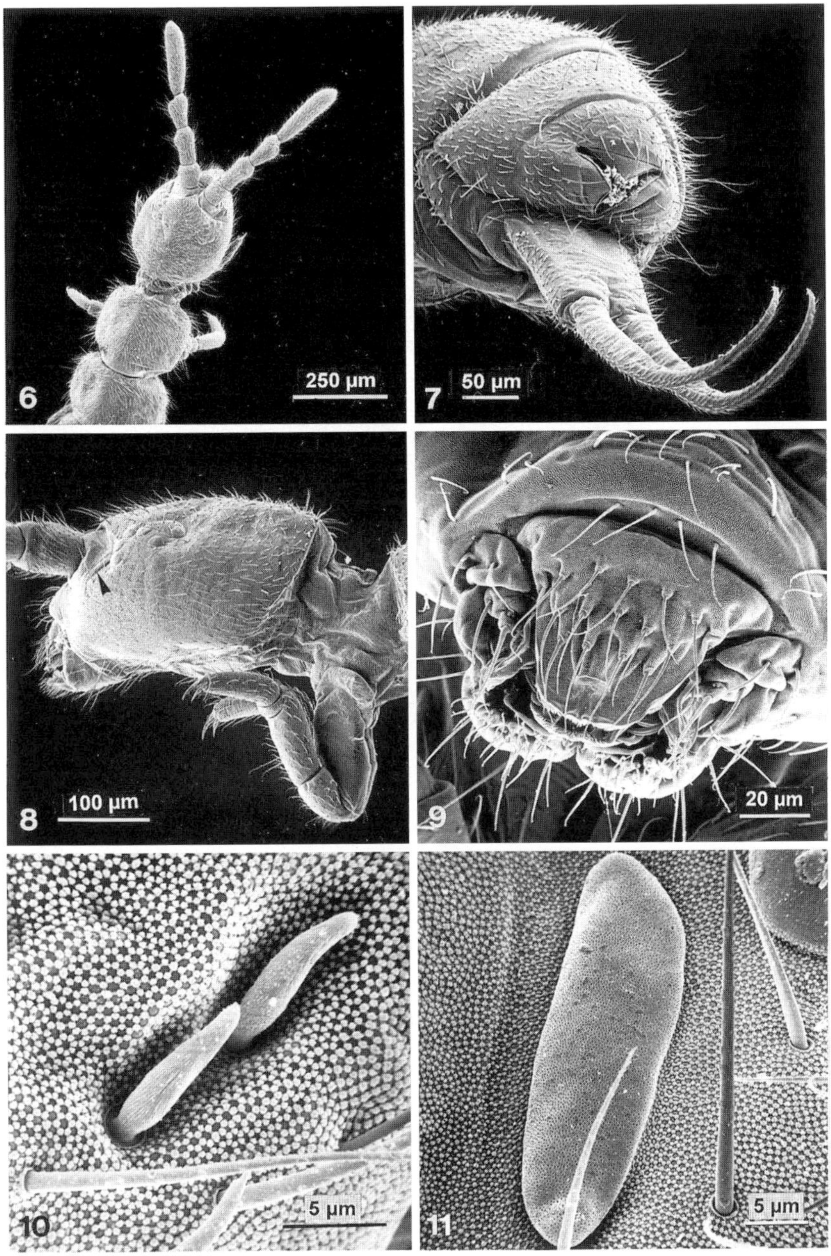

Figs. 6–11. Scanning electron microscope pictures of *Isotoma* 'sp. G'
(**6**) Anterior body dorsal view; (**7**) Abdomen caudal view with anal region and extended furca; (**8**) Head and prothorax lateral view with eye spot, mouth cone and fore legs, the *arrow* points to the postantennal organ; (**9**) Frontal view of mouthparts with labrum, tips of mandibles, maxillae, labium and tip of hypopharynx; (**10**) Postantennal organ III surrounded by a hexagonal pattern of micrograins and regular setae; (**11**) Area in front of the eye spot with the lobelike postantennal organ, look at the perforated cuticle of the organ

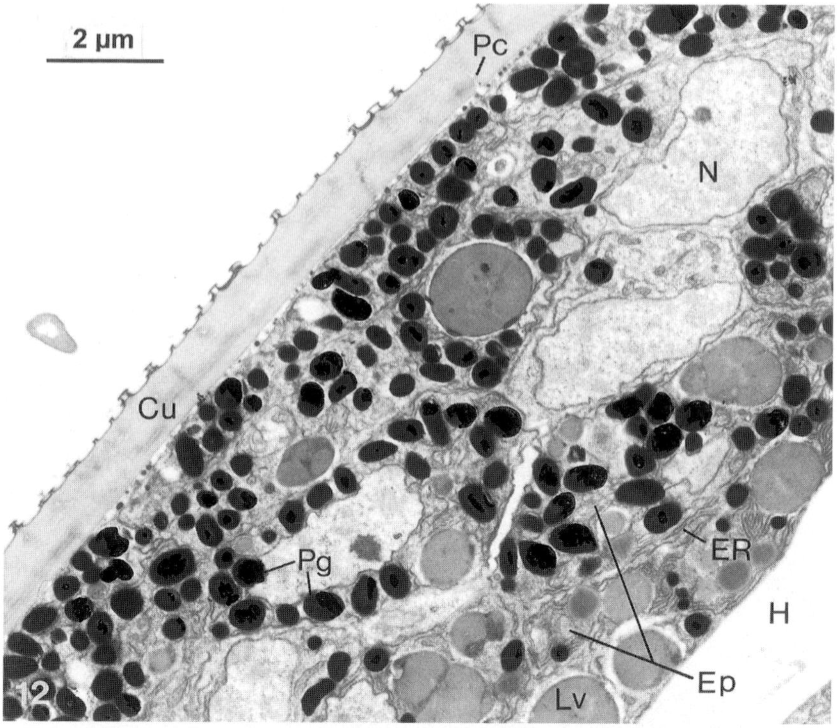

Fig. 12. Transmission electron microscope (TEM) section of the cuticle (*Cu*) and epidermis (*Ep*) of *Isotoma* 'sp. G'
Pc pore channels; *Pg* pigment granules, *ER* endoplasmic reticulum; *H* hemolymph space; *Lv* lipid vacuoles; *N* nucleus

on August 28th 1990. *I. palliceps* is not quite as intensively pigmented as *Isotoma* 'sp. G', especially not on its legs. Kopeszki[26] regards glacier Collembola as small "warming machines", said to even make the ice melt due to the radiation absorption being increased by their black pigmentation. This makes sense for the period they aggregate in the top 30 cm of the ice during the snow-free summer period, a hypothesis that certainly needs investigating.

In its eversible vesicles the ventral tube is equipped with a typical transport epithelium for the transportation of water and ions. A mitochondria-rich transport border with apical microvilli lies beneath an almost 1-µm-thick wettable cuticle (Figs. 13–14). This is followed by a mitochondria-rich intermediate zone with delicate pigment granules evenly distributed. Proximally the 4–5-µm-thick cell layer adjoins the hemolymph cavity. This zonation is in accordance with that of *Tomocerus flavescens* und *Orchesella villosa*, species occurring in temperate forest soils.[52] An epithelial part with a perikaryon containing a lobed active nucleus is shown in the bottom right hand corner of Figure 13. On this Figure the transport border contains no larger vesicles.

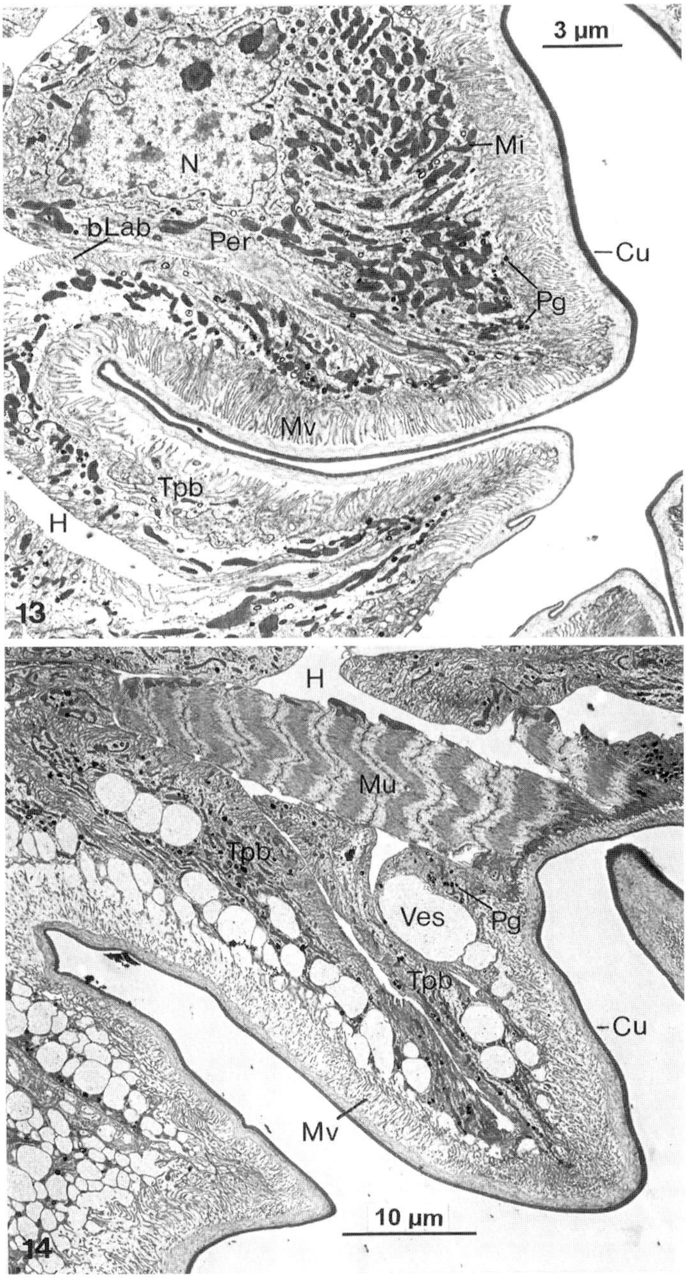

Figs. 13–14. TEM section of a vesicle of the ventral tube of *Isotoma* 'sp. G'
(13) Transport border (*Tpb*) with microvilli (*Mv*), mitochondria (*Mi*) and basal labyrinth (*bLab*), perikaryon (*Per*) of a transport cell with nucleus (*N*); **(14)** Transport border (*Tpb*) filled with vesicles (*Ves*)
Cu cuticle, *H* hemolymph, *Mv* microvilli, *Mu* retractor muscle, *Pg* pigment granules

The investigation of a second animal, however, revealed some different aspects concerning the tissue. The apical cytoplasma of the transport border was filled with a dense package of electron lucent vacuoles with a slightly flaky content (mucus), most probably forming larger units in due course (Fig. 14). From the fact that the smaller units are usually located near the brush border and the larger ones are situated in the intermediate part of the transport epithelium, we may assume that with regard to this cell aspect fluid is most probably transported from the outside via the epithelium to the hemolymph. Possibly a water-binding mucous primary substance is secreted into the subcuticular area by the transport cells. This, however, has to be proved in more detail.

With both species (*Isotoma* 'sp. G' and *I. palliceps*) Kopeszki[26] observed the formation of a droplet at the tip of the ventral tube, which was taken up by the animals with their legs and dispersed evenly over the cuticula. This is referred to as greasing the cuticle. Collembola have already been known to demonstrate this behavior.[53] This droplet is obviously excreted by the ventral tube. On the other hand, it is also possible that it is produced by the labial glands and transported onto the ventral tube via the ventral channel. Schaller[53] also reports similar observations. We must point out that the above-mentioned dehydration and absorption experiments revealed for certain that the animals gain weight when touching the tube vesicles onto the ground. Similar experiments were carried out with *Tomocerus flavescens* und *Orchesella villosa*, two species of temperate forest soils.[46,47] Marking experiments using tritiated water (THO) showed an increased uptake of THO into the body water whenever the ventral tube was used for the water uptake. Preventing the ventral tube from functioning reduces the uptake of THO significantly (Eisenbeis, unpubl. data). Previous authors regard the ventral tube as a multifunctional organ, so that we can assume that both secretion and absorption are involved.

Apart from the strongly pigmented epidermis and the ventral tube, the midgut epithelium is also of interest. As with all other Collembola, it is replaced by a new epithelium during the process of molting. Consequently, the gut demonstrates an additional excretory function. Figure 15 shows a section of the midgut of *I. palliceps* with a well-developed microvilli border. The ultrastructure of the midgut of *Isotoma* 'sp. G' is identical. In the apical cytoplasm, close below the microvilli border a fine electrondense band exists that has proved to be perforated when taking a closer look (Figs. 15–17). It is a special form of the cytoskeleton, called terminal web. It normally only appears in the intestinum of mammals, and is generally not known with insects. Dallai and Callaini[54] described comparable structures with *Isotoma palustris* and *Cryptopygus thermophilus* and Barra and Poinso-Balaguer[55] with *Isotoma viridis*.[27] With other Collembola families this structure has not been found, so that up to now it seems to be restricted to the Isotomidae. The sections clearly show that larger cell organelles like vacuoles and mitochondria are not able to reach the base of the microvilli, which demonstrates that the terminal web is involved in forming compartments in the apical cytoplasm. Furthermore, Eisenbeis and Meyer[27] discussed the possibilities that it (1) could function as a support structure within the apical cytoplasm, (2) possibly assisting the micro-pinocytosis along the

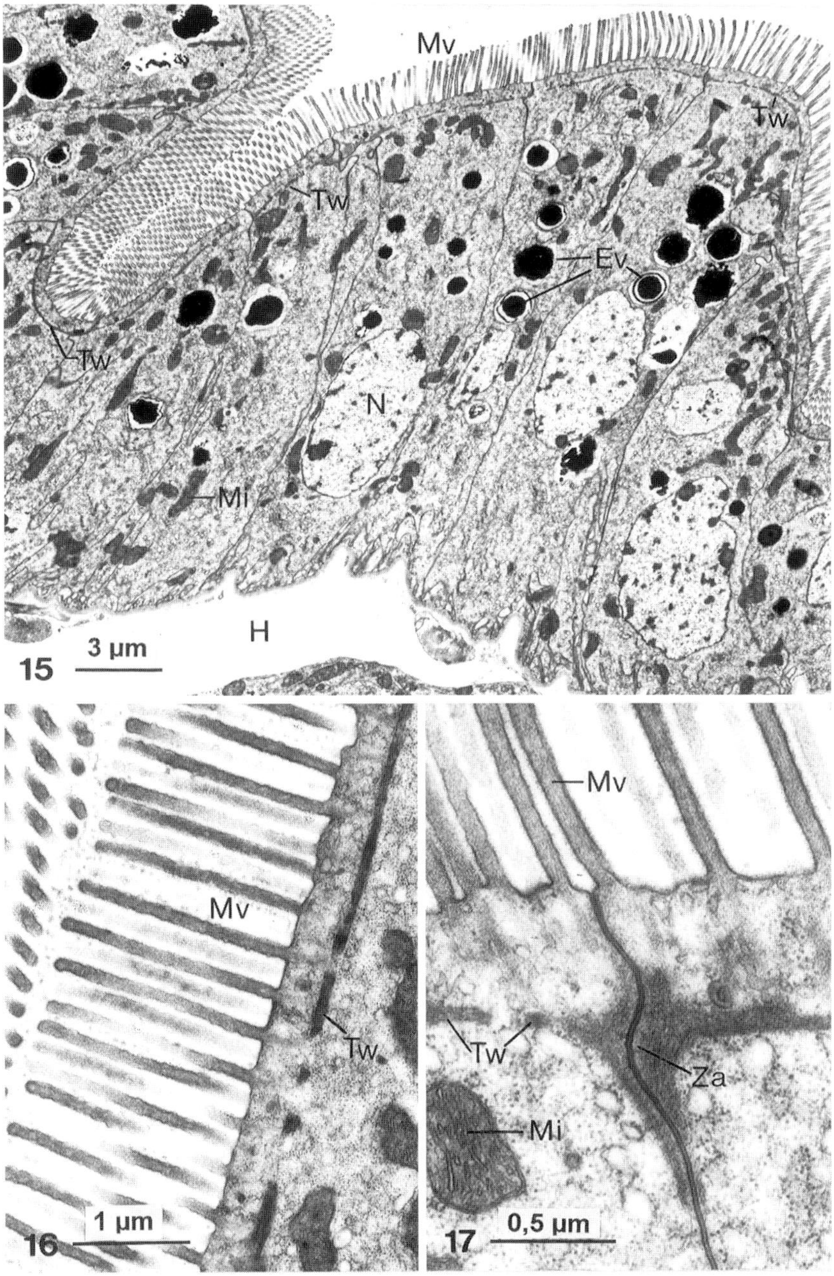

Figs. 15–17. TEM section of the midgut epithelium of *Isotoma* 'sp. G' and *Isotomurus palliceps* **(15)** Midgut epithelium of *Isotomurus palliceps* with microvilli brush border (*Mv*), terminal web (*Tw*) and excretory vacuoles (*Ev*); *N* nucleus; *H* hemolymph space; *Mi* mitochondria; **(16)** Microvilli brush border (*Mv*) and perforated terminal web (*Tw*) of *Isotoma* 'sp. G'; **(17)** Apical cell border between two midgut cells of *Isotoma* 'sp. G' showing a connection of zonula adhaerens (*Za*) and terminal web (*Tw*)

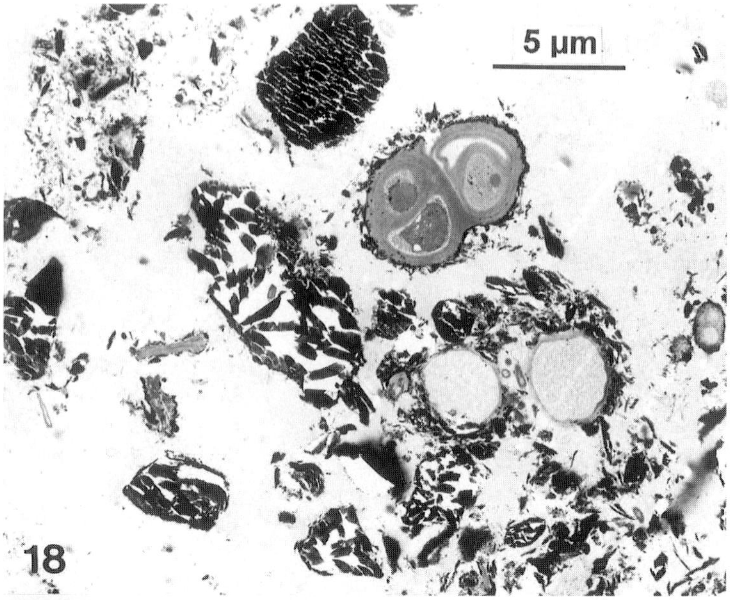

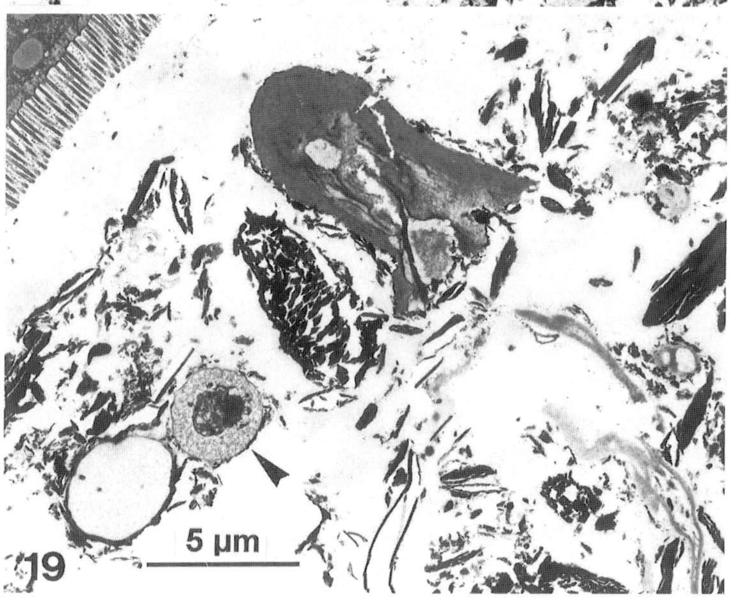

Figs. 18–19. Gut content of *Isotoma* 'sp. G' with formed food components (algae, pollen) and sharp-edged mineral particles. The *tip of the arrow* indicates a cross-section of the snow algae *Chlamydomonas nivalis*

microvilli, and (3) function as a protection of the epithelium against the sharp edges of the mineral components in the food.

The intestinal epithelium of insects is usually protected by a peritrophic membrane. Sections of the gut content of glacier Collembola have shown that a high percentage of sharp edged components of presumably mineral origin occur in the gut content (Figs. 18–19). The terminal web could possibly form a kind of inner barrier against intrusion by mineral particles, in case injuries of the cell surface occur. Collembola, especially soil-dwelling ones, are known to mix mineral and organic materials in their gut (formation of organomineral complexes). This accounts for the development of the characteristic sharp-edged pellets, which differ distinctly from those produced by mites.[56] Collembola dwelling on the glaciers basically feed on cryoconite, a blackish mixture predominantly consisting of organic debris and mineral particles, algae, pollen, microfauna and arthropod fallout. Figures 18–19 show several of these components, in particular the thick coat consisting of minute mica-type plates wrapped around the pellet-shaped food components. Additionally, sections of gregarina with their plasma densely filled with lipid vacuoles were found in midgut epithelium of *Isotoma* 'sp. G'.[27] With some of the sections, gregarina extruding into the lumen of the gut were observed. So far nothing is known about a possible mutual relationship.

4
Conclusions

- At least two Collembola species, *Isotoma* 'sp. G.', a deep black Isotomidae, and *Onychiurus* 'sp. G.', a salmon red Onychiuridae, dwell on glaciers in the Ötztal Alps. A further species, *Isotomurus palliceps*, is common to the marginal firn fields.
- Glacier Collembola tolerate a dehydration only up to 30%. The water loss rates of the glacier-dwelling species *Isotoma* 'sp. G.' are three times higher than those of *I. palliceps* living on the marginal firn fields. The very high water loss rates of *Isotoma* 'sp. G.' place this species within the group of extremely hygric arthropods. Therefore, Collembola living on glaciers remain only for a short time at the surface. They rapidly seek protection in the moist interstice of the snow layer or the glacier ice. When comparing the transpiration and absorption rates, the ability of *I. palliceps* to rapidly rehydrate via their ventral tubes becomes apparent.
- All structures of the body surface of glacier Collembola are also exhibited by other Collembola species. They do not constitute any specific form of adaptation to glacial conditions.
- The intensive epidermal pigmentation certainly functions as protection against the high radiation on the glacier surface. *I. palliceps* is not quite as intensively pigmented as *Isotoma* 'sp. G.', especially not on its legs.
- Apart from the strongly pigmented epidermis, the midgut epithelium is of interest too. In the apical cytoplasm of this epithelium a special form of the cytoskeleton, known as terminal web, can be observed. Besides other functions it may

protect the midgut epithelium against the intrusion of sharp edged mineral components in the food. Sections of glacier Collembola have shown that a high percentage such components occur in the gut content. Collembola dwelling on the glaciers basically feed on cryoconite, a blackish mixture predominantly consisting of organic debris, algae, pollen, microfauna, arthropod fallout and mineral particles.

Acknowledgements. We are indebted to Prof. Dr. G. Patzelt from the Alpine Research Station Obergurgl providing all research facilities of the station. The Departments of Zoology of the Universities of Mainz (Germany) and Innsbruck (Austria) and the Feldbauschstiftung of the Department of Biology at the University of Mainz provided support for our glacier campaigns, which is gratefully acknowledged. For technical support (electron microscopy) we thank E. Köddermann, E. Sehn and M. Ullmann.

5
References

1. Sømme L. Cold hardiness of winter active Collembola. Norw J Entomol 1976; 23:149-153.
2. Sømme L. Cold-hardiness of *Cryptopygus antarcticus* (Collembola) from Bouvetøya. Oikos 1978; 31:94-97.
3. Block W, Zettel J. Cold hardiness of some alpine Collembola. Ecol Entomol 1980; 5:1-9.
4. Miller K. Cold-hardiness strategies of some adult and immature insects overwintering in interior Alaska. Comp Biochem Physiol 1982; 73A:595-596.
5. Sømme L. Supercooling and winter survival in terrestrial arthropods. Comp Biochem Physiol 1982; 73A:519-5.
6. Convey P. Overwintering strategies of terrestrial invertebrates in Antarctica – the significance of flexibility in extremely seasonal environments. Eur J Entomol 1996; 93:489-505.
7. Convey P. The influence of environmental characteristics on the life history attributes of Antarctic terrestrial biota. Biol Rev 1996; 71:191-225.
8. Bale JS, Hodkinson ID, Block W, Webb NR, Coulson SC, Strathdee AT. Life strategies of arctic terrestrial arthropods. In: Woodin SJ, Marquiss M, eds. Ecology of Arctic Environments. Spec Publ Brit Ecol Soc 1997; 13:137-165.
9. Maclean SF Jr. Ecological adaptations of tundra invertebrates. In: Vernberg J, ed. Proc AIBS Symp Physiol Adaption to the Environment. Amhurst, MA, 1975:269-300.
10. Addison JA. Biology of *Hypogastrura tullbergi* (Collembola) at a high arctic site. Holarctic Ecol 1981; 4:49-58.
11. Aitchison CW. Winter-active subnivean invertebrates in Southern Canada I. Collembola. Pedobiologia 1979; 19:113-120.
12. Aitchison CW. Low temperature and preferred feeding by winter-active Collembola (Insecta, Apterygota). Pedobiologia 1983; 25:27-36.
13. Brummer-Korvenkontio M, Brummer-Korvenkontio L. Springtails (Collembola) on and in snow. Mem Soc Fauna Flora Fenn 1980; 56:91-94.
14. Leinaas HP. Activity of Arthropoda in snow within a coniferous forest, with special reference to Collembola. Holarctic Ecol 1981; 4:127-138.
15. Leinaas HP. Winter strategy of surface dwelling Collembola. Pedobiologia 1983; 25:235-240.
16. Christian E. Über massenhaft auftretende Schneeflöhe in Österreich. Burgenländische Heimatblätter 1977; 39:140-142.
17. Hagvar S. Long distance, directional migration on snow in a forest Collembolan *Hypogastrura socialis* (Uzel). Acta Zool Fennica 1995; 196:200-205.
18. Zettel J. The significance of temperature and barometric pressure changes for the snow surface activity of *Isotoma hiemalis*. Experientia 1984; 40:1369-1372.

19. Block W. Insects and freezing. Sci Progress 1995; 78:349-372.
20. Steinböck O. Der Gletscherfloh. Z Dt Alpenvereins 1939; 70:138-147.
21. An der Lan H. Zur Winter-Ökologie des Gletscherflohes. Die Pyramide 1961; 33-34.
22. An der Lan H. Tiere im Ewigschneegebiet. Umschau Wiss Technik 1963; 2:49-52.
23. An der Lan H. Neues zur Tierwelt des Ewigschneegebietes. Zool Anz Suppl Verh Dt Zool Ges 1963; 26:673-678.
24. Schaller F. Beobachtungen am Gletscherfloh *Isotoma saltans* (Nicolet 1841). Zool Anz Suppl Verh Dt Zool Ges 1963; 26:679-682.
25. Schaller F. *Isotoma saltans* und *Cryptopygus antarcticus*, Lebenskünstler unter Extrembedingungen (Collembola: Isotomidae). Entomol Gener 1992; 17:161-167.
26. Kopeszki H. Zur Biologie zweier hochalpiner Collembolen – *Isotomurus palliceps* (Uzel, 1891) und *Isotoma saltans* (Nicolet, 1841). Zool Jb Syst 1988; 115: 405-439.
27. Eisenbeis G, Meyer E. Some ulstrastrucutral features of glacier Collembola, *Isotoma* 'sp. G.'and *Isotomorus palliceps* (Uzel, 1891) from the Tyrolean Central Alps. In: Dallai R, ed. 2nd Int Seminar Apterygota. Siena: University of Siena, 1986:257-272.
28. Eisenbeis G, Meyer E, Maixner M. Leben im ewigen Eis – Anpassung an einen aquatischen Extrembiotop am Beispiel von Gletscher-Collembolen. Verh Westd Entom Tag 1989:75-77.
29. Cannon RJC, Block W. Cold tolerance of microarthropods. Biol Rev 1988; 63:23-77.
30. Lee RE, Denlinger DL, eds. Insects at Low Temperature. London, New York: Chapman & Hall, 1991.
31. Duman JG. Insect antifreezes and ice-nucleating agents. Cryobiology 1982; 19:613-627.
32. Sømme L. Overwintering ecology of Alpine Collembola and oribatid mites from the Austrian Alps. Ecol Entomol 1979; 4:175-180.
33. Duman JG. Thermal-hysteresis-factors in overwintering insects. J Insect Physiol 1979; 25:805-810.
34. Zettel J. Cold hardiness strategies and thermal hysteresis in Collembola. Rev Ecol Biol Sol 1984; 21:189-203.
35. Block W. Terrestrial arthropods and low temperature. Cryobiology 1981; 18:436-444.
36. Bauer R, Christian E. Adaptations of three springtail species to granite boulder habitats (Collembola). Pedobiologia 1993; 37:280-290.
37. Worland MR, Block W. Survival and water loss in some Antarctic arthropods. J Insect Physiol 1986; 32:579-584.
38. Block W. Cold or drought – the lesser of two evils for terrestrial arthropods? Eur J Entomol 1996; 93:325-339.
39. Schaller F, Zinkler D. Atmungsphysiologische Untersuchungen am Gletscherfloh. Naturwissenschaften 1963; 50:385.
40. Sømme L, Block W. Adaptations to alpine and polar environments in insects and other terrestrial arthropods. In: Lee RE, Denlinger DL, eds. Insects ad Low Temperature. London, New York: Chapman & Hall, 1991:318-359.
41. Block W, Webb NR, Coulson S, Hodkinson ID, Worland MR. Thermal adaptation in the Arctic collebolan *Onychiurus antarcticus*. J Insect Physiol 1994; 40:715-722.
42. Block W. Cold hardiness in invertebrate poikilotherms. Comp Biochem Physiol 1982; 73A:581-593.
43. Block W. Cold tolerance of insects and other arthropods. Phil Trans Roy Soc London B 1990; 326:613-633.
44. Zinkler D, Schaller F. High amounts of long-chain polyunsaturated fatty acids in the glacier-flea *Isotoma saltans* (Insecta, Collembola). In: Striganova BR, ed. Soil Fauna and Soil Fertility. 9th Int Coll Soil Zool, 6th Int Coll Apterygota. Nauka, Moscow, 1987:729-734.
45. Eisenbeis G, Wichard W. Atlas on the Biology of Soil Arthropods. Berlin, Heidelberg: Springer, 1987.
46. Eisenbeis G. Physiological absorption of liquid water by Collembola: Absorption by the ventral tube at different salinities. J Insect Physiol 1982; 28:11-20.

47. Jaeger G, Eisenbeis G. pH-dependent absorption of solutions by the ventral tube of *Tomocerus flavescens* (Tullberg, 1871) (Insecta, Collembola). Rev Écol Biol Sol 1984; 21:519-531.
48. Baehr B, Eisenbeis G. Comparative investigations on the resistance to desiccation in Lycosidae, Hahniidae, Linyphiidae and Micryphantidae (Arachnida, Araneae). Zool Jb Syst 1985; 112:225-234.
49. Meyer E, Eisenbeis G. Water relations in millipedes from some alpine habitat types (Central Alps, Tyrol) (Diplopoda). Bijdragen to de Dierkunde 1985; 55:131-142.
50. Winston PW, Bates DH. Saturated solutions for the control of humidity in biological research. Ecology 1960; 41:232-237.
51. Eisenbeis G. Kinetics of transpiration in soil arthropods. In: Lebrun P, André HM, De Medts A, Grégoire-Wibo C, Whauty G, eds. New Trends in Soil Biology. Louvain-la-Neuve: Dieu-Brichart, 1983:626-627.
52. Eisenbeis G. Licht- und elektronenmikroskopische Untersuchungen zur Ultrastruktur des Transportepithels am Ventraltubus arthropleoner Collembolen (Insecta). Cytobiologie 1974; 9:180-202.
53. Schaller F. Collembola (Springschwänze). In: Helmcke JG, Starck D, Wermuth H, eds. Collembola. Handb Zool 1970; 4(2)2/1:1-72.
54. Dallai R, Callaini G. Comparative analysis of the Collembola pyloric region (Insecta Apterygota). Monit Zool Ital NS 1979; 13:45-66.
55. Barra JA, Poinsot-Balaguer N. Un "terminal web" fenêtre dans le mésentéron de *Isotoma viridis* Bourlet (Collembola, Isotomidae). Rev Écol Biol Sol 1987; 24:75-84.
56. Rusek J. Die bodenbildende Funktion von Collembolen und Acarina. Pedobiologia 1975; 15:299-308.

Ecological features of antarctic fishes

M. Vacchi[1]*, E. Pisano[2] and G. La Mesa[1]

[1] ICRAM, Istituto Centrale per la Ricerca Scientifica e Tecnologica Applicata al Mare, Via di Casalotti 300, I-00166 Roma, Italy
[2] Istituto di Anatomia Comparata, Università di Genova, Viale Benedetto XV 5, I-16132 Genova, Italy

1 Introduction

1.1 Antarctic fishes: a unique fauna

Only 1% of the world's modern fish species live in antarctic waters, but the antarctic fish fauna display a number of different and unique features for large marine ecosystems.[1,2] Most of the antarctic fishes possess so many physiological and biochemical peculiarities that they have been considered an extreme in the spectrum of animal diversification and adaptation.[3] A well-known example are the icefish (family Channichthyidae), which are unique among vertebrates because they lack hemoglobin and oxygen is transported by the plasma.[4] From an evolutionary perspective, the hemoglobin loss was possible due to greater adaptability of a reduced blood viscosity coupled with an increased oxygen solubility in cold antarctic waters. Another remarkable molecular adaptation is the presence of glycoproteins in body fluids that prevents freezing in these fishes and allows them to survive at subzero temperatures.[5] The antarctic fish evolved the antifreeze protein at an estimated time of about 14 million years ago.[6] From this time, the presence of small ice crystals, thick ice on the surface and on the bottom was no longer an obstacle for their survival and success during the severe cooling of the Southern Ocean.

The capacity to produce antifreeze proteins is one of the few traits which arctic and antarctic polar fish have in common. This capacity is an example of convergence because the similar antifreeze proteins evolved from different genetic starting points.[6] Apart from antifreeze proteins and some other physiological aspects, the antarctic fish fauna is very different from the arctic fish fauna in both faunistic and ecological features.[7,8] It is also surprisingly different from both modern fish fauna of other southern continents and from the fossil antarctic ichthyofauna.[2]

* Corresponding author; e-mail: pisano@unige.it

Because of their evolutionary adaptation to extreme cold and their ecological position in south polar ecosystems, antarctic fish have been of considerable interest to both biologists and ecologists. Major reviews of the growing body of literature on antarctic fish include refs. 2 and 9–16.

1.2
Ecological studies

The response of antarctic fish to an extreme habitat has held the attention of ichthyologists since the first specimen was taken during the James Clark Ross's Expedition to the Southern Ocean (1839–1843). However, ecological research became common only after the 1950s when the possible existence of substantial, exploitable fish stocks in the Southern Ocean attracted the attention of fishing nations. The development of large-scale fishing further expanded studies, especially concerning harvested species, on growth, reproduction, feeding, and ecological position in the trophic webs of the Southern Ocean. After it was shown that antarctic fish are particularly susceptible to overfishing, the need for more effective management measures greatly increased ecological research, especially in light of the Convention on Conservation of Antarctic Marine Living Resources (CCAMLR).[15]

Among antarctic fish, the shallow-water or coastal modern species have evolved in a unique condition of isolation as a result of the evolution and paleoceanography of the Southern Ocean.[17] This isolation occurred after the formation of the Antarctic Polar Front in the Early Miocene, 22 million years ago, producing a major ecological barrier which is still operating today.[18] The continental shelf water depth is much greater than that for other oceans and there is no shallow-water connection from this shelf to the continents to the north.[19] The major component of this fauna is represented today by the perciform suborder Notothenioidei. In this chapter we consider some main features of this fish group, focusing on their ecological diversification. We take examples from our own published and unpublished observations especially concerning our research at Terra Nova Bay, Ross Sea, as well as from the literature. First, we provide some basic information on the origin and evolution of these teleosts in relation to tectonic and climatic change of the antarctic continent. In fact the ecological diversity that we observe today in antarctic fish, even more than in other taxa, has been influenced by the unique environmental conditions that have occurred during the course of their unique evolutionary history.

2
The notothenioid fishes: origin and diversification

Although some question remains open, especially concerning the ancestral stock that gave origin to the Notothenioidei, the major features of the fascinating evolutionary history of this suborder have been established.[2,20–22] A summary of the major events in this evolution follows.

The origin and the evolution of antarctic fish are strictly linked to the tectonic, oceanographic and climatic events in the history of the supercontinent Gondwana. After the breakup of Gondwana, Antarctica was separated from other land masses by seafloor spreading and deep water in the Drake Passage at about the Oligocene/Miocene boundary, 25 million years ago. These events isolated Antarctica and initiated climatic changes that probably eliminated some habitats previously available to fish fauna. As a result, the diversity of the fauna was reduced and the taxonomic composition began to change over geological time. The low water temperature by itself was not the only important factor in the process of faunal change. The new ecological conditions such as the inshore habitat destruction by ice, changing trophic conditions and the grounding of ice shelves were equally important in determining the local extinction of most of the early Tertiary non-notothenioid fauna.

The ancestors of notothenioids were benthic fishes probably associated with the shelf of the Weddell Province in the late Cretaceous/early Tertiary. They were trapped on the continental shelf and drifted with Antarctica as it separated from Gondwana. The diversification of Notothenioidei was facilitated by the thermal isolation of Antarctica and by the increasing productivity of the Southern Ocean beginning about 22 million years ago.[23] Another ecological factor, the absence of competition from non-notothenioids, was most likely one of the most important in determining both the ecological and phyletic diversification of these tolerant fishes as they began to fill the broad ecological space available in the developing Southern Ocean. Notothenioids underwent a process of diversification directed away from the ancestral benthic habitat toward pelagic or partially pelagic lifestyles.[2,24] The high degree of endemicity among Notothenioidei suggests that once the modern fauna was established, low water temperature, geographic isolation and lack of south flowing surface currents limited immigration of many other fish groups into waters south of the Antarctic Polar Front.

3 Taxonomic diversity

The perciform suborder Notothenioidei includes 8 families and 121 species (Table 1). Over 20 new species have been described during the past 15 years and the number of species will certainly increase as ichthyological research continues and sampling techniques improve. Cladistic analyses of new sets of data have subjected the taxonomy of the group to rapid revision. The taxa Pseudaphritidae and Eleginopsidae have been recently considered as new families in the suborder, according to morphological, cytogenetic and molecular analyses.[25–27]

As can be seen from Table 1, four families are important for the taxonomic diversity and account for 90% of number of genera and species (Nototheniidae, Artedidraconidae, Bathydraconidae, Channichthyidae). These are the taxa for which we have more biological and ecological data. A short description of them follows.

Table 1. Present taxonomic statement of the antarctic fish suborder Notothenioidei (families, number of genera and number of species)

Family	Genera	Number of species
Bovichtidae	2	10
Pseudaphritidae	1	1
Eleginopsidae	1	1
Nototheniidae	13	49
Harpagiferidae	1	6
Artedidraconidae	4	24
Bathydraconidae	10	15
Channichthyidae	11	15

3.1 Family Nototheniidae

Nototheniids (antarctic cod) are the most rich and diverse group. The species of this family are widely distributed in the Southern Ocean with a range extending to the north of the Antarctic Polar Front onto the Southern South American shelf and New Zealand. Nototheniids are basically associated with benthic habitats, but several lineages are adapted in varying degrees to pelagic life. Among the pelagic forms *Pleuragramma antarcticum* represents an ecological novelty in having a life cycle that is entirely free of the substrate. This species has become the major notothenioid component of the water column system of most shelf areas of the Antarctic Ocean. As we will see later, some other nototheniids are specialized for life just beneath the platelet ice in what has been termed the cryopelagic habitat by Andriashev.[28] Most of the species of commercial interest are nototheniids, such as *Notothenia rossii*, *Lepidonotothen squamifrons* and the largest antarctic fish *Dissostichus eleginoides*. This latter species, also known as Patagonian toothfish, represents today the most relevant fin-fish resource of the Southern Ocean, and greatly commands the attention of the CCAMLR because of the increase in illegal fishing in several antarctic areas.

3.2 Family Artedidraconidae

Artedidraconidae (plunderfish) are distinguished from all other notothenioids by the presence of a mental barbel used as both an attractive lure and somato-sensory device. They are very sedentary, spending almost all their time motionless on the substrate.[29] Among them *Histiodraco velifer* is the most common plunderfish in the shallow coastal community we studied at Terra Nova Bay. During diving activities it is not difficult to observe this species at a few meters depth, where it can be easily caught by hand.

3.3
Family Bathydraconidae

The bathydraconids (dragonfish) are all distributed south of the Antarctic Polar Front. They are generally numerous in the coldest, deepest shelf water at the highest latitudes. The species *Bathydraco scotiae* has been collected deeper than any other notothenioid, at 2950 m. Some species are, in contrast, common in shallow waters. For example *Cygnodraco mawsoni* is the most common dragonfish in the fish community at Terra Nova Bay. Here large specimens, often females, are caught near the bottom from 80 to 150 m together with the more common channichthyid species, *Chionodraco hamatus*. In the shallower waters the dragonfish *Gymnodraco acuticeps* is found. This species is characterized by a very specialized mouth, armed with long, sharp teeth adapted for ambush predation on fishes.

3.4
Family Channichthyidae

The channichtyids (icefish) are the more phyletically derived notothenioids.[25,27] They are widely distributed in the antarctic regions. All species lack hemoglobin in the blood and the muscles contain no myoglobin. This unique condition among vertebrates makes this fish one of the most unusual taxon in the world. Most of species are primarily benthic but spend considerable time in the water column, where they prey heavily on krill. Some species have large pelvic fins that can be used both for elevated perching and for enhanced movements in the currents. This is especially true for channichthyid larvae that appear to be pelagic in all species studied. Icefish include exploited species such as the mackerel icefish (*Champsocephalus gunnari*) that is abundant on the fishing grounds of the South Georgia Islands and Kerguelen Plateau.

4
Environmental constraints

Many authors agree that water temperature, perhaps the most characteristic feature of the antarctic marine ecosystem, is not the major factor for which antarctic fish can be compared. For marine organisms cold water is not an unusual environment and low temperature as such is not a limiting factor for biosynthetic processes. Therefore it is unlikely that this parameter was directly responsible for the faunal reduction over geologic time and for the lack of diversity among notothenioids in the modern fauna.[2,3,15] Further, the greatest danger, freezing, has been resolved by the evolution of antifreeze and associated physiology in these fishes. More important, ecological constraints in the modern marine antarctic environment appear to be the presence of continental ice and the trophic conditions which distinguish the antarctic from any other marine habitat.

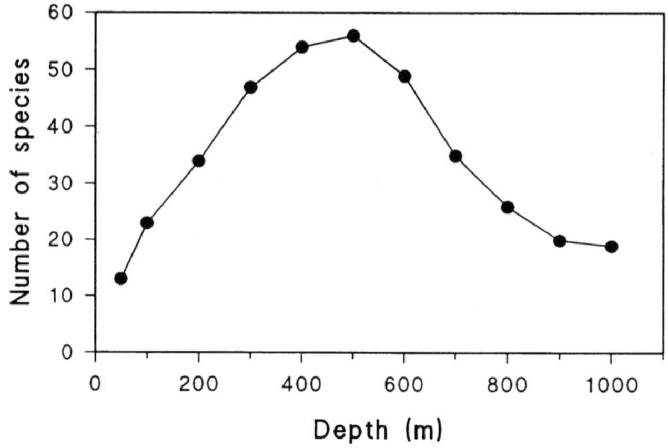

Fig. 1. Pattern of depth distribution of notothenioid species on antarctic shelves
The fish species diversity is greatest at 300–600 m in Antarctica while in temperate waters it is greatest at 100–200 m (redrawn after ref. 32)

4.1
Continental ice and coastal habitat

As a result of isostatic depression of the entire continent and the associated shelf regions by the cap ice since 25–15 million years ago, the antarctic continental shelf is unusually deep, several times greater than that of other continents, and twice as great as the Arctic.[31] Off the European coast the shelf break occurs at a depth of 200 m but in Antarctica it is 500 m on average. As a result of glacial erosion and ice sheets grounded below sea level, in most areas the shelf is narrow. Glaciation has eliminated archipelagos and a variety of shallow water habitats that characterize the coastal environment elsewhere in temperate and tropical coastal areas. Moreover ice shelves and fast ice at present occupy most of the continental margin. Anchor ice coats the bottom to a depth of 30 m and is a main disturbing factor for benthic communities at higher latitude. Destruction and deepening of the shelf habitat has caused the pattern of deep distribution typical of modern notothenioids living on the shelf.

When compared to the depth distribution of fishes living on temperate shelves, species diversity among antarctic notothenioids is greatest at 300–600 m rather than at 100–200 m (Fig. 1). This pattern of depth distribution, known as glacial submergence, may be attributable to the destruction of near-shore bottom habitat by ice shelves.[32]

4.2
Trophic resources

Although antarctic waters exhibit little seasonal variability in temperature, light and productivity show marked seasonal fluctuations. Antarctic light cycles impose strong seasonality on primary production and cause a marked seasonal oscillation in the food supply, especially in some areas. Both the biomass and the primary productivity of phytoplankton show seasonal and latitudinal variability in the Southern Ocean, requiring an adequate ecological response on the part of the feeders. Seasonality has been considered far more important than temperature in determining the character of polar ecosystems.[33]

Together with seasonality, productivity patchness is another relevant feature of antarctic waters. For example at Terra Nova Bay the biomass of some benthic mollusks varies greatly within just a few meters.[34] This distribution determines local feeding communities and can also lead to strict predator-prey relationships between the fish species and locally abundant benthic resources. An example is given in section 6 of this chapter describing the relationships between the scallop *Adamussium colbecki* and the nototheniid *Trematomus bernacchii*.

It could be objected that the seasonality would little affect species such as notothenioids which are mainly formed by benthic species because the benthos is a seasonally stable resource and theoretically available to the fish at all times of the year. However, the macrobenthic communities are mostly composed of scarcely edible items like sponges, sedentary polychaetes, sea urchins, sea stars and sea spiders and holothurians.[35,36]

As we will see below, the fish have developed several strategies in both their life history, niche breadth, and in their reproductive features to overcome seasonality and patchness of the trophic resources of antarctic waters.

4.3
Trophic resources and antarctic fish diversity

It is interesting to consider how purely trophic factors more than physical or oceanographic factors could have been important in conditioning the diversity of the teleosts in the Southern Ocean. In evaluating the pattern of diversification in fossil and recent antarctic fish, Eastman[2] follows Thomson[37] in suggesting that, as in other fish groups, in the evolution of antarctic fish there may have been a ceiling on diversity imposed largely by trophic resources. Although the high kinematic viscosity and related high cost of swimming at subzero temperatures could have been also important constraints, seasonal oscillation in the food supply is likely to have constrained the entry of certain groups of filter feeders into the ecosystem. For instance Clupeiforms, that were present in fossil antarctic fauna and are present in the waters off southern South America today, are absent from the modern antarctic fauna. Equally absent today are all phytoplanktonic feeders and obligatory mass consumers of planktonic crustaceans. Niches for filter feeders might be limited by the patchiness of the productivity in the summer or by the overall low productivity in the ecosystem during the austral winter.[2,15]

On the other hand, it is equally interesting to underline that the lack of fish competitors in the filter-feeding niche has probably permitted the ecological diversification that has given the opportunity to benthic species to use the trophic resources in the water column and to support the trend toward pelagization in many antarctic notothenioids.

5
Ecological diversity and life histories

If it is assumed that the ancestral notothenioid stock was benthic, it is not surprising that most of the notothenioid fishes are associated with the benthic habit. However, a clear trend towards morphological and ecological divergence to occupy pelagic, semipelagic, cryopelagic and epibenthic habitats is present in many groups. In some cases the adaptation of species to typical water column habitats has entailed considerable evolutionary modification of the original benthic body form, as we can see in the nototheniid *P. antarcticum* (antarctic silverfish) and in the channichthyid *C. gunnari* (mackerel icefish). At other times, species maintain a prevailing benthic morphology and habit but are able to capture actively planktonic prey in the water column, suggesting also for these species an evolutionary pathway from benthos-directed feeding to the more derived pelagic state.

Unique to polar ecosystems are the ice-associated communities. In Antarctica the undersurface of fast ice, pack ice, icebergs and platelet ice supports a rich community of cryopelagic organisms, composed of the adult and larval stages of invertebrates and some fish species. Some of these species feed directly on ice microalgae, such as amphipods, adult and larval euphausiids, and copepods. Fish enter the community mainly as secondarly consumers and are represented by larval stages of *P. antarcticum* and adults of a few other nototheniids of the genus *Pagothenia*.

These *Pagothenia* have strong adaptations to protect themselves against freezing as well as to modify their buoyancy, swimming behavior and color pattern. They feed on amphipods and copepods, cling to the underside surface of the sea ice when resting, and hide in ice crevices and holes when endangered.[28]

Among the other life histories described in notothenioids, the secondarily pelagic fishes have become permanent members of midwater communities, such as the species of the genera *Dissostichus* and *Pleuragramma*, after having achieved considerable evolutionary modifications in order to adapt to life in the water column. Among them *Pleuragramma* has a life cycle that is entirely pelagic, with pelagic eggs and spawning in the water column. Reduced competition from other fish groups, coupled with a great evolutive plasticity, led to the ecological success of this small fish. *P. antarcticum* is today the major nototheniid species by number and biomass in the water column of most shelves of the Antarctic; it has a circum-antarctic distribution at depths from 0 to 900 m and represents the key water column species in the food web of the antarctic shelf.[38,39] *Pleuragramma* diversified into a pelagic habitat by overcoming the absence of a swimming bladder through density lowering. The main morphological and physiological features of this species, including skeletal system and swimming, lipid storage, sensorial organs, blood and antifreeze, are summarized by Eastman.[40] The same author suggests that pedomorphosis may have played an important role in the evolution of this species toward the pelagic life style. Evidence from comparative analysis of morphogenesis and rate of development of various elements of the bony skeleton supports the hypothesis of a role of pedomorphosis in the diversification of other nototheniids, channichthyids and bathydraconids.[41,42]

6
The fish community at Terra Nova Bay, Ross Sea

Our knowledge of the fish fauna living along the continental coasts of Antarctica is far from being complete, and it is likely that differences in habitat structure, food availability and competition locally influence the ichthyofauna communities. However, some features common to fish coastal assemblages appear from studies undertaken in various continental areas.[43–48] Let us describe here, as an example, the shallow fish community of Terra Nova Bay, the Italian Antarctic base, where ecological studies on both benthos and necton have been carried out for 10 years.[49–52]

As in other High Antarctic coastal zones, at Terra Nova Bay, the benthic communities of the first meters are very poor and disturbed, due to the presence of drifting pack ice.

From 2–4 m down to 70 m depth, the algal community constitutes well defined belts: these represent shelter and food source for diversified and abundant vagile fauna, among which are echinoids and asteroids. Below, at 70-120 m depth, a complex community of sponges and anthozoans characterize the rocks, reaching higher values in biomass and species richness. The soft bottoms are characterized by bivalves and polichaetes with different species at different depths. The scallop *A. colbecki* reaches very high density and biomass values up to 70–80 m and constitutes characteristic discontinuous beds. Below 120–130 m depth, the hard bottoms become very sparse and the dominant serpulids and bryozoans contribute to a great heterogeneity in the microenvironment.

The fish community is dominated by notothenioids with only one non-notothenioid fish found as prey in the stomach of the deep-water notothenioid *D. mawsoni*.[53] The species belong to four notothenioid families and exibit different life styles with a clear vertical distribution. Figure 2 illustrates the species which we consider representatives of the fish life style types at Terra Nova Bay.

Nototheniids are the most common fish, in both specific diversity and abundance. Among them, *T. bernacchii* characterizes the community in the range depth from a few to 150 m constituting 60% of the individuals caught by gill nets. This species has typical benthic habit for most of his life cycle and as a result is strictly linked to the bottom community of invertebrates not only for feeding but also for egg deposition and nesting.

T. bernacchii is an ambush feeder on various benthic prey, usually eating available organisms in proportion to their abundance.[54] Although it can be considered a generalist, at Terra Nova Bay it mainly feeds on mollusks and at some sites appeared to forage somewhat selectively on the scallop *A. colbecki*, with a marked relationship prey-predator.[55] During ichthyological surveys, high concentrations of *T. bernacchii* with the stomach completely full of this prey were found at the same depth level of the scallop, indicating a locally preferential feeding on the dense mollusk beds (Fig. 3). Furthermore, the fish feeding is size-selective, and seemed to affect the scallop population structure. In fact, the analysis of size-frequency distribution during several austral summers showed a significative persistent deficiency of medium-sized scallops of the size classes 40–55 mm, representing the most vulnerable stage in the postsettlement life of the mollusk, the young scallop being just

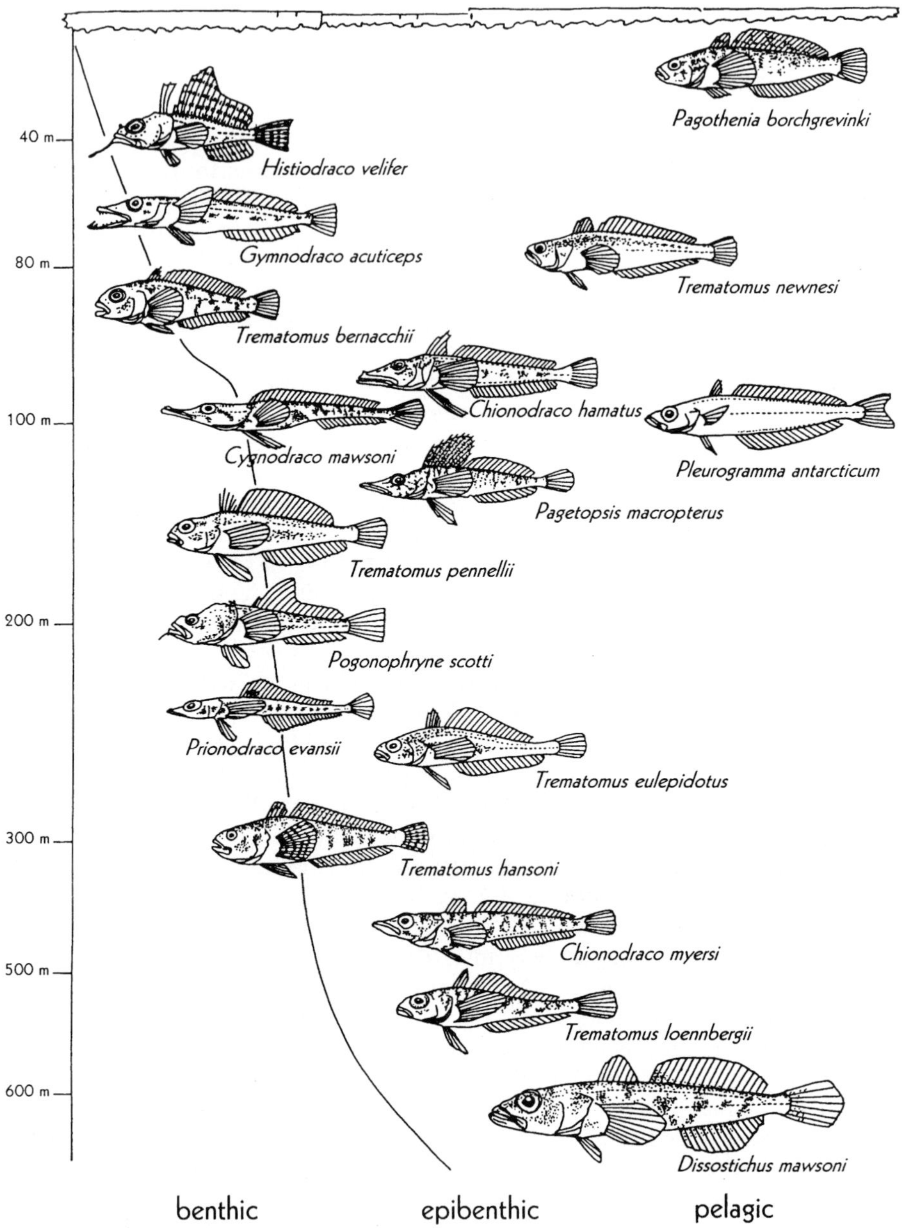

Fig. 2. The fish community at Terra Nova Bay, Ross Sea
The species representatives of the various life style types, at different depth, are illustrated

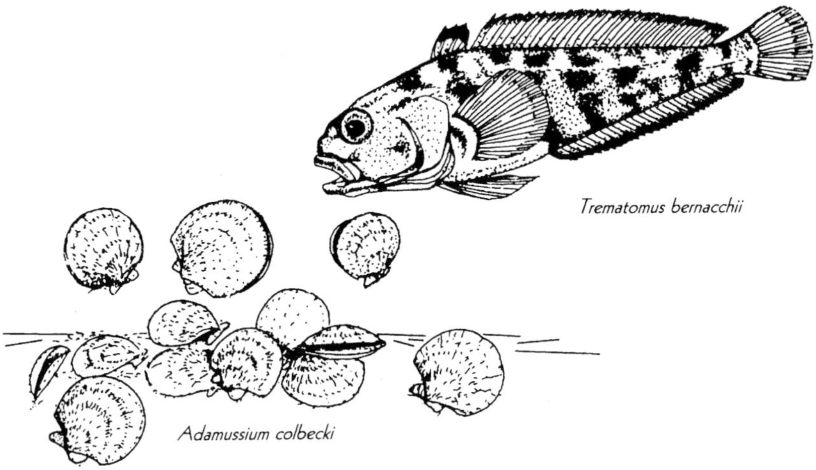

Fig. 3. Trophic relationships between the mollusk *Adamussium colbecki* and the antarctic fish *Trematomus bernacchii* at Terra Nova Bay
At sites where *A. colbecki* is more abundant (40–80 m) the fish preferentially feeds on the dense mollusk bed. The predation affects the population structure of the mollusk, being more effective to the young scallops which are not yet able to escape the predator by effective swimming

detached at this stage from the upper shell valve of adults and not yet able to escape predators by effective swimming performance.[56]

Like most of the notothenioid fishes, *T. bernacchii* has demersal eggs.[15] Underwater observations by Remote Operating Vehicle (ROV) at Terra Nova Bay showed this species on the top of large volcano sponges with aggressive behavior. In populations living in the Antarctic Peninsula this species exhibits a similar aggressive behavior during nesting and egg-guarding and it also undergoes a shift of diet from benthic to more pelagic organisms.[57] Other nototheniids of the genus *Trematomus* characterize the fish community at the different depth levels. Among them, *T. hansoni* inhabits shallow as well as deep waters, whereas *T. loennbergii* is commonly found at depths over 300 m. Both species are active swimming fish, probably sharing benthic and benthopelagic habits, but the feeding strategies adopted allow the two species to reduce interspecific competition. Fish and decapods are the main resources of *T. loennbergii*, which also feeds on epifaunal and tube-dwelling polychaetes. *T. hansoni* mostly relies on fish resources composed of juvenile stages of fish and eggs, the competition for food being mitigated due either to taking different prey or to taking different amounts of the same prey.[58] In other antarctic areas, *T. hansoni* showed great feeding plasticity, being able to vary its diet rapidly according to the availability of the prey in the environment.[59]

In the shallow water of Terra Nova Bay the semipelagic fish life style type is represented by the trematomid *T. newnesi*. This planktivorous feeder mostly utilizes zooplanktonic prey, showing sometimes a high level of cannibalistic behavior.[60] Several morphological and biochemical features in this fish are consistent with an active pelagic life and make this species quite different from others of the same group, like skeletal and lateral line, ocular structure and hemoglobin characteristics, as reported by Eastman.[61] Moreover, specimens studied at McMurdo by the same author revealed another unusual aspect of its biology because here *T. newnesi* was phenotypically plastic, consisting of a typical morph and a large-mouth/broad-headed morph. Because both morphs were caught at the same localities and times, Eastman suggests that the large mouth morph leads to a more benthic existence than the typical morph, probably living on the bottom as an ambush predator on both benthic and water column organisms, while the typical *T. newnesi* morph schools and feeds predominantly in the water column. The two differents morphs have been recently found at Terra Nova Bay.[62]

The small fish *Pagothenia borchgrevinki* is representative of the cryopelagic community at Terra Nova Bay. The ecology of this species and its role in inshore ecosystems have been studied at McMurdo, where it is easily caught on lures. As a typical zooplanktivor, this small fish feeds in the water column where it is able to preferentially select prey. *P. borchkrevinki* can recognize low vibration frequencies emitted by crustacean limb movements.[63] This indicates that, besides sight, the lateral line system may be important in prey detection during the winter months or at lightless depths. Flexibility and opportunism in diet was also observed in this species, which can include benthic foraging, and this can be considered as an ecological adjustment to seasonal variation because of varying ice cover, light availability and primary productivity.[64]

In the Terra Nova Bay community the Channichthtids are represented by the very common species *C. hamatus*, preferentially found at 100 m depth, with *Pagetopsis macropterus* and *Cryodraco* sp. while *Chionodraco myersi* is the deepest icefish. Dragon fish species showed a strict vertical distribution, with *Gymnodraco acuticeps* in very shallow water, *Cygnodraco mawsoni* at 100–200 m and *Prionodraco evansii* at a deeper level. Because of the small size of this latter fish, sampling by common fishing gears probably underestimates its actual abundance. In fact, large number of this bathydraconid were recorded by a ROV survey at a depth of 200 m. Among artedidraconids *Histiodraco velifer* is typical of shallower water and *Pogonophryne scotti* was found at deeper levels. Studies of specific trophic behavior and of other ecological parameters in these fish are underway to clarify how they are adapted to the different benthic and water column subhabitats and how they manage to partition the available resources for food, hiding and spawning.

On the basis of similar studies in high-latitude fish communities other than those in McMurdo Sound, we may expect the presence of a degree of planktivorous feeding in even most benthic species. For instance, *T. hansoni* and *T. nicolai* at McMurdo Sound also feed on plankton, indicating a trend toward the utilization of the water column resources, according to the greater feeding opportunities.[65] As a general rule, the overlapping of the benthic and pelagic habits seems a common ecological feature among notothenioids and can be considered adaptive to the trophic conditions of the antarctic ecosystem.

7
Reproductive strategies

Because of the importance of reproduction in the management of fish resources, most information on spawning, fecundity, sexual maturity and larval studies concerns subantarctic species, especially those living around Kerguelen, South Georgia and South Orkney, because of the relevant fishing grounds in these areas.[15] Instead, the species living at higher latitude have been studied more recently in the Weddell Sea, the Ross Sea and Mawson Sea.[52,66–70] Some reproductive features can be considered typical of an antarctic fish community. They are a prolonged gametogenesis, a biennial process of oocyte maturation, low fecundity and relatively large yolky eggs. Moreover, antarctic fish have an unusually long larval life if compared to species in temperate waters.

A comprehensive review of all the aspects of reproduction is contained in Kock's book,[15] that also reports data on larval development and fish growth, with special emphasis on management and fish stock strategies. Readers are referred to this volume for more detailed information and for additional references.

7.1
Ovarian cycle and spawning

Although spawning probably occurs every year, the cycle of maturation of ova lasts 2 years in the nototheniids and channichthyids studied. In the nature females, ova of two different stages of maturation and size are typically found in the ovaries: large yolky oocytes form the current season's spawn, and small ova in various stages of previtellogenic growth represent the next year's spawn. Given the presence of only one generation of large oocytes of very similar size, multiple spawning during the year seems to be unlikely and eggs are probably released in one batch.

Yolk deposition may occur throughout most of the year in both types of oocytes, although the greatest amount is deposited in the larger oocytes in the months immediately prior to spawning. It is still not clear why the fish take 2 years to lay down yolk in the oocytes. This can be linked to the process of synthesis and accumulation of a great amount of yolk in the eggs; but it has also been suggested that this feature is again linked to the seasonality in antarctic waters, which may permit rapid deposition prior to spawning. Existing information on spawning time of the antarctic fish indicates that the species living in the Seasonal Pack-Ice Zone or around the islands north of it are mostly autumn and winter spawners, while in the High Antarctic Zone a higher proportion of fish are spring and particularly summer spawners.

At Terra Nova Bay the reproductive features of the nototheniids *T. bernacchii* and *T. hansoni* and of the channichthyid *C. hamatus* were studied.[70] Comparison of reproductive periods and fecundity of these species in different antarctic areas showed a significative intraspecific variability. For example, specimens of *T. hansoni* caught off Adélie Land spawned in January/February but not those from the Weddell Sea. Considerable differences also occur between fishes from South Georgia, where they are late summer/early autumn spawners, and those from the Davis Sea, where they are spring spawners. At Terra Nova Bay this species spawned in summer between December and February, with a delay of 2 months compared to Casey Station. Moreover, egg size was higher and fecundity was lower in *T. hansoni* individuals of Terra Nova Bay than of Casey with a general trend, which is mostly confined to the family Nototheniidae, towards lower fecundity but larger eggs towards higher latitudes.[15]

The data on the channichthyid species *C. hamatus* again seem to indicate a shift of the spawning period, which occurs later for the population living at higher latitude (Terra Nova Bay versus Casey Station). However, several variables, dependent or not on latitude, are probably involved in determining the spawning season of a species; for example, Hureau found slight variations in the spawning period of *T. hansoni* sampled off Adélie Land in 2 consecutive years (1961-1962), probably due to variations in the onset of the austral summer.[71]

Although antarctic species are believed to spawn every year, some evidence from the study of the channichthyid *C. gunnari* indicates that 15–25% of the population around South Georgia and an unknown proportion of other populations in the Atlantic Ocean sector do not. The proportion of sexually mature fish at South Georgia which do not spawn in a particular season may be even higher in some years, as in 1990/91.[72] Such an intermission of spawning has also been observed in the population on the Kerguelen shelf, for example in 1983, when spawning concentrations and aggregations of sexually mature nonspawners were observed on the shelf at the same time.[73] Moreover an interruption of the annual gonadal cycle has been reported in *P. antarcticum*.[74] This phenomenon is probably more common than we know and merits further investigation. It could represent an adaptive strategy in the face of a shortage of food or a low energy content of available food. Under these conditions, if fish have been unable to refuel their energy stores sufficiently after the last spawning, egg production is sacrificed for the maintenance of somatic tissue.[75] This reproductive strategy is not uncommon in fish outside the Southern Ocean, such as in atlantic cod in the Baltic Sea or Greenland halibut.[76,77]

7.2 Reproductive effort

Information on the reproductive effort in antarctic fish is also reviewed in Kock's book.[15] Most of the research focuses on reproductive effort in females; let us just outline here some aspects in males. Small changes in lipid content of muscle and liver of males in the course of the maturation process suggest that males invest much less energy in maturation of the gonads than do females. This apparently minor reproductive effort in males may be partly balanced by the spawning maturity kept over an extended period of 1–2 months. Furthermore, secondary commitments like territoriality, courtship and parental care may considerably increase reproductive effort in males. This was most obvious in the nest-guarding *Lepidonotothen nudifrons* in aquaria, where nest guarding was carried out only by the male. Moreover, the earlier onset of spawning migration in males of a number of species, such as *Chaenocephalus aceratus* and *L. nudifrons*, may indicate that males establish territories prior to spawning.

7.3 Hatching and larval development

Although the spawning season for most antarctic fishes is relatively short, there is a considerable time lapse between spawning and hatching, ranging from 60 to 270 d.[78] However, the majority of species hatch in the spring and early summer when zooplankton production increases following the onset of the spring phytoplankton bloom. Spring-hatching larvae originate from eggs spawned in early autumn and exhibit a wide range of lengths at hatching (5–17 mm), indicating that different food size spectra are utilized by the feeding larvae, thus allowing for size-dependent resource partitioning by the abundant cooccurring larvae. A number of species follow a different reproductive strategy, hatching during the winter. They all have large larvae emerging from large and yolky eggs of 4–5 mm in diameter. These larvae have lower relative food requirements and large search volumes so that they can cope with food scarcity under the winter ice cover. However, the larvae of some species form part of the cryopelagic community feeding on various invertebrates associated with the ice. Besides the trophic niche segregation by larval size variation in spring, there exists a temporal sequence of hatching periods by species throughout the summer, probably related to the sequential occurrence of reproductive peaks of calanoid copepods.

In addition to the timing of spawning and hatching, the duration of the pelagic phase is a further important factor within reproductive strategies. In some species of artedidraconids, harpagiferids, and several nototheniids producing small eggs hatching in spring, the pelagic phase is completed before or at the onset of the following winter and pelagic development is confined to the period of elevated plankton production. Some other species, also producing small eggs, extend larval development over the following winter, depending for food on seasonally nonmigrating plankton component during winter. The delayed transition to demersal life in these fishes may allow for a settlement of the juveniles when competition on the bottom is lowest.

Most icefish generate large eggs and spawn throughout the year, producing larvae with spring, summer and winter hatching. In these species hatching occurs independently of the zooplankton production cycle and is linked instead to the early ontogeny of krill and nototheniid fish: in these species the pelagic phases are extended over most of the summer and winter months. The prolonged occurrence of early larvae in the plankton increases the chance of matching, the development of suitable food, and may compensate for the variability of their nearshore nutritional environment.

8 Antarctic fish biodiversity: ecological versus taxonomic diversity

In summary, focus upon antarctic fish biodiversity revealed that the antarctic fish fauna is small and limited in taxonomic diversity being dominated by a single perciform group; but if we consider the subzero shelf and slope waters, species of the suborder Notothenioidei occupy virtually all benthic and water column habitats and represent 90–95% of the biomass.[2,10,48] In some cases, as reported for the Weddell Sea, the only notothenioid family, Nototheniidae, accounts for more than 90% of the specimens captured.[29] In the shallow waters at the highest latitudes, such as McMurdo Sound and Terra Nova Bay, the diversity is further reduced.[45,52] Despite this low taxonomic diversity, notothenioids are unusual among marine fishes in their high ecological and morphological diversity. While there are other marine perciform groups that greatly exceed notothenioids in species richness, there is no other group that exibits similar ecological diversification and dominance of a continental shelf habitat. The reduced competitive pressure from other fish groups in antarctic waters probably enhanced the ecological opportunities of notothenioids. Therefore these fishes have been able to experiment a great field of morphoecological change allowing them to occupy many different niches that are filled by taxonomically unrelated fish in temperate and tropical oceans.

A number of life histories are described and a trend to diversification toward semipelagic and pelagic habitats is clear.[2,24] The trend toward a pelagic life can be considered the main relevant feature in both phyletic and ecological diversification of many notothenioid groups. As already reported, the antarctic silverfish *P. antarcticum* has become the key species of the antarctic pelagic coastal community after a series of morphological, behavioral and physiological adaptations. The high degree of planktonic feeding in most benthic species, as described for the nototheniids *T. hansoni* and *T. nicolai*, is another expression of the high ecological diversity in antarctic fish. The extended larval phase of notothenioids, in comparison with fish of temperate waters, may be considered another expression of the ecomorphological diversification toward the utilization of the water column resources, according to the greater feeding opportunities.

Their ecological diversity is further enhanced by the ontogenetic habitat shifts that are experienced by larvae and juveniles of many species.[20]

Moreover, the phenotypic plasticity described in populations of the nototheniid *T. newnesi* seems to indicate that another small-scale diversification has taken place within the larger radiation of notothenioidei, leading to different ecological morphs at the population level. In shallow water of the higher antarctic latitude, where faunistic diversity is strongly reduced, ecological morphs may represent another way to utilize available different resources of shallow inshore waters.[61] This, in turn, also means that the differentiation of ecological morphs could be another important source of antarctic fish diversity, until now unsufficiently known.

As research increases, it appears more and more evident that ecological diversity in antarctic fish is much more advanced than their taxonomic diversity. If and how much the ecological diversification of antarctic fish possibly precedes their phyletic diversification is one of the most exciting areas for future investigation.

Acknowledgements. Our research is founded by the Italian National Program for Antarctic Research (PNRA). It is also part of international cooperative projects for antarctic research between Italy (PNRA), France (IFRTP and Muséum National d'Histoire Naturelle, Paris) and Australia (Australian Antarctic Division, Tasmania). Figures 2 and 3 were drawn by Dr. Maurizio Wurtz.

9 References

1. Nelson JS. Fishes of the World, 3rd ed. New York: John Wiley & Sons, 1994.
2. Eastman JT. Antarctic Fish Biology. Evolution in a Unique Environment. San Diego: Academic Press, 1993.
3. Wilson E. The Diversity of Life. Cambridge, MA: Harvard University Press, 1982.
4. Ruud JT. Vertebrates without erythrocytes and blood pigment. Nature 1954; 60:848-850.
5. DeVries AL. Glycoproteins as biological antifreeze agents in Antarctic fishes. Science 1971; 172:1152-1155.
6. DeVries AL, Cheng C. Origin and mechanism of evolution of antifreeze glyproteins in polar fishes. In: di Prisco G, Pisano E, Clarke A, eds. Fishes of Antarctica. A Biological Overview. Milan: Springer, 1998:311-328.
7. Andriashev AP. Fishes of the Northern Seas of the USSR. Keys to the Fauna of the USSR. Zoological Institute of the USSR Academy of Sciences. 1954, no. 53.
8. Eastman JT. Comparisons of the Antarctic and Arctic Fish Faunas. Cybium 1997; 21:335-458.
9. Andriashev AP. A general review of the Antarctic fish fauna. In: van Oye P, van Mieghen, J, eds. Biogeography and Ecology in Antarctica. The Hague, The Netherlands: Junk, 1965:491-550.
10. DeWitt HH. Coastal and deep water benthic fishes of the Antarctic. Antarct Map Folio Ser Am Geophys Un 1971; 15:1-10.
11. Everson I. Fish biology. In: Laws RM, ed. Antarctic Ecology. London: Academic Press, 1984; 2:491-532.
12. Fischer W, Hureau JC. FAO species identification sheets for fishery purposes. Southern Ocean (Fishing Areas 48, 58 and 88) (CCAMLR Convention Area). Rome: FAO, 1985; 2:233-470.
13. Gon O, Heemstra PC. Fishes of the Southern Ocean. Grahamstown: JLB Institute of Ichthyology, 1990.
14. Miller RG. History and Atlas of the Fishes of the Antarctic Ocean. Carson City, Nevada: Foresta Institute for Ocean and Mountain Studies, 1993.
15. Kock KH. Antarctic Fish and Fisheries. Cambridge: University Press, 1992.
16. Knox GA. The Biology of the Southern Ocean. Cambridge: University Press, 1994.
17. Kennett JP. Cenozoic evolution of Antarctic glaciation, the circum-Antarctic Ocean, and their impact on global paleogeography. J Geophys Res 1977; 82:3483-3859.
18. Kennett JP, Houtz RE, Andrews PB, Edwards AR, Gostin VA, Hajos M, Hampton M, JenkinsDG, Margeolis SV, Overshine AT, Perch-Nielsen K. Cenozoic paleoceanography in the southwest Pacific ocean. Antarctic glaciation, and the development of the circum-antarctic current. Initial Reports of the Deep Sea Drilling Project, 1983; 29:1155-1169.
19. Adie RT. Geological history. In: Priestley R, Adie RJ, de Robin G, eds. Antarctic Research. A Review of British Scientific Achievement in Antarctica. London: Butterwords, 1964:118-162.
20. Eastman JT. The evolution of Antarctic fishes: questions for consideration and avenues for research. Cybium 1995; 19:371-389.
21. Clarke A, Johnston IA. Evolution and adaptive radiation of Antarctic fishes. Trends Ecol Evolution, 1996; 11:212-218.
22. Eastman JT, A Clarke. A comparison of adaptive radiation of Antarctic fish with those of non-Antarctic fish In: di Prisco G, Pisano E, Clarke A, eds. Fishes of Antarctica. A Biological Overview. Milan: Springer, 1998:3-28.
23. Kennett JP. Marine Geology. New Jersey, Prentice-Hall: Englewood Cliff, 1982.
24. Klingenberg CP, Ekau W. A combined morphometric and phylogenetic analysis of a ecomorphological trend towards pelagization in Antarctic fishes (Perciformes: Nototheniidae). Biol J Linnean Soc 1966; 59:143-177.

25. Balushkin AV. Classification, phylogenetic relationship, and origin of the families of the suborder Notothenioidei (Perciformes). J Ichthyol 1992; 32:90-110.
26. Pisano E, Ozouf-Costaz C, Prirodina V. Chromosomes diversification in Antarctic fishes (Notothenioidei). In: di Prisco G, Pisano E, Clarke A, eds. Fishes of Antarctica. A Biological Overview. Milan: Springer, 1998:275-286.
27. Bargelloni L, Lecointre G. Four years in Notothenioid systematics: a molecular perspective. In: di Prisco G, Pisano E, Clarke A, eds. Fishes of Antarctica. A Biological Overview. Milan: Springer, 1998:259-274.
28. Andriashev AP. Cryopelagic fishes of the Arctic and Antarctic and their significance in polar ecosystems. In: Holdgate MW, ed. Antarctic Ecology. London: Academic Press, 1970:297-304.
29. Hubold G. Ecology of notothenioid fish in the Weddell Sea. In: di Prisco G, Maresca B, Tota B, eds. Biology of Antarctic Fish. Berlin: Springer, 1991:3-22.
30. Clarke A. Temperature and evolution: Southern Ocean cooling and the Antarctic marine fauna. In: Kerry KR, Hempel G, eds. Antarctic Ecosystems. Ecological Change and Conservation. Berlin: Springer, 1990:9-22.
31. Elliott DH. Physical geography-geological evolution. In: Holm-Hansen O, Gilles R, eds. Key Environments: Antarctica. Oxford: Pergamon Press, 1985:39-61.
32 Andriashev AP. A general review of the Antarctic bottom fish fauna. In: Kullander SO, Fernholm B, eds. Proc 5th Congr Eur Ichthyol Stockholm. Stockholm: Swedish Museum of Natural History, 1987:357-372.
33. Dumbar MJ. Ecological Development in Polar Regions: A Study in Evolution. New Jersey, Prentice Hall: Englewood Cliff, 1968.
34. Cattaneo-Vietti R, Chiantore M, Albertelli G. The population structure and ecology of the Antarctic scallop *Adamussium colbecki* (Smith,1902) at Terra Nova Bay (Ross Sea, Antarctica). Sci Mar 1997; 61(Suppl 2):15-24.
35. Dell RK. Antarctic benthos. In: Russell FS, Yonge M, eds. Advances in Marine Biology. London: Academic Press, 1972:10:1-216.
36. White MG. Marine benthos. In: Laws RM, ed. Antarctic Ecology. London: Academic Press, 1984; 2:421-461.
37. Thomson KS. The pattern of diversification among fishes. In: Hallam A, ed. Pattern of Evolution as Illustrated by the Fossil Record. Amsterdam: Elsevier, 1977:377-404.
38. Hempel G. Antarctic marine food webs. In: Siegfried WR, Condy PR, Laws RM, eds. Antarctic Nutrient Cycles and Food Webs. Berlin, Heidelberg: Springer, 1985:226-270.
39. Hubold G, The early life history of the High Antarctic silverfish *Pleuragramma antarcticum*. In: Siegfried WR, Condy PR, Laws RM, eds. Antarctic Nutrient Cycles and Food Webs. Berlin, Heidelberg: Springer, 1985:445-451.
40. Eastman JT. Phyletic divergence and specialization for pelagic life in the Antarctic nototheniid fish *Pleuragramma antarcticum*. Comp Biochem Physiol 1997; 118A:1095-1101.
41. Iwami T. Heterochrony in the family Channichthyidae (Perciformes, Notothenioidei) and its phylogenetic implications. Antarctic communities, species structure and survival. Abstract 6th SCAR Biol Symp "Antarctic communties, species, structure and survival", Venice, 1994:134.
42. Voskoboinikova OS. Osteological development of the Channichthyidae (Teleostei, Notothenioidei). Cybium 1997; 21:369-379.
43. Hureau JC. Poissons antarctiques récoltés au cours de la onzième expedition française en terre Adélie (1960-1962). Bull Mus Nat Hist 1962; 34:22-238.
44. DeWitt HH. Coastal and deep-water benthic fishes of the Antarctic. Antarctic Map Folio Series 1971; 15:1-10.
45. Eastman JT, DeVries AL. Buoyancy studies of notothenioid fishes in McMurdo Sound, Antarctica. Copeia, 1982; 2:385-393.
46. Naito Y, Iwami T. Fish fauna in the northeastern parts of Lutz-Holm Bay with some notes on stomac contents. Memoirs Natl Inst Polar Res, Special Issue 1982; 32:64-72.

47. Williams R. The inshore fishes of the Vestfold Hills region, Antarctica. Hydrobiologia 1988; 165:161-167.
48. Ekau W. Demersal fish fauna of the Weddell Sea, Antarctica. Antarctic Sci 1990; 2:129-137.
49. Gambi MC, Lorenti M, Russo GF, Scipione MB. Benthic associations of the shallow hard bottoms off Terra Nova Bay, Ross Sea: zonation, biomass and population structure. Antarctic Sci 1994; 6:449-462.
50. Cattaneo-Vietti R, Chiantore M, Gambi MC, Albertelli G, Cormaci M, Di Geronimo I. Spatial and vertical distribution of benthic littoral communities in Terra Nova Bay. In: Guglielmo L, Ianora A, eds. Ross Sea Ecology. Italian Antarctic Expeditions (1987-1995). Berlin: Springer, 1999 (in press).
51. Vacchi M, Greco S, La Mesa M. Ichthyological survey by fixed gears in Terra Nova Bay (Antarctica). Fish list and first results. Memorie di Biologia Marina ed Oceanografia 1991; 19:197-202.
52. Vacchi M, La Mesa M, Greco S. Some ecological features of the coastal fish community of Terra Nova Bay. In: Guglielmo L, Ianora A, eds. Ross Sea Ecology. Italian Antarctic Expeditions (1987-1995). Berlin: Springer, 1999 (in press).
53. Vacchi M, Greco S. Capture of the giant nototheniid fish *Dissostichus mawsoni* in Terra Nova Bay (Antarctica): Notes on the fishing equipment and the specimens caught. Cybium 1994; 18:199-203.
54. Kiest KA. Relationship of diet to prey abundance and the foraging behavior of *Trematomus bernacchii*. Polar Biol 1993; 13:291-296.
55. Vacchi, Cattaneo-Vietti, Chiantore M, Dalù M. Prey-predator relationship between Antarctic scallop *Adamussium colbecki* and spotted trematom *Trematomus bernacchii* in Terra Nova Bay (Ross Sea). Abstract 7th SCAR Int Biol Symp, ChristChurch, New Zealand, 1998.
56. Chiantore MC, Cattaneo-Vietti R, Povero P, Albertelli G. The population structure and ecology of the Antarctic scallop *Adamussium colbecki* in Terra Nova Bay. In: Guglielmo L, Ianora A, eds. Ross Sea Ecology. Italian Antarctic Expeditions (1987-1995). Berlin: Springer, 1999 (in press).
57. Moreno CA. Observations on food and reproduction in *Trematomus bernacchii* (Pisces: Nototheniidae) from the Palmer Archipelago, Antarctica. Copeia 1980; 1:171-173.
58. La Mesa M, Vacchi M, Castelli A, Diviacco G. Feeding ecology of two nototheniid fishes, *Trematomus hansoni* and *Trematomus loennbergii*, from Terra Nova Bay, Ross Sea. Polar Biol 1977; 17:62-68.
59. Pakhomov EA. Feeding plasticity of the Antarctic fish *Trematomus hansoni* Boulenger, 1902 (Pisces:Nototheniidae): the influence of fishery waste on the diet. Polar Biol 1998; 19:289-292.
60. Vacchi M, La Mesa M. The diet of the Antarctic fish *Trematomus newnesi* Boulenger, 1902 (Nototheniidae) from Terra Nova Bay, Ross Sea Antarctic Sci 1995; 7:37-38.
61. Eastman JT, DeVries AL. Biology and phenotypic plasticity of the Antarctic nototheniid fish *Trematomus newnesi* in McMurdo Sound. Antarctic Sci 1997; 9:27-35.
62. Vacchi M, La Mesa M, Zunini-Sertorio T. Planktivory in *Trematomus newnesi* (Pisces: Nototheniidae) at Terra Nova Bay, Ross Sea: feeding plasticity related to environmental conditions. Abstract 7th SCAR Int Biol Symp, ChristChurch, New Zealand, 1998.
63. Montgomery JC, MacDonald JA. Sensory tuning of lateral line receptors in antarctic fish to the movements of plantonic prey. Science, 1987; 235:1195-196.
64. Foster BA, Cargil JM, Montgomery C. Planktivory in *Pagothenia borchgrevinki* (Pisces: Nototheniidae) in McMurdo Sound, Antarctica. Polar Biol 1987; 8:49-54.
65. Montgomery JC, Foster BA, Milton RC, Carr E. Spatial and temporal variation in the diet of nototheniid fish in McMurdo Sound, Antarctica. Polar Biol 1993; 13:429-431.
66. Ekau W. Reproduction in high Antarctic fishes (Notothenioidei). Meeresforsch 1991; 33:159-167.
67. Duhamel G, Kock KH, Balguerias E, Hureau JC. Reproduction in fish of the Weddell Sea. Polar Biol 1993; 13:193-200.

68. Pankhurst, NW. Growth and reproduction of the Antarctic nototheniid fish *Pagothenia borchgrevinki*. Polar Biol, 1990; 10:387-391.
69. Shandikov GA, Faleeva TI. Features of gametogenesis and sexual cycles of six notothenioid fishes from East Antarctica. Polar Biol 1992; 11:615-621.
70. Vacchi M, Williams R, La Mesa M. Reproduction in three species of fish from the Ross Sea and Mawson Sea. Antarctic Sci 1996; 8:185-192.
71. Hureau JC. Biologie comparée de quelques poissons antarctiques (Nototheniidae). Bull Inst Océanogr Monaco 1970; 68:11-250.
72. Kock KH. Reproduction of the mackerel icefish (*Chamspocephalus gunnari*) and its implications for fisheries management in the Atlanctic sector of the Southern Ocean. SC-CAMLR Selected Scientific Papers, 1990:51-68.
73. Duhamel G. Reproduction des Nototheniidae et Channichthyidae des Iles Kerguelen. In: Comité National Français des Recherches Antarctiques 577, Territoire des Terres Australes et Antarctiques Françaises, ed. Actes du Colloque sur l'Ecologie Marine des Iles Subantarctiques et Antarctiques, 1987:1-9.
74. Faleeva II, Gerasimchook VV. On the reproductive biology of the Antarctic sidestripe *Pleuragramma antarcticum* (Nototheniidae). Voprosy Ikthiologii,1990; 30:416-424.
75. Tyler AV, Dunn RS. Ratio, growth and measures of somatic and organ condition in relation to mean frequency in winter flounder, *Pseudopleuronectes americanus*, with hypothesis regarding population homeostasis. J Fish Res Board Can 1976; 33:63-75.
76. Thurow F. Über die Fortpflanzung des Dorsches *Gadus morhua* L in der Kieler Bucht. Berichte Dt Wiss Kommiss Meeresforsch 1970; 21:1170-192.
77. Fedorov KY. The state of the gonads of the Barents Sea Greenland halibut *Reinhardtius hippoglossoides* (Walb.) in connection with failure to spawn. J. Ichthyol 1971; 11:673-682.
78. Burchett MS, Sayers PJ, North AW, White GM. Some biological aspects of the nearshore fish population at South Georgia. Brit Antarctic Survey Bull 1983; 59:63-64.

Structure and function of hemoglobins from antarctic organisms: the search for correlations with adaptive evolution

G. di Prisco*, R. D'Avino and M. Tamburrini

Institute of Protein Biochemistry and Enzymology, C.N.R., Via Marconi 10, I-80125 Naples, Italy

1 Introduction

For the study of temperature adaptations, Antarctica, more than any other habitat on earth, is indeed a unique natural laboratory. After 15 years of extensive investigations on the molecular basis of cold adaptation in antarctic marine and terrestrial organisms, we are finally able to begin identifying firm guidelines for understanding the interplay among biochemical/ physiological processes of oxygen transport, ecology and adaptive evolution.

Oxygen carriers are one of the most interesting systems for studying the interrelationships between environmental conditions and molecular evolution. It is easy to imagine that hemoglobin (Hb), being a direct link between the exterior and body requirements and fulfilling its primary function under extremely variable conditions, has experienced a major evolutionary pressure to adapt and modify its functional features. Therefore, one of our research areas is the structure/function of Hb in fishes and birds.[1–4] This chapter summarizes the advances in the study of molecular adaptations in Hbs of antarctic organisms. In view of the role of temperature in modifying the oxygenation-deoxygenation cycle in respiring tissues, thermodynamic analysis deserves special attention. X-ray crystallography and molecular modeling are additional tools to correlate amino acid sequence, multiplicity and oxygen binding features of Hb with ecological constraints. Amino acid sequences are used as source of evolutionary information.

* Corresponding author; e-mail: diprisco@dafne.ibpe.na.cnr.it

2
Fish

2.1
Antarctic Notothenioidei

The perciform suborder Notothenioidei, mostly confined within antarctic and sub-antarctic waters, is the dominant component of the Southern Ocean fauna. Ninety-five of the 174 benthic species of the continental shelf and upper continental slope, and 120 (including nonantarctic species) of the 274 Southern Ocean species described to date[5] are notothenioids. There is lack of information on their site of origin, existence of a transition fauna and time of their radiation in the Antarctic. Indirect indications suggest that notothenioids appeared in the early Tertiary[6] and began to diversify in isolation on the antarctic continental shelf in the middle Tertiary, gradually adapting to progressive cooling. Bovichtidae, Nototheniidae (with the highest number of species, 34 antarctic and 15 nonantarctic), Harpagiferidae, Artedidraconidae, Bathydraconidae and Channichthyidae are the families of the suborder. Notothenioids are red-blooded, except Channichthyidae, the only known vertebrates whose pale whitish blood is devoid of Hb.[7]

Our studies on the oxygen-transport system have so far been addressed to 38 out of a total of 80 red-blooded notothenioid species, encompassing all families.

In the first stages of our research, the antarctic fish fauna seemed a rather uniform and simplified community. The fish[1,2,8] had a single major Hb (Hb 1) and often a minor one (Hb 2, generally sharing the β chain, and sometimes the α chain) both displaying – with some exceptions – strong Bohr[9] and Root[10] effects (Table 1). All these species are bottom dwellers, and this was a first important correlation. A general trend was found in the amino acid sequences of antarctic Hbs (Table 2). The sequences of major and minor Hbs cluster in two groups; in each group, identity is high (73–99 and 84–100%). However, the identity between major and minor Hbs is lower, ranging between 61 and 73%. The analysis of sequences, together with the similar functional features of major and minor Hbs in a given species, led us to

Table 1. Hemoglobin (Hb) components and oxygen-transport system of Notothenioidei

Family	No. of species examined	Major Hb components	Bohr and Root effects[a]
Bovichtidae/Pseudaphritidae	1	1 (95%)	Strong
Nototheniidae	16	1 (95–99%)	Strong[b]
Bathydraconidae	8	1 (90–99%)	Strong[c]
Artedidraconidae	9	1 (99%)	Weak
Harpagiferidae	1	1 (99%)	Not investigated

[a] Bohr coefficients ($\Delta \log P_{50}/\Delta pH$) higher than -0.6 and lower than -0.3 denote *strong* and *weak* Bohr effects, respectively. *Strong* and *weak* Root effects correspond to Hb oxygenation at atmospheric pressure, pH 6.0, lower than 40% and higher than 80%, respectively.

[b] Except in *Aethotaxis mitopteryx*.

[c] Except in *Gymnodraco acuticeps*.

conclude[8] that minor Hbs are vestigial (perhaps larval) remnants, devoid of physiological significance.

The most surprising findings came when we examined the Hb system of three nototheniids whose life style differs from that of the sluggish benthic species: *Trematomus newnesi*, *Pagothenia borchgrevinki* (two active cryopelagic species) and *Pleuragramma antarcticum* (a pelagic, sluggish but migratory fish). Each species had a unique oxygen-transport system, and each system appeared adjusted to the fish-specific mode of life.

The Hb system of *T. newnesi* is made of Hb C, Hb 1 (with the α chain in common) and Hb 2 (having the β chain in common with Hb 1).[11] This notothenioid is the only species having two major Hbs, only one of which (Hb C, a mere trace in all other notothenioids) displays pH and organophosphate regulation (Table 3). Temperature has little effect on the oxygen affinity of the most abundant component Hb 1 (the small ΔH of oxygen binding is constant in the pH range 8.0–6.5); thus no significant amounts of energy are required during the oxygenation-deoxygenation cycle. This Hb system can ensure oxygen binding at the gills (via Hb 1) and controlled delivery to tissues (via Hb C) also when active behavior produces acidosis. High levels of Hb C, conceivably redundant in other notothenioids (which count on Root and Bohr-effect Hb), balance the lack of proton/effector regulation of Hb 1 and Hb 2.

Pelagic, migratory *P. antarcticum* combines the more general adaptations of all notothenioids with specializations necessary for life in the water column. It has three major Hbs (Hb 1, Hb 2 and Hb 3), the highest multiplicity within notothenioids (Table 3). Hb 1 has the α chain in common with Hb 2 and the β chain with Hb 3. High sequence identity was observed between Hb 1 of *P. antarcticum* and Hb 1 of other species (Table 2). The chains of Hb 2 and Hb 3 which are not in common with Hb 1 have high identity with those of other minor Hbs.[12,13] The Hbs display strong, effector-enhanced, dependence of oxygen affinity on pH (alkaline Bohr effect). The Root effect is also displayed by all Hbs, and its amplitude is regulated by the physiological effectors organophosphates and chloride, indicating a strong pH-dependence of Hb oxygenation in air.

Although the oxygen-binding features are similar, these Hbs reveal important differences in thermodynamic behavior (Table 4). Hb 1 and Hb 3 show a very strong enthalpy change at pH 8.0, enhanced by chloride and organophosphates in the former, but drastically decreased in the latter; the heat of oxygenation of Hb 2, in the presence and absence of effectors, is much lower. A dramatic decrease is observed at lower pH in Hb 3 and Hb 2 (ΔH approaches zero); in contrast, Hb 1 retains high oxygenation enthalpy, especially when effectors are absent. These observations clearly indicate a stronger Bohr effect at physiological temperatures in Hb 1 (in the presence of effectors) and Hb 3 (in their absence). In addition, the moderate effect of temperature on Hb 2 in the pH range 7.0–8.0 and on Hb 3 at pH 7.0 is indicative of energy-saving mechanisms of oxygen loading and unloading. From a thermodynamic standpoint, the oxygen-transport system of *P. antarcticum* is one of the most specialized ever found in fish. It appears designed to fit its unusual pelagic life style through refined molecular adaptation in the thermodynamic features of each Hb. Rather than having to respond to acidosis, this Hb system

Table 2. α and β chain sequence identity (%) between major hemoglobins (Hbs) of antarctic and nonantarctic Notothenioidei (sequences are from refs. 11,12,26)

Species	P. *antar.* Hb 3	P. *urvillii** Hb 1,2	N. *angus.** Hb 1	P. *antar.* Hb 1, 2	P. *scotti*	A. *orianae*	R. *glacial.* Hb 1	G. *acut.*	C. *maws.* Hb 1, 2	A. *mitopt.*	P. *bern.* Hb 1	T. *new.* Hb 1, C
α chains												
N. coriiceps Hb 1	69	80	99	95	91	91	95	82	83	83	89	87
T. newnesi Hb 1, C	68	78	94	89	85	86	90	92	90	90	97	
P. bernacchii Hb 1	69	80	96	92	88	87	93	91	91	90		
A. mitopteryx	65	73	91	88	84	84	88	84	84			
C. mawsoni Hb 1, 2	70	78	89	87	85	84	89	93				
G. acuticeps	68	76	89	85	82	82	87					
R. glacialis Hb 1	72	77	94	94	91	91						
A. orianae	69	75	90	88	96							
P. scotti	70	76	90	88								
P. antarcticum Hb 1, 2	72	81	94									
*N. angustata** Hb 1	70	78										
*P. urvillii** Hb 1,2	65											

* Nonantarctic species.

Table 2 – continuation

	P. *antar.* Hb 2	T. *newn.* Hb C	P. *urvillii** Hb 1	N. *angus.** Hb 1,2	P. *antar.* Hb 1, 3	P. *scotti*	A. *orianae*	R. *glacial.* Hb 1	G. *acut.*	C. *maws.* Hb 1	A. *mitopt.*	P. *bern.* Hb 1	T. *new.* Hb 1, 2
β chains													
N. coriiceps Hb 1, 2	72	70	77	93	86	82	83	91	80	88	82	90	86
T. newnesi Hb 1, 2	69	68	77	85	84	77	79	86	80	84	83	93	
P. bernacchii Hb 1	72	70	82	91	90	79	80	90	83	87	86		
A. mitopteryx	69	67	77	84	84	75	76	84	77	80			
C. mawsoni Hb 1	68	70	75	87	82	83	81	92	85				
G. acuticeps	69	67	75	81	80	78	77	83					
R. glacialis Hb 1	72	70	80	92	86	84	84						
A. orianae	70	68	74	84	77	90							
P. scotti	68	65	73	82	76								
P. antarcticum Hb 1, 3	72	69	77	88									
*N. angustata** Hb 1,2	73	71	81										
*P. urvillii** Hb 1	70	69											
T. newnesi Hb C	91												

* Nonantarctic species.

Table 3. Hemoglobin (Hb) multiplicity and oxygen-transport system of *Trematomus newnesi*, *Pleuragramma antarcticum* and *Pagothenia borchgrevinki*

Hemoglobin component	% of total	Bohr and Root effects	
		No ATP	3 mM ATP
T. newnesi			
Hb C	20–25	Strong	Enhanced
Hb 1	70–75	Weak or absent	Not enhanced
Hb 2	3–5	Weak or absent	Not enhanced
P. antarcticum			
Hb 1	25–30	Strong	Enhanced
Hb 2	20–25	Strong	Enhanced
Hb 3	45–50	Strong	Enhanced
P. borchgrevinki			
Hb C	1	Strong	Enhanced
Hb 0	5–10	Strong	Enhanced
Hb 1	70–80	Weak	Not enhanced
Hb 2	5–10	Weak	Slightly enhanced
Hb 3	5–10	Weak	Slightly enhanced

responds to the need to optimize oxygen loading and unloading during seasonal migrations through water masses which can have different and fluctuating temperatures. Thus *P. antarcticum* relies on three Hbs, which differ functionally mainly in thermodynamic behavior rather than pH and organophosphate regulation.

The cryopelagic nototheniid *Pagothenia borchgrevinki* has a higher Hb concentration than other notothenioids;[14] the high oxygen capacity may correlate with the more active mode of life. It has five Hbs (Hb C, Hb 0, Hb 1, Hb 2 and Hb 3), which were purified and functionally characterized (Table 3).[15] Intact erythrocytes have a weak Bohr effect. Hb 1, Hb 2 and Hb 3 are functionally similar, with a slightly stronger Bohr effect (not significantly influenced by the effectors) and a weak Root effect, slightly enhanced by ATP in Hb 2. Hb 0 has a strong Bohr effect and (neglect-

Table 4. Heat of oxygenation of *Pleuragramma antarcticum* hemoglobins (Hb)

Hb	ΔH (kcal mol^{-1} oxygen)[a]			
	No effector		100 mM NaCl, 3 mM ATP	
	pH 7.0	pH 8.0	pH 7.0	pH 8.0
Hb 1	-12.8	-15.3	-8.6	-17.4
Hb 2	-3.6	-6.4	-1.8	-8.1
Hb 3	-0.1	-16.5	-4.1	-7.6

[a] The oxygenation enthalpy change ΔH (kcal mol^{-1}), corrected for the heat of oxygen solubilization (-3 kcal mol^{-1}), was calculated by the integrated van't Hoff equation:
$\Delta H = -4.574\ [(T_1.T_2)/(T_1-T_2)]\ \log P_{50}/1000$
(P_{50} partial pressure of oxygen required to achieve Hb half saturation)

ing Hb C) is the only component with a strong, effector-enhanced, Root effect. Differences were detected in the heat of oxygenation of each component, in the absence and presence of the effectors. The high multiplicity of functionally distinct Hbs indicates that also this active, cryopelagic species has a very specialized Hb system. The structure/function relationship in each Hb is currently under investigation.

2.2 Molecular modeling

The major Hbs of *Trematomus bernacchii* and *T. newnesi* differ only by four residues in the α chain and ten in the β.[11,16] Despite this high identity, the functional behavior of these Hbs is very different. *T. newnesi* Hb 1 has no Root effect and a weak Bohr effect, whereas both effects are very large in *T. bernacchii* Hb 1. Models of the quaternary structure of *T. newnesi* Hb 1 were built by homology modeling,[17] using the crystallographic coordinates of *T. bernacchii* Hb 1 in R and T state[16,18] as a template and BIOSYM/MSI Insight II, with direct residue replacements in the template structure (Fig. 1). *T. newnesi* and *T. bernacchii* Hbs were minimized at pH 8.0, after surrounding the structure with a 5-Å layer of water molecules, using the program Discover_3.

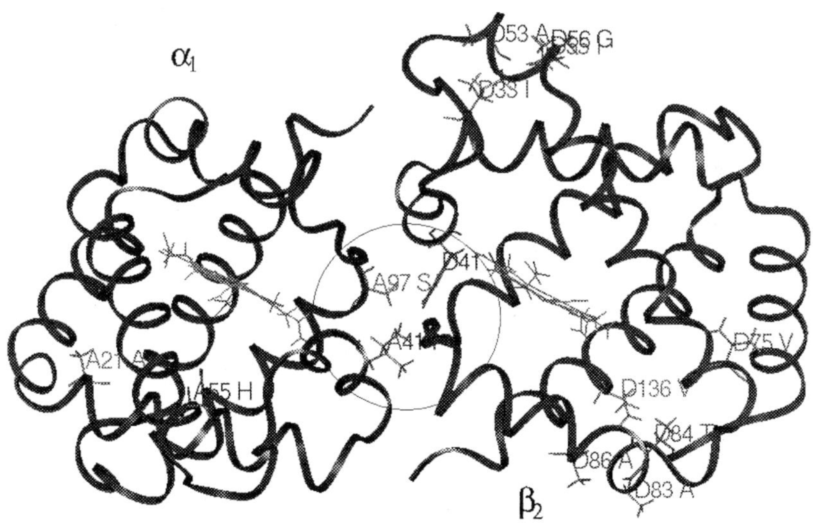

Fig. 1. The $\alpha_1\beta_2$ αC skeleton of T-state *Trematotus bernacchii* hemoglobin (Hb) 1, showing the 14 residues which have been substituted to build the T-state model of *T. newnesi* Hb 1
Labels denote the chain position, followed by the one-letter amino acid code of the residues in *T. newnesii* Hb 1. The *letters* preceding the sequence positions indicate the α chain (A) and the β chain (D); the region involved in the T-state stabilization[18] is shown in the *circle*

The analysis of the two structures did not evidentiate substantial structural differences in the R state, in accordance with the crystallographic structure of *T. newnesi* Hb 1 (Mazzarella et al., unpubl. data). A substitution at the $\alpha_1\beta_2$ interface (crucial in the R-T transition), where bulkier Ile in *T. newnesi* replaces Thr α41 in *T. bernacchii*, causes conformational modifications which may justify the loss of Root effect. In fact displacement of the βG helix away from the interface occurs in the T-state model of *T. newnesi* Hb 1, affecting the interaction between Asp β101 and Asp α95, considered responsible for T-state stabilization in *P. bernacchii.*[18] Antarctic fish (but not temperate fish) lacking the Root effect have this substitution[19] in α41.

2.3 Nonantarctic Notothenioidei: Bovichtidae/Pseudaphritidae

Comparison between cold-adapted and noncold-adapted nonantarctic notothenioids may help to understand their evolutionary history, as well as the molecular mechanisms of cold adaptation. Together with Nototheniidae, Bovichtidae is a notothenioid family inhabiting also temperate waters. Morphological[20,21] and karyological[22] studies indicate that Bovichtidae are the most primitive notothenioid family.[23,24] The family includes 11 species, all living in temperate waters except *Bovichtus elongatus*, found at the Antarctic Peninsula. A recent classification[25] separates the new family Pseudaphritidae from Bovichtidae.

Pseudaphritis urvillii, a bovichtid (or pseudaphritid) very common in estuaries and lower portion of Australian rivers, is considered a relict species. It is euryhaline and may migrate upstream as far as 120 km from the sea. It has no antifreeze glycoproteins.[6] No information was available on the Hb system of this primitive family.

Like most notothenioids, *P. urvillii* has a single major component (Hb 1), accounting for 90% of the total.[26] Hb 1 has a strong Bohr effect; the Root effect is ATP-induced. Although the Hb multiplicity closely resembles that of sedentary antarctic notothenioids, the oxygen affinity of Hb 1 is much higher; in fact, $\log P_{50}$ (partial pressure of oxygen required to achieve Hb half saturation) stays below zero in the pH range 8.0–7.0. This feature can be ascribed to the widely different constraints arising from the habitat of this nonantarctic species, and is likely to be reflected in changes in the primary structure. We have indeed found two substitutions which may alter the geometry of the invariantly hydrophobic heme pocket: in α87(F7) and α92(FG3) Glu and Met replace Leu. Glu α87(F7) is found only in this species; interestingly, all antarctic Notothenioidei have Gln, whose codon differs from that of Glu by a single base change at the first position.

Although *P. urvillii* has never developed cold adaptation, the globin amino acid sequences reveal high identity with those of the other notothenioids. Most of the residues which differentiate antarctic notothenioids from temperate fish are found in *P. urvillii* Hb. Following the general trend, the identity of Hb 1 with antarctic notothenioids (Table 2) is higher (73–81% for the α and 75–82% for the β chains) than with any temperate fish[8] (60–63 and 60–66%, respectively). However, the identity between *P. urvillii* and antarctic Hb 1 is close to, or lower than, the low extreme of the ensemble of values of antarctic notothenioids (82–99 and 77–93% for the α and β chains, respectively). These data argue in favor of a common origin of this species within notothenioids but also suggest that the major Hb has undergone modifications only to a limited extent. If sequence mutations in antarctic fish are indeed related to the development of cold adaptation, this may imply divergence during the first stages of the cooling process, and in any case before the event which gave origin to antifreeze glycoproteins. In contrast, the Hb sequences of the nonantarctic nototheniid *Notothenia angustata*,[27] as well as the finding of antifreeze genes in its genome (Cheng, personal communication) suggests that this fish was cold-adapted prior to its recent migration from antarctic to temperate waters.

We have initiated an investigation on the Hb system of the bovichtids *Cottoperca gobio* and *Bovichtus variegatus*. These species have higher Hb multiplicity. Moreover, preliminary phylogenetic evidence from the sequence of *C. gobio* Hb 1 seem to exclude monophyly of Bovichtidae, since in a tree of tandemly arranged chains *C. gobio* Hb 1 groups together with minor Hbs. These findings support the hypothesis that other bovichtids diverged from antarctic notothenioids before *P. urvillii* and that the latter is a nonbovichtid notothenioid.[25]

3 Birds

3.1 Emperor and Adélie penguins

The blood of 3–4-month-old Emperor penguin (*Aptenodytes forsteri*) has a major (Hb 1) and a minor Hb (Hb 2, approximately 10% of the total), the latter disappearing in the adult. The amplitude of the Bohr effect[4] is reduced in Hb 1 (Fig. 2). This evidence suggests that this Hb does possess peculiar structural features, important for the modulation of its function in relation with the extreme environmental conditions, but not easy to interpret. In fact, the alkaline Bohr effect finds its molecular basis in the preferential binding of protons to deoxyHb with respect to oxyHb; its amplitude is related to the number of protons bound per deoxyHb tetramer minus the number of protons bound to oxyHb. His β146 is a primary candidate as "alkaline Bohr group" because of a strong ionic bond with Asp β94 in the T state, whereas it is highly solvated in the R state. His β146 contributes the chloride-independent part of the Bohr effect of human Hb A (approximately 40%), and Val α1 and Lys β82 contribute the remaining chloride-dependent part. Our evidence show that, despite the presence of His β146, the Bohr effect (approximately 60% of that of Hb A) is abolished in the absence of chloride. This implies that the salt bridge between His β146 and Asp β94 is either not closed in deoxyHb or fails to open upon oxygenation. The latter case seems more likely. Structural investigations, including crystallographic analysis, will provide an unequivocal answer.

The blood of 2–3-week-old Adélie penguins (*Pygoscelis adeliae*) contains six Hbs in approximately equal amounts; five gradually disappear during development. Two-month-old birds retain a major (Hb 1) and a minor component (Hb 2, approximately 10% of the total); the latter disappears in the adult. We have begun

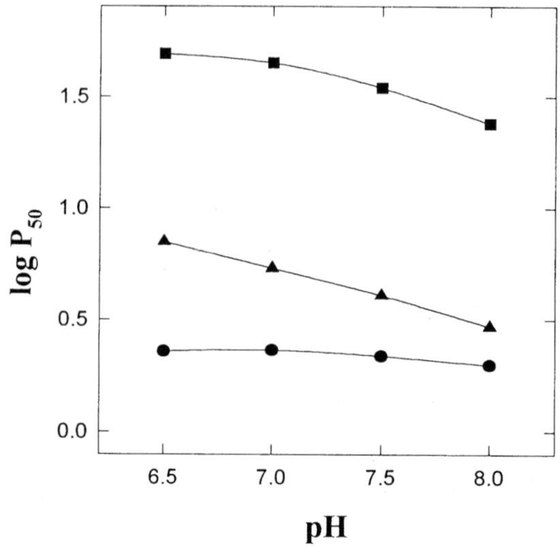

Fig. 2. Effect of pH on the oxygen affinity of Emperor penguin hemoglobin 1 (Hb 1) at 37°C in 0.1 M Hepes (*circles*) and 0.1 M bisTris-HCl or Tris-HCl buffers containing 0.1 M NaCl, in the absence (*triangles*) and presence (*squares*) of 3 mM inositol hexakisphosphate
P_{50} partial pressure of oxygen required to achieve hemoglobin half saturation

studying the amino acid sequence and functional properties of the six Hbs of the chick. A reduced Bohr effect was found in all Hbs, similar to Emperor penguin and south polar skua (see below). The six Hbs had different oxygen affinities, similarly reduced in the entire pH range by a temperature increase. Unlike Emperor penguin, Adélie penguin Hb 1 retained the Bohr effect also in the absence of chloride. Partial amino acid sequences of the three α and the two β chains which build up the six Hbs were elucidated. High identity was found between the α and β chains of the major or minor Hbs of Adélie penguin and the homologous chains of Hbs of Emperor penguin and skua (see below).

3.2
South polar skua

Reproduction of south polar skua (*Catharacta maccormicki*) occurs during the summer along the antarctic coastal regions. This bird has behavioral and physiological characteristics which differ from those of penguins. We elucidated the amino acid sequences and the oxygen-binding properties of the two Hbs, Hb 1 and Hb 2, of the adult.[28]

Unlike in *A. forsteri*, Hb 2 is not a fetal component. Similar to *A. forsteri*, the two Hbs have identical β chains. The α and β chains have 141 and 146 residues, respectively. The identity between the α chains of Hb 1 and Hb 2 is lower (approximately 60%) than that with the homologous chains (approximately 90%) of the two Hbs of *A. forsteri*. The identity between the β chains of *C. maccormicki* and

A. forsteri is 92%. Similar to *A. forsteri*, both Hbs display a small Bohr effect, enhanced by organophosphates in Hb 1. Hb 1 has a lower oxygen affinity than Hb 2, and shows a strong endothermic contribution of chloride binding at pH 6.5. In the pH range 7.5–8.0, the effectors lower ΔH of oxygenation in Hb 1 more than in Hb 2. The curves of oxygen affinity as a function of chloride and organophosphate concentration indicate a strong effect of phosphate, almost constant in the physiological pH range.

The two Hbs of *C. maccormicki* show overall similarities, but also structural and functional differences – probably acquired in order to meet special needs in relation to the environmental conditions – which need further analysis.

4
Conclusions

In antarctic waters, fish had to cope with temperatures below the freezing point of the body fluids and with high oxygen concentrations. They developed physiological and biochemical (often unique) specializations. Cold adaptation was developed in the last 20–30 million years during increasing isolation south of the Polar Front, a barrier to migration in both directions and thus a key factor for fish evolution.

The adaptive reduction in erythrocyte count and Hb content/multiplicity in the blood of antarctic fish counterbalances the increase in blood viscosity produced by subzero seawater temperature[29] with potentially negative physiological effects (i.e. higher demand of energy needed for circulation). The coexistence of Hb-less and naturally cytopenic red-blooded species suggests that the need for an oxygen carrier in a stable, cold environment is reduced also in red-blooded fish. Functional incapacitation of Hb and hematocrit reduction to 1–2% in *T. bernacchii*[29,30] caused no discernible harm in the absence of metabolic challenges. Like channichthyids, red-blooded fish can carry routinely needed oxygen dissolved in plasma.

Channichthyidae, the most phyletically derived notothenioid family, carry oxygen in physical solution at approximately 10% of the carrying capacity of red-blooded notothenioids, and have developed physiological adaptations that maintain adequate tissue oxygenation, e.g. enhanced gas exchange by highly vascularized gills and skin, and increased cardiac output, circulatory volume and heart size. Their genomes share retention of transcriptionally inactive DNA sequences closely related to the α-globin gene of red-blooded notothenioids and loss of the β-globin locus.[31] The most plausible mechanism leading to the Hb-less phenotype might be deletion of the β-globin locus in the ancestral channichthyid; the α-globin genes, no longer under positive selection pressure, would have accumulated mutations which caused loss of gene expression without complete loss of sequence information. The status of globin genes suggests that loss of gene expression is a primitive character (established in the ancestral channichthyid about 25 million years ago, prior to diversification within the clade), and opens promising pathways for evolutionary studies.

Our study has so far been addressed to 38 out of a total of 80 red-blooded antarctic species, a highly representative number indeed. It encompasses all notothenioid families, and has also been extended to four species of nonantarctic notothenioids, making Notothenioidei by far the most thoroughly characterized group of fish in the world. A wealth of data has been gathered on Hb structure and function (including amino acid sequence and X-ray crystallography). Thus, it has become possible to correlate sequence, multiplicity and oxygen-binding features (thermodynamics in particular) with ecological constraints, as shown by the Hb systems of *T. newnesi*, *P. borchgrevinki* and *P. antarcticum*, whose life style differs from that of the benthic species. Information on evolution has often been achieved, for instance from the oxygen-transport system of nonantarctic fish, such as Bovichtidae/Pseudaphritidae, which have been characterized for the first time. The available information on *P. urvillii* and *N. angustata* Hbs suggests a marked temporal difference in their divergence, occurring during the first stages of the cooling process for the most primitive notothenioid and much more recently (after cold adaptation) for the nototheniid. This hypothesis is supported by phylogenetic analysis of our sequences of the α and β chains of antarctic Hbs, together with those of seven nonantarctic species.[32] The phylogenetic trees indicate that Bovichtidae diverged before the formation of the Antarctic Polar Front; the grouping of *N. angustata* with two antarctic nototheniid species supports the hypothesis of recent migration from antarctic to temperate waters.[27] The trees are in agreement with those obtained by morphological analysis and other molecular approaches.[24] They give strong support to monophyly of antarctic notothenioids, with *P. urvillii* as ancestral taxon, and to prior divergence of bovichtids (e.g. *C. gobio*).

Antarctic fish Hbs display large differences in functional characteristics despite high identity in primary structure. Thus the ensemble of antarctic Hbs is a greatly simplified system, which makes it possible to apply molecular modeling as a useful approach to structure/function relationships, not only to tackle intriguing questions pertaining to the antarctic extreme environment (e.g. the lack of subunit cooperativity found in many antarctic Hbs; Tamburrini and di Prisco, unpubl. data), but also questions of general bearing (e.g. the molecular basis of the Root effect, an important functional feature of fish Hb). The sequences of *T. bernacchii* and *T. newnesi* Hb 1 are highly similar[11,16] (and were used in model building), yet their functional behavior is very different. The sequence differences include neither the residues considered responsible for the Root effect in temperate fish nor those involved in the binding of organophosphates.[33] Moreover, crystallographic analysis of the two structures in the R state does not evidentiate substantial structural differences. The limited number of residues involved makes these two functionally distinct Hbs the simplest system so far available for searching differences in the T state. Perhaps it is not surprising that the primary-structure characteristics which are at the basis of the absence or presence of the Root effect in antarctic fish Hb can be different from those of temperate fish, which have been exposed to totally different evolutionary constraints. Structure-function interplay may well entail very refined and often unpredictable mechanisms. Site-directed mutagenesis is likely to become a tool of choice, in combination with molecular modeling, to tackle the molecular basis of the Root effect, as well as to shed light on other open questions.

The oxygen-transport system of birds has been characterized in Emperor and Adélie penguins (during development) and in south polar skua. In Emperor pen-

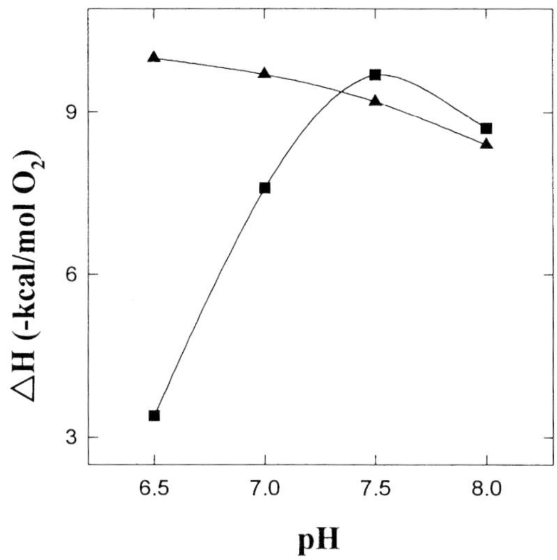

Fig. 3. Overall oxygenation enthalpy as a function of pH of Emperor penguin hemoglobin 1 (Hb 1), in 0.1 M bisTris-HCl or Tris-HCl buffers containing 0.1 M NaCl, in the absence (*triangles*) and presence (*squares*) of 3 mM inositol hexakisphosphate. Data are corrected for the heat of oxygen solubilization (-3 kcal mol^{-1} oxygen)

guin, the Hb features appear well suited to the diving behavior and winter reproduction. The shape of the Bohr effect curve of Emperor penguin Hb seems well adapted for gas exchange during prolonged dives. In fact, the pH dependence of the oxygen affinity, within the physiological pH range, is drastically reduced, preserving this Hb from uncontrolled stripping of oxygen in response to acidosis. Hence, during diving, oxygen delivery should be essentially modulated by the partial pressure of the gas at the specific tissue. Secondly, during dives the peripheral tissues are characterized by acid pH and by a lower temperature than that of the rest of the body; due to the low ΔH observed at low pH (Fig. 3), oxygen transport is not impaired at flippers and feet (which experience lower temperature and great muscular activity), allowing the bird to endure prolonged periods of anaerobiosis. Moreover, Emperor penguins reproduce in winter on the sea ice, and the skin temperature of the feet falls to near 0°C. Following egg laying, the male incubates the egg, keeping a standup position for over 60 d, holding it on the feet against the body to prevent it from freezing, and living on stored fat reserves. This would produce metabolic acidosis, which in turn may benefit tissue respiration at the feet, once again because of the low ΔH of oxygenation observed at acid pH.

Acknowledgements. This research is in the framework of the Italian National Programme for Antarctic Research.

5 References

1. di Prisco G. Physiological and biochemical adaptations in fish to a cold marine environment. In: Battaglia B, Valencia J, Walton DWH, eds. Proc SCAR 6th Biol Symp, Venice (Antarctic Communities: Species, Structure and Survival). Cambridge: Cambridge Univ Press, 1997:251-260.
2. di Prisco G, Giardina B. Temperature adaptation: molecular aspects. In: Johnston IA, Bennett AF, eds. Animals and Temperature. Phenotypic and Evolutionary Adaptation. Soc Exptl Biol, Seminar Series 59. Cambridge: Cambridge Univ Press, 1996:23-51.
3. di Prisco G, Tamburrini M, D'Avino R. Oxygen-transport systems in extreme environments: multiplicity and structure/function relationship in hemoglobins of Antarctic fish. In: Pörtner HO, Playle R, eds. Cold Ocean Physiology. Soc Exptl Biol, Seminar Series 66. Cambridge: Cambridge Univ Press, 1998:143-165.
4. Tamburrini M, Condò SG, di Prisco G, Giardina B. Adaptation to extreme environments: structure-function relationships in Emperor penguin haemoglobin. J Mol Biol 1994; 237:615-621.
5. Gon O, Heemstra PC, eds. Fishes of the Southern Ocean. JLB Smith Institute of Ichthyology, South Africa, 1990.
6. Eastman JT. Antarctic Fish Biology. Evolution in a Unique Environment. San Diego: Academic Press, 1993.
7. Ruud JT. Vertebrates without erythrocytes and blood pigment. Nature 1954; 173:848-850.
8. di Prisco G, D'Avino R, Caruso C, Tamburrini M, Camardella L, Rutigliano B, Carratore V, Romano M. The biochemistry of oxygen transport in red-blooded Antarctic fish. In: di Prisco G, Maresca B, Tota B, eds. Biology of Antarctic Fish. Heidelberg: Springer, 1991:263-281.
9. Riggs AF. The Bohr effect. Ann Rev Physiol 1988; 50:181-204.
10. Brittain T. The Root effect. Comp Biochem Physiol 1987; 86B:473-481.
11. D'Avino R, Caruso C, Tamburrini M, Romano M, Rutigliano B, Polverino de Laureto P, Camardella L, Carratore V, di Prisco G. Molecular characterization of the functionally distinct hemoglobins of the Antarctic fish *Trematomus newnesi*. J Biol Chem 1994; 269:9675-9681.
12. Tamburrini M, D'Avino R, Fago A, Carratore V, Kunzmann A, di Prisco G. The unique hemoglobin system of *Pleuragramma antarcticum*, an Antarctic migratory teleost. Structure and function of the three components. J Biol Chem 1996; 271:23780-23785.
13. Tamburrini M, D'Avino R, Carratore V, Kunzmann A, di Prisco G. The hemoglobin system of *Pleuragramma antarcticum*: correlation of hematological and biochemical adaptations with life style. Comp Biochem Physiol 1997; 118A:1037-1044.
14. Macdonald JA, Wells RMG. Viscosity of body fluids from Antarctic notothenioid fish. In: di Prisco G, Maresca B, Tota B, eds. Biology of Antarctic Fish. Heidelberg: Springer, 1991:163-178.
15. Tamburrini M, di Prisco G. The unique features of the hemoglobin system of the Antarctic teleost *Pagothenia borchgrevinki*. Abstract "Proteine '94", 1994; Abstr A25:43.
16. Camardella L, Caruso C, D'Avino R, di Prisco G, Rutigliano B, Tamburrini M, Fermi G, Perutz MF. Haemoglobin of the Antarctic fish *Pagothenia bernacchii*. Amino acid sequence, oxygen equilibria and cristal structure of its carbomonoxy derivative. J Mol Biol 1992; 224:449-460.
17. De Luca R, di Prisco G, D'Avino R. Homology modelling of the R and T structures of Hb 1 of the Antarctic fish *Trematomus newnesi*. Abstract "Proteine '96", 1996; Abstr B22:84.
18. Ito N, Komiyama NH, Fermi G. Structure of deoxyhaemoglobin of the Antarctic fish *Pagothenia bernacchii* with an analysis of the structural basis of the Root effect by comparison of the liganded and unliganded haemoglobin structures. J Mol Biol 1995; 250:648-658.

19. D'Avino R, di Prisco G. A substitution in the α chain of the main hemoglobin of an Antarctic fish may have a role in the Root effect. Abstract "Proteine '94" , 1994; Abstr A9:27.
20. Iwami T. Osteology and relationships of the family Channichthyidae. Mem Natl Inst Polar Res, Tokyo, 1985; Ser E, 36:1-69.
21. Balushkin AV. Morphological Bases of the Systematics and Phylogeny of Nototheniid Fishes. Leningrad: Acad Sci USSR, Zool Inst, 1984.
22. Prirodina VP. Karyotypes of *Cottoperca gobio* (Bovichthyidae, Notothenioidei) as Compared to Karyotypes of other Notothenioidei. Leningrad: USSR Acad Sci, Proc Zool Inst, 1986; 153:67-71.
23. Andriashev AP. A general review of the Antarctic bottom fish fauna. In: Kullander SO, Fernholm B, eds. Proc 5th Congr Eur Ichthyol, Stockholm. Stockholm: Swedish Museum of Natural History, 1987:357-372.
24. Bargelloni L, Ritchie PA, Patarnello T, Battaglia S, Lambert DM, Mayer A. Molecular evolution at subzero temperatures: mitochondrial and nuclear phylogenesis of fishes from Antarctica (suborder Notothenioidei), and the evolution of antifreeze glycopeptides. Mol Biol Evol 1994; 11:854-863.
25. Balushkin AV. Classification, phylogenetic relationships, and origins of the families of the suborder Notothenioidei (Perciformes). J Ichthyol 1992; 30:132-147.
26. D'Avino R, di Prisco G. The hemoglobin system of Antarctic and non-Antarctic notothenioid fishes. Comp Biochem Physiol 1997; 118A:1045-1049.
27. Fago A, D'Avino R, di Prisco G. The hemoglobins of *Notothenia angustata*, a temperate fish belonging to a family largely endemic to the Antarctic Ocean. Eur J Biochem 1992; 210:963-970.
28. Romano M, Tamburrini M, Romano M, di Prisco G. Amino acid sequence and oxygen-binding properties of the two hemoglobins of the Antarctic bird *Catharacta maccormicki.* Ital Biochem Soc Transactions 1996; 7:182.
29. Wells RMG, Macdonald JA, di Prisco G. Thin-blooded Antarctic fishes: a rheological comparison of the haemoglobin-free icefishes *Chionodraco kathleenae* and *Cryodraco antarcticus* with a red-blooded nototheniid, *Pagothenia bernacchii.* J Fish Biol 1990; 36:595-609.
30. di Prisco G, Macdonald JA, Brunori M. Antarctic fish survive exposure to carbon monoxide. Experientia 1992; 48:473-475.
31. Cocca E, Ratnayake-Lecamwasam M, Parker SK, Camardella L, Ciaramella M, di Prisco G, Detrich HW, III. Genomic remnants of α-globin genes in the hemoglobinless antarctic icefishes. Proc Natl Acad Sci USA 1995; 92:1817-1821.
32. Stam WT, Beintema JJ, D'Avino R, Tamburrini M, di Prisco G. Molecular evolution of hemoglobins of Antarctic fishes (Notothenioidei). J Mol Evol 1997; 45:437-445.
33. Perutz MF, Brunori M. Stereochemistry of cooperative effects in fish and amphibian haemoglobins. Nature 1982; 299:421-426.

Enzymology

Cold enzymes: a hot topic

C. Gerday*, M. Aittaleb, J. L. Arpigny, E. Baise, J. P. Chessa, J. M. François, G. Garsoux, I. Petrescu and G. Feller

Laboratory of Biochemistry, Institute of Chemistry, Bâtiment B6, University of Liège, Sart-Tilman, B-4000 Liège, Belgium

1 Introduction

Chemical reaction rates often show a strong temperature dependency and a decrease of 10°C from room temperature typically divides the rate by a factor oscillating between 1.5 and 4. The decrease of the rate constant k indeed obeys an equation proposed by Svante Arrhenius as early as in 1889:

$$k = A\ e^{-Ea/RT} \tag{1}$$

in which E_a is the activation energy, R the gas constant (8.31 kJ mol^{-1}) and T the temperature in Kelvin.

An important corollary to this law is that the greater the activation energy E_a, the stronger the temperature dependence of the reaction rate. Reactions displaying small activation energy around, for example, 5 kJ mol^{-1} will have rates that will decrease only slightly with temperature, in contrast to reactions showing usual activation energies around 50 kJ mol^{-1} which will have rates that will strongly depend on temperature (Fig. 1).

In the case of enzyme catalyzed reactions, the behavior towards the parameter temperature can be close to that of a noncatalyzed reaction or quite different, depending on the type of weak interactions involved in the formation of the enzyme-substrate complex and also on the substrate concentration. Indeed, if one considers only the case of enzymes following Michaelis-Menten kinetics, it is clear that at high substrate concentration the concentration of K_m becomes negligible, and at a fixed enzyme concentration the thermodependency of the reaction rate will be, as in the case of chemical reactions, dependent only on k_{cat}, the rate constant. The situation will be quite different at low substrate, nonsaturating substrate concentration, since in this case the reaction rate will also depend on the K_m and globally on the ratio k_{cat}/K_m, the so-called physiological efficiency. K_m can in most

* Corresponding author; e-mail: ch.gerday@ulg.ac.be

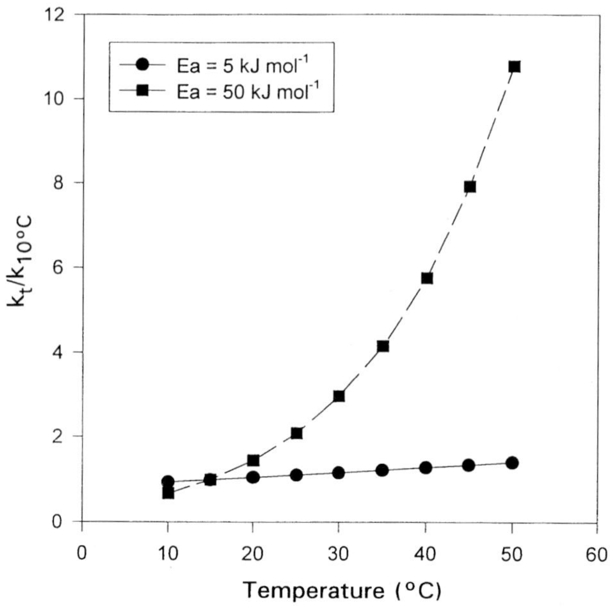

Fig. 1. Evolution of the ratios $k_t/k_{10°C}$ as a function of the temperature for two values of the activation energy: a typical one, 50 kJ mol^{-1} (*squares*), and a small one, 5 kJ mol^{-1} (*circles*)

cases be considered as a measure of the affinity of the enzyme for the substrate, in other words, the dissociation constant of the enzyme-substrate (ES) complex [E][S]/[ES]. The higher the K_m value, the lower the affinity of the enzyme for the substrate. This affinity is defined by the nature and strength of the weak interactions between enzyme and substrate. As there are only a few weak interactions involved, the affinity of the enzyme for the substrate will be strongly dependent on temperature. It is also known that there are two types of weak interactions: (1) the Van der Waals, hydrogen and electrostatic, which are formed with a negative modification of enthalpy and are therefore exothermic, and (2) the hydrophobic, occurring mostly with a positive modification of enthalpy and thus endothermic. The exothermic interactions will be destabilized by an increase in temperature whereas, on the contrary, endothermic interactions will tend to be stabilized. It follows that the affinity of an enzyme for the substrate will be differentially affected by temperature, depending on the relative contribution of each type of bond to the interaction between enzyme and substrate. If hydrophobic interactions prevail, elevated temperatures will have little effect on the K_m, and the Q_{10} value, expressing the ratio of reaction rates measured at an interval temperature of 10°C, will be kept constant; on the contrary at low temperatures, the interaction between the enzyme and the substrate can be strongly affected, leading to an increase in K_m and of Q_{10} values well over 3. These aspects have to be taken into account when considering the possible strategies used by extracellular and intracellular cold enzymes to fulfill adequate performances at low temperature.

2
Working hypotheses

It is clear that in terms of molecular adaptations, the selective pressure in low temperature environments, around 0°C, will be exerted towards the selection of enzymatic tools capable of achieving an adequate catalytic efficiency enabling the organism to display metabolic fluxes ideally comparable to those exhibited by mesophilic organisms at moderate temperatures. There are several theoretical ways to achieve such an objective, the simplest being to produce more enzyme-compensating low k_{cat} values by high concentrations of enzyme. Although not impossible, this strategy is rather improbable due to the energy cost of the enzyme synthesis. The second strategy would be to express temperature-controlled isotypes, but as stated by Hochachka and Somero,[1] this requires the availability of several more or less drastically modified genes. Quite common in eukaryotic cells, isoforms are rather unusual in microorganisms such as bacteria. This strategy would therefore be appropriate in the case of organisms occasionally displaying gene duplication processes giving rise to multiple copies of nucleotide sequences evolving subsequently by divergent evolution towards products adapted to different environmental conditions. This process is particularly useful for seasonal reversible adaptation mechanisms as demonstrated, for example, by early studies on fish enzymes,[2] and more recently on nematode enzymes.[3]

Organisms permanently exposed to low temperatures necessitate precisely tuned mechanisms conferring to the adaptation process an irreversible character. For the reasons exposed above, it is probably the strategy mainly used by organisms inhabiting low and constant temperature habitats, such as polar marine waters. When dealing with enzymes, it is probably also justified to distinguish between extracellular and intracellular enzymes, since the constraints are certainly sufficiently different to justify a specific strategy. First, it is reasonable to think that in most cases, the extracellular enzymes are confronted to high substrate concentrations enabling them to perform catalysis at maximum speed. If this assumption is correct, the strategy will involve only the adaptation of the catalytic constant k_{cat}. In the case of intracellular enzymes usually performing catalysis at low substrate concentration, the adaptation process can either concern k_{cat} or K_m or both and the constant to be taken into account is k_{cat}/K_m.

From numerous studies carried out on the structural characteristics of enzymes produced by thermophilic organisms, the thermostability required at the environmental temperature not only influences the resistance to other denaturing agents[4] but generally also impairs the catalytic efficiency so that, for a given chemical reaction, the catalytic constant never reaches the value which could be predicted by the Arrhenius law. If there is no general strategy of thermostabilization, one can say that it proceeds through the possible addition of several structural modifications such as the reinforcement of weak interactions, the improved compactness of the molecules or of constituting domains, the weakening of the interaction with the solvent and the stronger affinity of subunits for each other,[5] possibly leading to a higher state of association or to aggregated forms of monomeric proteins. It was notably observed in the case of a thermophilic xylanase that "sticky aromatic patches" at the surface of the protein favor the tendency to form dimers and tetramers which could be part of the strategy used to reach high thermostability.[6] These structural modifications lead in general to a moderate increase in the stabilizing energy ΔG_{stab} corresponding only to a few additional weak interactions. Nevertheless, the rigidity of the molecular structure is concomitantly enhanced when compared to mesophilic counterparts. In turn, the rigidity opposes serious constraints to the conformational changes necessary to reach the transition state enabling catalysis.

In summary, one can derive from this a statement not necessarily always correct but which could be quite useful for the discussion below: the higher the rigidity of the enzyme, the lower its catalytic efficiency. It follows that for cold-adapted enzymes, the reverse strategy could be considered as a working hypothesis although a net distinction has to be made. Indeed, in the case of thermophilic enzymes, the selective pressure is exerted towards the objective to select molecular structures characterized by a high thermostability at high temperature whereas in the case of cold enzymes, the environmental characteristics will favor molecular structures displaying high catalytic efficiencies at low temperatures. From the structural characteristics of thermophilic enzymes, one can predict that cold adaptation will tend to increase the plasticity of the molecular edifice in order to improve the accommodation of the substrate at low temperature. Depending on the type of three-dimensional (3D) structure involved, it seems clear that the molecular adaptation will be a more or less successful function of the generation time and of the relative importance of the enzyme for the viability of the organism.

3
Stability – flexibility – activity

The fact that a correlation does exist between the stability of proteins and the physiological temperature, which can possibly be defined as the temperature keeping the organism free of stress, has been known for a very long time thanks to the pioneer work of Rigby[7] carried out initially on skin collagens from various species. He stated at that time that the melting temperature of monomeric collagens was quite close to the upper limit of their environmental temperature. This idea was substantiated later by a rather exhaustive compilation of data carried out by Privalov.[8] This correlation was explained by the level of mobility necessary for efficient functioning. The mobility, plasticity or flexibility, being itself inversely related to stability. At room temperature, a thermophilic enzyme would be therefore stable, rigid, and poorly active whereas a psychrophilic enzyme would be rather unstable, flexible, and highly active. This statement is certainly an overstatement since the flexibility is not always *sensu stricto* incompatible with a high stability. The flexibility is only necessary to secure a good catalytic efficiency and can concern only the catalytic region which itself is certainly originally unstable.[9] Stabilization can be secured by other regions of the protein. It has, indeed, been shown that protein domains, the role of which was initially unknown, were in fact thermostabilizing domains contributing to the overall stability of the protein. This is, in particular, well documented in the case of xylanases.[10,11] Chimeric constructions made of highly efficient flexible catalytic domains and of stabilizing domains could therefore theoretically give rise to some sort of ideal enzymes associating best-possible catalytic properties and stability. In the case of cold enzymes, a high flexibility of the molecular edifice is possibly not always required especially for monospecific enzymes interacting with small-sized substrates which can be easily accommodated even at low temperatures. The problem is certainly quite different and more crucial in the case of macromolecular substrates since the rigidity of both the enzyme and the substrate will strongly depend on temperature.

How can we approach the flexibility of a protein structure? The danger is to use a technique giving an average measure of the flexibility which will not correctly reflect the local flexibility required for catalysis; this could be largely underestimated. The techniques which could provide some evaluation of the relative rigidity of a molecular edifice are:

1. the H-D exchange rates of amide protons using nuclear magnetic resonance or Fourier transformed infrared (FTIR) spectropscopy,
2. fluorescence quenching reflecting the accessibility of the tryptophan residues through the bimolecular rate constant κ_q,
3. fluorescence polarization giving access to the freedom of orientation,
4. possible enhanced activity in the presence of low denaturant concentrations,
5. relative resistance to proteolysis.

Provided that the 3D structure of the protein is known, the limited proteolysis technique associated with amino acid sequence determination can identify the more flexible parts of the proteins so that useful comparison between psychrophilic, mesophilic and thermophilic proteins can be undertaken. In favorable cases, also, significant differences in flexibility can also be assigned to specific regions using fluorescence techniques.

The relation between the thermostability of a protein and the kinetic of hydrogen exchange has been extensively studied in the case of lysozyme.[12] Denaturing agents such as pH and temperature were used, but also deforming agents such as LiCl. The data in fact made it possible not only to associate high hydrogen exchange rate to unfolded forms but also to define expanded conformations corresponding to the breaking of some of the 44 intramolecular peptide-petide hydrogen bonds stabilizing the lysozyme structure. This microunfolding giving rise to an increase in the rate of hydrogen exchange was related to the mobility of the macromolecular structure and it was found that collagens originating from various sources showed, at the same temperature, a broad range of hydrogen exchange rates from which Gibbs energies of microunfolding (ΔG_{mic}) could be derived. The ΔG_{mic} was regarded as a measure of the rigidity of a protein structure and was found to be also related to its macrostability.

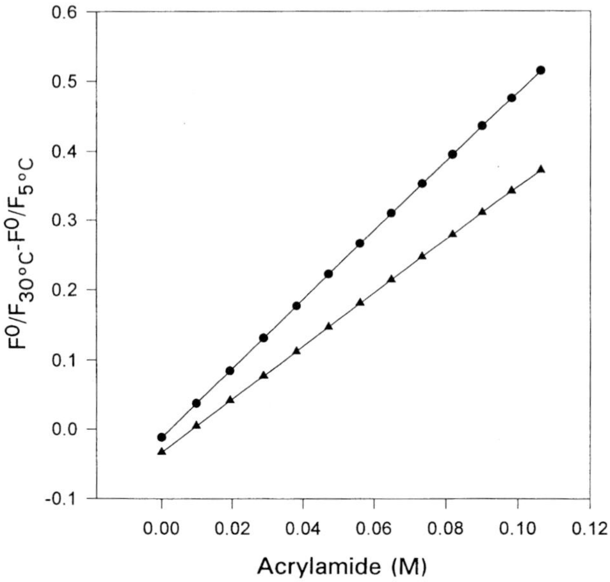

Fig. 2. Quenching of the tryptophane fluorescence of mesophilic (*triangles*) and psychrophilic (*circles*) Ca^{2+}-Zn^{2+} protease by acrylamide
In order to abolish the artefactual effect of the additional Trp present in the mesophilic enzyme, the differences of the fluorescence ratios at 30 and 5°C were plotted as a function of acrylamide concentration

The presumed increased flexibility of cold enzymes has not yet been demonstrated nor its inverse relation with thermostability. The hydrogen exchange rates have been measured at 20°C in the case of a psychrophilic α-amylase from *Alteromonas haloplanctis* and pig pancreatic α-amylases using FTIR spectroscopy (Goormagtigh et al., pers. communication). No significant difference could be found indicating either that there is no real difference or that the temperature was not appropriate. Maybe at 5°C, close to the environmental temperature of the psychrophilic enzyme, would it be possible to discriminate between the mesophilic and thermophilic enzymes. To our knowledge, only one experiment was successful in probably demonstrating the higher flexibility of cold enzymes. It concerns a Ca^{2+}-Zn^{2+} protease isolated from psychrophilic and mesophilic *Pseudomonas* sp. It was clearly shown, using fluorescence spectroscopy, that the quenching effect of acrylamide, when comparing the quenching constant at 5 and 30°C, was much more important in the case of the psychrophilic enzyme (Fig. 2). It must still, however, be demonstrated that an increased flexibility gives rise to a higher specific activity at low and moderate temperature, at least in the case of macromolecular substrates. Site-directed experiments carried out in the case of a psychrophilic subtilisin have indicated that the thermostability of the cold enzyme could be drastically improved by increasing the affinity of the enzyme for Ca^{2+} (mutation T85D).[13] Although more stable, the mutant showed in fact a higher (x1.5) specific activity when compared to the wild-type enzyme indicating that stability (rigidity) and specific activity are not systematically inversely related and that as stated above, the regions of the protein modulating the catalytic efficiency can be clearly distinct from the regions involved in stability (rigidity). Thus, more experiments are needed to establish clearly the relations, if any, between the stability, flexibility and activity of an enzyme.

4
Kinetic and molecular properties

At the present time, three crystallographic studies carried out on psychrophilic enzymes have been successfully achieved on a cold α-amylase,[14,15] a cold Ca^{2+}-Zn^{2+} protease[16] and a cold triosephosphate isomerase.[17] The details of the comparison between these refined structures and their mesophilic counterparts have not yet been published. One can possibly add to this the structures of salmon trypsin and elastase although the real psychrophilic character of this fish species can be questioned.[18,19] Several recent reviews have also been published,[20–28] covering all aspects of the psychrophilic world.

The essence of the knowledge accumulated on the molecular adaptation of enzymes from psychrophilic organisms is contained in papers describing 3D models of these enzymes which have been built for α-amylase,[29] subtilisins,[13,30] triosephosphate isomerase,[31] antarctic fish trypsin[32] and elastase,[33] β-lactamase[34] and lipase.[35] Other recent works although incomplete bring possibly new light on the properties of cold enzymes: they concern glucose-6-phosphate and glutamate dehydrogenase from antarctic fish[36] and citrate synthase from antarctic bacterium.[37] A cold methyltransferase from an antarctic psychrotroph has also been described[38] but has not been discussed in terms of molecular adaptation to cold. The data have to be analyzed as a function of the statements that, in microorganisms, the adaptation of extracellular enzymes can differ from that of intracellular enzymes and that the cold enzymes from cold-adapted fish have possibly molecular adaptations less pronounced than those of enzymes from microorganisms due to the large difference in generation time.

4.1
Extracellular enzymes from microorganisms

The extracellular enzymes which have been extensively studied are α-amylase, Ca^{2+}-Zn^{2+} protease, β-lactamase, lipase, subtilisin and xylanase. The refined 3D structures of α-amylase and Ca^{2+}-Zn^{2+} protease are available along with the computerized models of β-lactamase, lipase, subtilisin and xylanase. Note that all these enzymes display two common characteristics: they have a higher specific activity and higher thermosensitivity when compared to their mesophilic counterparts; the β-lactamase has a high but not the highest specific activity but it is a rather peculiar case which has been discussed by Feller et al.[34] and Feller and Gerday.[28] From the comparison of these structures with their equivalent originating from

Table 1. Possible structural parameters involved in cold adaptation in the case of extracellular enzymes from microorganisms investigated so far and deduced from the comparison of their 3D structures with those of mesophilic counterparts

Parameter	α-Amylase	Ca^{2+}-Zn^{2+} Protein	β-Lactamase	Lipase	Subtilisin	Xylanase
Salt bridges	F	ND	–	F	F	F
Hydrogen bonds	ND	ND	ND	ND	ND	ND
Aromatic-aromatic interactions	–	ND	F	F	F	–
Other X-aromatic interactions	F	ND	–	–	–	–
Hydrophobic clusters	F	ND	F	F	–	F
Glycine content	–	–	–	–	–	–
Proline content	F	–	F	–	F	–
R/(K+R)	F	–	F	F	R	R
Disulfide bonds	–	–	–	–	–	–
Helices dipoles	F	ND	–	–	–	F
Ca^{2+} affinity	F	F	–	–	F	/
Cl^- affinity	F	/	/	/	/	/
Other metals	/	–	/	/	/	/
Solvent interactions	–	–	F	F	F	F

(F) parameters are altered so that the plasticity of the molecular edifice is improved as well as its tendency to be more easily unfolded; (–) the situation is not significantly different in mesophiles; (R) a stabilization ion effect is expected; (/) the parameter is not involved; (ND) not determined

mesophilic counterparts, some structural parameters, which could be implicated in the adaptation to cold, have been deduced.

Table 1 summarizes the apparent strategies used by these enzymes in order to improve the catalytic efficiency at low temperatures. Although the comparison of the respective hydrogen bonds networks was not possible on model structures, one can see that the strategy appears specific to each enzymes. In the case of metallo-enzymes, the reduced cation-anion affinity seems to be a general strategy, but it must still be decided whether the lower affinity observed is an integral part of the mechanism of adaptation or a consequence of the increased plasticity of the molecular edifice. In the case of the psychrophilic α-amylase, although the affinity of the enzymes for Ca^{2+} and Cl^- is low when compared to the mammalian enzymes, it was found using site-directed mutagenesis and studying the apo-forms that neither Ca^{2+} or Cl^- were involved in the stability of the psychrophilic enzyme which has apparently reached a limit of stability and could not be further destabilized without losing its 3D structure.[39] It was indeed found that the difference in stabilizing energy with pig pancreatic α-amylase was about 10 kJ mol^{-1}.[29] Among the other strategies, the decrease in the number of salt bridges and in the strength of hydrophobic clusters, and the increase of the interaction with the solvent appear to be the more general ways to increase the flexibility of the molecular structure. In all cases, however, the structural modifications are rather marginal. The majority of amino acid substitutions occur within loops connecting secondary structure.[13,30] Some others affect the stability of α-helices.[31] The residues involved in catalysis or in substrate binding are strictly conserved.[13,29]

4.2 Intracellular enzymes from microorganisms

To our knowledge, seven intracellular or periplasmic enzymes from psychrophilic microorganisms have been studied or are still presently under investigation at the molecular level. They are a periplasmic alkaline phosphatase,[40] a L-lactate dehydrogenase,[41] a triosephosphate isomerase,[17,31] a citrate synthase,[37] a PspPI methyltransferase,[38] a phosphoglycerate kinase (Bentahir et al., pers. communication) and a β-galactosidase (François et al., pers. communication). All originate from antarctic bacteria. The methyltransferase has not been discussed in terms of adaptations to cold and will therefore fall outside the scope of this chapter. We also have no indication about the thermostability and specific activity of the cold lactate dehydrogenase. So only five incomplete investigations will constitute the basis of the discussion. These enzymes have at least one property in common with the other cold enzymes; this is their higher thermolability illustrated in Figure 3 in the case of the phosphoglycerate kinase from the antarctic strain TACII18 isolated from seawater. The high specific activity characterizing psychrophilic extracellular enzymes is not apparently a general feature of intracellular cold enzymes. Table 2 illustrates the data compared to those of mesophilic counterparts. With the exception of phosphoglycerate kinase, the intracellular psychrophilic enzymes mentioned in Table 2 have catalytic constants no higher than those of mesophilic enzymes even at their environmental temperature, the k_{cat}/K_m values also do not favor the psychrophilic enzymes. This is rather puzzling because clearly the thermostability, probably

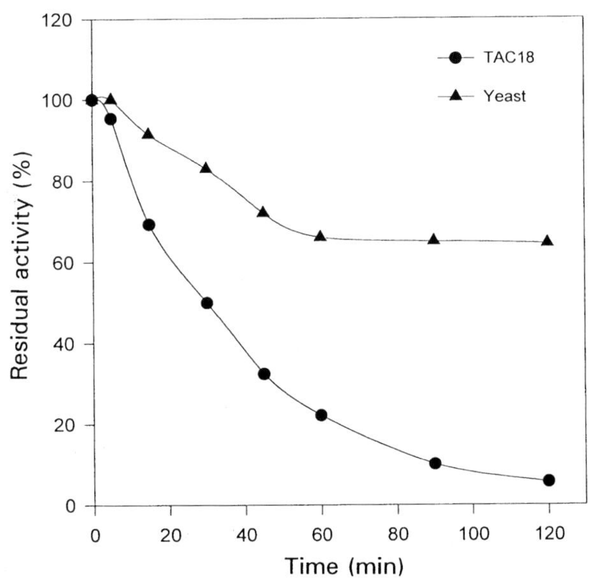

Fig. 3. Residual activity after exposure at 40°C of a psychrophilic phosphoglycerate kinase TACII18 (*circles*) and of the mesophilic enzyme from yeast (*triangles*) as a function of time

Table 2. Kinetic parameters of intracellular enzymes from psychrophilic microorganisms as compared to those of mesophilic enzymes

Enzyme	Temperature (°C)	k_{cat}	K_m
Citrate synthase			
Antarctic DS2-3R[a] (freshwater, Gram$^+$)	23	42 s^{-1}	230 μM (acetyl CoA)
Pig enzyme	23	61 s^{-1}	< 50 μM
β-Galactosidase			
Antarctic strain TAE79[b] (Gram$^-$)	5	30 s^{-1}	230 μM (ONPG)
	25	60 s^{-1}	350 μM
Escherichia coli	5	220 s^{-1}	1020 μM
	25	610 s^{-1}	650 μM
Phosphoglycerate kinase			
Antarctic strain TACII18[c] (marine water, Gram$^-$)	25	429 U mg^{-1} [f]	0.36 mM
Yeast	25	276 U mg^{-1}	1.33 mM
Triosephosphate isomerase			
Antarctic *Moraxella* TA139[d] (Gram$^-$, recombinant)	25	4142 U mg^{-1} [g]	–
Bacillus stearothermophilus	25	6349 U mg^{-1} [g]	–
Antarctic *Vibrio marinus*[e]	10	4.2×10^5 min^{-1}	1.9 mM
Escherichia coli	10	5.4×10^5 min^{-1}	1.03 mM

[a] Gericke et al.[37]
[b] François et al. (pers. communication).
[c] Bentahir et al. (pers. communication).
[d] Rentier-Delrue et al.[31]
[e] Alvarez et al.[17]
[f] One unit of enzyme converts 1 μmol of 3-phosphoglycerate min^{-1}.
[g] One unit of enzyme converts 1 μmol of D-glyceraldehyde 3-phosphate min^{-1}.
ONPG o-nitrophenyl β-D-galactopyranoside

inversely associated with plasticity, is, in all cases, much lower than that of mesophilic enzymes. In the case of citrate synthase, Gerike et al.[37] have argued that the low activity of the psychrophilic enzyme at low temperature is probably high enough to permit cell growth and function. This argument, of course, makes sense, but what is the relevance in selecting flexible and instable structure if it is not ultimately to achieve better performance at low temperature? In the case of the β-galactosidase, one can argue that lactose is not a very important substrate for an antarctic strain so the selective pressure does not probably exert a strong effect towards the enzyme. Moreover, the kinetic constants determined on the natural substrate could be significantly different from those obtained on o-nitrophenyl-β-D-galactoypyranoside. On the contrary, citrate synthase, triosephosphate isomerase and phosphoglycerate kinase appear to be crucial enzymes warranting adequate metabolic fluxes. Artifacts in the measurement of the specific activity of these rather unstable enzymes are always possible, since a more or less important part of the protein can be inactivated during the course of the purification. Maybe also some of these enzymes originating from rather exotic environments require unusual cofactors to optimally express their activity. The bacterial strains from which these enzymes originate have been isolated from various antarctic environments. DS2-3R (citrate synthase) was found in freshwaters, TAE79 on algae necrosis, TACII18 in marine waters and TA137 in the digestive track of the antarctic fish *Notothenia neglecta*. They are Gram-negative bacteria with the exception of DS2-3R, and all have temperature growth limits below 30°C. They are not in this sense true psychrophiles[20] but display, at least for the TA strains, generation times around 5 h at 4°C[42] illustrating their good adaptation to cold. TACII18 producing the well cold-adapted phosphoglycerate kinase also excretes the equally well-adapted extracellular Ca^{2+}-Zn^{2+} protease.[16]

The attribute common to the intracellular enzymes, citrate synthase, β-galactosidase and triosephosphate isomerase, displaying unexpectedly low specific activity at low temperatures, is their multimeric character, in contrast to phosphoglycerate kinase, which is monomeric. Could it be that the necessity to preserve the appropriate intermolecular interactions prevents or limits the molecular adjustments required to achieve a high catalytic efficiency at low temperature? The set of available data is not, of course, large enough to formulate a consistent answer.

4.3 Fish enzymes

We have postulated that the success of the molecular adaptation to cold is probably dependent on the generation time. When compared to microorganisms displaying at 4°C doubling time often well below 10 h, the generation time of multicellular organisms such as fish inhabiting cold waters like the Antarctic Ocean is a matter of months. In the Antarctic Ocean, a single suborder, the notothenioids, dominates the fish population and displays considerable diversity. They most likely arose during the Eocene about 50 million years ago, gradually adapting to cooling conditions estimated at an average value of 0.03°C per 100,000 years (15°C over a period of 50 million years).[43] The onset of markedly cold conditions, below 0°C, occurred only about 2 million years ago,[44] giving little time to the fish species to finely tune their adaptation to cold. Only a few studies have been carried out at the molecular level on enzymes from species inhabiting cold waters such as the Arctic and Antarctic Oceans. From the Northern Oceans, we will consider mainly the digestive enzymes such as trypsin from atlantic salmon,[17,45] trypsin from atlantic cod,[46,47] elastase from north atlantic salmon[19] and elastase from atlantic cod.[48] From the Southern and especially Antarctic Oceans, the structures of trypsin from the antarctic fish *Paranotothenia magellanica*,[32] of elastase from the antarctic fish *Notothenia neglecta*,[33] of glucose-6-phosphate dehydrogenase from the icefish *Chionodraco hamatus* and of glutamate dehydrogenase from *Chaenocephalus aceratus*[36] will be discussed.

A comparison of homologous cold enzymes from species from the northern and southern hemispheres could indeed allow the detection of a possible convergence in the strategy used to compensate for the low temperature of the respective environments, keeping in mind that the selective pressure is surely more intense in the south when compared to the north.

The crystal structures of trypsin and elastase from atlantic salmon have been solved,[18,19] whereas model structures of trypsin and elastase from two antarctic fish have been built.[32,33] All the above-mentioned enzymes also display the usual properties of cold enzymes, they have a higher catalytic efficiency and a higher thermosensitivity, reflecting their permanent adaptation to low temperature environments.

Starting with salmon trypsin, the structural differences with bovine trypsin appear rather discrete with the displacement in atomic position not exceeding 0.8 Å and very similar temperature factor profiles, possibly reflecting the same overall flexibility. Some loop regions, however, display a marked difference in dynamic properties. The strength and number of the usual weak interactions also appear comparable with the exception of the level of the interaction between the C-terminal helix and the rest of the molecule, leading to a possible weaker overall interaction between the two β-barrel domains of salmon trypsin. This region appears clearly more flexible than in bovine trypsin, favoring the idea that the interdomain motion is facilitated during substrate binding and catalysis. The internal residues in salmon trypsin also appear to be less hydrophobic than in bovine trypsin.

In the case of antarctic fish trypsin, most of the amino acid substitutions involve surface residues which carry more charges than in bovine trypsin. This can contribute to lowering the stability of the antarctic enzyme. The hydrophobicity index is also much lower than in the case of the bovine enzyme.[32] Another important difference is seen at the level of the interdomain region where Tyr151 of bovine trypsin is missing in the antarctic and atlantic salmon and atlantic cod trypsin. This gives more freedom to Gln192 and increases the accessibility of the binding pocket. In addition, the prolyl residue in position 152 of the bovine trypsin is replaced in *Notothenia* and atlantic cod trypsins by a glycyl and by a seryl residue in salmon trypsin, further supporting the idea that the catalytic cavities of the cold trypsins are more accessible than in the case of the mammalian enzyme and that a similar strategy has been used by the cold-adapted fish species to solve the problem of the specific activity at low temperatures.

The fact that there are at least four types of pancreatic elastases makes the comparison between elastases from different origins rather difficult, since in addition the different families also display more or less different substrate specificities related to the size, small, medium or large, of the amino acid residue occupying the P1 position.[49] The 3D structure of the elastase I from north atlantic salmon has been solved at 1.61 Å resolution[19] and compared to the homologous pig pancreatic elastase I showing 67% amino acid sequence identity. When superimposed, the main chain atoms of the two enzymes display an overall RMS deviation of 0.55 Å. Only nine buried residues are different but show compensatory substitution. The reported apparently interesting differences are first the deletion of Asp186, which belongs to the central core of the substrate binding site, giving rise to a large deviation in structure (± 3.9 Å) near the active Ser195. This modification does not appear, however, in antarctic fish nor in cod elastases. Second, the last three residues of the C-terminal helix are not visible, which can also be due to a modification of the interdomain substrate-binding regions as in the case of cold fish trypsins. The antarctic fish elastase differs from pig elastase by a lower molar ratio R/(R+K). It also shows, as in the case of cod elastase, a high number of Met. This residue is believed to increase the plasticity of the molecular edifice since it displays a free rotation around three single bonds.[48]

There is also a striking difference at the level of the catalytic cavity of the *Notothenia* elastase: indeed, in pig elastase, residues A99a, A99b and G100 limit the accessibility of the catalytic cavity, while in *Notothenia* elastase the deletion of residues A99a, G100 and the substitution A99bR make the catalytic cavity markedly more accessible. These structural modifications also occur in cod B elastase, suggesting again a convergent strategy. One cannot, of course, exclude a substrate-driven constraint rather than a cold-induced adaptation.

From this limited amount of cases, it can be said that the molecular adaptations of enzymes from cold-adapted fish are more discrete than in the case of enzymes from microorganisms and are possibly mainly concentrated in the case of digestive enzymes of the trypsin family on the interdomain regions, which is also the catalytic crevice.

5
Expression of psychrophilic genes in mesophilic hosts

The expression, under the form of active products, of genes coding for psychrophilic proteins in mesophilic hosts can, in theory, be impaired by a growing temperature far remote from that of the natural environment. This can possibly cause denaturation of the psychrophilic enzyme or inappropriate folding. Less than ten psychrophilic enzymes have been, up to now, cloned in *Escherichia coli* cells or in *Bacillus*[13] without apparently altering their specific properties provided that the culture temperature of the mesophilic host is low enough to avoid irreversible denaturation of the heterologous protein. *E. coli* RR1, HB101 or XL1-Blue grown at 18°C are quite suitable since the stationary phase is reached after 30 h.[50] Only one enzyme, a cold subtilisin from an antarctic *Bacillus*, has been expressed in a mesophilic *Bacillus*. In this case also, the growing temperature of the mesophilic host was kept at 18°C.[13] A systematic investigation of the expression of a psychrophilic α-amylase in *E. coli* cells as a function of the growing temperature of the mesophilic host has been carried out recently.[50] The study clearly shows that the maximal enzyme production was achieved below 18°C, whereas at higher temperatures a gradual decrease of the enzyme production occurs (Fig. 4) reaching 0 at 35°C. At 18°C, both the kinetic and spectroscopic properties of the recombinant enzyme are identical to those of the wild-type α-amylase. In the case of citrate synthase, the transformants were grown at 37°C but the expression of the recombinant enzyme was induced at 25°C,[37] a fairly high temperature if one takes into account that the apparent optimal temperature of the enzyme was found to be 30°C. The recombinant psychrophilic triosephosphate isomerase from *Vibrio marinus* was also produced in *E. coli* BL21 cells. The enzyme has a half-life of 10 min at 25°C[17] and surprisingly the cold enzyme was expressed at 37°C. The specific activity of this

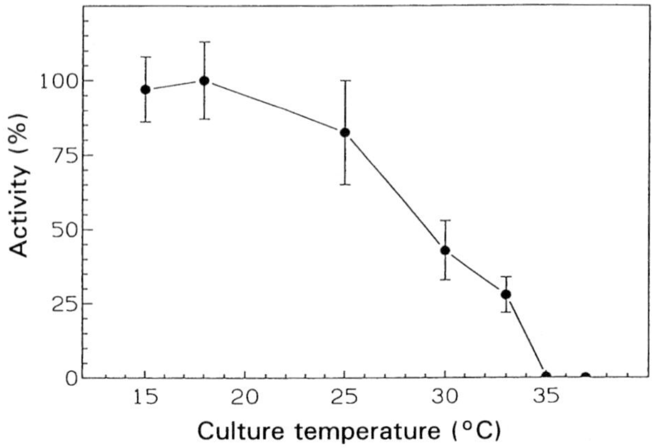

Fig. 4. Relative production in *E. coli* of recombinant α-amylase from antarctic strain *Alteromonas haloplanctis* as a function of culture temperature of the host bacterium
Data are expressed as percentage of the maximal activity recorded at 18°C

recombinant triosephosphate isomerase, measured at 10°C, is lower than that of the *E. coli* enzyme at the same temperature (Table 2). In view of the apparent inadequacy between the expression temperature and the enzyme stability, one can seriously question the physical homogeneity of the recombinant psychrophilic triosephosphate isomerase.

6 Site-directed mutagenesis

Three psychrophilic enzymes have been already submitted to site-directed mutagenesis experiments in order to check the hypotheses derived from the comparison of the 3D structure of the cold enzymes obtained either by protein modeling[13,39] or by X-ray crystallography,[17] with the refined structures of the mesophilic counterparts.

6.1 Alpha-amylase

In the case of the cold α-amylase from *Alteromonas haloplanctis*, it was found that the affinity of the enzyme for Ca^{2+} and Cl^- was lower than that observed in the case of pig pancreatic α-amylase. It was therefore suggested that the weak affinity of the enzyme for Ca^{2+} and Cl^- was part of the strategy used to increase the plasticity of the structure.[29] The Ca^{2+} and Cl^- binding sites of the cold α-amylase in fact differ from their equivalent found in pig pancreatic α-amylase by only one residue: in the Cl^- site an arginine is replaced by a lysine (R337K), and in the Ca^{2+} site an arginine is replaced by a glutamine (R158Q). As arginine is able to establish multivalent weak interactions with appropriate groups, it was thought that these substitutions were at the origin of the lower affinity of the cold enzymes for the ions. It was therefore decided to introduce in the cold enzymes the appropriate replacements K337R and Q158R in order to mimic the anion and cation sites found in pig pancreatic α-amylase. An additional mutation K337Q was selected in order to mimic the situation occurring in Cl^--independent α-amylases. The mutation K337R failed to restore the high affinity for Cl^- found in the case of the pig enzyme. On the contrary, the affinity was lowered by a factor[39] close to 10 and the catalytic efficiency by a factor close to 2. The mutation K337Q indeed renders the enzyme independent of Cl^- but the activity is lowered by a factor close to 10.[39]

At the level of the Ca^{2+} binding site, the mutation Q158R does not restore the high affinity for Ca^{2+} found in pig pancreatic α-amylase. The affinity for Ca^{2+} is further lowered by a factor close to 3 and the activity by a factor close to 4. The thermostability is not, however, affected.

Thus, clearly, the mutations observed at the level of the Ca^{2+} and Cl^- binding sites are not part of the strategy used by the cold α-amylase to increase its plasticity. The low affinity for the ions is the result of other structural modifications which have still to be defined.

6.2 Subtilisin

The objective of the site-directed mutagenesis experiments as applied to the psychrophilic subtilisin TA39[13] was mainly to try to improve the stability of the cold enzyme and see what would be the effect on the specific activity. Three successful mutants belonging to this category were produced in a mesophilic *Bacillus*; one, A181K-S210E, was supposed to restore a salt bridge present in mesophilic subtilisin BNP' and Carlsberg, the second, H121W, was supposed to restore an aromatic interaction F50/W113 stabilizing subtilisin Carlsberg, and the last, T85D, consisted on an exchange of a poor Ca^{2+} ligand T for a good ligand D enabling to restore a high affinity for Ca^{2+} in site 1 involved in protein stability.

Despite the use of serum albumin to limit the autolysis, the data illustrating the stability of the mutants were rather unclear in two cases. Indeed, contrarily to what was expected, the mutants carrying the additional salt bridge and aromatic interaction were respectively apparently less and equally stable when compared to the wild-type enzyme. As the thermostability was followed by measuring the residual activity, the results can be misleading due to the fact that both mutations unexpectedly increased the specific activity and also, of course, the rate of autolysis. A much more clear cut result was observed in the case of the third mutant. Indeed, the replacement of T85D increased the affinity of Ca^{2+} site 1 in the psychrophilic enzymes by a factor close to 1000 so that the overall thermostability of the cold enzyme was drastically improved to roughly reach that of mesophilic subtilisins. In this case, the hypothesis fitted the experiment but rather unexpected additional data were observed, since the more stable psychrophilic enzyme also displayed a specific activity increased by a factor of 1.5. This illustrated the fact that obviously stability and specific activity can be altered independently, at least within certain limits. This result is also important because it abolishes, at least for certain applications, the usual drawback of cold enzymes, i.e. their low stability on long time storage.

6.3
Triosephosphate isomerase

The 3D structure of the triosephosphate isomerase (TiM) from the psychrophilic strain *Vibrio marinus* (vTiM) has been solved at 2.7 Å resolution[17] and its structure has been compared to that of *E. coli* (eTiM) showing 66% amino acid sequence identity. In vTiM there is one insertion of two residues at the C-terminal end of helix 2, near the N-terminal end of the β-barrel. The two structures can be superimposed with an Cα RMS deviation of 0.59 Å. The vTiM sequence shows one unique feature when compared to the 45 known sequences, an alanine in position 238 rather than a serine in all the others. This amino acid is located at the beginning of the phosphate binding 3_{10} helix in loop 8 and the lateral chain (Ala in vTiM or Ser in other TiM) is oriented towards the protein interior. In other TiM the Oγ of the conserved serine is engaged in two hydrogen bonds with main chain oxygen atoms creating a hydrogen bond network with other residues. Thus the replacement of a serine by an alanine can obviously perturb the hydrogen bond arrangement and can be the origin of a decrease in stability. Indeed, the replacement of the Ala238 in the psychrophilic TiM by a serine causes a 5°C increase in the melting temperature of the mutated enzyme which approaches that of *E. coli* TiM (46 vs. 54°C). The specific activity is, in contrast, lowered by a factor close to 2, illustrating a clear inverse relation between stability and specific activity. The exchange of a serine for an alanine in the psychrophile enzyme appears therefore well related to the adaptation to cold. It must still, however, be explained why the specific activity at low temperature of the cold enzyme is not better than that of the *E. coli* enzyme. Possibly in this particular case the expression of the recombinant enzyme at a temperature as high as 37°C can give rise to a partially and irreversibly denatured enzyme.

7
Conclusions

From the limited amount of data available, one can say that enzymes produced by psychrophiles display a lower stability than their mesophilic counterparts. This sensitivity to various denaturing agents reflects an increased mobility or plasticity of the overall edifice or of selected parts of the enzyme structure. The improved plasticity is supposed to provide the appropriate flexibility of crucial regions involved in substrate binding and catalysis, so that a high specific activity is reached at low temperatures. This latter property is indeed encountered in all extracellular psychrophilic enzymes studied so far originating from either microorganisms or fish. In the case of intracellular enzymes, the data indicate that although they share with extracellular enzymes a high thermosensitivity, apparently not all of them display a higher specific activity or a higher physiological efficiency at low temperatures. It has been argued that the crucial question is to know whether the activity is high enough to allow appropriate metabolic fluxes at the environmental temperature. Everybody will agree with this, but why therefore has nature only selected psychrophilic extra- or intracellular enzymes of extreme marginal stability?

Acknowledgements. This research has been supported by the EU under the form of a network contract CT97-0131, a concerted action BIO4-CT95-0017 and a Biotech programme BIO4-CT96-0051. We also acknowledge the strong support of the Ministère de l'Education, de la Recherche et de la Formation (Concerted action ARC93/98-170) and of the Ministère de la Région wallonne - Direction générale des Technologies (Conventions 9613492 and 1828). The National Foundation for Scientific Research, FRSM contract no. 3.4502-97 to Charles Gerday has also been of great help. We also thank the Institut français de Recherche et de Technologie polaires for generous accommodation of our research fellows at the antarctic station J.S. Dumont d'Urville in Terre Adélie.

8
References

1. Hochachka PW, Somero GN. Temperature adaptation. In: Hochachka PW, Somero GN, eds. Biochemical Adaptation. Princeton Univ Press, 1984:355-449.
2. Baldwin J, Hochachka PW. Functional significance of isoenzymes in thermal acclimatization: acetylcholine esterase from trout brain. Biochem J 1970; 116:883-887.
3. Jagdale BG, Gordon R. Effect of temperature on the activities of glucose-6-phosphate dehydrogenase and hexokinase in ertopathogenic Nematodes. Comp Biochem Physiol 1997; 118A:1151-1156.
4. Cowan DA. Enzymes from thermophilic archae bacteria: current and future application in biotechnology. Biochem Soc Symp 1992; 58:149-169.
5. Jaenicke R. Stability and folding of ultrastable proteins: eye lens crystallins and enzymes from thermophiles. FASB J 1996; 10:84-92.
6. Harris GW, Pickersgill RW, Connerton I, Debeire P, Touzel JP, Breton C, Perez S. Structural basis of the properties of an industrially relevant thermophilic xylanase. Prot Struct Funct Gen 1997; 29:77-86.
7. Rigby JB. Amino acid composition and thermal stability of the skin collagen of the Antarctic fish. Nature 1968; 219:166-167.
8. Privalov PL. Stability of proteins. Adv Prot Chem 1982; 35:1-104.
9. Shoichet DK, Baase WA, Kuroki R, Matthews BW. A relationship between protein stability and protein function. Proc Natl Acad Sci USA 1995; 92:542-549.
10. Lee YE, Lowe SE, Henrissot B, Zeikus JG. Characterization of the active site and thermostability regions of endoxylanase from *Thermoanaerobacterium saccharolyticum* B6AR1. J Bacteriol 1993; 5890-5898.
11. Fontes GM, Hazlewood GP, Moraq E, Hall J, Hirst BH, Gilbert HJ. Evidence for a general role for non catalytic thermostability domains in xylanases from thermophilic bacteria. Biochem J 1995; 307:151-158.
12. Nakanishi M, Tsuboi M, Ikegami A. Fluctuation of the lysozyme structure. J Mol Biol 1992; 70:351-361.
13. Narinx E, Baise E, Gerday C. Subtilisin from psychrophilic Antarctic bacteria: characterization and site-directed mutagenesis of residues possibly involved in the adaptation to cold. Protein Eng 1997; 10:1271-1279.
14. Aghajari N, Feller G, Gerday C, Haser R. Crystallization and preliminary X-ray diffraction studies of α-amylase from the Antarctic psychrophile *Alteromonas haloplanctis* A23. Protein Sci 1996; 5:2128-2129.

15. Aghajari N, Feller G, Gerday C, Haser R. Crystal structures of the psychrophilic α-amylase from *Alteromonas haloplanctis* in its native form and complexed with an inhibition. Protein Sci 1998; 7:564-572.
16. Villeret V, Chessa JP, Gerday C, Van Beeumen J. Preliminary crystal structure determination of the alkaline protease from the Antarctic psychrophile *Pseudomonas aeruginosa*. Protein Sci 1997; 6:2462-2464.
17. Alvarez M, Zeelens JP, Mainfroid V, Rentier-Delrue F, Martial J, Wyns L, Wierenga RK, Maes D. Triose-phosphate isomerase (TIM) of the psychrophilic bacterium *Vibrio marinus*. J Biol Chem 1998; 273:2199-2206.
18. Smalas RO, Heimsad ES, Hordvik A, Willasen P, Mall R. Cold adaptation of enzymes: structural comparison between salmon and bovine trypsin. Protein Struct Funct Gen 1994; 20:149-166.
19. Berglund GI, Willassen NP, Hordvik A, Smalas AO. Structure of native pancreatic elastase from North Atlantic salmon at 1.61 Å resolution. Acta Cryst 1995; D51:925-937.
20. Morita RY. Psychrophilic bacteria. Bacteriol Rev 1975; 39:144-167.
21. Russell NJ. Cold adaptation of microorganisms. Phil Trans R Soc Bond 1990; 326:595-611.
22. Russell NJ, Hamamoto T. Psychrophiles. In: Horikoshi K, Grant WD, eds. Extremophiles: Microbial Life in Extreme Environments. New York: Wiley, 1997:25-45.
23. Gounot AM. Bacterial life at low temperature: physiological aspects and biotechnological implications. J Appl Bacteriol 1991; 71: 386-397.
24. Margesin R, Schinner F. Properties of cold-adapted microorganisms and their potential role in biotechnology. J Biotechnol 1994; 33:1-14.
25. Arpigny JL, Feller G, Davail S, Genicot S, Narinx E, Zekhnini Z, Gerday C. Molecular adaptations of enzymes from thermophilic and psychrophilic organisms. In: Gilles R, ed. Comparative Environmental Physiology. Berlin: Springer, 1994; 20:269-295.
26. Feller G, Narinx E, Arpigny JL, Aittaleb M, Baise E, Genicot S, Gerday C. Enzymes from psychrophilic organisms. FEMS Microbiol Rev 1996; 18:189-202.
27. Gerday C, Aittaleb M, Arpigny JL, Baise E, Chessa JP, Garsoux G, Petrescu I, Feller G. Psychrophilic enzymes: a thermodynamic challenge. Biochim Biophys Acta 1997; 1342:119-131.
28. Feller G, Gerday C. Psychrophilic enzymes: molecular basis of cold adaptation. Cell Mol Life Sci 1997; 54:830-841.
29. Feller G, Paysan F, Theys F, Qian M, Haser R, Gerday C. Stability and structural analysis of α-amylase from the Antarctic psychrophile *Alteromonas haloplanctis* A23. Eur J Biochem 1994; 222:441-447.
30. Davail S, Feller G, Narinx E, Gerday C. Cold adaptation of proteins. Purification, characterization and sequence of the heat-labile subtilisin from the Antarctic psychrophile *Bacillus* TA48. J Biol Chem 1994; 269:17448-17453.
31. Rentier-Delrue F, Mande SC, Moyens S, Terpstra P, Mainfroid V, Goraj K, Lion M, Hal WG, Martial J. Cloning and overexpression of the triose phosphate isomerase genes from psychrophilic and thermophilic bacteria. J Mol Biol 1993; 229:85-93.
32. Genicot S, Rentier-Delrue F, Edwards D, Van Beeumen J, Gerday C. Trypsin and trypsinogen from an Antarctic fish: molecular basis of cold adaptation. Biochim Biophys Acta 1996; 1298:45-57.
33. Aittaleb M, Hubner R, Lamotte-Brasseur J, Gerday C. Cold adaptation parameters derived from CDNA sequencing and molecular modelling of elastase from Antarctic fish *Notothenia neglecta*. Prot Eng 1997; 10:475-477.
34. Feller G, Zekhnini Z, Lamotte-Brasseur J, Gerday C. Enzymes from cold-adapted microorganisms. The class C β-lactamase from the Antarctic psychrophile *Psychrobacter immobilis* A5. Eur J Biochem 1997; 244:186-191.
35. Arpigny JL, Lamotte J, Gerday C. Molecular adaptation to cold of an Antarctic bacterial lipase. J Mol Catalysis 1997; 3:29-35.

36. Ciardello A, Camardella L, Carratore V, Di Prisco G. Enzymes in Antarctic fish: glucose-G-phosphate dehydrogenase and glutamate dehydrogenase. Comp Biochem Physiol 1997; 118A:1031-1036.
37. Gericke U, Danson MJ, Russell NJ, Hough DW. Sequencing and expression of the gene encoding a cold-active citrate synthase from an Antarctic bacterium strain DS2-3R. Eur J Biochem 1997; 248: 49-57.
38. Rina M, Caufrier F, Markaki M, Mavromatis K, Kokkinidis M, Bouriotis V. Cloning and characterization of the gene encoding Psp PI methyltransferase from the Antarctic psychrotroph *Psychrobacter sp.* strain TA137. Gene 1997; 197:353-360.
39. Feller G, Le Bussy O, Houssier C, Gerday C. Structural and functional aspects of chloride binding to *Alteromonas haloplanctis* α-amylase. J Biol Chem 1996; 271:23836-23841.
40. Kobori H, Sullivan CW, Shizuya H. Heat-labile alkaline phosphatase from Antarctic bacteria: rapid 5' end labelling of nucleic acids. Proc Natl Acad Sci 1984; 81:6691-6695.
41. Vckovski V, Schlatter D, Zuber H. Structure and function of L-lactate dehydrogenase from thermophilic, mesophilic and psychrophilic bacteria IX. Biol Chem Hoppe-Seyler 1990; 371:103-110.
42. Feller G, Narinx E, Arpigny JL, Zekhnini Z, Swings J, Gerday C. Temperature dependence of growth, enzyme secretion and activity of psychrophilic Antarctic bacteria. Appl Microbiol Biotechnol 1994; 42:477-479.
43. Clarke A. Life in cold water: the physiological ecology of polar marine ectotherms. Oceanogr Mar Biol Ann Rev 1983; 21:341-453.
44. Eastman JT. Hypotheses perceiving to origins and speciation of fauna. In: Eastman JT, ed. Antarctic Fish Biology. San Diego: Academic Press, 1993:125-142.
45. Male R, Lorens JD, Smalas AO, Torrissen KR. Molecular cloning and characterization of amino anionic and cationic variants of trypsin from Atlantic salmon. Eur J Biochem 1995; 232:677-685.
46. Gudmundsdottir A, Gudmundsdottir E, Oskarsson S, Bjarnason JB, Eakin AK, Croik CS. Isolation and characterization of CDNAs from Atlantic cod encoding two different forms of trypsinogen. Eur J Biochem 1993; 218:1091-1097.
47. Asgeirsson B, Fox JW, Bjarnason JB. Purification and characterization of trypsin from poikilotherm *Gadus moshua* (?). Eur J Biochem 1989; 180:85-94.
48. Gudmundsdottir E, Spilliaert R, Qin Yang, Croik CS, Bjarnasson JB, Gudmundsdottir A. Isolation and characterization of two CDNAs from Atlantic cod encoding two distinct psychrophilic elastases. Comp Biochem Physiol 1996; 113B:795-801.
49. Tani T, Ohsumi J, Mita K, Takiguchi Y. Identification of a novel class of elastase isozyme, human pancreatic elastase III by DNA and genomic gene cloning. J Biol Chem 1988; 263:1231-1239.
50. Feller G , Le Bussy O, Gerday C. Expression of psychrophilic genes in mesophilic hosts: assessment of the folding state of a recombinant α-amylase. Appl Environ Microbiol 1998; 64:1163-1165.

Psychrophilic enzymes: insights into cold adaptation and catalysis from the first high resolution crystal structures

N. Aghajari and R. Haser*

Institute of Biology and Chemistry of Proteins, Centre National de la Recherche Scientifique, 7 Passage du Vercors, F-69367 Lyon cedex 07, France

1 Introduction

A number of reviews have already addressed the diversity, physiology, enzymology, ecology, and methodology for isolation, purification and characterization of extremophile proteins, including those from psychrophilic organisms.[1–5] Enzymes from organisms adapted to cold environments function optimally at low temperatures, and have therefore considerable potential in biotechnological applications. Psychrophilic enzymes are characterized by a high catalytic efficiency at low temperature and an enhanced thermosensivity, these characteristics being proposed to be related to a higher flexibility of the corresponding three-dimensional (3D) structures when compared to that of their mesophilic and thermophilic counterparts.

Until recently, the structural basis of psychrophily was poorly understood; as the sole approach to the structure/function relationships of psychrophilic enzymes on a molecular level were homology modeling studies, based on the crystal structures of the mesophilic counterparts (for an example see ref. 6). With the crystallization of the first psychrophilic enzymes, α-amylase[7] and a Ca/Zn metalloprotease,[8,9] and the availability of the first experimentally obtained 3D structures of these psychrophilic enzymes[9,10] (Villeret et al., unpubl. data) one can now provide clues for a better understanding of psychrophily and of the high activity of these enzymes at low temperatures. As the α-amylase structures, in the native state as well as in complex with inhibitors, are presently the only psychrophilic structures known at high resolution (up to 1.7 Å), we will hereafter focus on this family of enzymes.

* Corresponding author; e-mail: r.haser@ibcp.fr

2 Crystallography: what do we expect from protein crystal structures?

Protein crystallography has greatly advanced in the past years and is becoming a more and more important and powerful method for understanding the structure, function and control of the components in the living cell where proteins are among the major elements of the cellular organization and metabolism. The high activity in the field is reflected by a growth rate of about three new structures per day, without taking into account numerous structures determined for homologous proteins, mutants, and protein-ligand complexes. The structural results have already made a crucial contribution to the understanding of fundamental biological processes and open new insights into aspects of molecular recognition and biological control, which can be of importance to industry, biotechnology and medicine. In the past decade tremendous advances have been performed in areas specific (or coupled to) to protein crystallography, i.e. molecular biology, protein engineering and design, faster computers, powerful methods and software (like the famous O, X-PLOR, and AMoRe programs),[11–13] new X-ray detectors (image plates, charge coupled devices), synchrotron X-ray sources, use of multiwavelength anomalous scattering techniques, new methods for phase improvement, crystal structure refinement, crystallization techniques and protocols. It is common to say that the structures of proteins resulting from conventional X-ray approaches provide only a static picture and cannot directly address the conformational changes and/or transitions that these proteins often undergo when they are in action. Nevertheless, when diffraction data at high (or even atomic) resolution can be collected, often with synchrotron radiation sources, much information can be gained not only with increasing accuracy in atomic positions but even in terms of molecular dynamics (B-factors, thermal motion) or of evolution of the system under study. For example, visualization of the intermediate steps and the associated structural changes in real time in an enzyme-catalyzed reaction is now becoming a reality with the use of Laue diffraction and time-resolved crystallography techniques, which require synchrotron X-ray beam lines and data collection preferentially at cryotemperatures (see for example ref. 14, and references therein).

Until recently it was convenient and realistic to make a clear division between macromolecular and small-molecule crystallography. While small-molecule crystals usually diffract to atomic resolution, diffraction to equal to or better than 2 Å is considered as high resolution for proteins. In order to observe individual atomic features in a crystal structure they should be distinguishable in the electron density. As the shortest distance between covalently bonded non-H atoms in organic structures is about 1.2 Å, this value is usually adopted for an empirical definition of atomic resolution.

In order to judge the accuracy of any X-ray structure determination, of derived atomic coordinates, two principal factors should be taken into account: the resolution of the diffraction data and the level of refinement. In general, for small molecules there are sufficient X-ray data, typically to 1.0 Å resolution or better, to refine the atomic parameters against these data alone. In fact, it is then possible to refine nine parameters for each atom, three positional and six thermal parameters required to describe its anisotopic thermal motion. The situation is often quite different for proteins, where the resolution is generally limited to less than atomic and where, in addition, the crystal contains about 50% solvent. Therefore, refinement of the crystal structures of proteins cannot be performed using the measured data alone, owing to the limited resolution of the data observed.[15] In order to increase the ratio observed data/refined parameters, the experimental data must be complemented by restraints and/or constraints, like stereochemical restraints based on the accurate structures of amino acids[16] and methods of modeling bulk solvent. Discrepancy between observed and calculated diffraction data (as given by the R-factors) is often the result of an inadequate description of disorder (for example side chains poorly ordered or present in multiple conformations), thermal motion and bulk solvent.

As already stated, water forms a large portion of the volume in crystals of proteins. Usually a tightly bound layer of water molecules surrounds the macromolecule and can be well described, even at medium resolution (around 2.8 Å). However, the remaining bulk water is mostly disordered but a quite detailed description of this solvent electron density can be achieved using bulk solvent refinement methods.[17] That the solvent molecules bound to the protein may be biologically relevant has been strongly suggested on the basis of the 3D structures of several enzymes in complex with substrate analogues, for example in the structures of a trapped phosphoenzyme during a catalytic reaction[18] or of an acyl-trypsin intermediate.[19] In the case of α-amylases, well-defined, structural water molecules are proposed to be essential in the amylolytic catalytic process.[10,20]

Indeed, as high intensity synchrotron radiation sources, efficient 2D area detectors and the use of cryogenic freezing techniques are widely available, recording of protein diffraction data to very high resolution has become tractable. For example, the first structure of a serine proteinase in complex with an inhibitor solved at atomic resolution (1.05 Å) shows all the active site hydrogens and reveals novel structural features which can be directly related to the biological role of the enzyme.[21] When protein crystals diffract exceptionally well, like those of crambin, recording of ultrahigh resolution data (in this case up to 0.67 Å) is possible, and should allow to study in great detail the distribution of electrons in the protein, including proper modeling of bonding electrons.[22] In some cases, one can even use *ab initio* methods (similar to those used in small-molecule crystallography) to solve protein crystal structures, provided that some heavier atoms such as iron or even sulfur are present, and provided the availability of diffraction data to atomic resolution (1.2 Å or better). A nice example is the determination of the crystal structure of a cytochrome c_6.[23]

3 Amylases

Alpha-amylases (α-1,4-glucan 4-glucanohydrolase, EC 3.2.1.1) are widely distributed in plants, fungi, mammalian tissues and microorganisms, where they catalyze the hydrolysis of the α-1,4 glycosidic linkages in starch components, such as amylose and amylopectin. They belong to family 13 in the classification of glycosyl-hydrolases which is the largest of these enzyme families also containing, besides α-1,4-hydrolases, enzymes with distinct substrate and product specificities, like cyclodextrin glycosyltransferases and enzymes specific for α-1,6 and α-1,1 glycosidic linkages.[24–27]

It is obvious from the sequence analysis of all the known amylases that the overall primary structure similarity is rather low (around 15%), but, surprisingly, the percentage of strict identity rises to 53% between *Alteromonas haloplanctis* α-amylase (AHA) and porcine pancreatic α-amylase (hereafter PPA).[28] The latter high homology is quite remarkable, owing to the evolutionarily distant relationship between these two enzymes. The available crystal structures of several α-amylases from

- fungi: *Aspergillus oryzae* (TAKA)[29,30] and *Aspergillus niger* (acid);[31]
- plants: barley isozyme 2 (AMY2)[32] and barley isozyme 1 (AMY1; Aghajari et al., unpubl. data);
- mammalians: porcine pancreas (PPA),[33,34] human pancreas (HPA)[35] and human salivary (HSA)[36]
- bacteria: *Bacillus licheniformis*[37] and the psychrophilic *Alteromonas haloplanctis* (AHA)[10]

clearly show a highly conserved structural core. It is now well established that these enzymes share the same fundamental 3D structure, even if they may differ in substrate specificity due to some sequence-dependent structural variations, but not in the mechanism employed in catalysis. Numerous localized structural differences are observed in the loops and on the surface of the enzymes and may have functional significance, perhaps contributing to the differences in specificity. Large variations are also observed in domain C, which is made up of loops and β-strands arranged in a Greek key motif, the number of these strands varying from five in barley α-amylase[32] to ten in human salivary α-amylase.[36]

4 Alpha-amylase fold, overall structure of psychrophilic α-amylase (AHA) and calcium binding site

The structure of AHA, like the other known α-amylase 3D structures, can be divided into three domains (Fig. 1). The largest one, domain A, including residues 1–86 and 147–356, has the classic $(\beta/\alpha)_8$ barrel fold first seen in the structure of triose phosphate isomerase.[38] This stable framework has been observed in glycosyl hydrolase families 1, 2, 10, 13, 14, 17 and 18 and has been predicted for seven other families,[26,39] and it is also found for a number of other enzymes with widely different and unrelated functions and with little or no sequence homology.[40] Domain A is the catalytic core unit and is linked to a C-terminal eight-stranded β-sheet compact unit (domain C) consisting of residues 357–453, the last five residues being absent in the 3D structure as no corresponding electron density was visible, most probably due to high thermal motion or disorder. Domain C is only loosely associated to

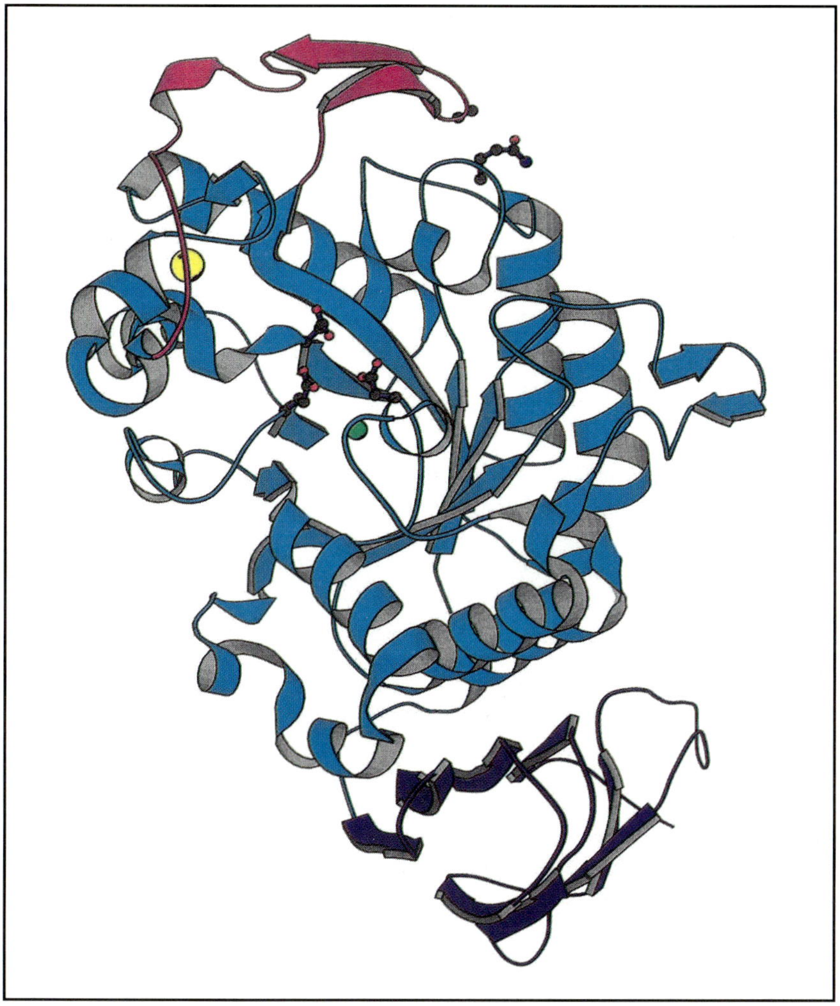

Fig. 1. Overall folding of the psychrophilic α-amylase from *Alteromonas haloplanctis* (AHA) The $(\beta/\alpha)_8$-barrel domain (domain A) is colored in *cyan*, domain B in *magenta* and the C-terminal domain in *dark blue*. In the top right side of the figure, the residues in AHA corresponding to the residues forming a disulphide bridge in mammalian α-amylases are highlighted. The three catalytic residues, Asp174, Glu200 and Asp264, situated in the $(\beta/\alpha)_8$-barrel domain are highlighted as well. The *yellow* ball situated between domains A and B corresponds to the calcium ion, and the *green* ball in the $(\beta/\alpha)_8$-barrel corresponds to the chloride ion (this Figure was generated using the program MOLSCRIPT[70])

the rest of the enzyme and the exact function of this domain has not yet been elucidated. Domain B is the smallest of the three domains (residues 87–146) and is inserted between the third β-sheet and the third helix of the $(\beta/\alpha)_8$ barrel; it consists essentially of loops, two short β-sheet elements and a short terminal helix. In the barley α-amylases, domain B has been shown to confer a number of isoenzyme specific properties.[41,42] Recently, the functional relevance of this domain has also been demonstrated on the molecular level with the determination of the crystal structures of complexes of various α-amylases bound to protein inhibitors.[43–45] Another characteristic of domain B in AHA, like in the other α-amylase 3D structures, is its contribution to the binding of a structurally conserved calcium ion which also interacts with domain A as seen in Figure 1.

Earlier studies showed that calcium is important for both stability and catalytic activity of α-amylases.[46] For example, in porcine pancreatic α-amylase, when the calcium was removed in the presence of a chelating agent, the activity of the enzyme is abolished, an observation which can be explained on the basis of the crystal structure of the enzyme.[33,34] The same calcium binding site is present in AHA, with eight ligands (five protein atoms and three water molecules); Asp144 (in AHA) appears to be an essential bidentate ligand strictly conserved among all known X-ray structures of α-amylases, including the fungal and plant enzymes. Among the five protein calcium ligands, four stem from domain B and one from domain A; such interactions between domains A and B mediated by the calcium ion are certainly important for enzyme stability, substrate binding and protein inhibitor recognition.

4.1
Psychrophilic α-amylase active site structure and the effector chloride ion

On the basis of sequence comparisons between the various α-amylases and the analysis of available α-amylase crystal structures, one could predict that in AHA most of the amino acids important for recognition and hydrolysis of a substrate would belong to the few regions which are highly conserved in all family-13 glycohydrolases. With the high resolution structures of AHA, in its native state and in complex with a Tris molecule acting as an inhibitor, the topology of the active site region could be described in fine details.[10] As is the case in all $(\beta/\alpha)_8$-barrel enzymes, the catalytic region is located on the C-terminal side of the barrel. Three carboxylate residues, previously identified in other α-amylase as strictly conserved and essential because of their implication in catalysis, are in AHA all involved in the binding of the Tris inhibitor. These residues, positioned at the end of a pocket located at the C terminus of the barrel (Fig. 2), form a catalytic trio with Asp174, Glu200 and Asp264 (AHA numbering), the two latter side chains being interconnected through hydrogen bonds via a water molecule most likely important in the catalytic process (see Sect. 5). Detailed crystallographic studies of several α-amylases have revealed the presence of this water molecule, which apparently is conserved in the 3D structures across this enzyme family. It is suggested that this specific water molecule is intimately linked to the catalytic process.

Close to the catalytic site, within the central cavity created by the $(\beta/\alpha)_8$ barrel, a chloride ion is located. All three protein ligands to this ion are common to those found in the structures of mammalian α-amylases, with the exception of Lys300 (in

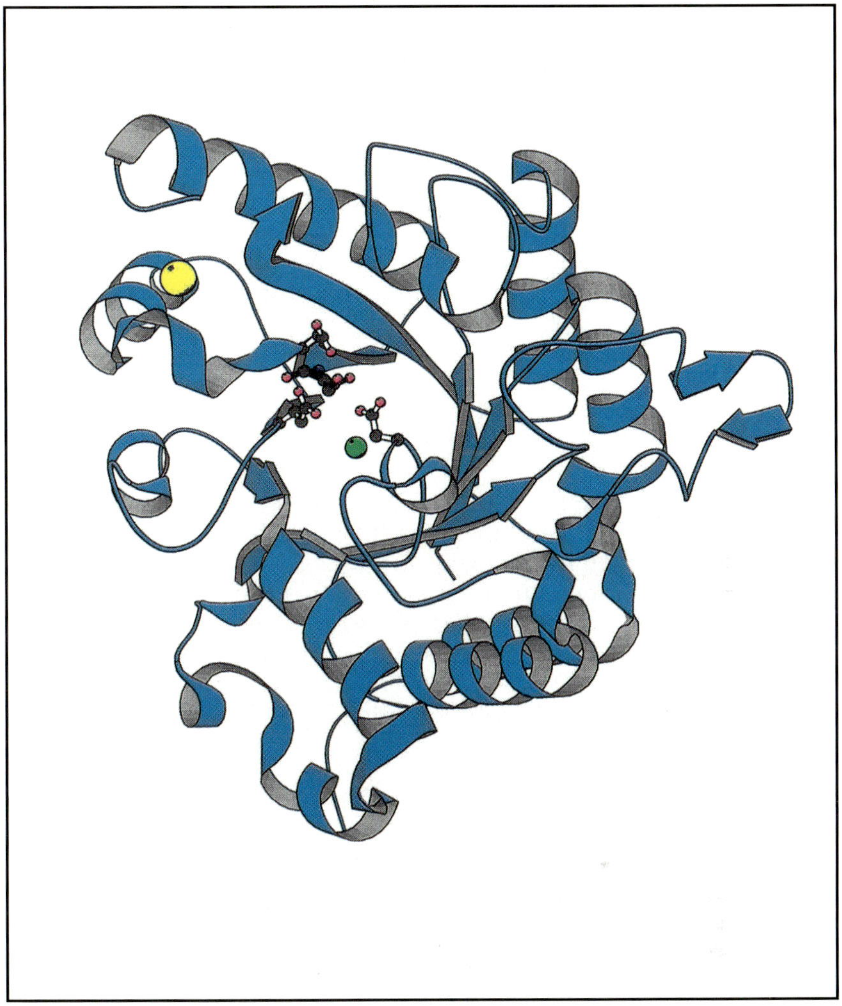

Fig. 2 Closeup on domain A of psychrophilic α-amylase in complex with the Tris inhibitor
The catalytic trio (Asp174, Glu200, Asp264) is *highlighted* as well as the Tris molecule (*black*). *Yellow* and *green* balls correspond to the calcium- and the chloride ion respectively (this Figure was generated using the program MOLSCRIPT[70])

AHA) which is an arginine in mammalian α-amylases (Arg337). The two other chloride ion ligands (Arg172 and Asn262) are adjacent to catalytically essential residues (Asp174 and Asp264). It is interesting to note that this anion binding site exists neither in the known structures of plant and fungal amylases, nor in that of the thermostable α-amylase from the mesophilic bacterium *B. licheniformis*. In all of these enzymes the residues corresponding to Arg172 and Asn262 are conserved, thus suggesting that the third protein ligand is critical for chloride binding. It apparently has to be a basic residue in chloride binding/dependent α-amylases, and site-directed mutagenesis studies in the chloride binding site of AHA[47] have shown that when lysine is mutated to Gln (as the situation in *B. licheniformis* α-amylase) a chloride independent but active enzyme is obtained. Chloride ions have been shown to activate the α-amylase reaction, to act as allosteric activators of catalysis[48] and this has until recently been thought to be characteristic only for mammalian α-amylases.[35] Now it is proved to be present in AHA as well, and sequence alignments have shown that the three amino acid residues critical for chloride binding have been found in some insect α-amylases as well.[10,47]

4.2
Is psychrophilic α-amylase a bifunctional enzyme or has it a vestigial serine protease-like active site?

At the surface of the $(\beta/\alpha)_8$ barrel, in close proximity to the C-terminal domain and approximately 22 Å from the active-site cleft, one finds a trio of residues, Glu19, His337, and Ser303 which superimposes almost perfectly onto the classical catalytic triad (Asp, His, Ser) found in the serine proteases and with that observed, for example, in the active site of some lipases,[10] where instead of a catalytic aspartic acid one also finds a glutamic acid residue, as in the lipase from *Geotrichum candidum*.[49] This striking structural feature observed in the $(\beta/\alpha)_8$ barrel domain of AHA is also a characteristic of the 3D structures of all known mammalian α-amylases, but not of those of the related plant and fungal enzymes. It is not clear whether this site is really active and biologically relevant, or if it corresponds to an ancient protease function. It is interesting to note that AHA in solution is slowly split into two fragments (G. Feller, pers. communication), suggesting that this apparent proteolysis could arise from the intrinsic activity of the above triad rather than from a minor contaminant.

5
Mechanistic implications: inhibition and activation of psychrophilic α-amylase

5.1
The catalytic residues

As mentioned before, α-amylases and related enzymes catalyze the cleavage of the 1,4 glycosidic bonds, which link the α-D-glucopyranoside units of the substrate. As the configuration of the anomeric center of the substrate is conserved after hydrolysis, α-amylases are retaining enzymes. They cleave at internal glycosidic linkages, therefore they fall into the endo-acting category of amylases, in contrast to the exoenzymes, such as glucoamylase and β-amylase, which remove saccharides from the nonreducing end with inversion of the anomeric configuration. A wealth of biochemical data on these systems have suggested that hydrolysis of these bonds proceeds via a double displacement reaction (or a two-step mechanism) in which a covalent glycosyl-enzyme intermediate is generated and then hydrolyzed by a water molecule, activated by a general base.[50–52]

Analysis of the structures of the active site of AHA in complex with a bound substrate analogue and with a reaction product (Aghajari and Haser, unpubl. data) or with an inhibitor like the Tris molecule,[10] readily reveals the molecular basis for the enzyme function. These enzyme complexes show that the trio Asp174, Glu200 and Asp264 plays a vital role in the catalytic mechanism. Mutagenesis studies of these residues in various α-amylases have already confirmed their importance to catalysis.[25,53,54]

5.2
Proposal for the catalytic mechanism of polysaccharide hydrolysis

The above three strictly conserved amino acids define the catalytic center in probably all α-amylases. Glu200 is proposed to be protonated and to act as a general acid protonating the interglycosidic oxygen atom of a polysaccharidic substrate and thus allows leaving group departure. In fact, this residue (and its homologue in other α-amylases) has been assigned the role of proton donor on the basis of its environment and the hydrogen bond between one of its carboxylate oxygen atoms and the interglycosidic nitrogen atom of acarbose (a pseudo-tetrasaccharide inhibitor of many glycosyl hydrolases) in the 3D structures of various glycosidase/acarbose complexes[20,55–57] (Aghajari and Haser, unpubl. data). In the presence of a true substrate, this interaction is believed to induce a distorsion of the adjacent nonreducing sugar ring (similar to that present in the acarbose substrate analogue), followed by the weakening and the subsequent breakage of the interglycosidic link. A transient positive charge on the leaving group will develop leading to an oxocarbonium ion which is stabilized by the negative charge of the nearby Asp174, acting by nucleophilic interaction with the C1 atom, with formation of a transient covalent intermediate. This aspartic acid has unequivocally been identified as the catalytic nucleophile in α-glucosidases.[58] The water molecule which bridges Glu200 and Asp264 (Fig. 3) is in an ideal position to attack the intermediate and to serve to regenerate the enzyme by reprotonating Glu200 (Fig. 3). In this scheme Glu200 behaves as the general base candidate for proton abstraction, and therefore has a dual function as acid/base catalyst. The geometry around the water molecule observed in all the α-amylase structures is in support of a proton transfer from the water to the carboxylate oxygen atom of the catalytic glutamic acid. Then, the resulting hydroxide ion (in the line of flight with the C1 atom) is now ready to move towards and bind to the C1 atom in order to complete the hydrolytic reaction.

5.3
A chloride ion acting as an effector of the activity of psychrophilic α-amylase

The chloride ion, close to (4.6 Å "behind") the water molecule (Fig. 3) which is proposed to be activated by Glu200 into a hydroxide ion, is in an ideal position to orient and enhance via electrostatic repulsion the movement of this hydroxide ion

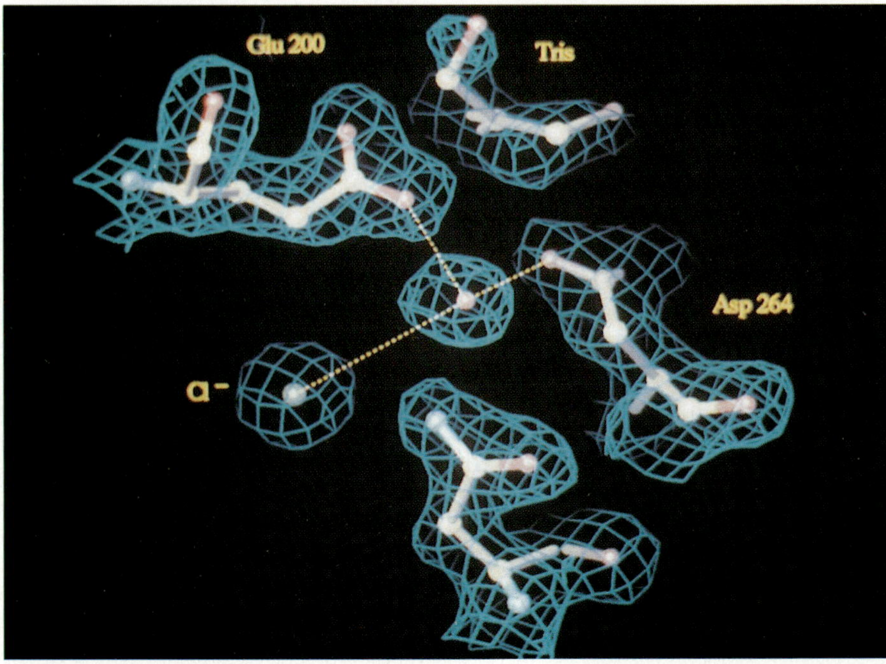

Fig. 3. The catalytic machinery of psychrophilic α-amylase, with the effector chloride ion and the water molecule (Wat 1004) proposed to perform the hydrolytic attack of the substrate, this water molecule is bridged between two of the catalytic residues, Glu200 and Asp264 (this Figure was generated using the program TURBO-FRODO[71])

towards the positively charged C1 atom. It is therefore quite obvious that the generated OH^- nucleophile encounters more electrostatic repulsion induced by the nearby carboxylate of Asp264 and by the chloride ion during its movement towards the C1 atom. The AHA structures (the substrate-free form and the enzyme/inhibitor complexes) allow understanding at a molecular level of this activation process, i.e. the role in modulating the enzyme activity with anions, like halogens, nitrate, etc. which also have been shown to act as activators for AHA.[47]

5.4
An essential structural water molecule in psychrophilic α-amylase?

The mechanism proposed above enlightens the role of a bound water molecule in the active site. Solving the crystal structures of several α-amylases has in all cases shown a water molecule bridging the proton donor Glu and an adjacent Asp. We have proposed that this water molecule is in a position that makes it likely to be able to act as the nucleophile in the hydrolysis reaction.[10,20] It is noteworthy that it is linked to a chain of well ordered water molecules forming a solvent channel which could act as a supply for further hydrolytic reactions[9] (Aghajari and Haser, unpubl. data).

6
Determinants of psychrophily in the light of the 3D structure of psychrophilic α-amylase

A variety of theories have been proposed to explain the structural basis for protein thermostability (for examples see refs. 59–62) as well as for enzyme cold adaptation (see refs. 5,63, and references therein). It is believed that comparing carefully 3D structures of homologous extremozymes from different organisms living in extreme environments (halophile, acidophile, alkalophile, thermophile, hyperthermophile, barophile, psychrophile) to their mesophilic (mesostable) counterparts should enable the specific structural features which distinguish these enzymes and which contribute to control their stability and activity to be evaluated.

The comparison of sequence and 3D structure of AHA with their mesophilic homologues (we will here restrict this comparison to the mammalian enzymes owing to their high sequence homology with AHA) is thought to clarify the reasons for greater thermolability, for high efficiency at low temperature and for cold adaptation of the former enzyme. This detailed analysis is in progress,[64] but several points can already be underlined. In accord with many other authors facing the problem of thermal stability, we come to the conclusion that there are several or even a great number of stabilizing structural features (rather than one single outstanding characteristic) necessary for optimal enzymatic activity under extreme conditions. In the case of AHA, we have selected a few features and significant differences with mammalian α-amylases which may be relevant for explaining cold adaptation and higher catalytic efficiency of this enzyme: loop conformational variations, disulfide bonds, interdomain flexibility and role of domain B, charged amino-acid distribution.

6.1
Loops

As already stated, the overall structure of AHA is very similar to the three known mammalian α-amylase structures. The superposition of AHA on PPA is shown in Figure 4. However, local variations primarily exist in loop regions, and 11 of these regions differ between the psychrophilic and mammalian α-amylases. Six out of the 11 differences correspond to deletions in AHA. Some of these changes occur in the immediate vicinity of the catalytic site, thereby they certainly influence the binding of substrates within the active site cleft (Fig. 4).

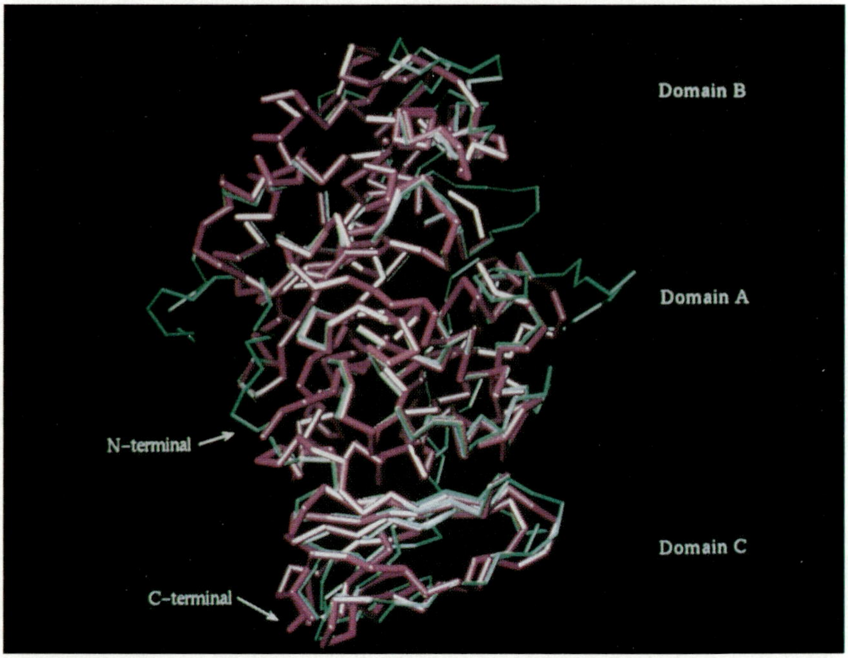

Fig. 4. Superimposition of the Cα backbone of AHA (*pink* and *thick-lined*) on the Cα backbone of pancreatic α-amylase (*green* and *thin-lined*), it is seen that the immediate biggest differences are found in the loop regions (this Figure was generated with TURBO-FRODO[72])

6.2
Disulfide bonds and interdomain interactions

That disulfide bonds reduce the unfolding entropy of proteins has been proved experimentally in the cases of trypsin and of pancreatic trypsin inhibitor,[65] and in many other proteins since. Five disulfide bridges are found and conserved in all mammalian α-amylases, whereas only four of them are common to AHA, the missing one being Cys70-Cys115; it is interesting to note that the latter in the mammalian enzymes contributes to link domain A [the catalytic $(\beta/\alpha)_8$-barrel unit] to domain B (Fig. 1 shows the corresponding residues in AHA). The absence of the interdomain disulfide bridge in AHA could be a factor contributing to cold adaptation, the enzyme gaining in overall mobility or overall conformational freedom in order to maintain high or at least adequate catalytic efficiency in the cold environment. As the missing cross-link in AHA apparently is not compensated by other interactions, one expects indeed additional flexibility and independence of domain B, compared to the mesophilic α-amylases. Therefore, the two domains A and B in AHA are less constrained and can move more easily relative to one another.

6.3 Role of domain B

At this point it is worth recalling that domain B plays an important role in substrate recognition, protein inhibitor binding and enzymatic activity: for example in pancreatic amylases when His101 (His89 in AHA) was mutated to Asn, the enzymatic activity was drastically affected.[66,67] In AHA, this residue participates in substrate binding (Aghajari and Haser, unpubl. data) and is also adjacent to one of the calcium ligands, Asn88, a network of interactions which clearly serves to orient His89 for proper binding to the substrate. This calcium ion therefore stabilizes the His89 side chain and guides it into the correct orientation for hydrogen bonding to the substrate. In mesophilic α-amylases domain B is also of crucial importance for binding proteinaceous inhibitors,[43–45] and tertiary structure and sequence variations in this domain can clearly prevent interaction with these inhibitors. These studies demonstrate that small changes in α-amylase structure can have drastic consequences on the structure/activity relationships of these enzymes.

6.4 Charged residue distribution and chloride binding site

The distributions of acidic and basic residues are evidently different in the two types of α-amylase, mesophilic and psychrophilic, as shown in Fig. 5. There are

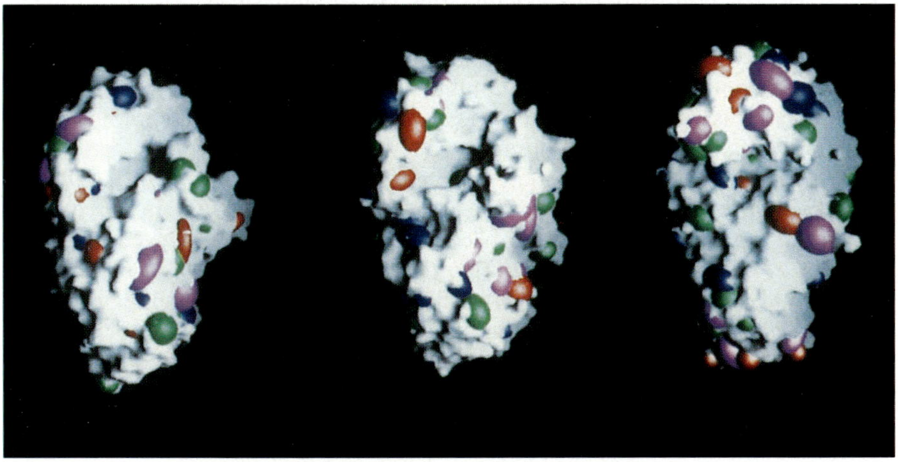

Fig. 5. Distribution of charged amino acids on the surfaces of psychrophilic α-amylase (*left side*) compared to the distributions on the mesophilic mammalian α-amylase HPA (*in the middle*) and the mesophilic thermostable bacterial α-amylase BLA (*right side*)
Color code: arginine residues are shown in *pink*, lysine residues in *blue*, aspartic acid residues in *green* and glutamic acid residues in *red*. The number of charged amino acids on the surface increases with the thermostability. All molecules are shown in the same orientation (this Figure was prepared with the program GRASP[73])

obviously less charges on the surface of AHA than in its mesophilic counterparts. The impact this has on intermolecular interactions is straightforward, with a very small number of those in the case of AHA. Such different distributions also correspond to significant differences in polarity between these structures, which may play a role in guiding substrates and inhibitors into the active site cleft. Among the basic amino acids, two of them (Arg172 and Lys200 in AHA) are essential for binding and partly neutralizing the chloride effector ion, a site which is absent in the mesophilic known 3D structures of fungal and plant α-amylases, as well as of the thermostable bacterial α-amylase from *B. licheniformis*. The fact that for the psychrophilic bacterial enzyme AHA nature has tailored a site optimized for binding an effector ion like chloride is probably not fortuitous; its presence contributes, with its associated electrostatic field, to speed up the catalytic reaction and to allow the enzyme to work efficiently despite the low temperature of its environment. Therefore, we suggest that the chloride ion binding site in the psychrophilic α-amylase appears as an integral part of the cold-adaptation system of this enzyme. That this site is also conserved in the homologous mammalian, and even an insect,[68] enzymes may not be surprising as the corresponding organisms have occasionally to cope with a wide range of conditions including low temperature media.

7 Conclusions and perspectives

The subtle structural variations found in the vicinity of the active site of various α-amylases including the psychrophilic one are certainly related to the observed differential substrate and cleavage pattern specificities between these enzymes. Indeed, the structure-based comparisons of AHA, complexed to various polysaccharide compounds, with its mesophilic counterparts (porcine pancreatic and barley α-amylases in complex with the same substrate analogues), indicate that, despite a very high structural homology between the corresponding active sites, AHA displays a drastically different behavior in terms of processing the substrate when comparing to the mesophilic enzyme (Aghajari and Haser, unpubl. data).

Cold-active enzymes had already been suggested to display enhanced abilities to perform conformational changes during the catalytic reaction in order to reduce the activation energy, and therefore to function with a reduced energy requirement.[69] The fact that in the psychrophilic α-amylase the disulfide bond linking domains A and B is missing, together with a number of structural variations in loops, especially in proximity to the active site cleft, gives higher overall flexibility to the enzyme and may allow it to interact with a broader range of substrates. The recently determined crystal structures of AHA in complex with various polysaccharide compounds are consistent with the above proposal (Aghajari and Haser, unpubl. data).

Our present results are consistent with the idea that there is no single obvious explanation for the molecular determinants of psychrophily, but that factors producing psychrophily correspond to small changes distributed throughout the protein structure and leading to an overall increased flexibility.

The crystal structures described here provide the highest resolution images of any psychrophilic enzyme to date. These structures allow now for the rational design of variants to try to further elucidate the mechanism of cold adaptation. A major issue will be to improve the stability of AHA without affecting its high catalytic efficiency at low temperature. Moreover, appropriate site-directed mutations guided by the 3D structures of AHA are expected to improve the specific activity of the enzyme. This approach was already successful with a subtilisin excreted by psychrophilic antarctic bacteria, as suitable mutants designed on the basis of a model were shown to display increased structural rigidity without decrease in catalytic efficiency, or even one of these mutants had an overall activity about 20 times higher than that of mesophilic subtilisin.[70] This kind of results suggests that mutagenesis programs based on experimentally obtained high resolution 3D structures, where for example a more precise description of hydrogen bond networks can be given, can be succesfully applicated to researches in the field of structure/stability/activity relationship of psychrophilic enzymes and open new perspectives in the use of these enzymes for industrial applications.

Other avenues of investigation can now be contemplated as a result of having the AHA structure to hand. Certainly, a very directed mutagenesis study can now be carried out to precisely define the importance of stabilizing features necessary for optimal activity under the extreme growth conditions. Finally, it is clear that only the cumulative biological, genetic, and structural information about a number of psychrophilic enzymes will further enhance our understanding of the mechanisms of cold adaptation and of their high catalytic efficiency at low temperature.

Acknowledgements. This work and N. Aghajari were supported by the Human Capital and Mobility Network CHRX-CT94-0521, the EUROCOLD Concerted Action and the Centre National de la Recherche Scientifique.

8
References

1. Morita RY. Psychrophilic bacteria. Bacteriol Rev 1975; 39:144-167.
2. Russel NJ. Cold adaptation of microorganisms. Phil Trans R Soc London 1990; B326:595-611.
3. Russel NJ. Physiology and molecular biology of psychrophilic microorganisms. In: Herbert RA, Sharp RJ, eds. Molecular Biology and Biotechnology of Extremophiles. London: Blackie, 1992:203-224.
4. Jaenicke R. Protein structure and function at low temperatures. Phil Trans R Soc London 1990; B326:535-553.
5. Feller G, Gerday C. Psychrophilic enzymes: molecular basis of cold adaptation. Cell Mol Life Sci 1997; 53:830-841.
6. Feller G, Payan F, Theys F, Qian M, Haser R, Gerday C. Stability and structure analysis of α-amylase from the antarctic psychrophile *Alteromonas haloplanctis* A23. Eur J Biochem 1994; 222:441-447.
7. Aghajari N, Feller G, Gerday C, Haser R. Crystallization and preliminary X-ray diffraction studies of α-amylase from the antarctic psychrophile *Alteromonas haloplanctis* A23. Protein Sci 1996; 5:2128-2129.

8. Villeret V, Chessa J-P, Gerday C, Van Beumenn J. Preliminary crystal structure determination of the alkaline protease from the Antarctic psychrophile *Pseudomonas aeruginosa*. Protein Sci 1997; 6:2462-2464.
9. Aghajari N. Crystallographic studies of psychrophilic enzymes with focus on structure/function relationships and cold adaptation. Université de la Méditerranée (Aix-Marseille II) and Université de Liège: PhD Thesis, 1998.
10. Aghajari N, Feller G, Gerday C, Haser R. Crystal structures of the psychrophilic α-amylase from *Alteromonas haloplanctis* in its native form and complexed with an inhibitor. Protein Sci 1998; 7:564-572.
11. Jones TA, Zou JY, Cowan SW, Kjeldgaard M. Improved methods for building protein models in electron density maps and the location of errors in these models. Acta Cryst 1991; A47:110-119.
12. Brünger AT. X-PLOR: Version 3.1. A System for Crystallography and NMR. New Haven, CT, USA: Yale Univ Press, 1993.
13. Navaza J. AMoRe: an automated package for molecular replacement. Acta Cryst 1994; A50:157-163.
14. Srajer V, Teng T-Y , Ursby T, Pradervand C, Ren Z, Adachi S-I, Shildkamp W, Bourgeois D, Wulff M, Moffat K. Photolysis of the carbon monoxide complex myoglobin: nanosecond time-resolved crystallography. Science 1997; 274:1726-1729.
15. Brünger AT. Crystallographic phasing and refinement of macromolecules. Curr Opin Struct Biol 1991; 1:1016-1022.
16. Engh RA, Huber R. Accurate bond and angle parameters for X-ray protein structure refinement. Acta Cryst 1991; A47:392-400.
17. Jiang J-S, Brünger AT. Protein hydration observed by X-ray diffraction: solvation properties of penicillopepsion and neuraminidase crystal structures. J Mol Biol 1994; 243:100-115.
18. Lee Y-H, Olson TW, Ogata CM, Levitt DG, Banaszak LJ, Lange AJ. Crystal structure of a trapped phosphoenzyme during a catalytic reaction. Nature Struct Biol 1997; 4:615-618.
19. Singer PT, Smalås A, Carty RP, Mangel WF, Sweet RM. The hydrolytic water molecule in trypsin, revealed by time-resolved Laue crystallography. Science 1993; 259:669-673.
20. Kadziola A, Søgaard M, Svensson B, Haser R. Molecular structure of a barley α-amylase-inhibitor complex: Implications for starch binding and catalysis. J Mol Biol 1998; 278: 205-217.
21. Rypniewski WR, von der Osten C, Ostergaard P, Dauter M, Wilson KS. Direct observation of catalytic hydrogen atoms in a serine proteinase: subtilisin/CI-2A complex at 1.05 Å resolution. In: Bartunik H, Laasch W, Lamzin V, Wilmanns M, eds. Hamburger Synchrotronstrahlungslabor Hasylab am Deutschen Elektronen-Synchrotron. Desy: Annual Report Part II,1997:405-406.
22. Teeter MM, Dauter Z, Wilson KS, Lamzin VS. Crambin at ultra-high resolution: experimental observation of bonding electrons in proteins. In: Bartunik H, Laasch W, Lamzin V, Wilmanns M, eds. Hamburger Synchrotronstrahlungslabor Hasylab am Deutschen Elektronen-Synchrotron. Desy: Annual Report Part II, 1997:435-436.
23. Frazão C, Soares CM, Carrondo MA, Pohl E, Dauter Z, Wilson KS, Hervás M, Navarro JA, De la Rosa MA, Sheldrick GM. *Ab initio* determination of the crystal structure of cytochrome c_6 and comparison with plastocyanin. Structure 1995; 3:1159-1169.
24. Henrissat B, Bairoch A. New families in the classification of glycosylhydrolases based on amino acid sequence similarities. Biochem J 1993; 293:781-788.
25. Svensson B. Protein engineering in the α-amylase family: catalytic mechanism, substrate specificity, and stability. Plant Mol Biol 1994; 25:141-157.
26. Henrissat B, Callebaut I, Fabrega S, Lehn P, Mornon JP, Davies G. Evolution: conserved catalytic machinery and the prediction of a common fold for several families of glycosyl hydrolases. Proc Natl Acad Sci 1995; 92: 7090-7094.
27. Davies G, Henrissat B. Structures and mechanisms of glycosyl hydrolases. Structure 1995; 3:853-859.

28. Feller G, Lonhienne T, Deroanne C, Van Beeumen J, Gerday C. Purification, characterization and nucleotide sequence of the thermolabile α-amylase from the Antarctic psychrotroph *Alteromonas haloplanctis* A23. J Biol Chem 1992; 267: 5217-5221.
29. Matsuura Y, Kunusoki M, Harada W, Kakudo M. Structure and possible catalytic residues of Taka-amylase A. J Biochem 1984; 95:697-702.
30. Swift HJ, Brady L, Derewenda ZS, Dodson EJ, Dodson GG, Turkenberg JP, Wilkinson AJ. Structure and molecular model refinement of *Aspergillus oryzae* (TAKA) α-amylase: an application of the simulated-annealing method. Acta Cryst 1991; B47:535-544.
31. Brady RL, Brzozowski AM, Derewenda ZS, Dodson EJ, Dodson GG. Solution of the structure of *Aspergillus niger* acid α-amylase by combined molecular replacement and multiple isomorphous replacement methods. Acta Cryst 1991; B47:527-535.
32. Kadziola A, Abe J-I, Svensson B, Haser R. Crystal and molecular structure of barley α-amylase. J Mol Biol 1994; 239:104-121.
33. Qian M, Haser R, Payan F. Structure and molecular model refinement of pig pancreatic α-amylase at 2.1 Å resolution. J Mol Biol 1993; 231:785-799.
34. Larson SB, Greenwood A, Cascio D, Day J, McPherson A. Refined molecular structure of pig pancreatic α-amylase at 2.1 Å resolution. J Mol Biol 1994; 235:1560-1584.
35. Brayer GD, Luo Y, Withers SG. The structure of human pancreatic α-amylase at 1.8 Å resolution and comparison with related enzymes. Protein Sci 1995; 4: 1730-1742.
36. Ramasubbu N, Paloth V, Luo Y, Brayer GD, Levine MJ. Structure of human salivary α-amylase at 1.6 Å resolution: Implications for its role in the oral cavity. Acta Cryst 1996; D52:435-446.
37. Machius M, Wiegand G, Huber R. Crystal structure of calcium-depleted *Bacillus licheniformis* α-amylase at 2.2 Å resolution. J Mol Biol 1995; 246:545-559.
38. Banner DW, Bloomer AC, Petsko GA, Phillips DC, Pogson CI, Wilson IA, Corran PH, Furth AJ, Milman JD, Offord RE, Priddle JD, Waley SG. Structure of chicken muscle triose phosphate isomerase determined crystallographically at 2.5 Å resolution using amino acid sequence data. Nature 1975; 255:609-614.
39. Jenkins J, Leggio LL, Harris G, Pickersgill R. β-glucosidase, β-galactosidase, family A cellulases, family F xylanases and two barley glycanases form a superfamily of enzymes with 8-fold β/α architecture and with two conserved glutamates near the carboxy-terminal ends of β-strands four and seven. FEBS Lett 1995; 362: 281-285.
40. Reardon D, Farber GK. The structure and evolution of α/β proteins. FASEB J 1995; 9:497-503.
41. Rodenburg KW, Juge N, Guo XJ, Søgaard M, Chaix JC, Svensson B. Domain B protruding at the third β strand of the α/β barrel in barley α-amylase confers distinct isozyme-specific properties. Eur J Biochem 1994; 221:277-284.
42. Juge N, Rodenburg KW, Guo XJ, Chaix JC, Svensson B. Isozyme hybrids within the protruding third loop domain of the barley α-amylase $(\beta/\alpha)_8$-barrel. Implications for BASI sensitivity and substrate affinity. FEBS Lett 1995; 363:299-303.
43. Wiegand G, Epp O, Huber R. The crystal structure of porcine pancreatic α-amylase in complex with the microbial inhibitor tendamistat. J Mol Biol 1995; 247:99-110.
44. Bompard-Gilles C, Rousseau P, Rougé P, Payan F. Substrate mimicry in the active center of a mammalian α-amylase: structural analysis of an enzyme-inhibitor complex. Structure 1996; 4:1441-1452.
45. Vallée F, Kadziola A, Juy M, Bourne Y, Rodenburg K, Svensson B, Haser R. Barley α-amylase bound to its endogenous protein inhibitor BASI: crystal structure of the complex at 1.9 Å resolution. Structure 1998; 6:649-659.
46. Vallée, BL, Stein EA, Summerwell, WN, Fischer, EH. Metal content of α-amylases of various origins. J Biol Chem 1959; 234:2901-2929.
47. Feller G, le Bussy O, Houssier C, Gerday C. Structural and functional aspects of chloride binding to *Alteromonas haloplanctis* α-amylase J Biol Chem 1996; 271:23836-23841.

48. Levitzki A, Steer ML. The allosteric activation of mammalian α-amylase by chloride. Eur J Biochem 1974; 41:171-180.
49. Schrag JD, Li Y, Wu S, Cygler M. Ser-His-Glu triad forms the catalytic site of the lipase from *Geotrichum candidum*. Nature 1991; 351:761-764.
50. Koshland DE. Stereochemistry and the mechanism of enzymatic reactions. Biol Rev 1953; 28:416-436.
51. Sinnott ML. Catalytic mechanism of enzymic glycosyl transfer. Chem Rev 1990; 90:1171-1202.
52. McCarter JD, Withers SG. Mechanisms of enzymatic glycoside hydrolysis. Curr Opin Struct Biol 1994; 4:885-892.
53. Svensson B, Søgaard M. Mutational analysis of glycosilase function. J Biotechnol 1993; 29:1-37.
54. Søgaard M, Kadziola A, Haser R, Svensson B. Site-directed mutagenesis of histidine 93, aspartic acid 180, glutamic acid 205, histidine 290, and aspartic acid 291 at the active site and tryptophan 279 at the raw starch binding site in barley α-amylase 1. J Biol Chem 1993; 268:22480-22484.
55. Qian M, Haser R, Buisson E, Duée E, Payan F. The active center of a mammalian α-amylase. Structure of the complex of a pancreatic α-amylase with a carbohydrate inhibitor refined to 2.2-Å resolution. Biochemistry 1994; 33:6284-6294.
56. Strokopytov B, Penninga D, Rozeboom HJ, Kalk KH, Dijkhuizen L, Dijkstra BW. X-ray structure of cyclodextrin glycosyl transferase complexed with acarbose. Implications for the catalytic mechanism of glycosidases. Biochemistry 1995; 34:2234-2240.
57. Brzozowski AM, Davies G. Structure of the *Aspergillus oryzae* α-amylase complexed with the inhibitor acarbose at 2.0 Å resolution. Biochemistry 1997; 36:10839-10845.
58. McCarter JD, Withers SG. Unequivocal identification of Asp-214 as the catalytic nucleophile of *Saccharomyces cerevisiae* α-glucosidase using 5-fluoro glycosyl fluorides. J Biol Chem 1996; 271:6889-6894.
59. Argos P, Rossmann MG, Grau UM, Zuber H, Frank G, Tratschin JG.Thermal stability and protein structure. Biochemistry 1979; 18:5698-5703.
60. Dill KA.Dominant forces in protein folding. Biochemistry 1990; 29:7133-7155.
61. Delboni LF, Mande SC, Rentier-Delrue F, Mainfroid V, Turley S, Vellieux FMD, Martial JA, Hol WGL. Crystal structures of recombinant triosephosphate isomerase from *Bacillus stearothermophilus*. An analysis of potential thermostability factors in six isomerases with known three-dimensional structures points to the importance of hydrophobic interactions. Protein Sci 1995; 4:2594-2604.
62. Russel RJM, Taylor GL. Engineering thermostability. Lessons from thermophilic proteins. Curr Opin Biotechnol 1995; 6:370-374.
63. Smalås AO, Heimstad ES, Hordvik A, Willassen NP, Male R. Cold adaption of enzymes: structural comparison between salmon and bovine trypsins. Proteins: Struct Funct Gen 1994; 20:149-166.
64. Aghajari N, Feller G, Gerday Ch, Haser R. Structures of the psychrophilic *Alteromonas haloplanctis* α-amylase give insights into cold adaptation at a molecular level. Structure 1998; 6:1503-1516.
65. Privalov PL. Stability of proteins. Adv Protein Chem 1979; 33: 167-241.
66. Ishikawa K, Matsui I, Honda K, Nakatani H. Substrate-dependent shift of optimum pH in porcine pancreatic α-amylase-catalyzed reactions. Biochemistry 1990; 29:7119-7123.
67. Ishikawa K, Matsui I, Kobajashi S, Nakatani H, Honda K. Substrate recognition at the binding site in mammalian pancreatic α-amylase. Biochemistry 1993; 32:6259-6365.
68. Strobl S, Gomis-Rüth FX, Maskos K, Frank G, Huber R, Glockshuber R. The α-amylase from the yellow meal worm: complete primary structure, crystallization and preliminary X-ray analysis. FEBS Lett 1997; 409:109-114.
69. Hochachka P, Somero GN. Temperature adaptation. In: Biochemical Adaptation. New Jersey: Princeton Univ Press, 1984:355-449.

70. Narinx E, Baise E, Gerday C. Subtilisin from psychrophilic antarctic bacteria: characterization and site-directed mutagenesis of residues possibly involved in the adaptation to cold. Protein Eng 1997; 10:1271-1279.
71. Kraulis PJ. MOLSCRIPT: A program to produce both detailed and schematic plots of protein structures. J Appl Crystallogr 1991; 24:946-950.
72. Roussel A, Cambillau C. In: Silicon graphics geometry partners directory 86. Mountain View, CA: Silicon Graphics, 1991.
73. Nicholls A, Bharadwaj R, Honig B. GRASP - graphical representation and analysis of surface properties. Biophys J 1993; 64:166-170.

Temperature adaptation in enzymes of antarctic fish

M. A. Ciardiello, L. Camardella and G. di Prisco*

Institute of Protein Biochemistry and Enzymology, C.N.R., Via Marconi 10, I-80125 Naples, Italy

1 Introduction

The study of the biochemical basis underlying the adaptation processes to different environments yields important information when investigating organisms living under extreme conditions. Antarctic organisms are a precious tool to investigate molecular properties acquired during adaptation to the lowest temperatures in the world. Antarctic fish are a taxonomically uniform group of teleosts, which evolved in geographic and climatic isolation and live at -1.97°C throughout the year. Because of the constant temperature of the seawater they developed a high degree of stenothermality and were defined as a paradox in being ectothermic homeotherms.[1] In fact, antarctic species such as *Pagothenia borchgrevinki*, *Trematomus bernacchii* and *Trematomus hansoni* do not survive at temperatures of only a few degrees higher than that of the habitat.[2] Due to their unique ecological characteristics, antarctic fish show unusual features. One of the most significant evolutionary trends is the decrease in erythrocyte number and hemoglobin concentration in their blood,[3,4] possibly reaching the extreme stage in the hemoglobin-less family Channichthyidae (icefish),[5] whose blood is virtually devoid of erythrocytes.

Comparative studies of structure/function relationship in enzymes from antarctic and mesophilic organisms may provide information on the biochemical mechanisms sustaining cold adaptation. Although the first studies on enzymatic cold adaptation in antarctic fish were undertaken many years ago,[1,6,7] the literature so far available is still insufficient to allow unequivocal understanding of molecular properties acquired during evolution. Recently, enzymes of psychrophilic bacteria received more attention than those from cold-adapted higher organisms, providing a wider literature, especially on secretory enzymes.[8–10] The comparative use of data on enzymes from bacteria and higher organisms, such as vertebrates, is useful, although it must be taken with some caution. In fact, bacteria can usually tolerate a wider temperature range than antarctic fish; in particular, the secretory enzymes may have developed distinct structural and functional adaptations linked to specif-

* Corresponding author; e-mail: diprisco@dafne.ibpe.na.cnr.it

ic requirements of the external environment. Therefore, in order to identify trends linked to cold adaptation in antarctic fish, we try, when possible, to compare the properties of intracellular enzymes from stenothermic organisms with each other. It was reported[1] that in some biochemical adaptations (e.g. lack of multiple hemoglobins and some properties of oxygen binding) antarctic fish are more similar to mammals than to other fish. These features were related to the stenothermality of these organisms. Therefore, we compare the properties of enzymes from antarctic fish (whose physiological temperature is constantly below zero) with those of enzymes from homeothermic vertebrates (whose physiological temperature is constantly near 40°C).

2
Effect of temperature on activity

Antarctic fish enzymes normally work at constant subzero temperatures since the cells of ectothermic species have the same temperature of the habitat. At low temperatures the rates of enzyme-catalyzed reactions are reduced by the low heat content of the cellular environment (Q_{10} effect). On the basis of this consideration, strategies of biochemical adaptation developed during the cold-adaptation process are expected, allowing antarctic fish to produce an adequate metabolic flux offsetting the Q_{10} effect.

Functional properties of two enzymes from antarctic fish, glucose-6-phosphate dehydrogenase (G6PD) and glutamate dehydrogenase (GDH), have been analyzed in comparison with the homologous enzymes of mesophilic vertebrates. G6PD was purified from the blood of two antarctic species, the red-blooded *Dissostichus mawsoni* and the icefish *Chionodraco hamatus*. In the blood, G6PD is usually found in the erythrocytes. The blood of icefish contains only few erythrocyte-like cells; the enzyme was found associated with these cells, at a concentration four- or fivefold higher than that present in the erythrocytes of red-blooded fish.[11–13] Despite the highly reduced number of icefish erythrocyte-like cells, the higher blood volume[14] allows the maintenance of a similar G6PD activity level in both families. These results suggest that the icefish blood cells may have acquired the function of storing G6PD,[13] ensuring the availability of the physiologically necessary amount of enzyme activity.

Table 1. Kinetic parameters of glucose-6-phosphate dehydrogenase at 5°C

Source	K_m (NADP) (μM)	K_m (G6P) (μM)	k_{cat} (min^{-1})	k_{cat}/K_m (NADP) (min^{-1} μM^{-1})	k_{cat}/K_m (G6P) (min^{-1} μM^{-1})
D. mawsoni	1.1	10	7700	7000	770
C. hamatus	0.8	13	7400	9300	570
Human	1.0	20	3400	3400	170

K_m Michaelis constant; k_{cat} catalytic constant; *NADP* β-nicotinamide adenine dinucleotide phosphate; *G6P* glucose 6-phosphate

Activity as a function of temperature of antarctic G6PD and GDH has been compared with that of homologous enzymes from human erythrocytes and from bovine liver, respectively. It should be noted that mesophilic G6PD and GDH are still active at 0°C, although with lower efficiency. Therefore the activity at 0°C is a necessary but not sufficient condition to define an enzyme as cold-adapted.

The shift of the apparent optimal temperatures towards temperatures lower than those of the mesophilic enzymes is considered a general feature of cold-adapted enzymes.[15] The behavior of G6PD and GDH from antarctic fish is in line with this feature. In fact, the apparent optimal temperatures of antarctic and mesophilic G6PD were 55 and 65°C, respectively; apparent optimal temperatures at 30 and 45°C for antarctic and bovine GDH, respectively, were observed.[16,17]

2.1 Kinetic properties

The affinity of antarctic and human G6PD for the substrate glucose 6-phosphate (G6P) was not significantly influenced by temperature variations; however, the apparent K_m (Michaelis constant) of the fish enzymes was lower than that of human G6PD. In contrast, the affinity for the coenzyme NADP increased at increasing temperatures, *D. mawsoni* G6PD displaying the largest variation. In the temperature range considered, k_{cat} (katalytic constant) and k_{cat}/K_m of the antarctic enzymes were higher than those of human G6PD. The higher catalytic efficiency of antarctic G6PD is mainly due to higher k_{cat}, although lower K_m values at some temperatures also give a positive contribution.[16] Table 1 shows the values calculated at 5°C. At low temperatures, the specific activity of antarctic GDH was fourfold higher than that of the bovine enzyme.[11]

Higher catalytic efficiency than that of homologous enzymes from mesophilic organisms seems to be a general feature of many antarctic fish enzymes. Higher catalytic efficiency due to both kinetic parameters (increased k_{cat} and decreased K_m) was described in trypsin from *Paranotothenia magellanica* when compared with the homologous enzyme from ox and trout.[9] In contrast, enzymes such as phosphofructokinase and glycogen phosphorylase[7] isolated from antarctic fish show little or no functional difference related with the low temperature. It should be noted that compensation for low temperature could be reached not only through higher catalytic rate per enzyme molecule, but also through other mechanisms such as higher number of enzyme molecules.[18,19] An increase in both enzyme concentration and specific activity can also be adopted, as reported for lactate dehydrogenase in the heart of the icefish *Channichthys rhinoceratus*.[20] These combined mechanisms were also found in G6PD in the blood cells of antarctic icefish.[11,12]

2.2 Thermodynamic parameters

The temperature effect on the catalytic rates of *C. hamatus*, *D. mawsoni* and human G6PDs were investigated in the range 0–70°C. The curves of activity as a function of temperature with *D. mawsoni* and *C. hamatus* G6PDs were almost identical. In the range 0–50°C antarctic G6PDs displayed an approximately twofold higher catalytic activity than that measured in the human enzyme. The temperature effect on the catalytic activity was analyzed using the Arrhenius equation; the values of activation energy (E_a) calculated in the low temperatures range were 9,300, 10,300 and 13,000 cal mol^{-1} for *D. mawsoni*, *C. hamatus* and human G6PD, respectively. The lower values of E_a of antarctic G6PDs are in line with the results reported for other enzymes from cold-adapted organisms,[6,7,9] indicating that the active complex formation is favored at the low temperature of the environment.

The thermodynamic parameters of the reaction catalyzed by antarctic fish G6PD were also investigated. At low temperatures, the enthalpy of activation (ΔH^*) and entropy of activation (ΔS^*) values were lower than the corresponding values of the human enzyme; the free energy of activation (ΔG^*) was relatively constant (once again, slightly lower in the antarctic fish enzymes). The same trend was described by Low et al.[21] and Hochachka and Somero[18] after comparative analysis of the thermodynamic activation parameters of the reaction catalyzed by muscle-type lactate dehydrogenase, D-glyceraldehyde-3-phosphate dehydrogenase and muscle glycogen phosphorylase *b* from cold- and warm-adapted vertebrates. This trend was correlated with the physiological temperature of the organism. Significant lower enthalpy and entropy values and slightly lower free energy values were reported for the cold-adapted enzymes, suggesting that catalysis is favored by a lower energy barrier and a more ordered enzyme-substrate complex.

3
Thermostability

We have investigated the thermostability of G6PD, GDH and glycogen phosphorylase *b* of antarctic fish.[17] The heat response of *C. hamatus* and *D. mawsoni* G6PD was very similar to that of the human enzyme and similar temperatures of half inactivation were calculated after 10 min incubation, i.e. 52 and 55°C for the fish and mesophilic enzyme, respectively (Table 2). The temperature of heat inactivation of *Chaenocephalus aceratus* GDH was also similar to that of the bovine enzyme; the half-inactivation temperatures, calculated after 10 min incubation, were 50 and 53°C, respectively. Phosphorylase *b*, purified from the muscle of *T. bernacchii*, also showed high temperatures of heat inactivation and half-inactivation (49°C, after 10 min incubation). The allosteric effector adenosine 5'-monophosphate (AMP) yielded protection and shifted the half-inactivation temperature to 55°C. Although the highest temperatures at which no heat inactivation of antarctic and mesophilic enzymes was observed were similar, a common feature is shown by antarctic enzymes, i.e. they were consistently heat-inactivated abruptly in a very narrow temperature range. In G6PD and GDH, where the comparison with the mesophilic homologues is available, this temperature range is clearly narrower than that of the mesophilic enzymes, suggesting that the molecular structure becomes highly sensitive to small temperature variations.

Feeney and Osuga[7] compared the heat inactivation of several enzymes from antarctic fish muscle with that of homologous mesophilic enzymes. These early

Table 2. Temperatures at which 100% and 50% residual activity was measured, after 10 min incubation, for glucose-6-phosphate dehydrogenase (G6PD), glycogen phosphorylase *b* and glutamate dehydrogenase (GDH)

Enzyme	Source	Temperature (°C) Residual activity	
		100%	50%
G6PD	*D. mawsoni*	40	52
	C. hamatus	40	52
	Human	40	55
Phosphorylase *b*	*T. bernacchii*	45	49
+AMP (1 mM)		48	55
GDH	*C. aceratus*	45	50
	Bovine	45	53

results already showed nonhomogeneous behaviors. In fact, the temperatures of half inactivation of glycogen phosphorylase from *D. mawsoni* and *P. borchgrevinki* did not substantially differ from those of phosphorylase from rabbit, trout and salmon and fell in the range 42–48°C. Glyceraldehyde-3-phosphate dehydrogenase from *D. mawsoni* and rabbit had nearly identical temperature stability curves. In contrast, the half-inactivation temperatures of *P. borchgrevinki*, *D. mawsoni* and rabbit muscle fructose-1,6-diphosphate aldolases clearly indicated lower thermostability of the antarctic enzymes. More recently, it was reported that trypsin from *P. magellanica* was more sensitive to temperature than the mesophilic counterparts.[9]

Although heat inactivation, occurring at temperatures far from the physiological ones, may not have a direct physiological meaning, it can provide useful information on the molecular structure evolution. Different behaviors suggest different degrees of adaptation. For example, the microtubule proteins of mammals and birds polymerize to yield microtubules at temperatures near 37°C, but these polymers are cold-labile and disassemble to their subunits at low temperatures (0–4°C).[22] Thus, in order to survive, cold-adapted organisms were driven to modify their own microtubule proteins so that microtubules can assemble at low temperatures. As mentioned before, mesophilic enzymes such as G6PD and GDH can still function at low temperature. Substantial alterations of enzyme structure, therefore, may not be necessary to allow survival of cold-adapted organisms.

4 Structural properties

A more flexible molecular structure with less internal binding than mesophilic enzymes has been previously suggested for cold-adapted enzymes on the basis of thermolability, susceptibility to chemical denaturation and catalytic rates at low temperatures observed in several enzymes.[1] Three-dimensional molecular struc-

	115					120					125					130
E. coli	N	A	L	T	T	L	P	M	G	G	G	**K**	G	G		S
S. cerevisiae	N	S	L	T	G	L	D	M	G	G	G	**K**	G	G	L	C
S. solfataricus	N	S	I	L	I	L	P	Y	G	G	G	**K**	G	G	V	R
Bovine	C	A	V	V	D	V	P	F	G	G	A	**K**	A	G	V	K
Chicken	C	A	V	V	D	V	P	F	G	G	A	**K**	A	G	V	K
Human	C	A	V	V	D	V	P	F	G	G	A	**K**	A	G	V	K
Rat	C	A	V	V	D	V	P	F	G	G	A	**K**	A	G	V	K
Mouse	C	A	V	V	D	V	P	F	G	G	A	**K**	A	G	V	K
Tuna	C	A	V	V	D	V	P	F	G	G	A	**K**	A	G	V	K
C. aceratus	C	A	V	V	D	V	P	F	G	G	A	**K**	A	G	V	R

Fig. 1. Alignment of the amino acid sequence region 115–130 of glutamate dehydrogenase from *Chaenocephalus aceratus* and the corresponding regions of the homologous enzymes from other organisms.[23,26] Identical residues are *shadowed*

tures are not yet available to allow the definition of structural properties related with cold adaptation in antarctic fish enzymes.

Conservation of higher orders of protein molecular structure has been observed in antarctic fish G6PD and GDH. The apparent molecular weight of the subunit of antarctic G6PDs is 61 kDa, slightly higher than that of human G6PD. The apparent subunit molecular weight of *C. aceratus* GDH is 57 kDa and the apparent molecular weight of the native protein indicates a hexameric quaternary structure. These values are very similar to those of tuna[23] and dogfish[24] and other vertebrate GDHs. Partial amino acid sequencing of *C. aceratus* GDH has been performed.[11] Alignment with the region of residues 115–130 of the bovine enzyme which contains Lys-126, an essential residue of the active site (Fig. 1), revealed that this region is identical in vertebrate GDHs (ox, chicken, human, rat, mouse, tuna), except for the substitution of Lys-130 with Arg in *C. aceratus*; in contrast, bacteria and yeast GDHs display several substitutions. The striking sequence identity with enzymes of temperate fish, birds and mammals highlights conservation of GDH molecular structure during evolution. Further evidence on primary and three-dimensional structures of antarctic fish enzymes are required in order to have information on structural modifications acquired during cold adaptation.

5
Conclusions

According to Gerike et al.,[25] parameters such as temperature optimum, thermostability and specific activity are not necessarily good indicators of cold adaptation of enzymes. In fact, higher catalytic efficiency and lower thermostability seem to be a trend rather than a general rule. The structural properties of G6PD and GDH from antarctic fish indicate that at least some levels of molecular structure have been conserved, as shown by the similarities in molecular weight, number of subunits and amino acid sequence of the active site. The extent of the structure/function modifications may vary depending on the degree of physiological correction necessary to have or maintain a suited function. G6PD and GDH from mesophilic organisms are not cold-labile, as mesophilic microtubules are, and can function at a temperature as low as 0°C, at which their catalytic activity is two- to fourfold lower than that of antarctic fish counterparts. These results, seen together with similar thermostability, suggest that the molecular structure of mesophilic and antarctic fish G6PD and GDH does not need gross alterations to work better at low temperatures. In some cases, it is likely that the physiology of the organism may have chosen to reach adaptation by modulating reaction rates through enhancing the synthesis of enzyme molecules or through regulation of allosteric effects, rather than by extensively modifying the molecular structure. The combination of many factors, each one giving a small and synergistic contribution, is also a conceivable mechanism to reach suitable activity levels. It will be of interest to assess whether the structure of allosteric enzymes needs to be altered less than that of nonallosteric enzymes in order to satisfy the physiological requirements of the cell at low temperatures. Furthermore, the process of enzyme cold adaptation must also take into account (and respond to) the different importance that a given metabolic pathway may have in organisms living in cold habitats.

Acknowledgements. This research is in the framework of the Italian National Programme for Antarctic Research.

6
References

1. Macdonald JA, Montgomery JC, Wells RMG. Comparative physiology of Antarctic fishes. Adv Mar Biol 1987; 24:321-387.
2. Somero GN, DeVries AL. Temperature tolerance of some Antarctic fishes. Science 1967; 156:257-258.
3. Everson I, Ralph R. Blood analyses of some Antarctic fish. Br Antarctic Surv Bull 1968;15:59-62.
4. Wells RMG, Ashby MD, Duncan SJ, Macdonald JA. Comparative study of the erythrocytes and hemoglobins in nototheniid fishes from Antarctica. J Fish Biol 1980; 17:517-527.
5. Ruud JT. Vertebrates without erythrocytes and blood pigments. Nature 1954; 173:848-850.
6. Somero GN, Hochachka PW. The effect of temperature on catalytic and regulatory functions of pyruvate kinases of the rainbow trout and the Antarctic fish *Trematomus bernacchii*. Biochem J 1968; 110:395-400.
7. Feeney RE, Osuga DT. Blood proteins and muscle enzymes of cold-adapted antarctic fishes. In: Quam LO, ed. Research in the Antarctic. Am Assoc Adv Sci, Washington DC, 1971: 227-257.
8. Feller G, Payan F, Theys F, Qian M, Haser R, Gerday C. Stability and structural analysis of *a*-amylase from the Antarctic psychrophile *Alteromonas haloplanctis* A23. Eur J Biochem 1994; 222:441-447.
9. Genicot S, Feller G, Gerday C. Trypsin from Antarctic fish (*Paranotothenia magellanica* Forster) as compared with the trout (*Salmo gairdneri*) trypsin. Comp Biochem Physiol 1988; 90B:601-609.
10. Davail S, Feller G, Narinx E, Gerday C. Cold adaptation of proteins. Purification, characterisation and sequence of heat-labile subtilisin from the Antarctic psychrophile *Bacillus* TA41. J Biol Chem 1994; 269:17448-17453.

11. Ciardiello MA, Camardella L Carratore V, di Prisco G. Enzymes in Antarctic fish: glucose-6-phosphate dehydrogenase and glutamate dehydrogenase. Comp Biochem Physiol 1997; 118A:1031-1036.
12. Ciardiello MA, Camardella L, di Prisco G. Enzymes in cold-adapted Antarctic fish: Glucose-6-phosphate dehydrogenase. In: Battaglia B, Valencia J, Walton H, eds. Proc 6th SCAR Biol Symp, Antarctic Communities: Species, Structure and Survival. Cambridge: Cambridge Univ Press; 1997:261-265.
13. di Prisco G, D'Avino R. Molecular adaptation of the blood of Antarctic teleosts to environmental conditions. Antarctic Sci 1989; 1:119-124.
14. Hemmingsen EA, Douglas EL. Respiratory and circulatory adaptation to the absence of hemoglobin in Channichthyid fishes. In: Llano GA, ed. Adaptation within Antarctic Ecosystems. Washington, DC: Smithsonian Institution, 1977:479-487.
15. Feller G, Narinx E, Arpigny JL, Aittaleb M, Baise E, Genicot S, Gerday C. Enzymes from psychrophilic organisms. FEMS Microbiol Rev 1996; 18:189-202.
16. Ciardiello MA, Camardella L, di Prisco G. Glucose-6-phosphate dehydrogenase from the blood cells of two Antarctic teleosts: correlation with cold adaptation. Biochim Biophys Acta 1995; 1250:76-82.
17. Ciardiello MA, Camardella L, di Prisco G. Enzymes of Antarctic fishes: effect of temperature on catalysis. Cybium 1997; 21:443-450.
18. Hochachka PW, Somero GN. Biochemical Adaptations. New Jersey: Princeton Univ Press; 1984.
19. Somero GN. Biochemical mechanisms of cold adaptation and stenothermality in Antarctic fish. In: di Prisco G, Maresca B, Tota B, eds. Biology of Antarctic Fish. Heidelberg: Springer, 1991:232-247.
20. Feller G, Pauly J-P, Smal A, O'Carra P, Gerday C. The lactate dehydrogenase of the icefish heart: biochemical adaptations to hypoxia tolerance. Biochim Biophys Acta 1991; 1079:343-347.
21. Low PS, Bada JL, Somero GN. Temperature adaptation of enzymes: roles of the free energy, the enthalpy and the entropy of activation. Proc Natl Acad Sci USA 1973; 70:430-432.
22. Detrich HW III. Polymerization of microtubule proteins from Antarctic fish. In: di Prisco G, Maresca B, Tota B, eds. Biology of Antarctic Fish. Berlin: Springer, 1991:248-262.
23. Veronese FM, Bevilacqua R, Boccù E, Brown DM. Glutamate dehydrogenase from tuna liver. Purification, characteristics and sequence of a peptide containing an essential lysine residue. Biochim Biophys Acta 1976; 445:1-13.
24. Corman L, Prescott LM, Kaplan NO. Purification and kinetic characteristics of dogfish liver glutamate dehydrogenase. J Biol Chem 1966; 242:1383-1390.
25. Gerike U, Danson MJ, Russell NJ, Hough DW. Sequencing and expression of the gene encoding a cold-active citrate synthase from an Antarctic bacterium, strain DS2-3R. Eur J Biochem 1997; 248:49-57.
26. Benachenhou-Lahfa N, Forterre P, Labedan B. Evolution of glutamate dehydrogenase genes: evidence for two paralogous protein families and unusual branching patterns of the archeobacteria in the universal tree of life. J Mol Evol 1993; 36:335-346.

Aspartate aminotransferase from *Moraxella* TAC125: an unusual psychrophilic enzyme

M. L. Tutino, L. Birolo, B. Fontanella, K. Mainolfi, F. Vinci, G. Sannia and G. Marino*

Department of Organic and Biological Chemistry, University of Naples Federico II, Via Mezzocannone 16, I-80134 Naples, Italy

1 Introduction

Life at low temperature requires molecular modifications that are not just the opposite to those adopted by thermophilic organisms. Hot-adapted proteins, indeed, have to face the problem of stability at high temperature, whereas the psychrophilic ones have to cope with the reduction of the enzymatic reaction rates at low temperature in order to maintain metabolic fluxes. A more flexible structure seems to be the simplest strategy of enzyme adaptation to catalysis at low temperature in order to undergo the rapid and reversible conformational modifications that are required by the catalytic cycle.

A question we have addressed ourselves after studying the aspartate aminotransferase from *Moraxella* TAC125 (*M*AspAT) is if all the enzymes from psychrophilic or psychrotrophic organisms should necessarily exhibit a truly psychrophilic or psychrotrophic behavior. In other words, can it be safely assumed that mutative adaptation of proteins to specific physiological conditions to maintain "corresponding states" regarding structure, flexibility, ligand binding and catalysis also applies to proteins coming from the cold? Although factors affecting the enhanced stability of thermophilic proteins indicate no single influence as producing the major effect, with individual proteins exhibiting different combinations of stabilizing effects,[1] it has been demonstrated that, in fact, molecular alteration responsible for the adaptation to "extreme" growth environments lead to shift "normal" characteristics to high temperature, so that the molecular properties, at their respective physiological conditions, turn out to be comparable.[2]

Aspartate aminotransferase (AspAT), an ubiquitous enzyme, is an interesting paradigm for studies concerning structure/function relationships. AspATs have

* Corresponding author; e-mail: gmarino@unina.it

been purified and characterized from many sources, and three-dimensional structures have been solved for mammalian (pig cytosolic),[3] avian (chicken mitochondrial)[4] and bacterial (*Escherichia coli*)[5] AspATs.

The thermophilic AspAT from a sulphur-dependent archaeon *Sulfolobus solfataricus* (*Ss*AspAT) proved to be a truly thermophilic and thermostable protein[6] with properties that are "normal" for a thermophilic protein belonging to the aspartate aminotransferase family,[7] despite the very low sequence identity (<15%) with any of the mesophilic AspAT.[8] In other words, *Ss*AspAT behaves as an orthodox thermophilic enzyme. The work on *M*AspAT started with the aim to extend to low temperatures the activity/stability relationships for this class of enzyme. As a result, we may say that *M*aspAT behaves quite differently from the psychrophilic enzymes so far investigated.

2
Gene structure and regulation

The AspAT gene was recently cloned from *Moraxella* TAC125, a new antarctic isolate, overexpressed in *E. coli*, and purified to homogenity in good yield (unpubl. data). Figure 1 shows the nucleotide sequence of TAC125 *MaspC* and the deduced amino acid primary structure of *M*AspAT and the flanking regions that were searched for the presence of putative transcription and translation regulation sequences. A putative Shine-Dalgarno-like sequence (AGAGA) can be observed 7 bp upstream of the ATG start codon. Two sequences potentially homologous to the -35 and -10 bp regions of bacterial promoters were identified upstream the *M*AspAT translation start site and they are marked in Figure 1. A region that could fold in a "stem and loop" RNA secondary structure was recognized a few bp downstream of the *MaspC* stop codon, followed by a sequence rich in pyrimidines, that resemble a canonical rho-independent transcription terminator. These observations suggested that *MaspC* could be transcribed as a monocystronic m-RNA.

Total RNA extracted from exponentially growing cultures of *Moraxella* TAC125 was analyzed by Northern blotting hybridization, using a *MaspC* fragment as probe. This experiment is shown in Figure 2; a single band of an estimated 1.3 kb was detected, demonstrating that the gene is really transcribed. As far as the codon usage is concerned, the low C+G content of *MaspC* gene justifies the occurrence of T (39%) and A (34%) as preferred 3rd codon position. It is also worth noting that the psychrotrophic bacterium severely restricted the genetic code degeneration in the cases of arginine and isoleucine. The CGT codon is generally preferred (82%) out of the six Arg coding triplets (CGN, $AG^{A}/_{G}$), whereas the occurrence of ATT codon for Ile is quite absolute (90%).

The AspAT gene is 1191 bp long and codes for a protein of 397 amino acids. The primary structure of *M*AspAT, deduced from the nucleotide sequence and checked by mass spectrometry on the recombinant protein, was aligned to those of other mesophilic [*E. coli* (*Ec*AspAT)] and thermophilic AspATs [*S. solfataricus* (*Ss*AspAT), *Bacillus* spp. (*Bac*AspAT), *Thermus thermophilus* (*Tt*AspAT)] (Fig. 3). *M*AspAT shows the highest identity with the *E. coli* enzyme (52%) and a quite low identity

```
   1   AAGGTAGTCTGAGGTGTGCTACCTTATTAATAGGAGTGGTGAATGCGATGGAATTATTGA
  61   TGCTGAGSGTTTCATAATGATTTTTTTGATGAAAAAGTACTCGCAGTCCTAATTGCATGC
 121   TGTATTAAAATTCCTCAATTTTTAGTTTAAATAACAAAAATAGACTATGTTGAGGTTTTT
 181   AATAATGATTTTTTGCATGCACAAGGTATTAACTGTATTAATGGCATGCTGTATTAAAAT
 241   TCCTCACTTTTTTGTACCAATAACACTAGTCAACCTAGCGTGTTTTTATGATAAGCTTAA
                                                      -35
 301   GCTCACACCAACCCCCTAAATAATGATAAGAGATAGTCAATGTTCTCAGTATTAAAACCA
                        -10          SD         M  F  S  V  L  K  P      7
 361   TTACCCACAGATCCTATTCTTGGCCTTATGGCGGCCTATAAGCAAGATACTAACCCTAAT
        L  P  T  D  P  I  L  G  L  M  A  A  Y  K  Q  D  T  N  P  N     27
 421   AAAATTGATTTAGGTGTGGGTGTCTATAAAGATGAGCTAGGCAATACACCGGTACTTAAA
        K  I  D  L  G  V  G  V  Y  K  D  E  L  G  N  T  P  V  L  K     47
 481   GCGGTAAAAAAAGCAGAAGCGTTTCGTTTAGAAAATGAAACCAGCAAATCTTATATTGGT
        A  V  K  K  A  E  A  F  R  L  E  N  E  T  S  K  S  Y  I  G     67
 541   TTAGCGGGTAATTTAGATTACTGTCAAAAAATGGAAAGCTTACTATTAGGTGAGCACAAA
        L  A  G  N  L  D  Y  C  Q  K  M  E  S  L  L  L  G  E  H  K     87
 601   ACATTACTAGCTAACCGTGTACGTACGGCGCAACGGCCAGGTGGTACTGGGGCATTACGT
        T  L  L  A  N  R  V  R  T  A  Q  R  P  G  G  T  G  A  L  R    107
 661   GTTGCTGCTGAATTTATTATGCGTTGTAATCCAAAAGCAACAGTGTGGGTAACTACGCCA
        V  A  A  E  F  I  M  R  C  N  P  K  A  T  V  W  V  T  T  P    127
 721   ACATGGGCTAATCATATTAGCTTATTTGAAGCAGCGGGCTTAACAGTAAAAGAATACCCT
        T  W  A  N  H  I  S  L  F  E  A  A  G  L  T  V  K  E  Y  P    147
 781   TATTACGATTACGAAAATAAAGACTTATTATTTGATGAGATGATCAACACACTTAAGCAA
        Y  Y  D  Y  E  N  K  D  L  L  F  D  E  M  I  N  T  L  K  Q    167
 841   GTGCCTAAAGGCGATGTGGTTTTACTACATGCGTGTTGTCATAACCCAAGTGGTATGGAT
        V  P  K  G  D  V  V  L  L  H  A  C  C  H  N  P  S  G  M  D    187
 901   TTAAACGAAGCACAGTGGAAGGTAGTTGCTGAGCTTGCAAAAGAAGTGGGCTTTACGCCG
        L  N  E  A  Q  W  K  V  V  A  E  L  A  K  E  V  G  F  T  P    207
 961   CTTGTTGATATTGCATACCAAGGTTTTGGCTCTTCACTAGAAGAAGATGCACGTGGTTTA
        L  V  D  I  A  Y  Q  G  F  G  S  S  L  E  E  D  A  R  G  L    227
1021   CGTATTTTAGCTGACGCAGTTGAAGAGTTAATTATTTGTTCTTCATGTTCTAAAAACTTT
        R  I  L  A  D  A  V  E  E  L  I  I  C  S  S  C  S  K  N  F    247
1081   GGTCTTTACCGCGAACGTATTGGTGCGTGTTCACTAATTGCAAAAGACAGCGCAACTGCT
        G  L  Y  R  E  R  I  G  A  C  S  L  I  A  K  D  S  A  T  A    267
1141   GATATTTCAAACTCAGTATTACTTAGTGTTGTACGTAGTATTTATTCAATGCCGCCAGCG
        D  I  S  N  S  V  L  L  S  V  V  R  S  I  Y  S  M  P  P  A    287
1201   CATGGTGCTGACATTGTAAATACAATTTTAAGCAGCACTGAGCTTACGCAAATGTGGCAT
        H  G  A  D  I  V  N  T  I  L  S  S  T  E  L  T  Q  M  W  H    307
1261   CAAGAGCTTGATGAAATGCGTAGCCGTATAAATGGTTTACGTACACAAATTAAAGAAACA
        Q  E  L  D  E  M  R  S  R  I  N  G  L  R  T  Q  I  K  E  T    327
1321   CTAGCGACTAAAGACATAGCGCAAGACTTCTCGTTTATTGAGCGTCAGGACGGTATGTTT
        L  A  T  K  D  I  A  Q  D  F  S  F  I  E  R  Q  D  G  M  F    347
1381   TCTTTCTTAGGAATTAATAAAGAGCAAATTACTCGCCTACAAAAAGAGTACGGCATTTAT
        S  F  L  G  I  N  K  E  Q  I  T  R  L  Q  K  E  Y  G  I  Y    367
1441   ATTGTTGGCTCAAGTCGTGTAAACGTAGCGGGCGTGAGTGATGCAAATATTGAATACTTT
        I  V  G  S  S  R  V  N  V  A  G  V  S  D  A  N  I  E  Y  F    387
1501   GCAAATGCAGTTGCCGACGTATGTAAATCATAATAGTTAACCTGAACTCTGGTTAAGAGA
        A  N  A  V  A  D  V  C  K  S                                  397
1561   TTTACTAACATATTGAATCATCGTTATATTTAATTTTATTTTCTCTAGCCAGAGGTTTTC
1621   AGTGAAAACAAGGCGAATTTACSCGSTCAATAGCTGGCCTATTGCAAGTAAATCAACCCA
1681   GTTAGCCCTGAAAACAGCTGCTTGAGATAAGTTTTATTATCCGGACCTCAGGTAGTTAAA
1741   TTATTTAA
```

Fig. 1. Nucleotide sequence of *Moraxella* TAC125 aspartate aminotransferase gene and the deduced primary structure of the protein
The TAC125 *MaspC* flanking regions were searched for the presence of transcription and translation regulative signals. A Shine-Dalgarno-like sequence is located 7 bp upstream the translation start site (*bold, double underlined*); a putative bacterial promoter was identified and the two constitutive elements, -35 and -10 bp boxes, are indicated in *bold*; an inverted repeated sequence (*bold, dotted underlined*) follows the end of the *MaspC* coding region and could represent the stem and loop secondary structure that characterizes the ρ-independent transcription terminator

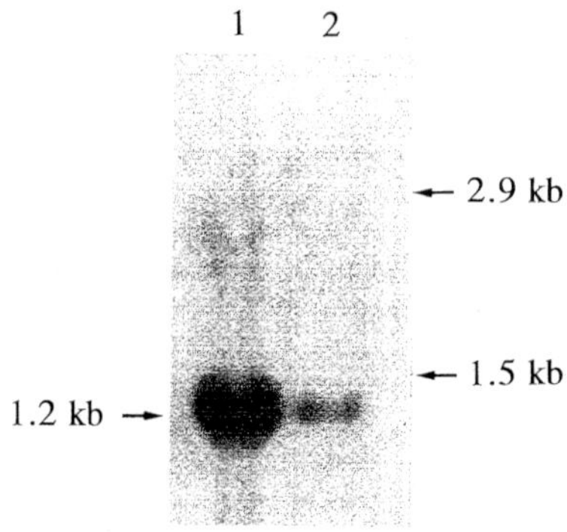

Fig. 2. Analysis of *MaspC* transcription
Two amounts of total *Moraxella* TAC125 RNA (*lane 1*: 15 μg; *lane 2*: 5 μg) were analyzed by Northern blotting hybridization using the *MaspC* gene as probe; *arrows* indicate the positions corresponding to the RNA molecular weight markers

with both the thermophilic proteins (about 14%), similar to the homology observed between the thermophilic and the mesophilic ones (Table 1). The 11 amino acid residues which are absolutely conserved in all the aminotransferases of subgroup I sequenced so far[9] are also present in *M*AspAT (Fig. 3). These residues are either involved in the binding of the coenzyme pyridoxal-5'phosphate and of the substrates or appear to be essential for the functional conformation of the enzyme.

In the analysis of the amino acid composition of these proteins, a higher cysteine and a lower proline content were observed in the psychrotrophic enzyme. When the amino acid sequence of *M*AspAT was compared with that of *Ec*AspAT the number of cysteinyl residues was found to have increased from five to eight. Moreover, the amino acid sequence of thermophilic AspAT from both *S. solfataricus* and *Bacillus* contains no cysteinyl residues, and that from *Th. thermophilus* presents only one residue. The proline content follows an opposite behavior: facing the 14 prolines of *M*AspAT and the 15 of *Ec*AspAT are the 16 of the *S. solfataricus* enzyme, the 20 of

Table 1. Percentage identity among aligned aminotransferases by pair-wise comparison of the sequences

	*M*AspAT	*Ec*AspAT	*Bac*AspAT	*Tt*AspAT	*Ss*AspAT
*M*AspAT	–	52.02	14.03	15.84	12.85
*Ec*AspAT	52.02	–	14.03	16.10	16.16
*Bac*AspAT	14.03	14.03	–	47.01	33.67
*Tt*AspAT	15.84	16.10	47.01	–	30.91
*Ss*AspAT	12.85	16.16	33.67	30.91	–

AspAT aspartate aminotransferase; *M Moraxella* sp.; *Ec Escherichia coli*; *Bac Bacillus* sp.; *Tt Thermus thermophilus*; *Ss Sulfolobus solfataricus*

```
                                                      50                           •
.....MFSVLKPLPTDPILGLMAAYKQDTNPNK...IDLGVGVYKDELGNTPVLKAVK..KAEAFRLENE TSKS.Y...I
.....MFENITAAPADPILGLADLFRADERPGK...INLGIGVYKDETGKTPVLTSVK..KAEQYLLENE.TTKN.Y...L
vslldFNGNMSQVTGETTLLYKEIARNVEKTKKIKIIDFGIG.....QPD...LPTFK..RIRDAAKEALDQGFTFYTSAF
mk..eLLANRVKTLTPSTTLAITAKAKEMKAQGIDVIGLGAG.....EPD...FNTPQ..NIMDAAIDSMQQGYTKYTPSG
mrglsRRVQAMKPSATVAVNAKALEL...RRQGVDLVALTAG.....EPD...FDTFTPEHVKEAARRALAQGHTKYAPPA

                                     100                                •      150
GLAGNLDYCQKMESLLLGEHKTLLAN-RVRTAQAPGGTGALRV..AAEFIMRCNPKAT.VWVTTPTWANHISLFEAAGL.TV
GIDGIPEFGRCTQELLFGKGSALINDKRARTAQTPGGTGALRV..AADFLAKNTSVKR.VWVSNPSWPNHKSVFNSAGL.EV
GIDELREKIAQYLNTRYGTD.....VKKEEVIVTPGAKPALFL..VFILY...INPSDEVILPDPSFYSYAEVVKLLGGKPI
GLPALKQAIIEKFKRDNQLE.....YKPNEIIVGVGAKHVLYT..LFQVI...LNEGDEVIIPIPYWVSYPEQVKLAGGVPV
GIPELREALAEKFRRENGLS.....VKKEEVIVTVGAKQALFTFNVFQAI...LDPGDEVIVLSPYWVSYQEMVRFAGG..V

                                                 •• •200                        • •
KEYP.YYDYENKDLLFDEMINTLK..QVAKGDVVLLHACCHNPTGMDLKEAQWKVVAELAKEVGFTPLVDIAYQGFGSS.LE
REYA.YYDAENHTLDFDALINSLN..EAQAGDVVLFHGCCHNPTGIDPTLEQWQTLAQLSVEKGWLPLFDFAYQGFARG.LE
YANLKWSREEGFSIDVDDLQSKIS..K..RTKMIVFNN.PHNPTGTLFSPNDVKKIVDISRDNKIILLSDEIYDNFVYEGKM
YIEA..TSEQNYKITAEQLKNATT..D..KTKAVIINS.PSNPTGMVYTREELEDIAKIALENNILIVSDEIYEKLLYNGAE
VVEVETLPEEGFVPDPERVRRAITPFT..RTKALVVNS.PNNPTGAVYPKEVLEALARLAVEHDFYLVSDEIYEHLLYEGEH

                              250        •       • •                            300
E...DAR.GLRILADAVEELIICSSCSKNFGLPRERIGACSLIAKDSATADISN.SVLLSVVRSIYSMPPAHG..ADIVNTI
E...DAE.GLRAFAAMHKELIVASSYSKNFGLYNERVGACTLVAADSETVDRAF.SQMKAAIRANYSNPPAHG..ASVVATI
RSTLED..SD.WRDF....LIYVNGFSKTFSMTGWRLGYIVA........KREIIQKMGILAANVYTAPTSFVQKAAVKAFD
MFSIAQISEE.VKAQ....TIVINGVSKSHSMTGWRIGYAAG........NADIINAMTDLASHSTSNPTTASQYAAAISAYN
FSPGRV..AF.TPEH....TLTVNGAAKAFAMTGWRIGYACG........PKEVIKAMASVSSQSTTSPDTIAQWAATLEALT

                                                  350
LSSTELTQMWHQELDEMRSRINGLRTQIKETLATKDIAQDFSFIERQNG.MFSFL.........GIN..KEQIVRLINEYSI
LSNDALRAIWEQELTDMRQRIQRMRQLFVNTLQEKGANRDFSFIIKQNG.MFSFS.........GLT..KEQVLRLQKEFGV
TFDE.........VNQMVSLFKKRRDVMYDELTKVKG....VEVSKPNGAFYMFPnvskilktsGFD.VKSLAIKLREEKGV
GPQD........SVEEMRKAFESRLETIYPKLSAIPG....FKVVKPQGAFYLLPdsveqqqktGFASVDEFASALLTEANV
NFTQEASRAF...VEMAREAYRRRRDLLLEGLTAL.G....LKAVRPSGAFYVLMdtspiapd.IIE.VRAAERFTLLEAGV

                                •            400
YIVGS...........SRVNVSGVSDANIEYFANTVADVCKS........
YAVAS...........GRVNVAGMTPDNMAPLCEAIVAVL..........
VTIPGevfplnigkefLRLSFA....VNEEVIKEGIQKIRefaeqmmnsr
AVIPGsgfg...apstIRISYA....TSLNLIEEAIERIDefv.......
AVVPGtdfa...afghVRLSYA....TSEENLRKALERFArvlgra....
```

Fig. 3. Sequence alignment of aspartate aminotransferases (AspAT)
The primary sequence of *Moraxella* TAC125 AspAT (TACAspAT) was aligned to those of mesophilic [*Escherichia coli* (*Ec*AspAT)] and thermophilic [*Sulfolobus solfataricus* (*Ss*AspAT), *Bacillus* spp. (*Bac*AspAT), *Thermus thermophilus* (*Tt*AspAT)] microbial AspATs. The 11 residues conserved in all AspATs sequenced so far are indicated by a *closed circle*; *numbering* is referred to the first and longest AspAT sequence, that of porcine cytosolic AspAT (cAspATp)[21]

the AspAT from *Bacillus* and the 25 of the *Th. thermophilus* protein. A decrease in proline residue content could contribute to increase local flexibility of the peptidic backbone because the pyrrolidine ring severely restricts the available dihedral angles of the preceding residue and decreases the backbone entropy of unfolding. Interestingly, among the proline residues of *Ec*AspAT that are not conserved in *M*aspAT, there is a proline, Pro141, that is in a β-turn located at the entrance of the active cleft. Its substitution for an alanine in the psychrophilic enzyme might be instrumental to increase the flexibility of that loop. A comment on the higher cysteine content has to be viewed from the opposite aspect: the cysteine side chain is indeed quite sensitive to high temperature and thus, from the thermoresistance point of view, only strictly necessary cysteines have to be retained in the hot-adapted proteins.

The capacity of arginine residue to form multiple hydrogen bonds to backbone carbonyl and side-chain oxygens and to interact favorably with the solvent have been proposed as the structural feature at the basis of the stabilizing effect of arginine and therefore to the observation that thermophilic proteins generally exhibit a higher R/(R+K) ratio than their mesophilic counterparts. This stabilizing feature could be also observed for AspATs if we could consider only *M*AspAT [(R/(R+K) is 0.41], *Ec*AspAT (0.55) and *Tt*AspAT (0.68). Nevertheless, the trend is inverted for the other two thermophilic proteins, *Ss*AspAT showing an arginine versus basic amino-acid ratio of 0.30 and *Bac*AspAT of 0.26, both even lower than the value calculated for *M*AspAT. Comparison of the amino-acid composition of AspATs shows that other parameters, such as the percentage of hydrophobic or charged residues, do not change significantly between psychrophilic, mesophilic and thermophilic enzymes.

3
Gene expression and protein purification

Cold-adapted enzymes can be successfully overexpressed in mesophilic hosts, as already demonstrated in a few cases such as the citrate synthase from an antarctic bacterium[10] and the α-amylase from *Alteromonas haloplanctis*.[11] The aspartate and tyrosine aminotransferase-deficient *E. coli* strain TY103[12] was chosen for overexpressing *M*AspAT, because of the high sequence homology between the psychrophilic enzyme and *Ec*AspAT. Special care was taken to choose the appropriate temperature for inducing the expression of the cold-adapted enzyme, and an optimal compromise between the likely thermolability of *M*AspAT and the growth needs of *E. coli* was found at 34°C. Two mg of homogenous recombinant *M*AspAT g^{-1} wet cells could be obtained in a reasonably fast purification procedure (unpubl. data).

4
Recombinant *M*AspAT characterization

The molecular mass of recombinant *M*AspAT was determined by electrospray mass spectrometry and the primary structure verified by peptide mapping. Physical parameters of *M*AspAT such as isoelectric point and hydrodynamic behavior in either native or denaturing conditions revealed no peculiar features of the psychrophilic enzyme in respect to *Ec*AspAT.

Michaelis constants (K_m) of *M*AspAT at 25°C for natural substrates (α-ketoglutarate and aspartic acid) and for the substrate analogue cysteinsulphinic acid (CSA) are reported in Table 2 and compared to the values determined for the mesophilic enzyme from *E. coli* (*Ec*AspAT). The apparent higher affinity of *M*AspAT for substrates is in agreement with a common strategy of the cold-adapted enzymes to overcome the lower intracellular metabolic rate and/or metabolite concentrations due to the low growth temperature.[13–15] Nevertheless, thermophilic *Ss*AspAT shows at 60°C an affinity for aspartate even higher than *M*AspAT (0.25 mM),[16] thus suggesting that this strategy is not a general rule in AspATs.

The apparent K_m value for *M*AspAT and *Ec*AspAT for the substrate analogue CSA has been determined at different assay temperatures in the range between 7 and 45°C (Fig. 4) in an assay method described by Marino et al.[7] The K_m increased with temperature for the psychrophilic enzymes, whereas it stayed almost constant for *Ec*AspAT. This has also been reported for other enzymes from psychrophilic organisms.[17] It can be commented in terms of a strategy of psychrophilic enzymes to resist the slowing down of the reaction velocity at lower temperature by increasing the affinity for substrates; but it can also be considered in terms of a more and more flexible active site as the temperature increases, resulting in a loosening of the interactions at the catalytic site which determine the affinity for substrates.

Taking care of the temperature dependence of K_m, the initial velocity towards CSA was then assayed at saturating concentrations of substrates at different temperatures in the range 5–70°C. The maximum of activity has been observed at 64 and 75°C for *M*AspAT and *Ec*AspAT, respectively. It can be argued that 60°C is a relatively high temperature for a psychrophilic enzyme, but the data are well in agreement with the general tendency towards thermophilicity and thermoresistance of AspATs,[18] that usually exhibit an optimum temperature for activity about 30–40°C above the optimum growth temperature of the organism.

Table 2. Michaelis constants (K_m) for transamination substrates

Enzyme	Temperature (°C)	α-Ketoglutarate (mM)	L-Aspartate (mM)	L-Cystein-sulphinate (mM)
*M*AspAT	15	0.053	1.12	3.36
*Ec*AspAT	25	0.6[a]	2.00[a]	8.17

[a] Values after ref. 22.
AspAT aspartate aminotransferase; *M Moraxella* sp.; *Ec E. coli*

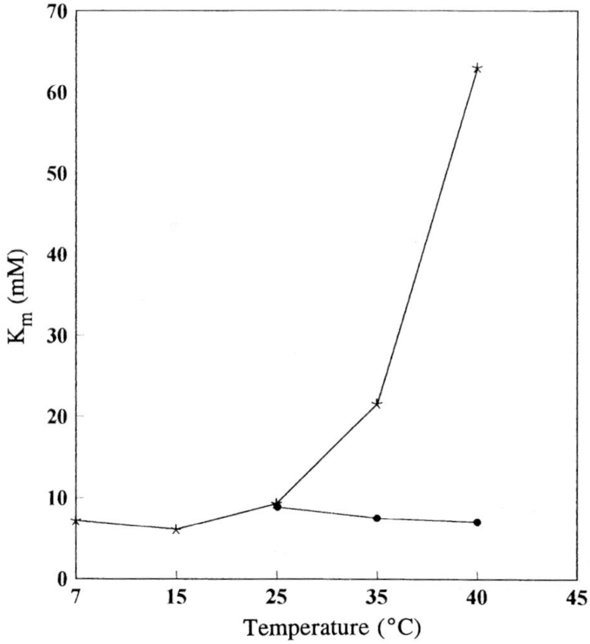

Fig. 4. Dependence of the K_m values for L-cysteinsulphinic acid on temperature
K_m values for the pseudo-substrate L-cysteinsulphinic acid have been determined for *M*AspAT (✻) and *Ec*AspAT (●) at different temperatures according to Marino et al.[7]

If we have to believe that one of the main characteristics of cold-adapted enzymes is their enhanced turnover number (k_{cat}) and catalytic efficiency (k_{cat}/K_m) at low temperature, then *M*AspAT cannot be considered as a truly cold-adapted enzyme: the k_{cat} values of *M*AspAT are indeed lower than those of *Ec*AspAT in the whole temperature range, as well as the k_{cat}/K_m values, even at 5°C. A similar feature has also been observed for the cold-active citrate synthase from the antarctic bacterium, strain DS2-3R, and suggests that comparison of factors such as temperature optimum, thermostability and specific activity are not enough to determine if an

Table 3. Kinetic and thermodynamic activation parameters of transamination at 20°C

Parameter	*M*AspAT	*Ec*AspAT
k_{cat} (s^{-1})	19.1	281.5
ΔH^* (kJ mol^{-1})	84.4	71.5
ΔS^* (J mol^{-1})	68.7	46.9
ΔG^* (kJ mol^{-1})	64.3	57.8

k_{cat} turnover number; ΔH^* enthalpy of activation; ΔS^* entropy of activation; ΔG^* free energy of activation; *AspAT* aspartate aminotransferase; *M Moraxella* sp.; *Ec E. coli*

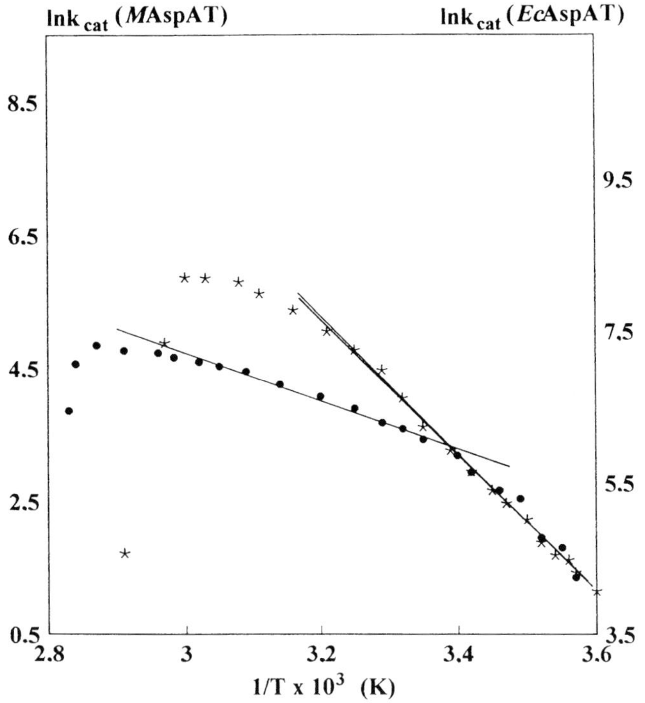

Fig. 5. Arrhenius plot for initial velocity of transamination of *M*aspAT (*) compared to *Ec*AspAT (•)
Initial transamination velocities were measured at increasing temperatures using L-cysteinsulphinic acid and α-ketoglutaric acid at saturating concentrations as substrates

enzyme is a psychrophilic one, while it is more crucial to understand whether the activity at low temperature is sufficient to permit cell growth and function.

The k_{cat} values measured at different temperatures have been used to construct Arrhenius plots (Fig. 5) and to calculate the activation energy parameters for the transamination reaction (Table 3). So far, all the AspATs examined for their thermodependence properties have shown a nonlinear Arrhenius plot of the k_{cat} which could be approximated by two incident lines and that was interpreted in terms of a temperature-dependent conformational change.[19] An interesting feature of *M*AspAT is the linearity of its Arrhenius plot. The energy of activation (E_a) of *M*AspAT is 86.8 ± 1.4 kJ mol^{-1}, close to the E_a that can be calculated in the low temperature range for *Ec*AspAT (73.9 ± 6.4 kJ mol^{-1}, whereas in the high temperature range an E_a of 26.0 ± 1.0 kJ mol^{-1} can be calculated). Nevertheless, its higher value and all the thermodynamic parameters reported in Table 3 are not in agreement with the common characterics of psychrophilic enzymes whose more flexible protein structure should be reflected in the thermodynamics of the enzyme-substrate complex formation.

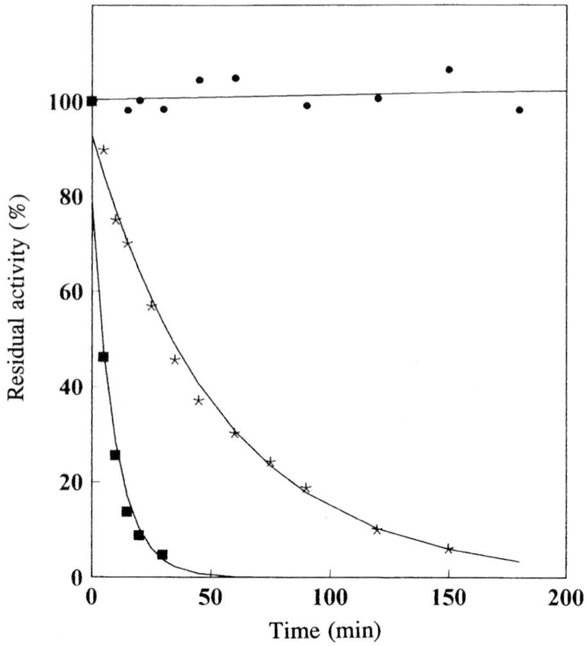

Fig. 6. Thermoresistance of *M*aspAT and *Ec*AspAT at 50°C
*M*aspAT (■) and *Ec*AspAT (•) were incubated at 50°C at 0.17 mg ml^{-1} in Na-TES 50 mM, pH 7.5, and the residual transamination activity measured at different times of incubation; (⁎) *M*aspAT incubated as above but in the presence of 2 mM α-ketoglutarate

*M*AspAT is actually much less thermoresistant than *Ec*AspAT. In Figure 6 the inactivation curve upon incubation at 50°C is reported for *M*AspAT either in presence or in absence of the substrate α-ketoglutarate. The half-time of inactivation ($t^1/_2$) of *M*AspAT at 50°C is 6.8 min whereas *Ec*AspAT can be incubated under the same conditions for more than 3 h without any sensible change in activity. It can be observed a strong protective effect of α-ketoglutarate against thermoinactivation of *M*AspAT ($t^1/_2$ at 50°C is 37.9 min when 2 mM α-ketoglutarate is added to the incubation mixture). It can be commented that the presence of substrate, which likely induces also in the case of *M*AspAT the conformational change that is at the basis of the mechanism of transamination (a crystallographic study of the two conformations for *Ec*AspAT is reported by Okamoto et al.),[20] can protect the active site from irreversible denaturation.

5
Conclusions

It is interesting to conclude by stressing that *M*AspAT has all the disadvantages of psychrophilic enzymes, such as the thermolability, without any of the advantages, for example the enhanced catalytic efficiency at low temperature. As far as cold adaptation is concerned, we can surely say that *M*AspAT has a high conformational flexibility, if we can judge from the results of the thermoinactivation experiment, but this cannot be considered as the drawback to an evolution towards increased catalytic efficiency in order to compensate for the reduction of reaction rates at low temperatures.[14] In other words, the mesophilic *Ec*AspAT is a better enzyme in the whole range of temperatures tested and it is also more stable. Care must be taken, therefore, when analyzing the temperature optimum, thermostability and specific activity of a psychrophilic enzyme because they can be not simply different aspects of cold adaptation, but they might differentially reflect cell needs for growth and function, such as a well-suited catalytic efficiency, but also a quite rapid turnover when needed.

6
References

1. Menèndez-Arias L, Argos P. Engineering protein thermal stability; sequence statistics point to residue substitutions in α-helices. J Mol Biol 1989; 206:397-406.
2. Jaenicke R. Protein stability and molecular adaptation to extreme conditions. Eur J Biochem 1991; 202:715-728.
3. Arnone A, Rogers PH, Hyde CC, Briley PD, Metzler CM, Metzler DE. Crystallographic studies of pig and cytosolic aspartate aminotransferase. In: Christen P, Metzler DE, eds. Transaminases. New York: John Wiley & Sons, 1985:138-155,
4. McPhalen CA, Vincent MG, Jansonius JN. X-ray structure refinement and comparison of three forms of mitochondrial aspartate aminotransferase. J Mol Biol 1992; 225:495-517.
5. Jäger J, Moser M, Sauder U, Jansonius JN. Crystal structures of *Escherichia coli* aspartate aminotransferase in two conformations. Comparison of an unliganded open and two liganded closed forms. J Mol Biol 1994; 239:285-305.
6. Arnone MI, Birolo L, Pascarella S, Cubellis MV, Bossa F, Sannia G, Marino G. Stability of aspartate aminotransferase from *Sulfolobus solfataricus*. Protein Engin 1997; 10:237-248.
7. Marino G, Nitti G, Arnone MI, Sannia G, Gambacorta G, De Rosa M. Purification and characterization of aspartate aminotransferase from the thermoacidophilic archaebacterium *Sulfolobus solfataricus*. J Biol Chem 1988; 263:12305-12309.
8. Cubellis MV, Rozzo C, Nitti G, Arnone MI, Marino G, Sannia G. Cloning and sequencing of the gene coding for aspartate aminotransferase from the thermoacidophilic archaebacterium *Sulfolobus solfataricus*. Eur J Biochem 1989; 186:375-381.
9. Mehta PK, Hale TI, Christen P. Aminotransferases: demonstration of homology and division into evolutionary subgroups. Eur J Biochem 1993; 214:549-561.
10. Gerike U, Danson MJ, Russell NJ, Hough DW. Sequencing and expression of the gene encoding a cold-active citrate synthase from an Antartic bacterium, strain DS2-3R. Eur J Biochem 1997; 248:49-57.
11. Feller G, le Bussy O, Houssier C, Gerday C. Structural and functional aspects of chloride binding to *Alteromonas haloplanctis* α-amylase. J Biol Chem 1996; 271:23836-23841.

12. Kamitori S, Hirotsu K, Higuchi T, Kondo K, Inoue K, Kuramitsu S, Kagamiyama H, Higuchi Y, Yasuoko N., Kusunoki M, Matsuura Y. Overproduction and preliminary X-ray characterization of aspartate aminotransferase from *Escherichia coli.* J Biochem (Tokyo) 1987; 101:813-816.
13. Somero GN. Temperature as a selective factor in protein evolution. The adaptational strategy of compromise. J Exp Zool 1977; 194:175-188.
14. Hochachka PW, Somero GN. Temperature adaptation. In: Hochachka PW, Somero GN, eds. Biochemical Adaptation. Princeton: Princeton Univ Press, 1984:355-449.
15. Feller G, Narinx E, Arpigny JL, Aittaleb M, Baise E, Genicot S, Gerday C. Enzymes from psychrophilic organisms. FEMS Microbiol Rev 1996; 18:189-202.
16. Birolo L, Arnone MI, Cubellis MV, Andreotti G, Nitti G, Marino G, Sannia G. The active site of *Sulfolobus solfataricus* aspartate aminotransferase. Biochim Biophys Acta 1991; 1080:198-204.
17. Hug DH, Hunter JK. Effect of temperature on urocanase from a psychrophile, *Pseudomonas putida.* Biochemistry 1974; 13:1427-1431.
18. Jenkins WT, Yphantis DA, Sizer IW. Glutamic aspartic transaminase. J Biol Chem 1959; 234:51-58.
19. Arnone MI, Birolo L, Giamberini M, Cubellis MV, Nitti G, Sannia G, Marino G. Limited proteolysis as a probe of conformational changes in aspartate aminotransferase from *Sulfolobus solfataricus.* Eur J Biochem 1992; 204:1183-1189.
20. Okamoto A, Higuchi T, Hirotsu K, Kuramitsu S, Kagamiyama H. Crystallographic study of pyridoxal 5'-phosphate-type aspartate aminotransferases from *Escherichia coli* in open and closed form. J Biochem (Tokyo) 1994; 116:95-107.
21. Ovchinnikov YA, Egorov CA, Aldanova NA, Feigina MY, Lipkin VM, Abdulaev NG, Grishin EV, Kiselev AP, Modyanov NN, Braunstein AE, Polyanovsky OL, Nosikov VV. The complete amino acid sequence of cytoplasmic aspartate aminotransferase from pig heart. FEBS Lett 1973; 29:31-34.
22. Inoue K, Kuramitsu S, Okamato A, Hirotsu K, Higuchi T, Morino Y, Kagamiyama H. Tyr225 in aspartate aminotransferase: contribution of the hydrogen bond between Tyr225 and coenzyme to the catalytic reaction. J Biochem 1991; 109:570-576.

Molecular Biology

Genes and enzymes involved in arginine and pyrimidine biosynthesis in psychrophilic *Vibrio* strains from the deep sea

Y. Xu[1], Z. Liang[1], C. Legrain[2], V. Villeret[2], J. Van Beeumen[3] and N. Glansdorff[1,2*]

[1] Free University of Brussels and Flanders Interuniversity Institute for Biotechnology, Avenue Emile Gryson 1, B-1070 Brussels, Belgium
[2] Research Institute Jean-Marie Wiame, Avenue Emile Gryson 1, B-1070 Brussels, Belgium
[3] Laboratory for Protein Biochemistry and Protein Engineering, University of Gent, Ledeganckstraat 35, B-9000 Gent, Belgium

1
Introduction

The physiologically significant temperature range for the activity of any enzyme appears quite limited when compared to the full range of temperatures compatible with life. This emphasizes optimization of function rather than stability as the driving force of evolution: whereas proteins working at high temperatures are a compromise between intermolecular forces conferring enough rigidity to ensure stability and the need for functional flexibility, cold-active enzymes must be flexible enough to make catalysis possible at a reduced energy cost.[1–5] As a corollary, enzymes of psychrophilic origin are expected to be relatively thermolabile. The same considerations relating function and stability could in principle apply to regulatory proteins.

Studies by several laboratories (see for example refs. 6–11) indicate that "strategies" for adaptation at extremes of temperature may be brought to light and therefore become amenable to experimental analysis, when one compares structures and structural models of homologous proteins from organisms representing the widest possible range of temperatures. Carbamoyltransferases, a group of enzymes already familiar to our group,[12,13] was chosen to embark on such a comparative study.

Ornithine carbamoyltransferase (OTC) catalyzes the sixth step of *de novo* arginine biosynthesis or, in the reverse direction, the phosphorolysis of citrulline in the catabolic arginine deiminase pathway. Aspartate carbamoyltransferase (ATC) catalyzes the committed step of pyrimidine biosynthesis (Fig. 1). *Escherichia coli* ATC is one of the best-known allosteric enzymes.[14] Because of the essential character of their anabolic function, both carbamoyltransferases are ubiquitous. They are also

* Corresponding author; e-mail: ceriair@ulb.ac.be

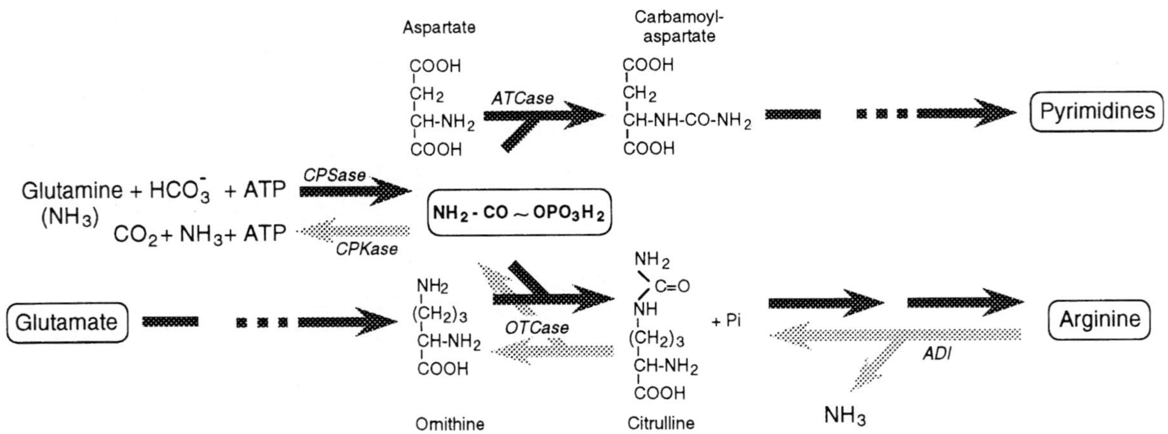

Fig. 1. Metabolism of carbamoylphosphate
Black arrows indicate biosynthetic pathways and *grey arrows* the arginine deiminase catabolic pathway

paralogous and it is very likely that the last common ancestor to all life on this planet already possessed redundant genes for either of them.[15] A large number of amino acid sequences is known (35 for OTC, 31 for ATC);[15] crystal structures are available for the OTCs of *Escherichia coli*, *Pseudomonas aeruginosa* and *Pyrococcus furiosus*,[11,16–18] and for the ATCs of *E. coli* and *Bacillus subtilis*.[19,20] Several other OTCs and ATCs are being investigated, including a number of thermophilic ones.[21] Until recently, however, no information was available on psychrophilic transferases.

An additional reason to study these enzymes is their intracellular character; they are therefore not submitted to the same molecular constraints as extracellular proteins and may reveal different strategies of adaptation. At the time of writing, the only two psychrophilic enzymes for which crystallographic data were available were secreted proteins: the α-amylase of *Alteromonas haloplanctis*[22] and a protease from an antarctic strain of *P. aeruginosa*.[23] Very recently the crystal structure of a cold-active citrate synthase was determined.[24]

Whereas the study of cold-active enzymes is gathering momentum, information on genetic regulatory mechanisms in strict psychrophiles is almost completely lacking. It is therefore interesting to compare the functional requirements of proteins or other elements involved in the regulation of arginine and pyrimidine biosynthesis over the full range of temperatures compatible with life. Both pathways have been extensively studied in this respect in a variety of organisms and therefore provide an excellent frame of reference.[12,13,25]

We had to select a prototrophic organism as subject matter in order to investigate genes and enzymes involved in the *de novo* biosynthesis of arginine and pyrimidines. Considering also that close relatedness with the best characterized homologues (i.e from Enterobacteriaceae) would be a serious advantage for comparative studies, we directed our choice towards psychrophilic strains presenting features of the genus *Vibrio*. These organisms were isolated by H. Rüger (Alfred-Wegener Institute for Marine and Polar Research, Bremerhaven, Germany) from the surface sediment layer of the deep Atlantic at a depth of 2815 m and a temperature of 2°C. This chapter summarizes the first data obtained on carbamoylphosphate (CP) metabolism and arginine biosynthesis in these organisms. The details of these studies and the characterization of the organisms are published elsewhere.[26,27]

2
Salient properties of the organisms

The three *Vibrio* strains (2674, 2693 and 2825) do not grow above 14°C and display their maximal growth rate around 6°C, with a doubling time of about 6 h in marine broth, 8.5 h in rich synthetic, arginine- and uracil-free medium, and 16.5 h in minimal salt medium.[28] The effect of pressure on these parameters is not yet known. Since the depth at which they were isolated overlaps the barophilic range,[29] it is not excluded that increasing pressure would increase their growth rate.

Enzyme assays revealed the presence of a glutamine-dependent carbamoylphosphate synthetase (CPS) as in most prokaryotes tested so far (see, however, refs. 30,31) and of both an OTC and an ATC (Table 1). Including arginine in the growth

Table 1. Specific activities of aspartate carbamoyltransferase (ATC) and ornithine carbamoyltransferase (OTC) from *Vibrio* strain 2693 in the presence of arginine and uracil

Addition to arginine- and uracil-free synthetic medium	Specific activity (units mg^{-1} protein) ATC	OTC
None	9.3	60.4
Arginine (5 mM)	7.8	7.0
Uracil (5 mM)	8.1	79.0
Arginine (5 mM) + Uracil (5 mM)	8.7	3.3

medium resulted in repression of OTC synthesis whereas addition of uracil did not affect the ATC level but increased OTC specific activity. This increase is probably due to a slightly excessive inhibition or repression of CPS as originally observed in *E. coli*.[13]

2.1 Arginine biosynthesis

The OTC genes of *Vibrio* 2674 and 2693 were cloned by complementation of the appropriate *E. coli* auxotroph. Determining the sequence of these genes and of flanking DNA segments revealed that most of the arginine biosynthetic genes were clustered, the positions of *argA* and *D* being still uncertain. The cluster is a large divergent operon, one wing of which is *argE* and the opposite one *argCBFGH*. This divergent pattern is reminiscent of Enterobacteriaceae, where *ECBH* and *ECBGH* operons have been described.[12] Considering the phylogenetic relationship of Vibrionaceae and Enterobacteriaceae in the γ-3 subgroup of the Proteobacteria,[32] these data suggest that the *Vibrio* pattern is ancestral. A parallel can be drawn with the evolution of the tryptophan operon in bacteria.[33–35] The presence of an *argE* gene (encoding an acetylornithinase) rather than an *argJ* gene (encoding an acetylornithine-glutamate acetyltransferase) is noteworthy. Arginine biosynthesis is indeed one of those rare instances where the enzymatic pathway is not universal: whereas most prokaryotes have *argJ*[12,13] and therefore recycle the acetyl group of acetylornithine on glutamate, Enterobacteriaceae and, as shown here, *Vibrio* also, have *argE*.

Preliminary experiments (unpubl. data) indicate that the DNA region between the *argE* and *argC* transcription startpoints contains a segment of about 40 bp which is recognized by the *E. coli* arginine repressor. The prospects of isolating a homologous psychrophilic arginine repressor are thus favorable. It will be interesting to compare this protein to its counterparts from *E. coli*[36,37] and *B. stearothermophilus*.[38]

2.2 Ornithine carbamoyltransferase

The OTCs from *Vibrio* strains 2674, 2693 and 2825 all appear distinctly psychrophilic, with apparent optima around 17 and 20°C. Even if these temperatures are above the functional range of the three organisms they are still lower than most of the optima reported for enzymes of psychrophilic origin.[2,39–43] Moreover, the enzyme is still considerably active at 0°C. Determination of the intrinsic catalytic efficiency awaits studies on pure enzyme. Figure 2 compares the temperature-dependent activity profiles of OTCs from *Vibrio* 2693, *E. coli* and *P. furiosus*,[45] a hyperthermophilic archaeon.

The thermostability of OTCs from *Vibrio* 2674 and 2693 (Fig. 3) proved almost identical (whether assayed on native or recombinant enzymes) and, as expected from studies on other OTCs, the enzymes were more stable in the presence of phosphate.[44] With a half life of 6.5 min at 60°C (phosphate being present), the enzyme is clearly more labile than its *E. coli* counterpart (half-life of 55 min at the same temperature). The OTC from strain 2825 is considerably more labile, being not stable above 25°C (Fig. 3).

OTC from strain 2693 was shown to have approximately the same molecular mass as in *E. coli* (98,500 vs. 105,000) on a Native Phast gel where enzyme activity was visualized by precipitation of the phosphate produced by the reaction as the lead salt.[28] It is therefore a trimeric enzyme, like most anabolic OTCs examined so far with the exception of *Pyrococcus* OTC.[11,45] The genes from both *Vibrio* 2674 and 2693 have a coding potential of 301 amino acids and differ at only five positions. The OTC gene from *Vibrio* 2825 has not yet been cloned.

In the absence of structural data and of published coordinates for the only trimeric homologue to have been studied by X-ray diffraction (*E. coli* OTC),[16,17] comments on the molecular basis for the psychrophily of *Vibrio* OTCs must remain limited. However, if we restrict ourselves to a comparison between Vibrio OTCs and others of the bacterial domain, and if we select organisms with GC% between 40 and 53% (*E. coli* and *Thermotoga maritima*) in order to avoid DNA compositional bias, it can be noted that there are more glutamines and asparagines (thermolabile residues) in *Vibrio* or in *E. coli* than in *Thermotoga*. The total number of charged residues susceptible to provide stabilizing interactions is lower in *Vibrio* than in *E. coli* and much lower than in *Thermotoga*. This is particularly so as regards arginine, which can play an important role in protein thermostability by its high hydrophilicity and hydrogen bonding capacity.[46,47] Structural comparisons between different trimeric OTCs will be essential to understand OTC psychrophily since comparing the dodecameric OTCs from *P. aeruginosa* and *P. furiosus* strongly suggests that the extreme thermostability of the archeal enzyme rests chiefly on interactions *between* trimers.[11] *Vibrio* OTC differs from all other OTCs by presenting substitutions in both of the motifs found to be the most conserved in this family of enzymes:[15,48] the ST57RT sequence involved in CP binding, where Thr57 is replaced by a leucine, and the HCL274P sequence, involved in ornithine binding, where leucine is replaced by a glutamine.[15] Restoring either consensus motifs or both in *Vibrio* 2693 OTC by site-directed mutagenesis did not increase stability or the comparatively low affinity of this enzyme for CP and ornithine (respectively 4.5 and 2.5 times less than in *E. coli*). The most obvious effect of these mutations was to decrease the activity of the enzyme.

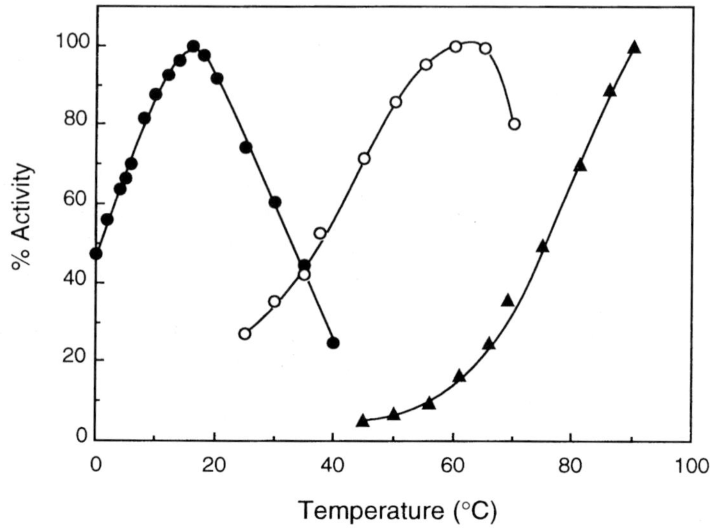

Fig. 2. Temperature activity profiles of ornithine carbamoyltransferases in crude extracts from cells of (●) *Vibrio* 2693, (o) *Escherichia coli*, (▲) *Pyrococcus furiosus*

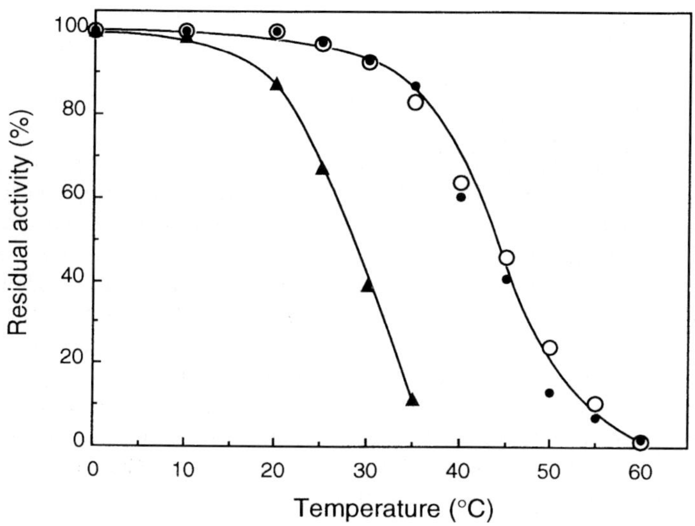

Fig. 3. Thermal inactivation profile of ornithine carbamoyltransferase in crude extracts of cells from *Vibrio* strains

Samples were incubated for 15 min at the temperature indicated in 30 mM Tris-HCl buffer, pH 8.0; (o) strain 2674; (●) strain 2693; (▲) strain 2825.

2.3
Aspartate carbamoyltransferase

In *Vibrio*, ATC is presently the only enzyme of the pyrimidine pathway to have been characterized at the genetic level.[27] ATC genes from both *Vibrio* 2674 and 2693 complement the corresponding *E. coli* auxotroph. Sequencing revealed an operon presenting striking similarities with the cognate determinants in *E. coli* and *Salmonella typhimurium*. Like other class B ATCs,[25,49] the enzyme is encoded by two genes: one is homologous to *E. coli pyrB*, the structural determinant of the catalytic chain, the other to *E. coli pyrI*, determinant of the regulatory chain. Percentages of amino acid sequence identity are quite high: 74 and 56, respectively (Fig. 4).

The two genes are preceded by a control region which extends 141 nucleotides upstream from the putative *pyrB* translation initiation codon (Fig. 5). Primer extension experiments located a predominant transcription start at a G nucleotide located at the onset of this region. The latter includes a 21-codon leader open reading frame containing a pyrimidine-rich stretch; the leader terminates at a TAA codon located within a putative rho-independent attenuation site. In principle, these elements are all that is required to control the operon by a pyrimidine-monitored attenuation mechanism as first proposed by Roof et al.[50] for *E. coli* and later established by others.[51] As already mentioned above, ATC synthesis is not repressible in *Vibrio* by an exogeneous source of pyrimidines and we were not yet able to demonstrate whether the operon can be derepressed. However, the region is functional in terms of gene expression since transcription starts at the onset of it. It is interesting that the sequence elements of the *Vibrio* control region, although they display a similar arrangement as in Enterobacteriaceae, are very dissimilar in composition and in length (20 amino acids for *Vibrio*, 44 for *E. coli*) whereas the structural genes themselves are very similar. Even between the two *Vibrio* strains, whose *pyrB* and *I* genes are more than 95% identical, the percentage of amino acid identity between the leader peptides (both of equal length in this case) is only 45%. These data can be regarded as a further illustration of the relatedness between Vibrionaceae and Enterobacteriaceae and this at two levels of function: the control region, where the significance of observed similarities resides in the layout of sequence elements which may extensively diverge in composition, and the structural genes, where functional constraints operate in a much more stringent way.

Class B ATCs consist of two trimers of catalytic chains held together by three dimers of regulatory chains.[14,49] For such an architecture and from the coding potential of *Vibrio pyrB* and *pyrI*, we would predict a M_r of 30,800. This is compatible with the molecular mass actually determined by molecular filtration (320 ± 20 kDa) and with the value of 300 kDa reported for *V. natriegens* ATC.[52]

Vibrio ATC is allosteric;[27] the aspartate saturation curve indicates homotropic interactions with respect to this substrate and CTP inhibits the reaction to about the same extent as in *E. coli* but no UTP-CTP synergy[49] could be observed. Moreover, ATP, an activator of *E. coli* ATC, is without effect in *Vibrio* (Fig. 6). Multiple sequence alignment of several class B ATCs revealed that all the critical residues involved in carbamoylphosphate and -aspartate binding are conserved in identical positions in *Vibrio* ATC (Fig. 4). Likewise, residues involved in interactions at the

A

```
                   10        20        30          40        50
VIBRIO   MANPLFRKHIVSINDISRNELELIVKTAAKLKKQP----QPELLKNKVIASCFFEASTRT
ECOLI    -ANPLYQKHIISINDLSRDDLNLVLATAAKLKANP----QPELLKHKVIASCFFEASTRT
SALTY    -ANPLYQKHIISINDLSRDDLNLVLATAAKLKANP----QPELLKHKVIASCFFEASTRT
SERMA    MANPLYHKHIISINDLSRDDLELVLATAAGLKANP----QPELLKHKVIASCFFEASTRT
PYRAB    --MDWKGRDVISIRDFSKEDIETVLATAERLERELKEKGQLEYAKGKILATLFFEPSTRT
              . ..** * *. ..  .. **  *         * *  * *..*. *** ****

             60        70        80        90       100       110
VIBRIO   RLSFETAIQRLGGTVIGFDNASNTSLAKKGETLADSISVISSYVDAFVMRHPQEGAARLA
ECOLI    RLSFETSMHRLGASVVGFSDSANTSLGKKGETLADTISVISTYVDAIVMRHPQEGAARLA
SALTY    RLSFETSMHRLGASVVGFSDSANTSLGKKGETLADTISVISTYVDAIVMRHPQEGAARLA
SERMA    RLSFETSMHRLGASVVGFADGSNTSLGKKGETLADTISVISTYVDAIVMRHPQE-GARMA
PYRAB    RLSFESAMHRLGGAVIGFAEASTSSVK-KGESLRDTIKTVEQYCDVIVIRHPKEGAARLA
         *****....***  .*.**    ...*.  ***.* *.*  .   * *  *.***.*  **.*

            120       130       140       150       160       170
VIBRIO   SEFS-NVPVINGGDGSNQHPTQTLLDLFSIYETQGCLDNLNIALVGDLKYGRTVHSLA-Q
ECOLI    TEFSGNVPVLNAGDGSNQHPTQTLLDLFTIQETQGRLDNLHVAMVGDLKYGRTVHSLT-Q
SALTY    TEFSGQVPVLNAGDGSNQHPTQTLLDLFTIQETQGRLDNLHIAMVGDLKYGRTVHFAKPR
SERMA    SEFSGNVPVLNAGDG-NQHPTQTLLDLFTIQETQGRLSNLSIAMVGDLKYGRTVHSLT-Q
PYRAB    AEVA-EVPVINAGDGSNQHPTQTLLDLYTIKKEFGRIDGLKIGLLGDLKYGRTVHSLA--
         .*  .  ***.* *** ***********..*    *  . .*  . ..**********

              180       190       200       210       220       230
VIBRIO   ALAKFSGCKFYFIAPDALAMPEYICDELDEHNVSYACYNSIEEVVPEIDVLYMTRVQKER
ECOLI    ALAKFDGNRFYFIAPDALAMPQYILDMLDEKGIAWSLHSSIEEVMAEVDILYMTRVQKER
SALTY    TLAKFSGNRFYFIAPDALAMPQYILDMLDEKGMAWSLHGSIEEVMADVDILYMTRVQKER
SERMA    ALAKFEGNRFYFIAPDALAMPAYILKMLEEKGIEYSSHGSIEEVVPELDILYMTRVQKER
PYRAB    EALTFYDVELYLISPELLRMPRHIVEELREKGMKVVETTTLEDVIGKLDVLYVTRIQKER
          *      * *.*. * **  *    * *...         ..*.*.  .*.**.**.****

              240       250       260       270       280       290
VIBRIO   FD-ETEYQHMKAGFILSASSLKHAKDNLKVLHPLPRVDEIAVDVDKTPYAYYFQQAENGV
ECOLI    LD-PSEYANVKAQFVLRASDLHNAKANMKVLHPLPRVDEIATDVDKTPHAWYFQQAGNGI
SALTY    LD-PSEYANVKAQFVLRP-DLNGARENMKVLHPLPRIDEITTDVDKTPHAWYFQQAGNGI
SERMA    LD-PSEYANVKAQFVLAA--DLAGAANLKVLHPLPRIDEIATDVDKTPHAYYFQQAGNGI
PYRAB    FPDEQEYLKVKGSYQVNLKVLEKAKDELRIMHPLPRVDEIHPEVDNTKHAIYFRQVFNGV
          **   .*  .  .            ....*****.***  .** *  * **.*  **.

             300       310
VIBRIO   YAREALLALVLNATIEG
ECOLI    FARQALLALVLNRDLVL
SALTY    FAAQALLALVLNSELSL
SERMA    FAR-SALALVVNADLAL
PYRAB    PVRMALLALVLGVI---
          . ****..
```

B

```
                  10        20        30        40          50
VIBRIO   -MKSNHMQVEAICNGYVIDHIPSGQGVKILRLFSLTDTKQR--VTVGFNLPSHDGTTKDL
ECOLI    -THDNKLQVEAIKRGTVIDHIPAQIGFKLLSLFKLTETDQR--ITIGLNLPSGEMGRKDL
SALTY    -THDNKLQVEAIKCGTVIDHIPAQVGFKLLSLFKLTETDQR--ITIGLNLPSGEMGRKDL
SERMA    MTHDNKLQVEAIKCGTVIDHIPAQIGFKLLTLFKLTATDQR--ITIGLNLPSNELGRKDL
PYRAB    ---MAELKVSAIKEGTVIDHIPAGKGLKVIEILKLGKLTNGGAVLLAMNVPSKKLGRKDI
            ..* **  * ******.  * *.. .  *    .  .  . *.**       **.

            60        70        80        90       100       110
VIBRIO   IKVENTEITKSQANQLALLAPNATVNIIENFKVTDKHSLALPKEVENVFPCPNSNCITHG
ECOLI    IKIENTFLSEDQVDQLALYAPQATVNRIDNYEVVGKSRPSLPERIDNVLVCPNSNCISHA
SALTY    IKIENTFLTEEQVNQLALYAPQATVNRIDNYDVVGKSRPSLPERINNVLVCPNSNCISHA
SERMA    IKIENTFLTEQQANQLAMYAPKATVNRIDNYEVVRKLTLSLPDHIDGVLTCPNGNCISRS
PYRAB    VKVEGRFLSEEEVNKIALVAPNATVNIIRDYKVVEKFKVEVPDVIEGILRCGNPNCITNH
         .*.*.  ..   .  ..*. ** **** *   .  *  *     .*  .  . ..  * * ***.

            120       130       140       150
VIBRIO   EPVISSFTIKMIKGNIGLKCKYCEKTFSKEIVTAQV
ECOLI    EPVSSSFAVRKRANDIALKCKYCEKEFSHNVVLAN-
SALTY    EPVSSSFAVKKRANDIALKCKYCEKEFSHYVVLAN-
SERMA    EPVRSSFSVKSRGGEVHLKCRYCEKEFEHQVVLQAD
PYRAB    EYVTTKFYVISRE-PLKVRCHYCERTMEEEEILANL
         * *  . *  .        .  ..*.***.       .
```

Fig. 4. Primary sequence alignement for the catalytic (**A**) and regulatory subunits (**B**) of several aspartate carbamoyltransferases (ATC) from *Vibrio* 2693, *Escherichia coli*, *Salmonella typhimurium*, *Serratia marcescens* and *Pyrococcus abyssi*

Residues, which in *E. coli* ATC are involved in CTP binding are in *bold*, those involved in ATP binding are *underlined*. *Asterisks* indicate identical residues and *dots* similar residues conserved in all protein sequences (reprinted with permission from ref. 27)

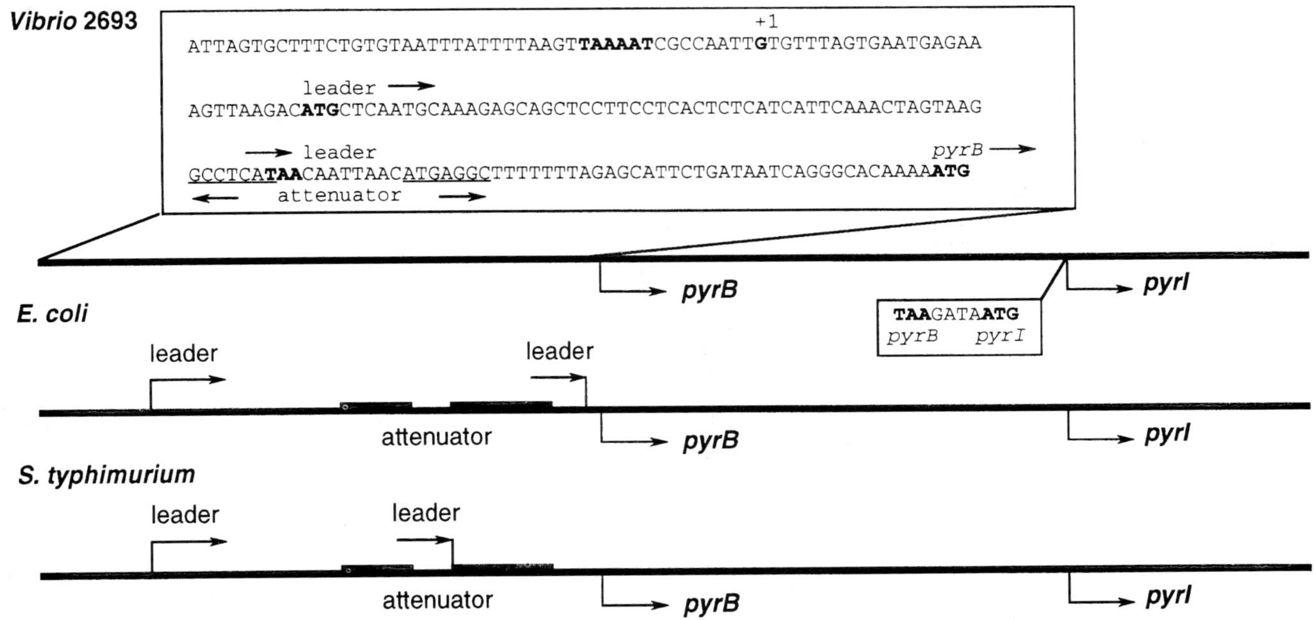

Fig. 5. Comparison of the promoter regions of the *pyrBI* operons in *Vibrio* 2693, *Escherichia coli* and *Salmonella typhimurium*
In the *Vibrio* sequence, *bold* characters indicate the -10 promoter element (TAAAAT), the experimentally determined transcription start point (+1), and the start and stop codons of the leader open reading frame, of *pyrB* and *pyrI*. Inverted repeats are *underlined* (reprinted with permission from ref. 27)

c1–c2 interface (between subunits of one catalytic trimer) and at the c1–c4 interface (between two catalytic trimers) are highly conserved in *Vibrio* ATC. In the regulatory chain most of the residues which constitute the binding site for nucleotides in *E. coli* ATC are also present, except Tyr88, which is replaced by a Phe and Arg95 by a Ser. Val17 and Ile86, both specific for CTP binding, are conserved. On the other hand, residues Asp19 and Leu58, both specific for ATP binding to *E. coli* ATC, are also conserved in *Vibrio*. Thus, even though *Vibrio* ATC does not appear to be sensitive to ATP, the site for this nucleotide effector is almost completely conserved. It is therefore likely that the insensitivity to ATP is due to substitutions in residues playing a role in the transmission of the regulatory signal. The four cysteines involved in the binding to the Zn atom of the regulatory chain are conserved.

Vibrio ATC displays an apparent optimum between 30 and 35°C, with still considerable activity in the actual temperature range of the host organisms (Fig. 7). Though higher than for OTC, this value is not uncommon for an enzyme of psychrophilic origin.[2,39,40,42,43] However, the protein is much more thermolabile than *E. coli* ATC: whereas the latter loses only 25% of its activity after 6 h incubation at 60°C,[53] *Vibrio* ATC is almost totally inactivated after 15 min at the same temperature. It is therefore interesting to examine whether the differences in composition already noted between *E. coli* ATC and its hyperthermophilic homologue from *Pyrococcus abyssi* also exist in *Vibrio* ATC and whether or not they are accentuated in this enzyme. The most notable features of the mesophilic ATC with respect to the thermophilic one are a lower number of charged residues susceptible to participate

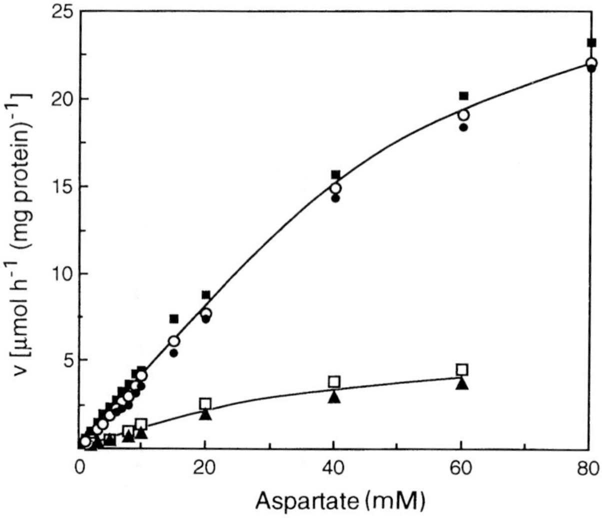

Fig. 6. Effect of various ligands in the presence of L-aspartate on the activity of aspartate carbamoyltransferase (ATC) in crude extracts from cells of a *pyrBI* deletion in strain of *Escherichia coli* expressing the cloned *Vibrio* ATC genes
(o) no effectors; (■) ATP 5 mM; (●) UTP 2 mM; (❑) CTP 2 mM; (▲) CTP 2 mM + UTP 2 mM (reprinted with permission from ref. 27)

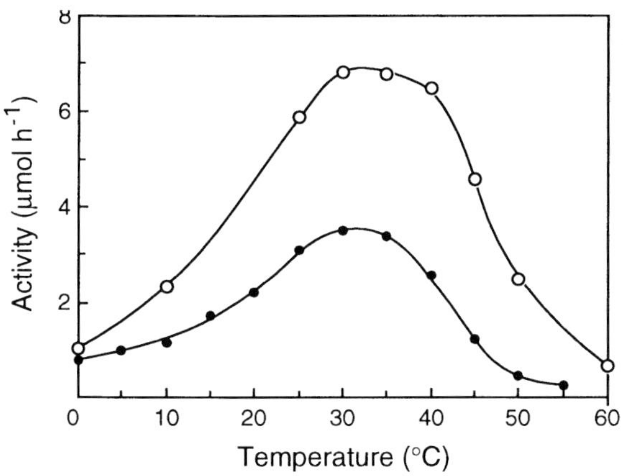

Fig. 7. Temperature activity profile of aspartate carbamoyltransferase in crude extracts from (●) *Vibrio* 2693 cells and (o) *Escherichia coli* cells expressing the corresponding cloned genes in a *pyrBI* deletion background

in stabilizing interactions (Arg, Glu and Lys), an excedent of the thermolabile residues $AspNH_2$ and $GluNH_2$, and an excedent of serine residues.[54] The same characteristics indeed show up in *Vibrio* ATC but, with respect to *E. coli*, the arginine content [and also the R/(R+K) ratio] undergoes a further and remarkable decrease (14 residues instead of 23) whereas the number of cysteine residues, susceptible to enhanced oxidation at higher temperature, rises from 5 to 10. This is a striking increase, considering that four of these cysteines are already required to bind the Zn atom of the regulatory subunit and are thus expected to be conserved in all class B ATCs. On the whole, therefore, *Vibrio* ATC presents characteristics which are in keeping with its psychrophilic character.

The high sequence identity observed between *E. coli* and *Vibrio* ATCs is propitious to molecular modeling. This allowed us to carry the analysis of the protein a step further. The details will be published elsewhere but the most salient features are summarized here. We have already mentioned above the high degree of conservation at the c1–c2 and c1–c4 interfaces. However, at the interfaces between r chains or between c and r chains, the model suggests that certain salt bridges shown to be important in the R form of *E. coli* ATC[14,55] are actually absent in *Vibrio*. At the r1–r6 interfaces (between the two chains of a regulatory dimer) the double interactions between *E. coli* residues Asp39 and Arg55 appear to be absent, since Asp39 is replaced by a Lys and Arg55 by a Thr. This could decrease the stability of the regularory dimer and facilitate the allosteric transition at low temperature. At the c1–r1 interface, Asp200c in *E. coli* interacts with Arg128r and Arg130r; in *Vibrio* ATC Arg128r is replaced by a Lys and Arg130 by an Ile. Moreover, Glu144r which in *E. coli* also interacts with Arg128r and Arg130r, becomes a Thr in *Vibrio*. It is therefore tempting to suggest that these rather striking changes are related to adaptation at low temperature.

3
Conclusions

Though crystallographic data on psychrophilic enzymes are still very limited, not much time will elapse before we become able to compare the molecular architecture of several families of homologous enzymes over the whole range of temperatures compatible with life. The observations summarized in this chapter show OTCs and ATCs to be particularly adequate material to investigate molecular strategies for extremophily. Besides, the value of psychrophilic ATC as a model system is enhanced by its allosteric character; indeed, the transmission of long-range intramolecular signals emphasizes the need for structural flexibility. We have seen above that modeling studies point out substitutions that could be responsible for increased flexibility and enhanced thermolability.

A recurrent observation concerning enzymes of psychrophilic origin is that their apparent optimal temperature of activity is much higher than the temperature at which the maximal growth rate of their host is recorded (the so-called optimal temperature). *Vibrio* OTC and ATC are no exceptions in this respect. Even the comparatively low apparent optimum of OTC (17°C) still exceeds considerably the optimum of its host (6°C). One can wonder whether this gap does not correspond to a limit of stability that most enzymes would be unable to cross, at least not without the intervention of extrinsic factors. In other words, to shift the apparent optimum towards even lower temperatures might require such an increase in flexibility that the protein would become too unstable. Another factor that may contribute to set a lower limit to the flexibility of psychrophilic enzymes is the necessity to resist cold denaturation. It may be that many psychrophilic enzymes actually approach the lowest possible intrinsic stability.

There are very little observations concerning the molecular genetics of psychrophiles. The genetic data reported in this chapter certainly contribute to our understanding of the evolutionary relationships between members of the γ-3 subgroup of the Proteobacteria, but it is still too early to say whether investigations of this kind will disclose novel features specifically related to psychrophily. However, we can expect interesting developments concerning the analysis of the *Vibrio* arginine repressor and its comparison with mesophilic and thermophilic homologues.

Acknowledgements. This work was supported by the EEC-sponsored programmes Eurocold and Coldzymes, by the Flanders Foundation for Scientific Research and by the Flemish Actionprogramme Biotechnology.

4
References

1. Daniel RM, Dines M, Petach HH. The denaturation and degradation of stable enzymes at high temperatures. Biochem J 1996; 317:1-11.
2. Feller G, Narinx E, Arpigny JL, Aittaleb M, Baise E, Genicot S, Gerday C. Enzymes from psychrophilic organisms. FEMS Microbiol Rev 1996; 18:189-202.

3. Gerday C, Aittaleb M, Arpigny JL, Baise E, Chessa JP, Garsoux G, Petrescu I, Feller G. Psychrophilic enzymes: a thermodynamic challenge. Biochim Biophys Acta 1997; 1342:119-131.
4. Jaenicke R. Protein folding and association: Significance of in vitro studies for self-organization and targeting in the cell. Curr Topics Cell Regul 1995; 34:209-314.
5. Shoichet BK, Baase WA, Kuroki R, Matthews BW. A relationship between protein stability and protein function. Proc Natl Acad Sci USA 1995; 92:452-456.
6. Mainfroid V, Mande SC, Hol WGC, Martial JA, Goraj K. Stabilization of human triosephosphate isomerase by improvement of the stability of individual alpha-helices in dimeric as well as monomeric forms of the protein. Biochemistry 1996; 35:4110-4117.
7. Rentier-Delrue F, Mande SC, Moyens S, Terpstra P, Mainfroid V, Goraj K, Lion M, Hol WGJ, Martial JA. Cloning and overexpression of the triosephosphate isomerase genes from psychrophilic and thermophilic bacteria: structural comparison of the predicted protein sequences. J Mol Biol 1993; 229:85-93.
8. Rice DW, Yip KSP, Stillman TJ, Britton KL, Fuentes A, Connerton I, Pasquo A, Scandurra R, Engel PC. Insights into the molecular basis of thermal stability from the structure determination of *Pyrococcus furiosus* glutamate dehydrogenase. FEMS Microbiol Rev 1996; 18:105-117.
9. Van de Casteele M, Legrain C, Demarez L, Chen P, Piérard A, Glansdorff N. Molecular physiology of carbamoylation under extreme conditions: what can we learn from extreme thermophilic microorganisms? Comp Biochem Physiol 1997; 118A:463-473.
10. Van de Casteele M, Chen P, Roovers M, Legrain C, Glansdorff N. Structure and expression of a pyrimidine gene cluster from the extreme thermophile *Thermus* strain ZO5. J Bacteriol 1997; 179:3470-3481.
11. Villeret V, Clantin B, Tricot C, Legrain C, Roovers M, Stalon V, Glansdorff N, Van Beeumen J. The crystal structure of *Pyrococcus furiosus* ornithine carbamoyltransferase reveals a key role for oligomerization in enzyme stability at extremely high temperatures. Proc Natl Acad Sci. USA 1998; 95:2801-2806.
12. Cunin R, Glansdorff N, Piérard A, Stalon V. Biosynthesis and metabolism of arginine in bacteria. Microbiol Rev 1986; 50:314-352.
13. Glansdorff N. Biosynthesis of arginine and polyamines. In: Neidhardt F, ed. *Escherichia coli* and *Salmonella typhimurium*, Molecular and Cellular Biology. Washington: ASM Press, 1996:408-433.
14. Lipscomb WN. Aspartate transcarbamylase from *Escherichia coli*: activity and regulation. Adv Enzymol 1994; 68:67-152.
15. Labédan B, Boyen A, Baetens M, Charlier D, Chen P, Cunin R, Durbecq V, Glansdorff N, Hervé G, Legrain C, Liang Z, Purcarea C, Roovers M, Sanchez R, Thia-Toong TL, Van de Casteele M, Van Vliet F, Xu Y, Zhang Y. The evolutionary history of carbamoyltransferases: a complex set of paralogous genes was already present in the last universal common ancestor. J Mol Evol 1998 (in press).
16. Ha Y, McCann MT, Tuchman M, Allewell NM. Substrate-induced conformational change in a trimeric ornithine transcarbamoylase. Proc Natl Acad Sci USA 1997; 94:9550-9555.
17. Jin L, Seaton BA, Head JF. Crystal structure at 2.8 Å resolution of anabolic ornithine transcarbamylase from *Escherichia coli*. Nature Struct Biology 1997; 4:622-625.
18. Villeret V, Tricot C, Stalon V, Dideberg O. Crystal structure of *Pseudomonas aeruginosa* catabolic ornithine transcarbamoylase at 3.0-Å resolution: a different oligomeric organization in the trancarbamoylase family. Proc Natl Acad Sci USA 1995; 192:10762-10766.
19. Ke H, Lipscomb WN, Cho Y, Honzatko RB. Complex of *N*-phosphonoacetyl-L-aspartate with aspartate carbamoyltransferase: X-ray refinement, analysis of conformational changes and catalytic and allosteric mechanisms. J Mol Biol 1988; 204:725-747.
20. Stevens RC, Reinisch KM, Lipscomb WN. Molecular structure of *Bacillus subtilis* aspartate transcarbamoylase at 3.0-Å resolution. Proc Natl Acad Sci USA 1991; 88:6097-6091.
21. Cunin R. Carbamoylphosphate metabolism in thermophiles and hyperthermophiles – a perspective. Path to Pyrimidines, Int Newslett 1997; 5:37-49.

22. Aghajari N, Feller G, Gerday C, Haser R. Crystallization and preliminary X-ray diffraction studies of alpha-amylase from the antarctic psychrophile *Alteromonas haloplanctis* A23. Prot Sci 1996; 5:21-28.
23. Villeret V, Chessa JP, Gerday C, Van Beeumen J. Preliminary crystal structure determination of the alkaline protease from the antarctic psychrophile *Pseudomonas aeruginosa*. Protein Sci 1997; 6:2462-2464.
24. Russell RJM, Gerike V, Danson MJ, Hough DW, Taylor GL. Structural adaptation of the cold-active citrate synthase from an antarctic bacterium. Structure 1998; 6:351-361.
25. Neuhard J, Kelln RA. Biosynthesis and conversions of pyrimidines. In: Neidhardt F, ed. *Escherichia coli* and *Salmonella typimurium*, Molecular and Cellular Biology. Washington: ASM Press, 1996:580-599.
26. Liang Z, Xu Y, Roovers M, Legrain C, Van de Casteele M, Rüger H, De Kegel D, Glansdorff N. Ornithine carbamoyltransferase from two new deep-sea *Vibrio* strains. Microbiology 1999 (in press).
27. Xu Y, Zhang YF, Liang ZY, Van de Casteele M, Legrain C, Glansdorff N. Aspartate carbamoyltransferase from a psychrophilic deep-sea bacterium, *Vibrio* strain 2693. Properties of the enzyme, genetic organisation and synthesis in *Escherichia coli*. Microbiology 1998; 144:1435-1441.
28. Liang Z. Physiology and molecular biology of enzymatic carbamoylation in marine psychrophilic bacteria. Brussels, VUB: PhD Thesis, 1997.
29. Yayanos AA. Microbiology to 10,500 meters in the deep sea. Ann Rev Microbiol 1995; 49:777-805.
30. Legrain C, Demarez M, Glansdorff N, Piérard A. Ammonia-dependent synthesis and metabolic channelling of carbamoyl phosphate in the hyperthermophilic archaeon *Pyrococcus furiosus*. Microbiology 1995; 141:1093-1099.
31. Durbecq V, Legrain C, Roovers M, Piérard A, Glansdorff N. The carbamate kinase-like carbamoyl phosphate synthetase of the hyperthermophilic archaeon *Pyrococcus furiosus*, a missing link in the evolution of carbamoyl phosphate biosynthesis. Proc Natl Acad Sci USA 1997; 94:12803-12808.
32. Baumann P, Schubert RHW Family II. Vibrionaceae. In: Krieg NK, Holt JG, eds. Bergey's Manual for Determinative Bacteriology, vol 1. Baltimore: Williams & Wilkins, 1984:516-517.
33. Glansdorff N. On the origin of operons and their possible role in evolution towards thermophily. J Mol Evol 1998 (in press).
34. Jensen RA. An emerging outline of the evolutionary history of aromatic amino acid biosynthesis. In: Mortlock RP, ed. The Evolution of Metabolic Function. Boca Raton: CRC Press, 1992:205-236.
35. Sterner R, Dahm A, Darimont B, Ivens A, Liebl W, Kirschner K. (beta alpha)(8)-barrel proteins of tryptophan biosynthesis in the hyperthermophile *Thermotoga maritima*. EMBO J 1995; 14:4395-4402.
36. Sunnerhagen M, Nilges M, Otting G, Carey J. Solution structure of the DNA-binding domain and model for the complex of multifunctional hexameric arginine repressor with DNA. Nature Struct Biology 1997; 4:819-826.
37. Van Duyne GD, Ghosh G, Maas WK, Sigler PB. Structure of the oligomerization and L-arginine binding domain of the arginine repressor of *Escherichia coli*. J Mol Biol 1996; 256:377-391.
38. Dion M, Charlier D, Wang H, Gigot D, Savchenko A, Hallet JN, Glansdorff N, Sakanyan V. The highly thermostable arginine repressor of *Bacillus stearothermophilus*: gene cloning and repressor-operator interactions. Mol Microbiol 1997; 25:385-398.
39. Davail S, Feller G, Narinx E, Gerday C. Cold adaptation of proteins. Purification, characterisation and sequence of the heat-labile subtilisin from the antartic psychrophile *Bacillus* TA41. J Biol Chem 1994; 269:17448-17453.
40. Morita RY. Psychrophilic bacteria. Bacteriol Rev 1975; 39:144-167.

41. Morita RY. Low-temperature environments. In: Lederberg J, ed. Encyclopedia of Microbiology, vol 2. New York: Academic Press, 1992:625-637.
42. Russell NJ. Cold adaptation in microorganisms. Phil Trans Royal Soc London 1990; B326:535-553.
43. Russell NJ. Physiology and molecular biology of psychrophilic microorganisms. In: Herbert RA, Sharp RJ, eds. Molecular Biology and Biotechnology of Extremophiles. New York: Chapman and Hall, 1992:203-224.
44. Legrain C, Stalon V. Ornithine carbamoyltransferase from *Escherichia coli* W. Purification, structure and steady-state kinetic analysis. Eur J Biochem 1976; 63:289-301.
45. Legrain C, Villeret V, Roovers M, Gigot D, Dideberg O, Piérard A, Glansdorff N. Biochemical characterisation of ornithine carbamoyltransferase from *Pyrococcus furiosus*. Eur J Biochem 1997; 247:1046-1055.
46. Merkler DJ, Farington GK, Wedler FC. Protein stability, correlation between calculated macroscopic parameters and growth temperatures for closely related thermophiles. Int J Peptide Res 1981; 18:430-442.
47. Mrabet NT, Van den Broeck A, Van den Brande I, Stanssens P, Laroche Y, Lambeir AM, Matthijssens G, Jenkins J, Chiadmi M, van Tilbeurgh H, Rey F, Janin J, Quax WJ, Lasters I, De Maeyer M, Wodak, SJ. Arginine residues as stabilizing elements in proteins. Biochemistry 1992; 31:2239-2253.
48. Sanchez R, Baetens M, Van de Casteele M, Roovers M, Legrain C, Glansdorff N. Ornithine carbamoyltransferase from the extreme thermophile *Thermus thermophilus*: analysis of the gene and characterization of the protein. Eur J Biochem 1997; 248:466-474.
49. Wild JR, Wales ME. Molecular evolution and genetic engineering of protein domains involving aspartate transcarbamylase. Ann Rev Microbiol 1990; 44:193-218.
50. Roof RD, Foltermann KF, Wild JR. The organization and regulation of the *pyrBI* operon in *E. coli* includes a rho-independent attenuator sequence. Mol Gen Genet 1982; 187:391-400.
51. Liu C, Turnbough CL Jr. Multiple control mechanisms for pyrimidine-mediated regulation of *pyrBI* operon expression in *Escherichia coli* K12. J Bacteriol 1989; 171:3337-3342.
52. Kenny MJ, Mc Phail D, Shepherdson M. A reappraisal of the diversity and class distribution of aspartate transcarbamoylases in Gram-negative bacteria. Microbiology 1996; 142:1873-1879.
53. Kerbiriou DC, Hervé G. An aspartate transcarbamylase lacking catalytic subunit interactions. I. Disconnection of homotropic and heterotropic interactions under the influence of 2-thiouracil. J Mol Biol 1972; 64:379-382.
54. Purcarea C, Hervé G, Ladjimi MM, Cunin R. Aspartate transcarbamylase from the deep-sea hyperthermophilic archaeon *Pyrococcus abyssi*: genetic organization, structure and expression in *Escherichia coli*. J Bacteriol 1997; 179:4143-4157.
55. Stebbins JW, Kantrowitz ER. The importance of the link between glu-204 of the catalytic chain and arg-130 of the regulatory chain for the homotropic and heterotropic properties of *Escherichia coli* aspartate transcarbamoylase. J Biol Chem 1989; 264:14860-14864.

Plasmids from antarctic bacteria

M. L. Tutino, B. Fontanella, M. A. Moretti, A. Duilio, G. Sannia and G. Marino*

Department of Organic and Biological Chemistry, University of Naples Federico II,
Via Mezzocannone 16, I-80134 Naples, Italy

1 Introduction

In recent years a growing interest has been addressed to cold-adapted microorganisms, because of their biotechnological applications.[1,2] The knowledge of the structural and functional features that justify the unusual activity of psychrophilic enzymes at low temperature represents the main goal of many comparative analyses;[3–6] indeed, this information is instrumental toward an exploitation of their biotechnological potential as suitable catalysts in future applications, like those in the dairy industry,[7,8] in "cold washing" or in any situation where high activity at low temperature and an easy inactivation of added catalysts by moderate heating are required.

However, this "finalized" approach to the problem of life at low temperatures resulted in a dramatic lack of information about genetic and molecular biology of cold-loving organisms. Although a small number of genes from psychrophilic organisms have been isolated and characterized in the past, nothing is virtually known about bacterial genetic elements, like plasmids or viruses.

Recently, two papers have been published on conjugative plasmid distribution in marine bacterial communities, and it is evident that the genetic diversity observed cannot be interpreted using the presently available information on mesophile-derived replicons.[9,10] These results strengthen the idea that cold environments are colonized by a not yet fully explored biodiversity, in which the strategies for DNA replication and genetic information flow are quite different from those observed in temperate organisms.

Therefore, the interest in the isolation and molecular characterization of plasmids from cold-adapted bacteria could fulfill several different scientific requirements:

- it could contribute to fill the gap in knowledge of basic molecular processes and their adaptation to low temperatures, providing easy model systems on which to

* Corresponding author; e-mail: gmarino@unina.it

study fundamental mechanisms like DNA replication, recombination and repair;
- the collected information could be finalized to set up cloning/expression systems in cold-adapted bacteria.

In particular, the construction of such genetic systems would greatly enhance the opportunities of genetic manipulation of the bacterial hosts, with the aim of clarifying the physiologic processes that allow them to thrive under such critical thermal conditions. Furthermore, an expression system in cold-adapted bacteria could represent the solution to the problems sometimes occurring during the heterologous production of psychrophilic proteins in mesophilic hosts, due to the limited stability of cold products at the expression temperatures.[11,12] This chapter reports some results from our laboratory on the molecular characterization of plasmids from antarctic bacteria.

2 Plasmid occurrence in antarctic bacteria

2.1 Screening of terrestrial isolates

Some terrestrial environments in Antarctica are among the most hostile permanently cold habitats in the world. However, complex microbial communities are able to thrive under these conditions, made harsh by the dramatic scarcity of humidity and nutrients. The evolution of peculiar physiological processes or the possibility of utilizing alternative nutrients via new metabolic pathways could have represented the successful strategy in the adaptation to these habitats.[1] Because plasmids represent one of the most important means by which bacteria exchange genetic materials and information, the distribution of these genetic elements could give an idea for spreading and natural selection of advantageous phenotypes in bacterial communities.

Table 1. Geographical information on the sampling sites from which the terrestrial bacterial strains analyzed were collected

Location	Grid ref.	Altitude (m)	Comment
Mount McGee	74°03'S 164°33'E	1410	Exposed fine scree at summit
Kay Island	74°04'S 165°19'E	about 80	Opening of snow petrel nesting site
Harrow Peaks	74°06'S 164°51'E	about 200	Exposed fine scoria/soil
Crater Circe	72°38'S 169°22'E	500	Raised beach/pond line soil

Table 2. Plasmid distribution in antarctic bacterial communities

Location	Analyzed strains	Plasmid-containing strains	Comment
Mount Mc Gee	12	10	Two plasmid subfamilies
Kay Island	10	–	No plasmid
Harrow Peaks	9	7	Several plasmids per strain, great plasmid complexity
Crater Circe	8	4	One type of high molecular weight plasmid
Total	39	21	

Thirty nine bacterial strains collected in Antarctica by N. Russell and co-workers (University of London, Wye College, UK) were analyzed for the presence of extrachromosomal elements. The antarctic isolates came from four different locations, whose geographical information is summarized in Table 1. Looking at the distribution of extrachromosomal elements (Table 2), they appear to be widespread in three of the four bacterial communities analyzed. Indeed, none of the ten strains collected from the opening of snow petrel nesting sites at Kay Island contains genetic elements, whereas the seven plasmid-containing isolates coming from exposed fine scoria/soil at Harrow Peaks exhibit a great genetic complexity. In general, they contain at least two plasmids, whose molecular weights range between a few kb to 20 kb or even more. A much simpler situation has been registered in the Mount McGee community, where it was possible to identify two bacterial subfamilies, each characterized by the same genetic element and with no more than two extrachromosomal elements for each bacterial strain. As far as the bacteria collected in the raised beach/pond line soil at Crater Circe are concerned, four out of eight strains analyzed contain a high molecular weight element, which characterizes this community.

In conclusion, the prevalence of high molecular weight genetic elements (>20 kb) can be observed in three of the four terrestrial bacterial communities analyzed. Due to their dimensions, these plasmids could represent automobilizing elements. However, their wide distribution confirms the existence of intense genetic exchanges between the bacteria living in the same habitat, and suggests the possibility that these large plasmids contain some advantageous genetic markers for surviving in the harsh polar environments.

2.2
Screening of marine isolates

The distribution of extrachromosomal elements in antarctic bacteria collected from the seawater was investigated in some strains kindly provided by C. Gerday (University of Liège, Belgium) and isolated in the surroundings of the French Antarctic Station Dumont d'Urville in Terre Adélie (66°40'S, 140°01'E). Four iso-

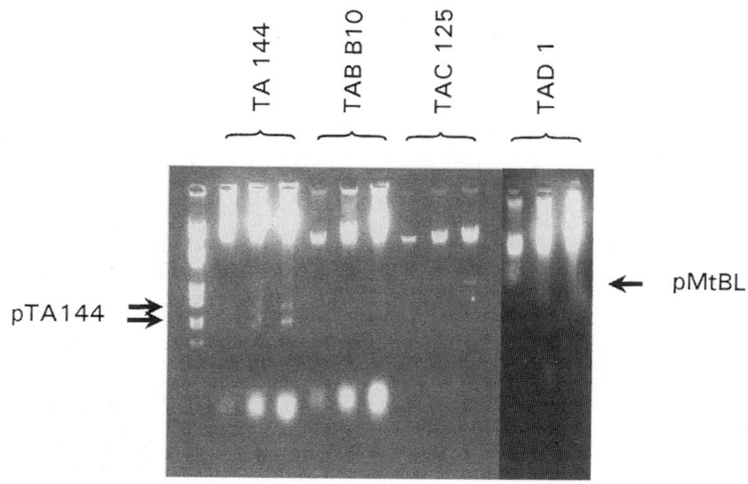

Fig. 1. Genomic analysis of four marine bacterial strains from antarctic seawater
Three different amounts of total DNA from four antarctic strains were analyzed by agarose gel electrophoresis. *Arrows* indicate that low molecular weight genetic elements are present in TA144 (pTA144 Up and Dw) and TAC125 (pMtBL)

lates (TA144, TAB B10, TAC125 and TAD1) were subjected to a genomic analysis and the results are presented in Figure 1. Two of them, TA144 and TAC125, contain small plasmids, whereas no extrachromosomal elements seem to be present in the other two.

The plasmids are in a different copy number inside their respective host; in fact, when the fluorescence intensity of plasmid signals is compared to that corresponding to the genomic DNA, it turns out that pMtBL plasmid, harbored by the TAC125 strain, is a medium-copy number element (about 20 copies cell^{-1}), whereas pTA144 plasmids from TA144 are less represented.

Due to the usefulness of a higher copy number, pMtBL was the first "cold" genetic element characterized to any extent; however, the preliminary characterization of the pTA144 plasmid system opened the way to investigate the cold adaptation of a different DNA replication mechanism, the rolling circle.

3
The TAC125 strain and pMtBL plasmid

Before starting the molecular characterization of pMtBL plasmid, some aspects of TAC125 biology were characterized. It is a Gram-negative pink-pigmented bacterium, tentatively classified as a *Moraxella* species (G. Feller, pers. communication). The dependence of TAC125 growth from incubation temperature was analyzed. Although this isolate comes from a permanently cold environment, characterized by an average temperature of -1.8°C,[1] its growth was checked in a wide temperature range, 4–25°C, and the lowest observed doubling time was at 20°C (31 min). The above results suggest that although this antarctic isolate was originally selected during a screening for true psychrophiles, it can be properly defined as a psychrotrophic rather than psychrophilic strain, according to Morita and Russell's definitions,[1,13] since it thrives either at very low temperature or under quite "mesophilic" conditions (optimum temperature above 15°C). It is worth noting that at 4°C the upper stationary phase is reached in about 20 h of incubation, whereas at 20 or 25°C liquid cultures require about 5 and 9 h of growth, respectively, to reach a higher stationary phase. This behavior is due essentially to the different length of the first lag phase, because the doubling time values determined in the exponential phase are comparable (unpubl. data).

Purification of pMtBL was achieved by means of classical methods for plasmid isolation,[14] while restriction analysis revealed the presence of several unique restriction sites that proved to be extremely useful in the following plasmid cloning (unpubl. data). The complete nucleotide sequence was determined and is shown in Figure 2. Searching for evidence about a possible pMtBL nature and function, its 4082-bp-long nucleotide sequence was compared to the entries contained in the GenBank/EMBL database, but no significative results were obtained. None of the sequences stored in the database appeared to be related to the TAC125 plasmid. Moreover, none of all the putative open reading frames possibly coded by pMtBL showed an appreciable identity to the known protein sequences stored in the SWISS PROT databank. Taken together, the above results straightened the idea that pMtBL represents a novel autonomous replicating element.

To collect some hints about the molecular mechanism of plasmid duplication, an exponentially growing TAC125 culture was treated with chloramphenicol and the effect of this addition on pMtBL copy number was analyzed (Fig. 3). In fact, some *Escherichia coli* plasmids, like ColE1 and pMB1, are specifically amplified by treatment with several drugs or antibiotics that inhibit protein synthesis.[15] The plasmid amplification, resulting in an increase of copy number of about 100–150 times, has been related to the peculiar replication control mechanism. The formation of an RNA molecule to be used as primer for the enzyme DNA polymerase I is the first step in duplication of the ColE1 plasmid family. Negative regulation is obtained by interaction between the precursor of the primer and RNA I, a plasmid-encoded small RNA. The RNA-RNA interaction between the two regulatory molecules is modulated by the Rop protein.[16] Inhibition of protein synthesis results in the absence of Rop molecules and in continuous plasmid replication, because it relies entirely on long-lived enzymes supplied by the host.[17]

As shown in Figure 3, a chloramphenicol treatment induces the enhancement of pMtBL copy number of approximately five times. Although this result could be interpreted under the aspect of a negative control of pMtBL copy number by a regulative protein, it demonstrates that the plasmid replication is insensible to the ongoing protein synthesis because it does not require any plasmid-encoded function. Furthermore, it is unlikely that the regulator factor, if it exists, is encoded by the plasmid itself, because transcriptional analyses demonstrated the absence of pMtBL transcriptional activity (unpubl. data).

```
TCTAGATAGG ATTTTTAAAA GCCTATCTGT ACCACTCTAA GGGGAAGTTA ATTCTTCCCC   60
TTTTTCCAAA TTATTGAGGG TTTTGATGCT GATAGGATAT CGGAGTTCAT TTATGCAGCG  120
GATTCAGGAA GCTGCGAATC ATGGGTATAC CAGTGTAGTT ACAGGTACCG TTAAAATTGA  180
AAAATACGAT AATTTATATC GTAAATTTAA TGCGATGTAC GATGTAGATA TGCAGAAATC  240
AAAACGCAGC CGGCAAAAAG CCCAGGGCGT TAGTTTAGCT AAATTTTTAG CGCATCCAGA  300
AGAGGGGGAT ATGGTTCGCT GGGTTTTAAT GTGCACCAGT GGTACCGGAA CATTTTTTGA  360
GGAAGAGAAG CGCGCCATTA TACCTGCAAA AAAAATAGAG CTGTTTGGCC AGTACCGTTT  420
AACCAGAGAG ACAACAAAAA GACACGGAAA AGCCGTTTGG ACGTTTTATT TAACAACCGA  480
ATATAAAAAC ACAGTTAAGA AGCTCGTTCA TAGTTACGTT CATCACAAAA ATGACGCTGG  540
ACTGAGAAAA ATCATTCACG AAACAAACCA GCTAATCCCA TTTTCAGGAG TTCGCCAGGA  600
CCTGTATGAG ATACATAAAC ATATTCGACA TGAGTGGCAT CGCTATCGAG GAAAAGCCGA  660
CATGCCGGAA TTGCCATTTA TACGTTACCA CCGCCGCCTA GCGACCTCTC CAGTTCATGC  720
TGATAACGTA ATACGTAGAA TGCAGCGAAC AGGTGAAACA GCCCTAGAAG CCGCTAGAAA  780
CATAGCACAA AACAGAAAGG TTAAACCTAA ACCAGCCTGA TAATATTGGT AATTTCGGTA  840
ATACTGAAAT TACCAATATT ATCAATTTTA CCGAAATTAC CGAAATTATT TTTATTCTGC  900
TACACTCAGA TTTACCTAAA ATAATTCAGT GTCTAATTTG ATGAAAATAT CAATTTACAG  960
CCCCAAAGGG GGATGCGGGA AAACACCACT GGCCATAAAC TTTGCCGTTG ATAAAGGTTA 1020
TGCCCTGGCG ACTAACGAAA GCTTTGATGC ATTCTCAGGT TTTTTTGATG ATGATCTTTT 1080
ATTGTTTGTG GCTGCAGATG AGGAATTTCC AGAAATCCCC AGTGATATCG ATGTTGTTTT 1140
CGATTTGGCG GGTACCATGT CTAAAAATGA CTACTCAACG CTCTCAGCTA TACAGCAGAG 1200
CGATGTTGTA ATTGTGCCTA TATGGAATAA TAGATTGTCT ATTTTAGGCG GCCTGAGAGC 1260
CATACAGAGT GTTCATGCGA TAGGTAGGCC TGTGATAGTT GTAGCAACTA AATTGACTGC 1320
AAAAACAAAA AAGAAAGGCC AGAATTGGAT TGAGTCAGAA GATTATCAGT GCGTTTGTAA 1380
TGCTGTTTCA TCTTTAAAAG ATGAGCTGGG CTATATGCCG GCTGTATTGC CATTGAAATA 1440
CTCTGAGGTA TTTGAAACCG TGCTAGGTGA ATGCGTCAGT GTTGATGTTG TGGCCAGTAA 1500
GTCTGGTTTG GCTAAATACA CACATAGAGA GCTGTTAGGG CAATTTAAGG ATTTGTATCA 1560
GGAGGTTGAA AAATATGGCT AAAAACTCAT TAGCGGCAAT TAAAAACGCG GATAACAGTA 1620
AAAATAGCCT ACCGTTAACG CCTAAATCTA AAGCCGGTCG TAAAGCTAAA CCAGCATCAG 1680
AGAAAGTAAG CCGGCCGGTA ACTCTAAAAT TTACAGAGAG TGAGTTAGAG TTAATAGAAA 1740
AGCGGAGTGG TTTAGTGCCG AACGCAACAT ACCTAAAACA ATTATTGATA GAAAACACGG 1800
ATTTATTTGG TAGTTAAAGT GCCCTAAAAG GCACTTTAAC TATGTTTTTG GCCCTTAAAG 1860
TGCCCAAAAG GTAGTGTTAA TAAAGGTGAC ATATGGAAAA TTTTATATTT GTAATACCAG 1920
CTATAGTTTT AACTGTAAGA CTCGGTTATT TGCTCTGGAA AACAAGAAAC AAATAAAGAA 1980
TGGGTGTTAT TTAGGCAATA GAAGTCTTTT TGATAGGGGG GTAACCCGTA TTACGGCATT 2040
ACCCTAGCTA TCATTAATGT TTGGCGTTGT TAAAAGAAAA ATCTCATTGT CAGGTGTTAT 2100
TTCTGTGCCG GATTAAGGGG GGCATAGACA CGCAGCTAAC CAAACTGTCC GACCTGTGGA 2160
GCATCAAACC AATAGATAAA AATCTATTCA CTCAGGCGTT TTTTAATGTT ACTCAAACCT 2220
ATTTTAAAAC CGTCCCATCT AACGGTGTAT TTTTCTTTTA GTTGTTTTTT ACTTTTTAAT 2280
TTCCAATGGA GAGATTAAGC CCTGCTTTTT TTTAATTGGT TGCTGGTTGG TTTGCCCGTA 2340
ATCCAATATC AATAGAAACA ACAAAAAATG TTTAAATACT TGCTTAGGTG AGTTCTTGAA 2400
ATATACTGAT TACGTGGTAG AATTGCAGAC GTATTGATGT ACAGTGTATT TTCAGTGATC 2460
TTGAACTCAT CTGAAAGTAT TCTTCAAAAA AGCCGCAACG CCAATTGCGG CTTTTTTAAT 2520
TCCTGCAAGC ATACTGTGAT CTCTAATTCT GATCAATGAT TGTGTTTATT TTTCTAGATC 2580
GAAAATTGCT AAAATTGCTA AATTCACAAA ACACTCTAAC TTCTTCGAAA ACAAACCTAA 2640
AATAATGGCA TATTATGCCA TCTTGCTCCC ATCTTGCTCC CATCTTGCTT TTAACCAGCT 2700
ACTCTTTCAG TGTTTCTATC AAAATCAACG TATAGCCGAT AACGGTTTTT ATGTTAAATA 2760
AGCTTGCCTG TGCATATCAC CTCTAAAAAC AATCTATAAA CTAACCTTTT AAGCATTACT 2820
TTGTTAAATA GACATAGAAG TGGCAATAAC TGCAATTAAG CTTATTTATC TATTCTAAAC 2880
GAAATGCAAA AAAAAAAGAA AATTCAGAAA AATGTGCAAC TTTGTATATA AAAAACACAG 2940
TGACGAGGTG AAACGCAATA ATAGGTAAGG TTTCTTCGAA ACCTTATATA AAAGATTCCT 3000
GTCAAATGCC CCCCCGATGT CCTTACTTAC TTTTGTCAGA TAGACAAACA CGTCATCTGC 3060
GAAGTTTTTA TCAGCACGTC AATGCTGATT GTTCACGATG AATTTCGCCA AAAATTCATT 3120
AGCTGCCGAA GCGATATGCA CGTAATTGTG CTCGTATCGC TACGACGGGC GCTAGAGGTT 3180
CACCCCTCTT TCGACCCAAT CCGATGACTG GCCAAGGTGA CTTGGCGAGT AATCAAACTT 3240
GAGAACGGGC GACAGGGACA CGGGAGCTGT GGTGACACAG TAAACCCTGT TTTATTTGAA 3300
CCAGTGTAAA TCTGGGTTAC GGGATGCTAG AAATGGTATG ACGTAAACCT CACACGTGGG 3360
TGTCAGCAGG TGGAGAAATC CATCTGAGCT ATCAGGGAGA AATCTGGCGA TAGCGGTACT 3420
GTCTGGGGCA ACCTGGGCAG ATCGGCGAAG AGCTGAAGCG TGTAACGGCA GCCAACTGGG 3480
CTGCTGTGGG GTAAGTTTGT TAAAGCTAGT GCTCAACATT AGTTGAGAAG CACTATAAAC 3540
GTATGATTTA AATAGACAGT AAGCGATCCG TTCGGGATCT TGGGACGCTG AAATTCTGGT 3600
TACACGCGCA TTTAAAAATT CTTTTTCGTG TAATTGGTTG AGGCTCATAG AGCTAGATAT 3660
ATTCTATTTA ATGACGTTGC TAGCGTTAAC ATTCGACAAA TAAAAATAAC CTGGGAAGCA 3720
CTAAGGGCTG TACTGTTTGG CAACAGTACA TTCGGTCTAT GTGTGGGAGA TTGGTTTAGT 3780
AGATCCGATT TAGTTCGGAG CATAACAGTG ATATAGGCGA AGCTGTTATG GTCAATTAGA 3840
TTTTGACGAA TCTGGTTGGT ATACGAAAGA CCGAGGTAAG CAAAAGCCCC CACTTTCGTG 3900
GGGGCCTGTG AACATTTAAA CTTCAATGAT TTATTCGAAA GGCATTGCTT GGCGGTAATG 3960
TTTAGTAGAA ACGTATGATG GAAATTTTGA CAATAAGGAC AAAATCTTAG CACTGCAATA 4020
ATTAGGTTAG AGGGATGAAA CCCACCAGAA AATACAGTCC TCTACAGAGG TCATCGTATT 4080
TT                                                                4082
```

Fig. 2. pMtBL nucleotide sequence

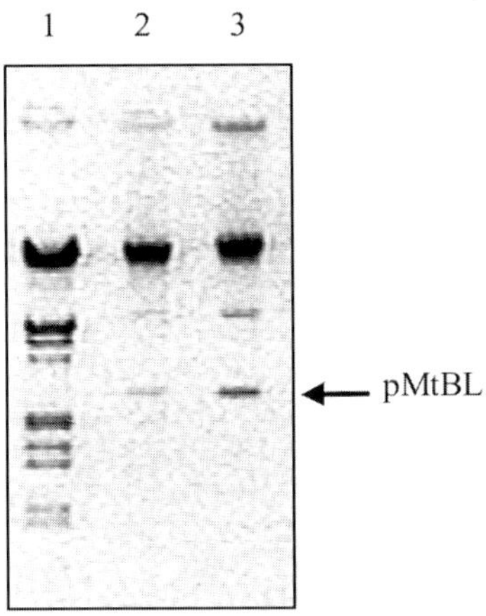

Fig. 3. The effect of chloramphenicol treatment on pMtBL copy number
Total TAC125 DNA was extracted from saturated cultures, previously treated with (*lane 3*) or without (*lane 2*) chloramphenicol (final concentration 80 μg ml^{-1}), and analyzed by agarose gel electrophoresis; the effect of pMtBL amplification is evident when the intensity of the plasmid signal is compared to the genomic DNA loaded

Indirect evidence of the efficiency of pMtBL replication process has been provided by some curing trials; TAC125 cultures were treated with several chemical and physical agents (acridine orange dye, coumermicin A1, novobiocin and electroporation) to induce the loss of pMtBL. None of the tested curing protocols leads to the isolation of TAC125 colonies without the plasmid.

4
Moraxella TA144 and pTA144 plasmids

Moraxella TA144 is one of the most studied antarctic isolates from the Gerday collection, because three lipolytic enzymes have been characterized from this strain, at both biochemical and genetic levels.[11,18–20] Although it is able to grow in a wide temperature range, 0–25°C, the secretion of exoenzymes into culture medium is maximal at a temperature close to that of its environment (-2 to 4°C) and is strongly inhibited at higher temperatures.[18] This result has been interpreted as proof of the real psychrophilicity of TA144, despite the thermal behavior exhibited by its growth.

The genomic analysis presented in Figure 1 revealed that this strain contains two small plasmids, called pTA144 Up and pTA144 Dw, respectively. A restriction analysis on a plasmid preparation from TA144 cells was carried out and it turned out that the two plasmids could be selectively linearized by one restriction endonuclease, *Eco*RV for pTA144 Up and *Spe*I for pTA144 Dw, leaving the other one in supercoiled topology (unpubl. data). These results have been instrumental in the

A ATCAACGAATCTTGGCTACTTGCTAATAGGTTTGAGGTTTTGGCTTGGATACGCTCCGCA 60
I N E S W L L A N R F E V L A W I R S A
Rep C-term
GAACAAAAAATAAGGCTTAAACATCGTGGTACTGGTCAGTTTAAAGGTTCAACGCTTTAT 120
E Q K I R L K H R G T G Q F K G S T L Y

TGGGGTAAAGGTTCTACTCGTTGGTTCTTAAAGTGTTACTCCAAAGGAGAAGAAATTAAC 180
W G K G S T R W F L K C Y S K G E E I N

CGTAAAAATAGTGGTTTCCCTGACTCTTTACGCACCCCTGACATGCTTTCATATGCTGAT 240
R K N S G F P D S L R T P D M L S Y A D

AGAGCCTTACGCTTAGAGCTTACAATGAAGTCTAATGCCCTCAGACAAGAGCAATTACAC 300
R A L R L E L T M K S N A L R Q E Q L H

ATACCCGCCAATTGGAACGCAGAAACTGCTAAAATGTTACTCTTAAACTCTATTCAAGGT 360
I P A N W N A E T A K M L L L N S I Q G

TTAGAGATGAGTAATAATTTTTCGCTTACCAATGATGTTTTGGCTACCTTGCCCTCTCGT 420
L E M S N N F S L T N D V L A T L P S R

GTCAGGCTTGCTTATGCTGCTTGGTATAATGGTGATGATTTAAGACAAATATTGTCGCGT 480
V R L A Y A A W Y N G D D L R Q I L S R

CCTACCTTCTACCGCTACCGCAAGCATTTGAAAGAGTATGACATAGACATTAGTATCCTT 540
P T F Y R Y R K H L K E Y D I D I S I L

AGAGACGTAGATAGAGAACCTAGCAACGTTATACCCCTTATTCGTGTTCTTGAAGCAGTA 600
R D V D R E P S N V I P L I R V L E A V

CCTGCCGGAATACCTGCTTGGGCATACGAGCAAGGATTAGTTGCTTGCTGATAAGATACA 660
P A G I P A W A Y E Q G L V A C

TGAGTATGAAATGACCTCGTTTTACACGGGGCTTTTTATTGCCTGTAATTCCTTCACCTC 720
CAGTTAACAGCTTTTTGCTGATAATGTCGTCATAGTTTTCTGTGCCCGTAACTGACGGAC 780
TATGACGCTGCCGTGAGAACCTATAAGAGCGAGTTTTCGACTTGGGAACAAATCCAAGTC 840
CCACTCTAATATTTGCGCGCGTGGTGGCTTGCGCTCCCCCGACCGCCAACGCACACGCGC 900
GCAAATAAGGGAGAGCTTAGTAATACCTCCCTTTTGTCTCAATTTGATAACACCTATACA 960
CAAACACCACTGCCAGACCTAACGAAATGAACCCTTTGGGGGTTATTCCGTACCCATCCC 1020
TTTTGTCTCACTTATTGCTTGACTTCACGCATTTATGGAGTTACATTGTATTTTCAAACG 1080
TTACTCCAAAGGAATATAACATGGCTACTCACAGTATAAATCTTCAACTTACCACCGAAC 1140
AATTAGTGCATTTAAATATGGCAATGAACGATAGTTATGACTGTTATTCTGATGCATTAG 1200
ACAGCATTGGTTTTGAAACTGAACGTCCGTCCCTCTTAGCCCAATGTAATGCTATTCAAG 1260
TGTTTATGGATACCATTAAACATAAATTATCAGAGATAAACCATGCCTAGTAAAAAAACC 1320
CGAATTAATCTAACCGTCGATGAAGATATGGATAATCTTTTAATTGAGCTTGCTGCGCTT 1380
CAAGGTATCCCTAAGACAACCCTAGTCTATGAATATTTAGAAGCAATGCGCCCTCACATG 1440
ATAGAGATTAGAGACGCTATTCGCATGGTCAACGAGAAAAAAGACCCCTCGCCACACCTC 1500
CGCAAAATGCTCCTTGATGGCCAACAAGACTTTCTTGATGTTCTCAAAG**AGGA**ATTTAAA 1560

AATGATTGATTGGGTTACGATGAAAATAGCTTGCGACCACGTTGGTATCATATCTAATGG 1620
M I D W V T M K I A C D H V G I I S N G
Rep N-term
TTCAGTCGTCAAAGTCTCTAAAGATAACGAAATAGAATGGACATCGTTGTCATGGCTGCA 1680
S V V K V S K D N E I E W T S L S W L Q

AGTTGCTGGTTCACATGATTCAAATATTGTAATTCGATCGCTTTCGGATAGTCAGTTAGA 1740
V A G S H D S N I V I R S L S D S Q L E

AATATCTGGTAATCCTGCTAAATGGCTTCAAGGTCATAATCTTTTTGGAACAAACGATTT 1800
I S G N P A K W L Q G H N L F G T N D L

AAACGCCCTCTGTTGGTTGTTTTTTTCAAAATTAACCAGTATCAAGGAATTATGCTTAAC 1860
N A L C W L F F S K L T S I K E L C L T

GCCCTCTGTAGAACAGATTAGGCTTATTAAAGATGGTCTATTCACGGTATCAAGAGTTGAT 1921
P S V E Q I R L I K D G L F T V S R V D

```
B    1 AGTGGGGGAATTTGCCCAATTTCGCTTTTTTCCTGAATTTGTCCTACTTCCTAACCCTTA
    61 CTATACATACGTTTGCAACGTTCCGCAAGAGCTAACGCGCTTCGCTTGTTGGGTCAATCA
   121 TTATTTTGCAATTCAAAATAGCCTAAGGGCTATTTGTGAATCACAAAAGTAATGATTCGA
   181 CCATCTGTTAATCGACTGATTTGTTTTAATTCTTTTGATTGCCCAGAGGGAAGGGAGTTG
   241 AGGAAAGGGCAATACTCACCCTAAGAGCTTGTAAGTACAAGCCCTTAGCGTTGCCCTCAA
   301 TGCTTGCCCGTCGGTACGGTCGCATCGTCAAGGGAGGGTATCACTGGTAGCAGCTACCCC
   361 TGCATAGATTACGGTACTACACACCGCCACAAGACTGTCTATGCCTTCAAGGCGCTGTTA
   421 TGCTAGCAAGGGCTTACAACCCAATATACCGCAACACAGACACACTCAAACACTCATTCA
   481 GCCGCCATCTTTCTTGTGTAACCCCAGTAGATGGCTTATAATCACGTTATGCAGGGCTGG
   541 CGTCTCTAATTGCCTTCCACCCCCTGCACCCGCCCCCAGTGAGTTATCGCTCATGTGGGG
   601 GTTTTTTAATGCTTGAAACTTGAGCAAGATTAGCACATGTGCACACACAGCACAATAGCC
   661 AAGTGCCCACCGTGCAAGCAAACGGGGTTTGCCCTTGCATCGGTTGGGCACTTCTGCTTT
   721 TTCTAAGTAATTTAGATCAGATTCTCTGTCATAGTTTATCGCAGGCGAAAAAATAGCCCT
   781 CATAGGCTTTTAACGGCTCATGAAGGCTATTCCCAAGGAGTTGTTATGCATATCTAGCTT
   841 CGAACTGTCAAAACGGCTTAAATCGCATTCTAGCGACTTTCAGCATCAAACATGTCACTG
   901 GTATTATCACATTTGCTAACTTTTTCTGCTTGACGGCTATCGGTTCTCACTCTTGTATCA
   961 CTATCAGCGATTTTACCGCGCCAATTGTCTTGTAGAGCCTTCTGCAACTCGCTCACCATG
  1021 AACGCGCGTATCTCATCCGATGCTACCACGCCCCTAGCCTTGCACAGCCCTAGGTAAGCC
  1081 CCTTTGAGTTCTTCCGTAGTCCGTACTGTCAGCCTCGTAGTGTAGTAATTGTCCAAACCT
  1141 CCACCCCTTAACCTAAGTGCGCACTTACGACTTGTCCCAAATGGGTACAATTCACAATGC
  1201 TAAAATATTATTTCTTAGCCCACATATGCACATAGACTAATTTTACTTCACAACATTAAT
  1261 CTTTACCCGCGCTATGGCGACCCTCTGTCGGGGTACTTTACCCCCGTATTACT
```

Fig. 4. Nucleotide sequences of TA 144 plasmids
(A) 1921 bp nucleotide sequence of pTA144 Up as determined after linearization by *Eco*RV, the translation of the putative Rep-encoding open reading frame is also shown, a putative ribosome binding site sequence (AGGA, *bold*) is located 9 bp upstream from the *rep* start codon; **(B)** Nucleotide sequence of pTA144 Dw (1313 bp long), the sequence starts with a half *Spe*I site used for the cloning strategy

following cloning of the genetic elements, whose nucleotide sequences are shown in Figure 4.

As was observed during the screening of terrestrial antarctic isolates, the presence of two or more plasmids in the same cell is not unusual. However, this result implies that their replication mechanisms are different, otherwise incompatibility problems could induce the loss of one or more of them during the following duplication.[21] Consequently, the absence of any relation between the two plasmids at the nucleotide sequence level does not surprise.

```
                 10        20        30        40        50
Rep TA   MIDWVTMKIACDHVGIISNGSVVKVSKDNEIEWTSLSWLQVAGSHDSNIVIRSL-SD---
         ::::::  . : :: . . : :..:. :. .:: :... .:.:: .:.: .::  ::
Rep I2-2 MIDWVTAVLPCLHVPV-DAGRVLSVAPDGSVEWESVKFSRVTGSFQSSISVRSQGSDGNG
                 10         20        30        40        50

             60        70        80        90       100       110
Rep TA   --SQLEISGNPAKWLQGHNLFGTNDLNALCWLFFSKLTSIKELCLTPSVEQIRLIKDGLF
           ..: ..:::.:::::::. :..:::.:   :.... :.  : .   .:. : . .: .
Rep I2-2 KATHLYVDGNPSKWLQGHNIVGSDDLNGLMIAFYARMLSL--LNIPHHLESYRQVLSGQY
        60        70        80        90         100       110

            120       130       140       150       160       170
Rep TA   TVSRVDINESWLLANRFEVLAWIRSAEQKIRLKHRGTGQFKGSTLYWGKGSTRWFLKCYS
          ..:::::  . : .  ..:.:...:: : . .:   .  :: :::.::.: :: .: ::
Rep I2-2 ELKRVDINYMFELPTLIDVRSWLHAAEFKAKTRHGRPATAKG-TLYFGKNSRRWSIKAYS
         120       130       140       150        160       170

            180         190       200       210       220       230
Rep TA   KGEEIN--RKNSGFPDSLRTPDMLSYADRALRLELTMKSNALRQEQLHIPANWNAETAKM
         : .:.:  .:  : :. ..   .. ..   ::::::...  : . .:..  ::..:::
Rep I2-2 KYDEVNCGKKAHGVPEEIKKTGLVEWSKNKLRLELTLRALQLTDINLNFAKNWSTETAYT
          180       190       200       210       220       230

              240       250       260       270       280       290
Rep TA   LLLNSIQGLEMSNNFSLTNDVLATLPSRVRLAYAAWYNGDDLRQILSRPTFYRYRKHLKE
         .. . .  .:::.:  ::.  . .::: .:..:. : .:  . ...:: :..:.:: :::
Rep I2-2 VFKEYMGRIEMSGNTLLTDTQVINLPSALRMTYVCWKQGICVTDMVSRATYFRHRKVLKE
          240       250       260       270       280       290

              300        310       320       330
Rep TA   YDIDISILRDVDR-EPSNVIPLIRVLEAVPAGIPAWAYEQGLVAC
         . :::..   ::: . :::.:::::::: ::.::.
Rep I2-2 FGIDIAVT--VDRVDNSNVVPLIRVLEAKPASIPSQYDNLIFSSNRVSSF
          300         310       320       330       340
```

Fig. 5. Amino acid alignment of the putative Rep protein from pTA144 Up and the homologous product from I2-2 bacteriophage
A *single dot* represents conservative amino acid substitutions; *double dots* indicate identical residues; the overall identity degree between the two proteins is 40%

A search for open reading frames (ORFs) potentially encoded by both plasmid strands suggested that the Dw element (1313 bp long) contained small ORFs, but none of them showed significative identity to the protein sequences stored in the SWISS PROT data bank. As far as pTA144 Up (1921 bp long) is concerned, a 1008-bp-long ORF was identified, whose potentially encoded product showed high identity degree with the family of Rep proteins from filamentous bacteriophages. In particular, the highest identity score (about 40%) was recorded when the putative cold-adapted Rep was aligned with the I2-2 phage protein (Fig. 5).

The identification of such a putative encoded protein suggests that the replication of pTA144 Up plasmid might occur through the well-characterized rolling circle mechanism.[22] Rep proteins, indeed, exhibit an origin-specific nicking-closing activity, involved in both initiation and termination steps in the rolling circle process.[23] The plasmid-encoded Rep protein nicks the leading strand origin to generate a free 3' OH end and becomes covalently attached to the 5' phosphate of the nick through a phosphotyrosine linkage. A host-encoded helicase displaces the strand and the 3' OH end serves as primer for the extension synthesis by DNA polymerase III. After the leading strand is completely synthesized, the Rep protein recognizes a subregion of the origin as the termination signal and carries out a second nicking event, with the concomitant ligation of the released single-stranded DNA corresponding to the leading strand. This single-stranded product is finally converted to a double strand form utilizing host-encoded factors.

The high identity degree recorded between the putative pTA144 Up Rep and the homologous protein from the I2-2 bacteriophage suggests the possibility that the psychrophilic plasmid could represent a genetic relict, derived from an abortive recombination of the original single-stranded phage genome. This event leaded to the deletion of all viral functions, producing a small auto-replicating element that contains only the information required for its maintenance, i.e. the so-called plus and minus strand origins besides the Rep protein. The characterization of pTA144 Up replication will allow us to investigate the adaptation of the rolling circle mechanism to the low temperature conditions.

5
Conclusions and future perspectives

The plasmids pMtBL and pTA144 Up represent two attractive model systems for the study of DNA replication mechanisms in cold-adapted bacteria. However, their molecular characterization represents only the first committed step in view of their use as new genetic tools for antarctic bacteria. Indeed, the setting up of suitable cloning systems for psychrophilic prokaryotes requires the solving of two main problems: the choice of suitable cold-adapted bacteria to be used as host for the replication of the new engineered cold plasmids, and the setting up of a convenient protocol for host transformation.

As far as the first problem is concerned, the lack of a phylogenetic classification of the bacterial isolates collected in Antarctica makes it difficult to identify a recipient strain closely related to the donor strain. From this point of view, the pMtBL system appears to be more useful than the pTA144 Up one, because the predicted absence of pMtBL-encoded factors suggests the possibility that no incompatibility mechanisms should work in TAC125; as a consequence, pMtBL-derived shuttle vectors could replicate into the same strain, without the need of curing it by the wild-type plasmid. Furthermore, the TAC125 ability of growing fast in a wide temperature range makes this strain a perfect host for a cold gene-expression system in which the heterologous production of proteins could be better set, varying also the growth conditions.

In the near future, cold-adapted gene expression systems could be used as alternative systems for the large-scale production of commercially or medically useful proteins, whose yield as soluble and active molecules in mesophilic hosts is not very efficient. In fact, many heterologous polypeptides fail to fold in their native state when expressed in *E. coli*; very often they accumulate in insoluble form, as inclusion bodies, because partially folded species or incorrectly folded intermediates contain exposed hydrophobic "sticky" surfaces which promote self-association.[24,25] To overcome these problems, it has been reported that the lowering of the expression temperature can facilitate the correct folding of a difficult product.[26] Several factors could explain this temperature effect on protein folding, including a decrease in the driving force for protein aggregation and changes in the folding kinetics of the polypeptide chain.

However, there is a physiological limit to the lowering of growth temperature in *E. coli*; below 15°C also specifically selected *E. coli* strains are unable to grow. Therefore, the setting up of a cold expression system could give the possibility to test this effect when the temperature is as low as 0°C.

6 References

1. Russell NJ. Physiology and molecular biology of psychrophilic micro-organisms. In Herbert RA, Sharp RJ, eds. Molecular Biology and Biotechnology of Extremophiles. New York: Chapman & Hall, 1992:203-221.
2. Gounot AM. Bacterial life at low temperature: physiological aspects and biotechnological implications. J Appl Bacteriol 1991; 71:386-397.
3. Aittaleb M, Hubner R, Lamotte-Brasseur J, Gerday C. Cold adaptation parameters derived from c-DNA sequencing and molecular modelling of elastase from Antarctic fish *Notothenia neglecta*. Protein Engin 1997; 10:475-477.
4. Davail S, Feller G, Narinx E, Gerday C. Cold adaptation of proteins. Purification, characterization, and sequence of the heat-labile subtilisin from the antarctic psychrophile *Bacillus* TA41. J Biol Chem 1994; 26:17448-17453.
5. Feller G, Payan F, Theys F, Qian M, Haser R, Gerday C. Stability and structural analysis of alpha-amylase from the antarctic psychrophile *Alteromonas haloplanctis* A23. Eur J Biochem 1994; 222:441-447.
6. Feller G, Zekhnini Z, Lamotte-Brasseur J, Gerday C. Enzymes from cold-adapted microorganisms. The class C β-lactamase from the antarctic psychrophile *Psychrobacter immobilis* A5. Eur J Biochem 1997; 244:186-191.
7. Burgess K, Shaw M. Dairy. In: Godfrey T, Reichelt J, eds. Industrial Enzymology. The Application of Enzymes in Industry. London: Macmillan, 1983:260-283.
8. Sharp RJ, Munster MJ. Biotechnological implications for micro-organisms from extreme environments. In: Herbert RA, Cod GA, eds. Microbes in Extreme Environments. London: Academic Press, 1986:215-295.
9. Dahlberg C, Linberg C, Torsvik VL, Hermansson M. Conjugative plasmids isolated from bacteria in marine environments show various degrees of homology to each other and are not closely related to well-characterized plasmids. Appl Environ Microbiol 1997; 63:4692-4697.
10. Sobecky PA, Mincer TJ, Chang MC, Helinski DR. Plasmids isolated from marine sediment microbial communities contain replication and incompatibility regions unrelated to those of known plasmid groups. Appl Environ Microbiol 1997; 63:888-895.
11. Feller G, Thiry M, Arpigny JL, Gerday C. Cloning and expression in *Escherichia coli* of three lipase-encoding genes from the psychrotrophic antarctic strain *Moraxella* TA144. Gene 1991; 102:111-115.
12. Gerike U, Danson MJ, Russell NJ, Hough DW. Sequencing and expression of the gene encoding a cold-active citrate synthase from an Antarctic bacterium, strain DS2-3R. Eur J Biochem 1997; 248:49-57.
13. Morita RY. Psychrophilic bacteria. Bacteriol Rev 1975; 39:144-167.

14. Sambrook J, Fritsch EF, Maniatis T. Molecular Cloning: a Laboratory Manual, 2nd ed. Cold Spring Harbor NY: Cold Spring Harbor Laboratory Press, 1989.
15. Clewell D.B. Nature of Col E1 plasmid replication in *Escherichia coli* in the presence of the chloramphenicol. J Bacteriol 1972; 110:667-676.
16. Cesareni G, Banner DW. Regulation of plasmid copy number by complementary RNAs. Trends Biochem Sci 1985; 10:303-306.
17. Scott JR. Regulation of plasmid replication. Microbiol Rev 1984; 48:1-23.
18. Feller G, Narinx E, Arpigny JL, Zekhnini Z, Swings J, Gerday C. Temperature dependence of growth, enzyme secretion and activity of psychrophilic Antarctic bacteria. Appl Microbiol Biotechnol 1994; 41:477-479.
19. Feller G, Thiry M, Gerday C. Nucleotide sequence of the lipase gene *lip2* from the Antarctic psychrotrophic *Moraxella* TA144 and site-specific mutagenesis of the conserved serine and histidine residues. DNA Cell Biol 1991; 10:381-388.
20. Feller G, Thiry M, Arpigny JL, Mergeay M, Gerday C. Lipases from psychrotrophic antarctic bacteria. FEMS Microbiol Lett 1990; 66:239-244.
21. Novick RP. Plasmid incompatibility. Microbiol Rev 1987; 51:381-395.
22. Gruss A, Ehrlich D. The family of highly interrelated single-stranded deoxyribonucleic acid plasmids. Microbiol Rev 1989; 53:231-241.
23. Koepsel RR, Murray RW, Rosenblum WD, Khan SA. The replication initiator protein of plasmid pT181 has sequence-specific endonuclease and topoisomerase-like activities. Proc Natl Acad Sci USA 1985; 82:6845-6849.
24. Wetzel R. Mutations and off-pathway aggregation of proteins. Trends Biotechnol 1994; 12:193-198.
25. Georgiou G, Valax P, Ostermeier M, Horowitz PM. Folding and aggregation of TEM β-lactamase: analogies with the formation of inclusion bodies in *Escherichia coli*. Protein Sci 1994; 3:1953-1960.
26. Georgiou G, Valax P. Expression of correctly folded proteins in *Escherichia coli*. Curr Opin Biotechnol 1996; 7:190-197.

Molecular responses in cold-adapted plants

M. Hiilovaara-Teijo* and E. T. Palva

Department of Biosciences, Division of Genetics, POB 56, University of Helsinki, FIN-00014 Helsinki, Finland

1 Introduction

Plants differ in their responses and resistance to chilling and freezing temperatures because they originate from different areas and have evolved in different environments. The two mechanisms for freezing resistance are avoidance and tolerance.[1] Freezing-avoidant plants may avoid freezing by overwintering as a seed with little freezable water, or they may completely avoid any freezing in the tissues by supercooling. Freezing tolerance is the capacity to withstand extracellular ice formation and to avoid intracellular ice formation.[2] Levitt classified plants according to their degree of freezing resistance.[1] Tender or completely unhardy plants have no avoidance of freezing and these plants are seriously damaged even at temperatures above freezing point, whereas slightly hardy plants survive freezing at about -5°C. Moderately hardy plants accumulate sugars and other solutes, and are able to survive freezing temperatures within -5 to -10°C. Very hardy plants have developed mechanisms leading to a marked avoidance of both intracellular freezing and extracellular freeze-dehydration resulting from removal of water from the cytoplasm to the growing ice crystals. These plants survive freezing temperatures within -10 to -20°C. Extremely hardy, the most freezing-tolerant plants are able to survive temperatures lower than -20°C by possessing a complete tolerance of freeze-dehydration-induced membrane disruption. The degree of freezing tolerance is different in woody and herbaceous perennials (e.g. conifers and *Vaccinium*), in overwintering annuals and perennial plants (e.g. winter cereals, winter rape, alfalfa, bromegrass) and in annual plants (e.g. *Arabidopsis thaliana*, spinach, cabbage).[3] Woody plants stop growing during fall and become dormant, whereas winter cereals only reduce their growth in fall, continue to grow slowly under favorable conditions and overwinter as green plants. *A. thaliana*, a widely used model plant for cold acclimation research, achieves rapidly, within a few days, the maximum freezing tolerance (the temperature at which 50% of the cells are injured, LT_{50}); i.e. the plants can survive freezing temperatures of -7 to -10°C,[4–7] or even -12°C when grown in soil in our

* Corresponding author; e-mail: mervi.hiilovaara-teijo@helsinki.fi

laboratory. By contrast, the most freezing-tolerant cereal winter rye (LT_{50} -27°C)[8] requires at least several weeks to become fully hardened.

During past decades, plant responses to low temperatures and mechanisms leading toward increased freezing tolerance have been studied extensively and published in several review articles,[4,9–19] books[1,3,20] and congress proceedings.[21–24] In this chapter we review the molecular mechanisms of cold acclimation process in cold-adapted plants. Examples are taken primarily from freezing-tolerant cereals which are exposed not only to freezing temperatures but also to other abiotic (e.g. high light, dehydration, ice encasement) and biotic (pathogens) stresses during cold acclimation and overwintering, and therefore serve as excellent model plants to study stress resistance. We will summarize the current knowledge on low temperature-induced proteins and discuss the possible roles of these proteins in cold acclimation and freezing tolerance. Moreover, we will report the recent advances in plant cold hardiness research obtained with transgenic plants and specific mutants to elucidate the regulation and role of low temperature-responsive genes in freezing tolerance.

2
Ice formation

The plant cell consists of the cell interior referred to as the intracellular space, and the apoplast which surrounds the living substance of the cell. In plant tissues ice formation can occur inside the cell. This intracellular ice formation is considered to be lethal to the cell because the plasma membrane is irreversibly ruptured resulting in the membrane destabilization and the loss of semipermeable characteristics.[25] Freezing-tolerant plants must have evolved mechanisms that allow them to avoid intracellular ice formation.[2] In hardy plant species, ice formation occurs first in the large vessels.[1] Ice crystals spread rapidly through the vessels and intercellular spaces to all other parts of the plant, because in living cells the cell wall together with hydrophobic plasma membrane form a barrier which prevents the growth of ice into the cytoplasm, and because effective intracellular ice nucleators are absent.[25,26] Gas-filled spaces provide room for crystal growth in leaves where the cell walls are also less rigidly structured.[27] Therefore, the extracellular ice formation in hardened leaves is not directly disruptive since there is no direct contact between ice and the protoplasm.[1] However, extracellular freezing leads to a secondary dehydration stress resulting from the water potential gradient with cellular water diffusing out of the cells to the growing ice crystals in the intercellular spaces.[1] Freeze dehydration, in turn, produces two other injurious strains, i.e. increased concentration of the cell sap and cell collapse.

The distribution of ice crystals in plant tissues is not uniform. In cold-acclimated winter rye (*Secale cereale* L.) leaves ice crystals form in xylem vessels and in intercellular spaces around the vascular bundles, below the epidermis and between the adjoining mesophyll cells, indicating several local sites for initiation of ice crystal growth within the plant.[28] In wheat (*Triticum aestivum* L.) leaves frozen in the field and then freeze-fixed, the extracellular ice is located in intercellular spaces below epidermis and in substomatal cavities.[29] These results suggest that ice formation in the field is initiated either by nucleators located on cell walls surrounding subepidermal spaces or by ice that initially forms on the leaf surface and then propagates into the leaf through stomata. In winter rye leaves ice nucleation was reported to occur over a range of -5 to -12°C.[30] During cold acclimation, winter rye leaves produce intrinsic ice nucleators that have been shown to initiate the formation of extracellular ice during freezing under controlled conditions. These heterogeneous ice nucleators present in rye leaves were characterized and shown to be complexes of proteins, carbohydrates, and phospholipids, in which both disulfide bonds and free sulfhydryls were important for activity.[30]

The formation of extracellular ice is essential for the survival of freezing-tolerant plants. The ability to restrict ice formation to extracellular areas and to prevent physical damage to the cells from ice crystals is a complex process characteristic to cold-adapted plants. Both structural and metabolic changes that occur in response to low temperature ultimately contribute to the increased cryostability of cellular membranes and to the maintenance of the structural integrity of the plasma membrane which is a prerequisite for survival during freeze–thaw cycles.

3 Freezing injury

There is a general consensus that the plasma membrane is a primary site of freezing injury because of its central role in cellular behavior during a freeze–thaw cycle.[25,31,32] However, there are controversial opinions on the nature and cause of injury. The visual symptoms of freezing-induced injury are a water-soaked appearance of the tissue, leakage of ions and organic solutes, and the inability to regain turgor during recovery.[1,31] These symptoms suggest that the cell membrane is a site of damage. Steponkus and his coworkers have studied freezing injury in winter rye using isolated protoplasts,[33,34] plasma membranes[35] and leaves.[36] Membrane damage and loss of membrane semipermeability have been explained to result from cell dehydration induced by extracellular freezing.[1,25] In nonacclimated rye protoplasts, freeze-induced dehydration results in two different forms of injury to the plasma membrane.[25,37] First, when nonacclimated rye protoplasts are cooled to -5°C, the predominant form of injury is lysis of protoplast during osmotic expansion after thawing of the suspending medium. This form of injury occurs because of irreversible endocytotic vesiculation of the plasma membrane during freeze-induced osmotic contraction.[38] Second, when nonacclimated protoplasts are frozen to lower subzero temperatures (e.g. -10°C), dehydration is more severe, resulting in the close apposition of the plasma membrane with endomembranes and the injury is manifested as a complete loss of osmotic responsiveness.[39] This form of injury is associated with dehydration-induced formation of aparticulate domains in the plasma membrane and subtending lamellae, and lamellar-to-hexagonal-II phase transition which is probably preceded by lipid-lipid demixing between the membranes.[40] In cold-acclimated protoplasts the surface area contraction is achieved by exocytotic extrusions of membrane material.[41] These membrane extrusions remain in association with the plasma membrane and are reincorporated during rehydration thus allowing the protoplasts to swell to their original size. The loss of osmotic responsiveness is a predominant form of injury in cold-acclimated protoplasts frozen to -25 to -35°C; however, the plasma membranes do not undergo lamellar-to-hexagonal-II phase transitions.[42] Rather, freezing injury is associated with a phenomenon referred to as the fracture-jump lesion between the plasma membrane and closely appressed cytoplasmic membranes.[40,43] Most frequently the plasma membrane is brought into close apposition with the chloroplast envelope during freeze-induced dehydration.[44,45] In conclusion, destabilization of the plasma membrane is a primary cause of freezing injury in winter rye protoplasts, and the manifestations of the injury are different during osmotic contraction and expansion, and severe freeze-induced dehydration. The differences in the cryobehavior of the plasma membrane of nonacclimated and cold-acclimated protoplasts have been proposed to reflect the alterations in plasma membrane composition or cryoprotective mechanisms following the cold acclimation process.[34]

The use of protoplasts (i.e. cells without cell walls) as a model system to investigate the mechanisms of freezing injury has been questioned in several reports because the effects of extracellular freezing and dehydration are reproduced by osmotic manipulation of the external solution.[31,46–49] In plants, ice formation in the extracellular space withdraws water from the cell resulting in collapse of the cell walls.[1,49] This process of dehydration and stresses that result from it are similar for water stress and extracellular freezing stress, but not for osmotic stress. In cells exposed to osmotic stress the plasma membrane separates from the cell wall, which does not occur during extracellular freezing or water stress.[31] When cold-acclimated frozen rye leaves were observed by low-temperature scanning electron microscopy (SEM), no gas-filled spaces or sublimable ice between cell walls and enclosed protoplast were detected, confirming that the cell wall and plasma membrane remain in contact during freezing and that ice grows extracellularly.[28] Thus the use of protoplasts does not take into account the membrane alterations in the tonoplast and the important role of the cell wall, which may function as a barrier to the propagation of intercellular ice.[26,46,47,50] Nevertheless, Webb and Steponkus confirmed by freeze-fracture electron microscopy (EM) that identical freezing injury observed in winter rye protoplasts also occurs in the leaves from which the protoplasts are isolated.[36] They conclude that the formation of the hex-II-phase transition in nonacclimated rye leaves requires a significantly longer equilibration times than in nonacclimated protoplasts. Therefore, the absence of the hex-II phase in frozen whole leaf cells may reflect the differences between the isolated protoplasts and intact cells, or differences in the experimental design, e.g. exposure times to freezing temperatures.[47,51]

Alternatively, the active transportation system of the plasma membrane may be the primary site of freezing injury.[31] Palta and coworkers have studied freezing injury in potato (*Solanum commersonii*) and observed that increased ion leakage was not associated with changes in photosynthetic and respiratory activity or the ultrastructure of cells following reversible freeze–thaw injury.[49] Instead, they showed that freezing injury results in the increase in K^+ transport properties following injury, and suggested that it could be explained in terms of freeze–thaw-induced perturbation of the membrane-associated H^+-ATPase.[31] This hypothesis was supported by alterations in the specific activity of the plasma membrane ATPase following freeze–thaw stress, i.e. an increase in specific activity following reversible injury and progressive decrease in specific activity following irreversible freezing damage.[52] These findings indicate that the activity of plasma membrane ATPase is needed for the uptake of leaked solutes and thus for recovery of cells.[53] The mechanism explaining how the plasma membrane ATPase is altered is not yet known, but changes in membrane lipids may play an important role in modulating the plasma membrane ATPase during cold acclimation.[54]

Freezing injury may originate during freezing, during thawing or during the postthaw period.[1] The enhanced ion leakage following a freeze–thaw cycle is indicative of the damage to the membranes. Pihakaski-Maunsbach and Harvey used X-ray microanalysis to study the effects of lethal freezing on K^+ distribution in compartments within nonacclimated and cold-acclimated winter rye cells to evaluate membrane leakage caused by freezing injury.[55] In nonacclimated cells the K^+ concentration decreased in the cytoplasm and increased in the cell walls at -4°C, suggesting that the plasma membrane was damaged already during freezing and not by subsequent thawing. In cold-acclimated rye cells the concentration of K^+ increased in the cytoplasm indicating the intactness of the plasma membrane at -12°C. Moreover, Uemura and Steponkus did not detect any substantial inactivation or change in plasma membrane ATPase activity after freezing of nonacclimated winter rye leaves, suggesting that the inactivation of plasma membrane ATPase is not a primary cause of freezing injury in winter rye; instead, injury is associated with an increase in the permeability (leakiness) of the plasma membrane.[35] The release of electrolytes from injured tissues was explained to occur because of lesions in the plasma membrane caused by the osmotic stress resulting from freezing.

In summary, it is widely accepted that the plasma membrane plays a key role in freeze–thaw stress. Different, sometimes controversial, results obtained by several research groups may reflect the different level of freezing injury which is influenced by numerous variables, like the type of plant and growth conditions prior to freezing experiments, the degree of hardiness, ice nucleation temperature, freezing (cooling) rate, lowest temperature reached, duration of exposure to ice, thawing rate and postthaw conditions. Freezing results in a multitude of stresses, and it is therefore more reasonable to suppose that injury is the result of the interaction of several stresses.[25] Several factors may be responsible for membrane damage. These factors may involve a physical effect of low temperature per se, mechanical effects of the accumulation of ice within tissues, freeze-induced reduction in the surface area and solute concentration effects, freeze-dehydration of the protoplasm, molecular packing of membrane constituents induced by cell shrinkage, as well as changes in pH and ionic strength.[3] Depending on the extent of the stress, the injury is either reversible or irreversible. If the cells are to survive, their plasma membranes must be able to withstand stresses associated with freezing. Membranes from cold-acclimated and nonacclimated plants have been shown to behave differently during freezing stress. Therefore, it is the process of cold acclimation that increases the stability of plasma membrane and its tolerance to extracellular ice formation.

4 Cold acclimation

4.1 Structural and morphological changes induced by cold acclimation

Growth at low temperatures affects several morphological characteristics. Cold-grown winter rye leaves are thicker than those of nonacclimated plants, which reflects an increase in mesophyll cell length.[56] The epidermal cell walls are thicker and stomatal frequence is lower in cold-acclimated leaves than in those of nonacclimated plants. Moreover, plant growth at low temperature results in an increased asymmetric thickening of the mestome sheath cell wall.[56] The thickening of epidermal and mestome sheath cell walls during low temperature growth of winter rye is accompanied by an increase in cuticle thickness and deposition of lipid polymers (lignin and suberin) in the epidermal and mestome sheath cell walls.[57,58] Water-soluble cell wall polymers in cold-acclimated winter rye have been analyzed and found to be mainly polysaccharides.[27,59] These polymers affect freezing, acting as competitive freezing inhibitors.[27] Deposition of cell wall materials and lipid-derived polymers was proposed to decrease ice propagation from xylem tissue to mesophyll tissue and to maintain the water status of rye plants.[58] In sum, low temperature-induced cell wall modifications including the accumulation of (1) antifreeze proteins (see Sect. 5.2), (2) polysaccharides, e.g. arabinoxylans,[27,59,60] (3) glycoproteins,[61] and (4) lipid polymers[58] have an important role in preventing secondary ice nucleation through cell walls and plasma membrane, and in limiting the growth of extracellular ice in freezing-tolerant cereals.

Once ice forms in xylem vessels, it can spread rapidly throughout the vascular system of the plant.[3] Anatomical adaptations that moderate the rate of propagation of ice throughout the plant and that prevent inoculation of sensitive tissues may be an important characteristic of freezing-tolerant plants. The xylem may be modified to slow down the growth of ice in the vascular system by vascular segmentation so that xylem vessels are not continuous throughout the plant but are interrupted by zones of tracheid differentiation. In freezing-tolerant cereals, the xylem vessels of both seminal and adventitious roots are separated from the vessels of shoots by the differentiation of tracheids in the crown.[62] This anatomical structure was shown to reduce the rate of ice propagation from roots to shoot.[63]

The role of microtubule cytoskeleton in plant cold hardiness is not clear. Cross-bridges between microtubules and plasma membrane have been proposed to contribute to the stability of the plasma membrane under stress.[64] In winter rye, cold acclimation induces cold-stable microtubules in leaf cells as well as in protoplasts isolated from leaves. These long cortical microtubules may protect and keep the plasma membrane intact and responsive to osmotic changes caused by subzero temperatures.[65]

4.2 Metabolic and biochemical changes

Freezing-tolerant plants tolerate ice formation within their tissues only after cold acclimation (hardening) which is induced by low nonfreezing temperatures as well as by drought (water stress), phytohormone abscisic acid (ABA) and short daylength. During cold acclimation, a series of metabolic and biochemical changes adjust the cell metabolism to new growth conditions at low temperatures and increase cell resistance to freezing.[10] Freezing-tolerant cereals require exposure to low temperatures for the induction and maintenance of a cold-acclimated state.[66]

A cessation or a marked retardation of growth during hardening has led to a general conclusion that freezing tolerance and growth are inversely related. However, in winter rye, growth at low temperatures is significant with growth parameters such as leaf dry weight, fresh weight, and area, but not with plant height.[67,68] The stage of leaf development influences its growth response to an altered environment.[67] Rye leaves fully expanded at 20°C undergo chlorosis and senescence when subjected to prolonged exposure to 5°C.[69] Thus growth and development at low temperatures is a prerequisite for cold hardening in freezing-tolerant cereals.[70] This developmental process requires energy, which comes from photosynthesis.

4.2.1 Changes in photosynthesis

Winter rye is able to modulate photosynthetic rates during growth at low temperatures such that photosynthetic capacity is increased with no change in photosynthetic efficiency.[71] Öquist et al. suggested that this low-temperature modulation of the photosynthetic apparatus may be an important factor of freezing resistance in cereals.[72] The increase in photosynthetic capacity has been shown to be associated with increases in the total activity and activation of enzymes responsible for the flow of fixed carbon through the Calvin cycle and through the sucrose synthesis.[73] Correspondingly, *Arabidopsis* leaves developed at low temperatures showed an increase in photosynthesis and upregulation of photosynthetic genes enabling production and accumulation of soluble sugars.[74] Mitochondria were proposed to contribute to increased photosynthetic capacity by using the reducing equivalents generated by noncyclic photosynthetic electron transport in cold-acclimated winter rye leaves.[75] To become cold-acclimated winter cereals and other overwintering herbaceous annuals must photosynthesize under suboptimal conditions of low temperature and moderate to high light which expose the plants to potential photoinhibition of photosynthesis.[76] Winter rye has been shown to acquire increased resistance to photoinhibition during growth at cold acclimation temperatures due to an increased capacity to maintain a greater fraction of the photosystem (PSII) reaction centres in an open state under given light and temperature conditions.[77] Recently, it was reported that low temperature-induced tolerance to photoinhibition in winter rye is not a growth temperature effect per se, but reflects photosynthetic adjustment to high PSII excitation pressure, which may be modulated by either temperature or irradiance.[78] Resistance to this kind of photoinhibition may be a general phenomenon in cold-tolerant plants.[76]

A strong correlation between photosynthetic capacity at low temperatures and freezing tolerance results from photosynthesis providing the energy necessary for the cellular changes that are required for the increased freezing tolerance.[72] Increased sucrose and other carbohydrate levels reflect an increased capacity for photosynthesis. Huner and coworkers believe that the carbohydrate accumulation during growth at low temperatures reflects specific adjustment of carbon metabolism to elevated PSII excitation pressure.[76,78] The increase in photosynthesis enables cold-acclimated winter rye to maximize the synthesis of compounds which have cryoprotective functions and which are necessary for basic metabolism during overwintering.[73]

4.2.2 Carbohydrate accumulation

Carbohydrate accumulation during cold acclimation has been traditionally interpreted to result from the excess of photosynthesis over respiration or from hydrolysis of starch to increase freezing tolerance by cryoprotection.[1] This accumulation has been thought to occur when demand for assimilates is reduced due to slower growth at lower temperatures. However, Huner et al. have emphasized that the accumulation of carbohydrates reflects important regulatory adjustments in photosynthetic carbon metabolism required for metabolic changes leading toward increased freezing tolerance.[76] They suggested that cold acclimation of cereals includes accumulation of sufficient energy stores at low temperature through the modulation of photosynthetic capacity and carbon metabolism, which is coupled with the allocation of photosynthates away from vegetative growth to cold acclimation of the crown for maximum winter survival and early spring growth. Overwintering cereals accumulate carbohydrates as the main carbon source for spring growth.[79] This accumulation has been assumed to be associated with an increase in resistance to freezing. Cold acclimation often results in an increased cellular osmolality and the major changes are due to increase in sugars.[1] During extracellular freezing, water moves out of cells, resulting in dehydration of the cell. As this freeze-dehydration proceeds, the cell contents become more and more concentrated.[1] Increased concentration of the cell sap also increases membrane-toxic compounds such as salts. Carbohydrates protect membranes against the deleterious effect of high concentrations of electrolytes during freezing.[80] The mechanism of this protection can be explained by colligative action of sugars, i.e. they nonspecifically dilute the concentration of toxic compounds. However, the protection of membranes during freezing in the presence of carbohydrates may involve more specific solute-membrane interactions.[81,82] Trehalose, sucrose and glycerol were shown to interact with membrane phospholipid headgroups, allowing them to remain associated with the membrane for the duration of the freezing process.[83]

Sucrose accumulation during cold acclimation is positively correlated with freezing tolerance.[1,79,84] The accumulation of sucrose is accompanied by the increase in amount and activity of sucrose synthase in winter wheat and sucrose phosphate synthase in spinach.[85,86] This increase in enzyme amount and activity may be needed to maximize synthesis of sucrose with important cryoprotectant properties.[86] If sugars play a role in freezing tolerance, they must accumulate near the primary site of freezing injury, the plasma membrane. Soluble sugars, mostly sucrose and raffinose, were shown to accumulate outside the vacuole, where they may contribute to cryoprotection.[84]

The vacuolar location of fructans and their ability to adjust the osmotic potential suggest that fructans are involved in freezing tolerance.[87] Suzuki and Nass studied the association of fructan accumulation with freezing resistance in hardy field-grown winter wheat, triticale and winter rye.[88] They found that fructans with a high degree of polymerization (DP) were more closely associated with freezing resistance than low-DP fructans, sucrose, and fructose. When winter cereals were exposed to freezing, fructan was converted to sugars, which may alleviate adhesive freezing stress in crowns.[89] Moreover, the energy of hydrolysis may be useful for translocating sugar into tissue sites where its cryoprotective activity is most effective. Significant changes in the concentration of individual sugars occur in winter rye crowns after a light freezing treatment.[90] Freezing induces conversion of fructan to the potential cryoprotective sugars, fructose and sucrose, and deposition of these sugars in the intercellular space which probably promotes fluid barriers to adhesion between hydrophilic plant components and ice.[90,91]

The role of accumulating sugars in cold-acclimating plants is still unclear. At the early stages of cold acclimation, sugar accumulation may be involved in perception of change in temperature. Sugars may act as a signal to genes controlling metabolic and morphological changes involved in cold acclimation of cold-adapted plants. Metabolic regulation of plant gene expression has been demonstrated in numerous cases, e.g. sugars induce expression of genes encoding storage proteins and regulate gene expression upon seed germination.[92–95] A supply of excess carbohydrate to cells has been shown to cause an increase in cytosolic Ca^{2+} concentration which may mediate "cross-talk" between signals related to carbohydrate metabolism and other stimuli in the regulation of gene expression.[94] Calcium signaling has been shown to play a role in cold acclimation and cold-induced gene expression.[96,97] In potato, the sugar accumulation was proposed to be associated with the decrease in osmotic potential, which by increasing the endogenous ABA level, may induce synthesis of proteins responsible for the development of freezing tolerance.[98] Moreover, the rapid increase in soluble sugars at the beginning of the cold acclimation of winter rye may be responsible for a special energy status enabling a plant to survive through the cold stress and to adjust metabolism to new growth conditions.[99] This energy may be required to increase capacity for protein synthesis which, in turn, produces cryoprotective compounds necessary for increased freezing tolerance and membrane stabilization.[100]

4.2.3
Alterations in plasma membrane lipid and protein composition during cold acclimation

Analyses of the individual lipid molecular species of the winter rye plasma membrane showed substantial differences after cold acclimation.[101] However, these lipid changes were only quantitative, not qualitative because no lipid species were unique to the plasma membrane of either nonacclimated or cold-acclimated leaves.[37] By altering the lipid composition of nonacclimated plasma membranes by fusion with liposomes isolated from the plasma membrane of cold-acclimated rye leaves, cryostability of the plasma membrane and freezing tolerance of the protoplasts were significantly increased.[102] The liposomes were composed of either mono- or di-unsaturated species of phosphatidylcholine. As a result, the changes in lipid composition following cold acclimation increase the plasma membrane cryostability. More evidence for this statement was obtained by comparing the plasma membrane lipid composition of the least freezing-tolerant cereal, spring oat, to that of winter rye.[103] Differences in the freezing tolerance of spring oat and winter rye may, in part, result from a vast difference in the lipid composition of the plasma membrane.[103]

By comparison, Uemura and Yoshida detected only slight changes in the degree of fatty acid unsaturation as well as in the proportions of phospholipid classes between the isolated plasma membranes from cold-acclimated and nonacclimated winter rye.[104] These results suggest that there is no direct relation between fatty acid unsaturation and the development of cold hardiness in winter rye. The lipid composition of the plasma membrane differs from other cellular membranes, and the plasma membrane lipid composition also varies within different organs from the same plant as well as between different plant species.[45] Therefore, changes in fatty acid unsaturation in lipid components of other membranes may play a role in cryoprotection. In fact, the genetic manipulation of fatty acid unsaturation in chloroplast membranes increased the chilling resistance of tobacco.[105] Moreover, Uemura and Steponkus showed that the lipid composition of the inner and outer membranes of the chloroplast envelope is altered during cold acclimation of rye leaves, increasing the cryostability of the membranes.[106]

Very little information on plasma membrane protein changes during cold acclimation of winter cereals has been presented so far. Uemura and Yoshida have reported some plasma membrane polypeptides as disappearing or decreasing while several new polypeptides appear during cold acclimation of winter rye, suggesting active turnover of membrane proteins.[104] Similar results were obtained with winter wheat plasma membrane proteins.[107] These results suggest that the plasma membrane may be structurally organized during cold acclimation to cope with the freezing-induced damage to the membrane, which may be mediated by changes in both protein and lipid components of the membrane.

4.2.4
Protein metabolism at low temperatures

Qualitative and quantitative differences in protein content between nonacclimated and cold-acclimated plants have been reported in various plants. A general increase in protein concentration and changes in protein composition in cold-acclimated plants suggest an increased capacity for protein synthesis. Overall changes in protein synthesis and gene expression can be monitored by analyzing alterations in the polyribosome populations.[108] At the beginning of cold acclimation of winter rye, the total polysome content (cytosolic and membrane-bound) increased when compared to that of control plants. Also the proportion of large polysomes containing more than five ribosomes was higher in cold-treated plants when compared to non-treated plants.[109] These results are indicative of the increase in protein synthesis in winter rye leaves developed at 25°C and shifted to cold acclimation conditions of 5°C. On the other hand, decreased termination of protein synthesis could cause a buildup of larger polysomes at low temperatures. It seems likely that the changes in the protein synthesis machinery of winter rye reflect the adjustment of the plant's metabolism to the low temperature by which winter rye ensures its survival during the first days of cold treatment by increasing the polysome content and especially the amount of large polysomes that are involved in protein synthesis. Correspondingly, fully cold-acclimated rye leaves which have developed and grown at low temperature for 9–10 weeks have higher polysome content and show an increase in the formation of large polysomes when compared to nonacclimated plants.[110] Similar results were observed in hardened winter wheat from which the polysome-bound mRNAs were translated *in vitro*. The polypeptides translated from these cold-induced mRNAs were proposed to play a role in the development of freezing tolerance.[111]

The role of protein metabolism on cold acclimation has been reviewed in refs. 11 and 112. It is generally accepted that cold acclimation requires altered gene expression and de novo protein synthesis, as first proposed by Weiser.[113] The first direct evidence to support Weiser's hypothesis of altered gene expression during cold acclimation was the observation that low temperature induced newly translatable mRNAs in spinach leaves.[114] The induction of low temperature-induced proteins and corresponding changes in translatable mRNA populations have been demonstrated in numerous reports (for reviews see refs. 11,13,15,17,115,116). Low temperature-induced proteins are referred to as COR-polypeptides (cold-regulated proteins)[117] or CAPs (cold acclimation proteins).[118] These polypeptides may be involved in metabolic adjustment to growth at low, nonfreezing temperature or they may be required for induction and maintenance of freezing tolerance. Cold-induced mRNAs have been shown to accumulate in many plants. In winter wheat, cold acclimation induced and increased at least 60 mRNAs as demonstrated by using *in vitro* translation assay combined with two-dimensional SDS-PAGE.[119] One group of mRNAs was induced and expressed at high levels during cold acclimation of freezing-tolerant wheat cultivars, suggesting a possible role in freezing tolerance. The second group of mRNAs expressed at similar levels during hardening of both freezing-sensitive and freezing-tolerant wheat cultivars may code for polypeptides that are required for cellular adjustment during growth at low temperature. The accumulation of mRNAs and the polypeptides which they encoded has been shown to be paralleled by the acquisition of freezing tolerance.[118] Several research groups working with different plants have examined low temperature-induced genes by extracting cold-induced mRNAs, constructing cDNA libraries and isolating cold-induced genes corresponding to cDNA clones. Analysis of the cold-induced gene sequences and comparison to the existing gene sequences may or may not reveal sequence homology. To date, over 100 cold-induced genes have been isolated and characterized but only a few have been shown to be related to freezing tolerance.

5
Low temperature-induced proteins

Little information is available about the identity or function of low temperature-induced proteins.[11,120] To characterize the genes encoding proteins associated with cold acclimation, many cDNAs corresponding to cold-regulated RNAs have been cloned.[17,18,116] The nucleotide sequences and the deduced peptide sequences have also been reported, and putative functions of these gene products have been proposed based on similarity with previously reported sequences of known DNAs or proteins. The putative functions of CORs include cryoprotection,[121,122] altered lipid metabolism,[123] protein protection,[124] desiccation tolerance,[125–127] and sugar metabolism.[85,128]

5.1
Cryoprotective proteins

Most promising candidate genes which may have a direct effect on freezing tolerance, are genes encoding polypeptides with potential cryoprotective properties. A gene product encoded by one of the *cor* genes from *A. thaliana* was shown to have cryoprotective activity in a standard *in vitro* assay, and was proposed to play some role in cryoprotection.[122] Recently, cold-regulated genes encoding COR15am and COR6.6 polypeptides were expressed in *Escherichia coli* and purified to near homogeneity.[129] Neither of these proteins had direct cryoprotective effect on the cryostability of membranes during freezing or dehydration stress.[130,131] Nevertheless, constitutive expression of *COR15a* that encodes a chloroplast-targeted polypeptide enhanced the *in vivo* freezing tolerance of chloroplasts in nonacclimated plants, as shown by chlorophyll fluorescence experiments.[132] Correspondingly, the proteins extracted from cold-acclimated spinach and cabbage leaves were shown to protect isolated thylakoids against mechanical membrane rupture during an *in vitro* freeze–thaw cycle.[121,133,134] Just recently, Hincha and colleagues have purified a boiling-stable protein (cryoprotectin) from the leaves of cold-acclimated cabbage and shown that the cryoprotective activity is associated with a cold-induced 7-kDa glycoprotein. So far, no reported genes encoding spinach or cabbage cryoprotective proteins have been isolated.[135]

5.2
Antifreeze proteins

A different approach to elucidate the identity, function and location of cold-induced proteins was chosen by Griffith and coworkers.[136] They started to look for proteins associated with increased freezing tolerance in the intercellular spaces where ice formation occurs. Proteins were extracted from the apoplast of cold-acclimated winter rye leaves by intercellular washing technique, which has been widely used by plant pathologists to obtain proteins secreted by plants in response to pathogen invasion. This protein extract from the apoplast was shown to possess the two characteristics unique to previously purified antifreeze proteins, i.e. to modify the normal growth pattern of ice crystals and to depress the freezing temperature of water.[136] Accordingly, the apoplastic proteins were named antifreeze proteins (AFPs). Seven polypeptides accumulate in the apoplast of cold-acclimated winter rye leaves.[137–139] Six of the polypeptides eluted from SDS-polyacrylamide gel showed a high degree of antifreeze activity.[138] The winter rye AFPs have been isolated and characterized by aminoterminal sequence comparisons, immuno-cross-reactions, and enzyme activity assays, and these polypeptides were found to be similar to members of three classes of pathogenesis-related (PR) proteins, namely, endochitinases, endo-β-1,3-glucanases and thaumatin-like proteins.[138,140] Two of AFPs (35 and 32 kDa) are glucanase-like (GLPs), two (35 and 28 kDa) are chitinase-like (CLPs) and two (25 and 16 kDa) are thaumatin-like proteins (TLPs).[139,140] The amino acid compositions of these polypeptides were similar in that they were all relatively enriched in the residues Asp/Asn, Glu/Gln, Ser, Thr, Gly, and Ala; they all lacked His (except one polypeptide), and they all contained up to 5% Cys residues.[138] The smaller apoplastic polypeptide with a molecular mass of 13 kDa exhibited very little antifreeze activity.[138] This polypeptide may therefore have other roles in cold acclimation than controlling ice formation. In fact, the 13-kDa apoplastic polypeptide has 80% identity to a barley lipid transfer protein as shown by amino-terminal sequence comparison (M. Griffith, pers. communication) suggesting a role in cutin deposition in cell wall modification.[141]

Apoplastic proteins with antifreeze activity produced by cold-acclimated winter rye leaves were shown to be distinct from antifreezes produced by fish and insects.[138] The lack of shared features reinforces the concept that ice-modifying proteins have arisen independently in different organisms in response to ice formation at freezing temperatures. Although antifreeze activity was reported to be fairly common in overwintering plants from diverse phylogenetic groups, the apoplastic protein secretion and antifreeze activity comparable to that found in winter rye is not common in all freezing-tolerant plants, but is limited to the *Triticum* and *Avena* groups in the Poaceae family.[8,142,143] Plant AFPs have been purified from a dicot *Solanum dulcamara* and a monocot winter rye, but these proteins show no similarity.[140,144] Despite the antifreeze activity observed in the extracts from different plants, such different proteins with the ability to modify ice growth accumulate in these plants. These results suggest that monocots and dicots respond differently to cold temperatures.

The term antifreeze protein is problematic because it leads readers to believe that ice is not formed in the presence of these proteins. However, the role of these proteins is to bind to and modify the growth of ice crystals, whether they are present in a fish that avoids freezing or in a plant that tolerates freezing.[2,145–147] Sicheri and Yang solved the crystal structure of an AFP from winter flounder and shown that there is a rigid structural requirement for ice binding.[147] The protein has to be rigid in the ice-binding area and has to have "ice-binding motifs" (IBMs), consisting of Thr/Asp or Thr/Asn/Leu, spaced at appropriate intervals to bind to ridges on the ice surface. The amino acids have free -OH, -C=O, and/or $-NH_2$ groups available to hydrogen bond to the ice crystal surface. These IBMs have to form a very flat surface to maximize the opportunity for forming H-bonds, and this occurs sterically hindering the movement of the hydrophilic groups to optimize their positions.[147]

AFPs, also referred to as thermal hysteresis proteins (THPs),[148] have been reported in many different organisms exposed to low temperatures, including terrestrial arthropods, bacteria, fungi and plants, which leads us to conclude that these proteins indeed play a role in modifying the growth of ice and have evolved independently.[143,149,150] So far, we do not know precisely how plant AFPs interact with extracellular ice *in vivo*. AFPs lower the freezing point without affecting the temperature at which the crystal melts.[145] The function of AFPs in freezing-tolerant organisms may be to inhibit ice recrystallization during freezing and thawing cycles.[151]

Kurkela and Franck proposed that cold-regulated hydrophilic alanine-rich KIN1 polypeptide from *A. thaliana* might have antifreeze properties based on the sequence similarity to fish AFP.[152] Thomashow and coworkers isolated a *cor6.6* gene (a homologue of *kin1*) from *A. thaliana*, overexpressed the polypeptide in *E. coli* and purified the protein to near homogeneity.[116] This protein was assayed for antifreeze activity, but no such activity was detected. Moreover, the low-temperature-induced protein BN28 in *Brassica napus* showing similarity to KIN1 and COR6.6 proteins was purified and tested for antifreeze activity using recrystallization inhibition assay.[153] No activity could be detected with BN28 further suggesting that, in spite of structural similarity to the type I AFP from winter flounder, low-temperature-induced plant proteins KIN1, COR6.6 and BN28 proteins do not possess antifreeze activity similar to that observed in cold-adapted plants or fish.

Before AFPs were identified as endogenously produced proteins in plants, most research was focused on transforming plants with fish AFPs to improve plant freezing tolerance. Cutler et al. vacuum-infiltrated winter flounder AFPs into the extracellular space of potato, canola and *A. thaliana* leaves.[154] After AFP infiltration, the freezing temperature was slightly decreased, thus supporting the feasibility of manipulating the freezing tolerance by transferring fish AFP genes to plants. Georges et al. designed and cloned a synthetic gene for the winter flounder AFP and electroporated the fusion gene into corn protoplasts.[155] AFP was expressed as an N-terminal extension of chloramphenicol acetyl transferase protein, but only 25% of the protein was exported to the extracellular space of the cell. Transgenic tomato plants have been shown to express genes encoding AFP at the mRNA level; however, the antifreeze activity as measured by inhibition of ice recrystallization was detectable only in transgenic plants containing a gene encoding a fusion protein between a truncated staphylococcal protein A and an AFP.[156] On the other hand, the accumulation of type I fish AFP without a fusion protein component has been shown to be cold-specific in transgenic tobacco.[157] Recently, fish AFPs and antifreeze glycoproteins (AFGPs) were shown to be cryotoxic to spinach thylakoid membranes.[158] Hincha et al. concluded that the expression of AFP genes in transgenic plants could lead to severe membrane damage during freezing, especially if the proteins are targeted to the chloroplast. These results indicate that AFPs should be secreted to the extracellular space of transgenic plants if they are to protect membranes from freeze-induced damage. This is also supported by Wallis et al., who transformed potato with a synthetic AFP fused to a signal peptide directing it to the apoplast. Transformants expressing this AFP were more resistant to freezing than the wild type, as determined by electrolyte leakage test.[159]

The different forms of ice crystals observed in apoplastic protein extracts during cold acclimation of winter rye may reflect the accumulation of different classes of AFPs. Hon et al. proposed that complexity of the faceted surfaces of bipyramidal ice crystals may indicate the presence of multiple AFPs, each of which binds onto a different plane of the ice crystal lattice.[140] The apoplastic proteins accumulate to high levels in leaves which have developed and grown at low temperatures.[8] Thus, accumulation of winter rye apoplastic proteins may also be related to developmental changes which occur in leaves during cold acclimation, e.g. modification of cell walls. Extraction of apoplastic proteins from rye leaves does not significantly change ice nucleation capacity, suggesting that the ice nucleators are tightly bound or are too large to be extracted.[30,137] However, the extraction of the apoplast reduces the amount of AFPs present in the leaves and leads to increased injury to the leaves when they are frozen, indicating that AFPs may well have a role in modifying the growth of extracellular ice *in vivo*.[137,139]

Winter rye apoplastic β-1,3-glucanases and chitinases that accumulate during cold acclimation have both antifreeze and enzyme activity. In contrast, purified pathogen-induced tobacco endochitinases and β-1,3-glucanases do not exhibit any ice-binding activity.[140] These results suggest that winter rye AFPs may have evolved from PR-proteins by accumulating mutations resulting in the ability to bind to ice. PR-proteins are also induced in plants by other conditions unrelated to pathogen stress such as ethylene, UV light, ozone, high salt, heat, phytotoxic metal, and wounding.[160] For these developmentally or environmentally induced PR-proteins, no specific roles could be defined, but a role in general stress response has been proposed.[161] Low-temperature induction of PR-proteins has been reported not only in rye but also in barley, bermuda grass, potato, cabbage and spinach.[127,162–165] Antikainen and Griffith tested whether low-temperature treatment induces secretion of proteins into apoplast in different freezing-tolerant and freezing-sensitive dicots and monocots which have been previously shown to produce PR- or PR-like proteins under pathogen or other stress.[8] They showed that apoplastic proteins do not increase in all tested plants as a response to long-term low-temperature exposure. Thus, the protein secretion into the apoplast accompanied by an increase in antifreeze activity is not a general stress response but a more cold-specific response associated with freezing tolerance of cereals. This was supported by the cold-induced apoplastic accumulation of polypeptides cross-reacting with antisera to winter rye AFPs in the *Triticum* group in the Poaceae family. Evidently, the freezing tolerance in monocots and dicots involves different proteins.

To ascertain the function of proteins in cold-adapted plants, it is important to know in which organs and tissues of the plant these proteins are located. Such studies alone will not provide the functional answer, but they will provide critical clues, and functional assignment will ultimately have to be consistent with the results of the localization studies. Thus, the protein localization identifies the place where to look for the function. So far, little information exists concerning the location and function of proteins preferentially synthesized at low temperatures.[11,17] The localization of AFPs in winter rye tissues by tissue printing and immuno-EM showed that the GLPs, CLPs and TLPs with antifreeze activity accumulate in all organs of cold-acclimated rye.[139,166] AFPs were localized in the epidermis and in mesophyll cells surrounding intercellular spaces in cold-acclimated winter rye leaves. Although GLPs, CLPs and TLPs were constitutively present in nonacclimated plants, they were found in different locations and did not exhibit antifreeze activity.

There are several hypothetical roles for AFPs in the apoplast of cold-acclimated rye plants. First, these AFPs are associated with the epidermis and with mesophyll cells lining intercellular spaces.[139] If the AFPs are not released from the cell walls, then the AFPs may function as barriers to ice propagation into the plant or into individual cells. Epidermal AFPs, coupled with cutinized cell walls and increased epicuticular waxes, may provide barriers for both ice propagation and penetration of pathogens from outside the plant. Furthermore, GLPs and CLPs associated with cells around vascular bundles may be important in slowing secondary ice nucleation from xylem and they may also prevent inoculative freezing of cells. Immunogold localization of GLPs in the cell walls of mesophyll cells and secondary thickenings of xylem vessels, i.e. facing the sites where extracellular ice formation occurs, support the role of AFPs functioning as a barrier to ice propagation.[166] AFPs have been localized in certain animal groups, as well. Fish skin contains AFPs which may provide an effective barrier to inoculative freezing.[149] Equally in insects, AFPs have been immunolocalized on the cell membranes of the epidermal cells, suggesting that they form a barrier to inoculative freezing just under the wax-coated cuticle.[149]

Second, localization of AFPs in the close vicinity of pathways for free water within the cold-acclimated rye leaves suggests that loosely bound AFPs act by binding to ice crystals and have important roles in controlling the crystallization and/or recrystallization of water.[139,166] In this case, the AFPs may not be tightly bound to the cell walls so that they are free to bind to the surfaces of extracellular ice crystals. This hypothesis is supported by Rohringer et al., who reported that intercellular washing technique removes substances dissolved in the free water of intercellular space or those weakly bound to the cell wall.[167] In addition, tissue printing technique has been successfully applied to localize cellular and extracellular proteins. With $CaCl_2$-pretreated nitrocellulose paper, it is possible to localize all soluble and salt-soluble proteins, including cell wall proteins.[168] If the cell wall proteins are cross-linked to each other or with other wall components, they cannot be transferred to the paper and detected by this method. Antifreeze activity was found to be higher in apoplastic extracts obtained with salts ($CaCl_2$ or $MgCl_2$) than in extracts obtained with ascorbic acid alone, indicating that more AFPs are released with salts.[138] When apoplastic proteins were extracted, the freezing injury to winter rye leaves was increased indicating that substantial amount AFPs were removed.[137] In fact, the 35-kDa GLP, 35-kDa CLP, 28-kDa CLP and 16-kDa TLP were effectively removed by apoplastic extraction indicating that they are secreted proteins dissolved in the apoplastic solution or loosely bound to the cell walls.[139] By contrast, the 32-kDa GLP and 25-kDa TLP were not completely removed, suggesting that these proteins accumulate intracellularly or that the proteins are more firmly cross-linked to the cell wall. The latter possibility was supported by the immuno-EM localization studies, which showed no GLP accumulation in vacuoles, but instead clearly demonstrated the secretion of GLPs via an exocytotic bulk-flow pathway and accumulation of GLPs in the cell walls.[166]

The location of GLPs in small vesicles close to the plasma membrane and over the plasma membrane suggests that AFPs may stabilize the outer surface of the plasma membrane to preserve the structural integrity during extracellular freezing and/or freeze-induced dehydration.[166] Correspondingly, THPs were immunolocalized on the plasma membranes of certain cell types in the stem of *Solanum dulcamara*.[142] Antifreeze glycopeptides from antarctic fish were proposed to interact with and stabilize cell membranes.[169] However, the mechanism by which AFGPs, THPs or AFPs might stabilize cellular membranes is not yet understood. To summarize, the GLPs, CLPs, and TLPs with antifreeze activity are localized in the apoplast of cold-acclimated winter rye tissues. The distribution of these proteins in the epidermis and surrounding intercellular spaces is correlated with the location of both epiphytic and extracellular ice suggesting that these proteins have a function in modifying the growth of ice. Moreover, the GLPs, CLPs, and TLPs present in nonacclimated plants lacked antifreeze activity and were generally localized in different tissue types suggesting that the AFPs may be different isoforms of PR-proteins.

5.3
Proteins involved in dehydration tolerance

Freezing and drought have been considered related stresses because plant cells become dehydrated during extracellular freezing, and therefore plants may exhibit at least partly overlapping responses to these stresses. In recent years, plant molecular responses to cold and drought stress have been studied extensively. The linkage between the responses to freezing and drought is also supported by the fact that freezing tolerance is also induced by desiccation stress or by exogenous application of ABA.[170-172] Moreover, the majority of the low temperature-induced genes characterized so far are also responsive to ABA and drought (for a review see refs. 17,173).

Dehydrins are one common group of proteins which are known to accumulate in response to both low temperature and drought stress.[174] Dehydrins or immunologically related proteins have been characterized in a wide range of photosynthetic organisms, including plants, algae and bacteria. Plant dehydrins, also known as the late embryogenesis abundant (LEA) proteins, are hydrophilic, boiling-soluble and characterized by the highly conserved Y, S and K segments.[174] Dehydrins may help stabilize cytoplasmic proteins against denaturation or the cytoplasmic surface of membranes against deformation in synergy with compatible solutes.[175]

Cold acclimation-induced dehydrin gene expression has been reported in freezing-tolerant winter cereals.[176–181] In winter wheat, the *Wcs120* gene family encodes a group of proteins which are specific to the Poaceae, highly abundant and coordinately regulated by low temperature.[182] The hydrophilicity, stability upon boiling, and structural similarity with the LEA proteins suggest that WCS120 protein family may have an indirect effect on freezing tolerance by protecting cells against dehydration caused by extracellular freezing.[176,180] By immunolocalization, the WCS protein family was shown to be highly expressed in the vascular transition zone but not in mature xylem, shoot apical meristem or lateral root primordia.[180] The WCS protein family was not found in cell walls or other organelles, but was present in both cytoplasm and the nucleoplasm, further supporting a general protective role in the soluble fraction of the cell.

Another protein showing similarity to dehydrins accumulates in spinach during exposure to low temperature and/or water stress.[183] Cell fractionation studies of cold-acclimated spinach tissue indicated that the CAP85 is predominately found in the cytosol.[183] Similar to CAP85 or a closely related cold-regulated protein, COR85, was purified and characterized from cold-acclimated spinach leaves.[184] This protein accumulated in leaves, petioles and stems, but not in roots of cold-acclimated spinach. On the contrary, COR85 accumulated in all tissues in response to drought and salt stress. COR85 existing as a 350-kDa oligomer in the cytosol is analogous to dehydrins and was proposed to be involved in cryoprotection of intracellular molecules against freezing-induced dehydration.

Several genes encoding dehydrin/LEA-like proteins have been found in *Arabidopsis*. The cold-induced *cor47*, *rab18*, *lti29* (formerly *lti45*) and *lti30* are all responsive to low temperature, drought and exogenous ABA, and their gene products show structural similarity to conserved serine and lysine repeats characteristic of dehydrin/LEA proteins.[125,185–187] These similarities support the view that there is a genetic link between freezing and dehydration tolerance.

5.4
Proteins involved in membrane/lipid modifications

The level of fatty acid unsaturation determines the membrane fluidity and function at decreasing temperatures (for a review see refs. 188,189). A decrease in growth temperature leads to a decrease in membrane fluidity, which needs to be compensated by the increase in the membrane lipid unsaturation. Chilling-sensitive and chilling-resistant plants have been shown to differ in the level of membrane unsaturation. A high level of saturated fatty acids is characteristic to the chloroplastic membranes of chilling-sensitive plants and is proposed to cause a low temperature-induced phase transition.[190] During cold acclimation, cold-adapted plants may change the membrane lipid composition and the level of lipid unsaturation to stabilize the membrane structure and preserve the semipermeable nature of the membrane. The role of fatty acid unsaturation in low-temperature-induced injury was shown by genetic manipulation of glycerol-3-phosphate acyltransferase (GPAT), the chloroplast enzyme which is involved in the phosphatidylglycerol fatty acid unsaturation.[105] Freezing-sensitive tobacco was transformed with cDNA encoding GPAT from chilling-sensitive squash and chilling-tolerant *Arabidopsis*. Transgenic tobacco plants expressing *Arabidopsis* gene were more resistant to low temperature stress (1°C for 10 d) than transformants with squash gene which were severely damaged. This difference correlated with the extent of fatty acid unsaturation in phosphatidylglycerol; i.e. being increased in transformant with *Arabidopsis* GPAT gene and decreased with squash gene.[105] The unsaturation on thylakoid membrane lipids was proposed to facilitate recovery of photosynthesis from low-temperature photoinhibition.[191]

Recent studies with plant and cyanobacteria mutants with specific alterations in membrane lipid composition and the level of fatty acid unsaturation have elucidated the role of membranes in low temperature injury (for a review see refs. 188,192). Both plant and cyanobacterial mutant systems have demonstrated the importance of the unsaturation of thylakoid lipids in protecting the photosynthetic machinery from photoinhibition under cold conditions. The *fab1* mutant (defective in fatty acid biosynthesis) of *A. thaliana* with the increased level of saturated fatty acids showed a decline in phosynthetic capacity (F_v/F_m) and destruction of chloroplasts grown for extended periods at 2°C.[193] This decline was concluded to result from a reduction in the function of PSII centers. In cyanobacteria mutants with inactivated desaturase genes, the degree of membrane lipid unsaturation was shown to affect the turnover of PSII reaction-center protein D1. The low level of fatty acid unsaturation impaired D1-protein turnover at lowered temperatures, thereby preventing recovery from photoinhibition.[194] Murata and Los proposed a model of membrane fluidity regulating the expression of genes for desaturases.[189] The decrease in temperature is sensed as the change in membrane fluidity by a putative sensor locating in the plasma membrane. This hypothetical sensor protein was proposed to be a Ca^{2+} channel or resemble a His kinase. The signal (decrease in temperature or in membrane fluidity) is mediated by an unknown pathway to the desaturase genes which are upregulated, the desaturases are synthesized and transported to the plasma and thylakoid membranes to increase the desaturation of the fatty acids in membrane lipids and to compensate for the decreased membrane fluidity.

Lipid-transfer proteins (LTPs) have been thought to participate in the transfer of lipids between membranes within the cell. However, the extracellular location of the LTPs does not support the essential intracellular role.[141] The search for stress-induced genes has revealed that genes encoding LTPs have been shown to respond to ABA and various stresses including cold, drought, salt and pathogens.[141] In winter barley, *blt4* gene family is induced by cold and drought stress.[123,195] Barley *blt4* gene family was shown to possess a consensus N-terminal signal sequence for extracellular transport.[195] The epidermal location of barley *blt4.9* expression in cold-treated meristematic and older leaf tissue was demonstrated by *in situ* hybridization.[196] *Blt4.9* expression localized in the epidermis is consistent with a proposed role in cuticle formation.[197] The location of LTPs in the extracellular space of epidermal cells and association with cuticular waxes suggest that LTPs may be involved in wax synthesis and/or secretion.[18,198,199] Another possible role of plant LTPs could be a general role in plant defense.[197,200] The analysis of the regulatory elements of barley *LTP* gene family showed putative ABA-responsive (ABRE) and low-temperature responsive elements (ACCGACA) which further supports the protective role of LTPs in low temperature and related stresses.[195]

6
Low-temperature control of gene expression in cold-adapted plants

The complex genetics of cold acclimation has been reviewed.[11,13,17,18,116] The capacity for freezing tolerance in cereals is a quantitative trait controlled by several genes, e.g. in winter rye it is controlled by genes with mainly additive effects, but other factors may also be involved.[13,201] The number of genes involved in low temperature-induced cold acclimation is not known; however, it is generally accepted that cold hardiness is controlled by many genes.

Studies with ABA-insensitive and ABA-deficient mutants have indicated that the development of freezing tolerance in *Arabidopsis* is controlled by genes which either require ABA for their low temperature induction, or respond to ABA but do not require ABA for low temperature-induction.[125,187,202] Most of the low temperature-induced genes characterized so far are also responsive to ABA and water stress.[17] Recently, it has been shown that low-temperature-induced and drought-induced genes in *A. thaliana* are regulated by at least two separate ABA-independent and ABA-responsive signal transduction pathways.[173,203–205] It has been suggested that dehydration or low temperature triggers the production of ABA, which induces gene expression.[98,206] Initial perception and transduction between different stresses could be similar, but they could diverge at the metabolic level, where specific responses are elicited.[207] This may be seen from the analysis of several stress-responsive promoters.[205,208,209] Several *Arabidopsis* genes which are induced by both cold and drought are arranged in tandem in the genome; namely *lti78* (or *cor78* or *rd29A*) and lti65 (or *rd29B*), *kin1* and *kin2* (or *cor6.6*), *cor15a* and *cor15b*, *lti29* (or *lti45*) and *cor47*, *RCI1A* (or *RCI1*) and *RCI1B* (or *RCI2*), and *RCI2A* and *RCI2B*.[185,186,204,205,210–213] These structurally related and closely linked gene pairs have been shown to exhibit differential response to cold and dehydration stress. The differential regulation within gene pairs was proposed to indicate that proteins with similar functions may have specialized to different stress situations.[186]

Gene expression can be regulated at the transcriptional, posttranscriptional or translational level. Evidence for the transcriptional regulation comes from the analyses of the promoter regions which have revealed that different cis-acting elements function in the gene expression under different environmental stresses.[205,208,209,214] To date, several responsive elements of cold-induced genes have been identified. The alcohol dehydrogenase (*Adh*) gene from *Arabidopsis* was shown to be induced by low oxygen, dehydration and low temperature.[209] Two promoter elements which interact differently under different stress conditions were identified: one element consists of a GT/GC-rich motif and functions as a general stress responsive element (GSRE) whereas the second element, G-box-1 sequence, is required for cold and dehydration induction. Three signal transduction pathways were proposed to be required for *Adh* gene induction by environmental stresses and exogenous ABA.[214] ABA could play the role of a stress signal, but other tissue-specific components may also be involved, e.g. transcription factor might bind to the GT/GC motif and mediate stress-responsive *Adh* expression through interactions with upstream regulatory elements such as G-box-1. The role of ABA in signal transduction has recently been reviewed by Merlot and Giraudat.[215]

In addition to the general stress responsive element found in *Arabidopsis Adh* gene, other regulatory elements have been identified in low-temperature-induced genes. Nordin et al. identified the sequence TACGTGTC of *lti78* and *lti65* to be similar to the ABA responsive element (ABRE) found in several ABA- and low-temperature-responsive genes, e.g. in *rab18*.[125,204] Moreover, comparison of the promoter sequences of these *ltis* suggests the presence of a putative low temperature-responsive element (LTRE) with the consensus sequence ACCGACA which has also been found in other cold-induced *Arabidopsis* genes *rab18*, *kin1* and *kin2* as well as in barley *blt4.6*.[195,204] The sequence TACCGACAT of *Arabidopsis rd29A* (or *lti78* or *cor78*) referred to as dehydration-responsive element (DRE) was reported to be involved in dehydration and low temperature stress, but not in ABA response.[205] Interestingly, this *cis*-acting element seems to contain the LTRE. Baker and coworkers identified a cold-responsive element (TGGCCGAC) termed as C-repeat in *Arabidopsis cor15a* gene.[216] The transcripts for *cor15a* increase in response to low temperature, drought and exogenous ABA.[211] All these three cold-responsive elements, LTRE, DRE and C-repeat, share an identical core sequence of CCGAC. Recently, an *Arabidopsis* cDNA that encodes a CRT/DRE (C-repeat/drought responsive element) binding factor, CBF1, was cloned and isolated.[217] The product of this gene, the 24-kDa protein, has a nuclear localization signal and a DNA binding domain. Therefore, CBF1 was proposed to bind to the *cis*-acting CRT/ DRE element of the *cor15a* and act as a transcriptional activator. The role of CBF1 in cold- and dehydration-regulated gene expression was examined by creating transgenic *Arabidopsis* plants that overexpress *CBF1*.[218] *CBF1* overexpression induced *COR* (*COR6.6*, *COR15a*, *COR47* and *COR78*) gene expression without low temperature treatment and increased plant freezing tolerance as determined by electrolyte leakage test. In comparison, plants expressing constitutively only *COR15a* were less freezing-tolerant than plants overexpressing *CBF1*.[218] These results suggest that the *Arabidopsis* transcription factor CBF1 induced the expression of *COR* gene family by binding to the *cis*-acting CRT/DRE sequence.

How plants sense the change in temperature is not yet known, although evidence for the involvement of Ca^{2+} in temperature-sensing and signal-transduction has been presented.[96,97] Studies with Ca^{2+}-chelators and channel blockers have demonstrated the involvement of Ca^{2+} in both ABA-mediated and ABA-independent signal pathways.[219,220] Both intra- and extracellular Ca^{2+}-stores appear to be involved in the ABA-independent pathway while only intracellular Ca^{2+} is utilized in triggering gene expression through the ABA-mediated pathway. Thus, ABA and low-temperature signaling both seem to be mediated by calcium, and may be coupled by calcium-dependent protein kinases (CDPKs) as was shown for cold-induced *kin*-genes.[220] Plasma membrane-associated protein kinases are proposed to act as transducers of extracellular signals, whereas protein kinases associated with the membranes of organelles may be involved in communication between the organelles and the cytoplasm.[221] Hong et al. have recently isolated a cDNA clone for a transmembrane receptor-like protein kinase gene (*RPK1*) from *Arabidopsis*.[222] This gene is rapidly induced by ABA, dehydration, high salt and low temperature, and is suggested to function in transmitting these extracellular signals into intracellular target molecules. The *RPK1* gene may be induced by an osmotic stress which is a common stress response in cold, salt and dehydration stresses as well as being known to involve ABA.[222] Several models for low temperature-responsive gene expression have been proposed (for details see refs. 18,96,182,187). To summarize these models, a decline in temperature decreases the membrane fluidity which in turn activates plasma membrane calcium channels. Calcium influx either activates or represses cold-specific protein kinases/phosphatases leading to a change in the phosphorylation status of cold-specific proteins. This change in protein conformation may trigger the cold-dependent gene expression which leads to increased freezing tolerance.

7 Interactions with other stresses

Cold acclimation increases not only the plant's tolerance to low temperatures but also to other abiotic and biotic stresses. There are several examples of interactions with cold and other stress responses in cold-adapted plants. Pathogenic infection impairs the plant's carbon metabolism, and pathogens may damage the plant by limiting its growth and the capacity to cope with abiotic stresses. Plants exposed to pathogens are more susceptible to drought stress, suffer from mineral deficiency and are prone to the effects of winter stresses.[223] However, many overwintering, freezing-tolerant plants are more resistant to plant diseases, e.g. winter cereals are more resistant to fungal diseases after cold acclimation.[224–226] Infection of a plant by a pathogen can trigger the plant's resistance mechanisms. Induced resistance is associated with induction of pathogenesis-related proteins, β-1,3-glucanases, chitinases and thaumatin-like proteins.[227] Similar proteins have been shown to accumulate in response to low temperature in winter rye, barley, bermudagrass, cabbage and spinach.[140,162,163,165] In fact, the same PR-proteins were produced during cold hardening that were found in barley inoculated with powdery mildew.[162] Tronsmo proposed that chitinase induction may induce resistance to fungal diseases in cold-hardened plants. Cold-acclimated winter barley was shown to accumulate apoplastic AFPs, i.e. GLPs, CLPs and TLPs with the ability to control ice formation.[8] Therefore, it is possible that AFPs similar to PRs may function in freezing tolerance and pathogen resistance in overwintering grasses. The localization of pathogen-induced PRs and the low temperature-accumulated GLPs, CLPs, and TLPs are similar, and suggest a defensive role against penetration and propagation of pathogens.[139,166,228,229] The putative role for winter rye AFPs/PRs in pathogen stress is further supported by the fact that GLPs, CLPs and TLPs which lack antifreeze activity accumulate as a response to infection by snow mold (*Microdochium nivale*) at non-hardening temperatures (Antikainen, Griffith, Hannukkala and Pihakaski-Maunsbach, unpubl. data). These proteins may play a pre-emptive role in the protection of cold-acclimated cereal leaves in the late autumn and early spring, when pathogen infection is most likely to occur.

One CLP was purified to homogeneity from cold-acclimated winter rye leaves and this chitinase was shown to exhibit both antifreeze and endochitinase activities.[140] In contrast, an endochitinase purified from tobacco plants infected with tobacco mosaic virus has a similar level of enzymatic activity to the low-temperature-accumulated rye chitinase but has no effect on ice crystal growth. Class I β-1,3-glucanase and class I chitinase were purified from tobacco and their cryoprotective effect was investigated in the *in vitro* assay.[165] Beta-1,3-glucanase protected spinach thylakoids against freeze–thaw injury, whereas chitinase had no effect in the same assay. Immunologically cross-reactive polypeptides with tobacco β-1,3-

glucanase antibody were detected in total soluble protein fraction of cabbage and spinach leaves cold-acclimated under field conditions.[165] The cellular localization and the target membranes of the cold-induced 41-kDa β-1,3-glucanase is not yet known. To sum up, low temperature and pathogen infection seem to induce the accumulation of related proteins with different functions. Understanding overlap of function and control of cold- and pathogen-responsive AFPs/PRs may improve our understanding of the relationship between resistance to cold and resistance to pathogens. This may lay a foundation for future exploitation of these resistance mechanisms.

Freezing tolerance is induced by low temperature as well as by drought. In common with these two stresses, plants accumulate dehydrins. Along with the potential cryoprotective role for dehydrins (discussed in Sect. 5.3), the preliminary studies with transgenic plants ectopically expressing dehydrin genes indicate that dehydrins are involved in tolerance to heavy metals as well as to oxidative stress.[202] Probably the sequestration of metals is one of the functions of dehydrins which is relevant to cold acclimation. Freeze-induced dehydration may increase the cellular concentrations of metals to toxic levels which could be removed by binding to dehydrins. Therefore, dehydrins were proposed to function as intracellular, stress-induced metal chelators in plants during cold, drought, and salt stress in order to maintain the metal homeostasis in dehydrating cell.[202]

Freezing tolerance may be induced under salt stress. Nonacclimated spinach leaves treated with 300 mM NaCl solution showed increased freezing tolerance which was reflected in reduced *in vitro* freeze–thaw damage to thylakoids.[230] High salt, along with ABA, low temperature and pathogen infection increased the expression of osmotin-like protein genes in potato.[127,231] The transcriptional activation of two osmotin-like protein genes by different abiotic stresses and by fungal pathogen was examined in transgenic potato.[232] Both genes were induced by ABA, NaCl, salicylic acid, wounding, and fungal infection. Low temperature activated these genes only slightly. Transgenic potato plants expressing high levels of osmotin-like proteins showed an increased resistance to fungal infection whereas expression of osmotin-like protein genes had no effect on freezing tolerance as determined by electrolyte leakage test after ABA- and low temperature treatments.[164] Therefore, osmotin-like protein may not be a major determinant of freezing tolerance despite the accumulation of osmotin-like mRNAs as a response to ABA and cold.[127] The complex interaction between different stresses was also demonstrated for salt-stressed tobacco cells which accumulated and secreted several antifungal chitinases.[233] Because both salt stress and pathogen attack cause osmotic stress, the possibility of "cross-talk" in the signal cascades was proposed. Thus, the signal transduction pathway that leads to osmotic adaptation may also share common intermediates with that leading to plant pathogen resistance.

Numerous reports have demonstrated that different abiotic and biotic stresses produce similar effects at the cellular level. Foyer suggests that plants may obtain cross tolerance to several stresses.[234] Therefore, induction of tolerance to specific abiotic stresses involving oxidative stress may also increase the tolerance to other abiotic and biotic stresses, e.g. cold-acclimated winter cereals were shown to increase the tolerance of activated oxygen.[235] Most environmental stresses stimulate excess production of active oxygen species which are harmful to plant cells (for a review see ref. 234). To eliminate these reactive oxygen species such as superoxide, hydrogen peroxide, and hydroxyl radicals, plants have evolved antioxidative defence mechanisms composed of low molecular weight antioxidants (glutathione, ascorbate, α-tocopherol) and protective enzymes (superoxide dismutase, ascorbate peroxidase, monodehydroascorbate radical reductase, dehydroascorbate reductase, glutathione reductase) (reviewed in refs. 236,237). On the other hand, in pathogen resistance hydrogen peroxide may restrict the pathogen growth or induce phytoalexins and PR-proteins.[234] Hydrogen peroxide was believed to serve as a putative signal (secondary messenger) during induction of plant defence genes.[237]

Oxidative stress is a common disorder in several stresses associated with winter including freezing, desiccation, ice-encasement, flooding and diseases. Therefore, activated oxygen may be a common link among these stresses. Freezing and oxidative stress cause similar injury symptoms in plant membranes probably due to accumulation of oxygen free radicals.[238] Superoxide dismutase (SOD) has a crucial role in detoxifying oxygen radicals because it catalyzes the dismutation of superoxide to hydrogen peroxide and oxygen. Transgenic alfalfa (*Medicago sativa*) overexpressing Mn-SOD cDNA showed enhanced tolerance to freezing stress and water deficit.[239,240] The product of SOD, hydrogen peroxide, was proposed to act as a mediator for disease resistance in transgenic potato by activating a number of host defence mechanisms.[241] Elevated levels of H_2O_2 may lead to salicylic acid accumulation which induces defence genes, or perhaps H_2O_2 itself activates defence genes including peroxidases and chitinases to contribute to plant disease resistance.[241] Hydrogen peroxide treatment of *Arabidopsis* caused an increase in cytosolic Ca^{+2} concentration similar to the one associated with cold acclimation.[97] These findings suggest that cold acclimation and hydrogen peroxide act through a similar pathway and that calcium functions as a second messenger in this signaling process.

Oxidative stress may be defined as an imbalance in cellular redox homeostasis.[242] The imbalance between energy supply and energy consumption is sensed as an over-reduction of the photosynthetic apparatus.[243] Many components contribute to the redox balance of the cell among which the thiol/disulfide exchange reactions and the accumulation of hydrogen peroxide provide central signals leading to the control of metabolism and changes in gene expression during environmental and biotic stresses. It is possible that these common redox signals are involved in the induction of acclimatory responses to abiotic and biotic stresses, and thus explaining the cross-resistance plants show to different stresses.[234] In support of this hypothesis, cold-induced transcript *Wcs19* was shown to accumulate in wheat exposed to high light stress (irradiance of 800 $\mu mol\ m^{-2}\ s^{-1}$) at 20°C without any significant increase in freezing tolerance.[244] The transcript level and the rosette growth habit were similar to those of the cold-acclimated plants. This was the first report on the induction of the previously cold-stimulated gene in response to a chloroplastic redox signal, i.e. the change in PSII excitation pressure.[244] Huner and coworkers propose that the regulation of freezing tolerance involves a complex interaction of low temperature, light, and redox-dependent signals.[243,244] Consequently, the applications of genetic manipulation to increase the freezing tolerance by transferring low-temperature-induced genes may not give desirable results because of the complex interactions between abiotic and biotic stresses which need to be taken into consideration.

8 Conclusions

Transfer of low temperature-induced genes to freezing-sensitive plants (e.g. tobacco) is believed to provide more evidence on their involvement in freezing tolerance. So far transformation of tobacco with cold-induced genes has not improved its freezing tolerance; e.g. a gene involved in carbohydrate metabolism,[245] *Arabidopsis rab18* gene[246] and *Arabidopsis kin1* gene (M. Borg-Franck, pers. communication) had no effect on freezing tolerance, as evaluated by the electrolyte leakage test. Evidently, these genes have no direct effect on freezing tolerance of the freezing-sensitive plant tobacco. The validity of the electrolyte test has been questioned because of the variability due to different plants, tissue types and size, growth conditions prior to testing, and cooling rates (for a review see ref. 247). Moreover, this test measures only the leakage of ions from injured cells, which does not necessarily reflect the true function of transferred genes in freezing and dehydration tolerance other than effect on membrane stabilization. Other viability tests (e.g. regrowth after freezing treatment) and specific tests for gene function are also needed to get a reliable estimation on freezing tolerance in transgenic plants. On the other hand, transfer of a single cold-responsive gene is unlikely to have a major effect on freezing tolerance which is already known to be a very complex multigenic process. Furthermore, the targeting of the encoded gene product may not be correct in transgenic plant and the freezing tolerance is not improved. As for the function of low temperature-responsive proteins, overexpression and antisense inhibition studies in transgenic plants will certainly elucidate their role in the freezing tolerance. Due to the complexity of the cold-acclimation process in cold-adapted plants, it may not be possible to produce genetically engineered plants with improved freezing tolerance by transferring low-temperature-induced genes.[120] However, one interesting strategy to enhance plant stress tolerance could involve manipulation of the signaling pathways leading to expression of tolerance genes. Very promising candidates for manipulation could involve alteration of the level of stress responsive transcription factors which regulate the genes required for improved stress tolerance. In our laboratory, we are currently (1) characterizing the signal processes involved in expression of low temperature-responsive genes, (2) ascribing the function of corresponding gene products, and (3) developing strategies to genetically engineer the freezing as well as dehydration and pathogen stress tolerance in plants.

9 References

1. Levitt J. Responses of Plants to Environmental Stresses: Chilling, Freezing, and High Temperature Stresses, vol 1, 2nd ed. New York: Academic Press, 1980.

2. Griffith M, Antikainen M. Extracellular ice formation in freezing-tolerant plants. Adv Low-Temp Biol 1996; 3:107-139.
3. Sakai A, Larcher W. Frost Survival of Plants: Responses and Adaptation to Freezing Stress. Berlin, Heidelberg: Springer, 1987.
4. Thomashow MF. *Arabidopsis thaliana* as a model for studying mechanisms of plant cold hardiness. In: Meyrowitz EM, Somerville CR, eds. *Arabidopsis.* New York: Cold Spring Harbor Laboratory Press, 1994:807-834.
5. Gilmour SJ, Hajela RK, Thomashow MF. Cold acclimation in *Arabidopsis thaliana.* Plant Physiol 1988; 87:745-750.
6. Kurkela S, Franck M, Heino P, Lång V, Palva ET. Cold induced gene expression in *Arabidopsis thaliana* L. Plant Cell Rep 1988; 7:495-498.
7. Lång V, Heino P, Palva ET. Low temperature acclimation and treatment with exogenous abscisic acid induce common polypeptides in *Arabidopsis thaliana* (L.) Heynh. Theor Appl Genet 1989; 77:729-743.
8. Antikainen M, Griffith M. Antifreeze protein accumulation in freezing-tolerant cereals. Physiol Plant 1997; 99:423-432.
9. Junttila O. Physiological responses to low temperature. Ann Sci For 1989; 46:604-613.
10. Kacperska A. Metabolic consequences of low temperature stress in chilling-insensitive plants. In: Li PH, ed. Low Temperature Stress Physiology in Crops. Boca Raton: CRC Press, 1989:27-40.
11. Guy CL. Cold acclimation and freezing stress tolerance: role of protein metabolism. Annu Rev Plant Physiol Plant Mol Biol 1990; 41:187-223.
12. Hällgren J-E, Öquist G. Adaptations to low temperatures. In: Alscher RG, Cumming JR, eds. Stress Responses in Plants: Adaptation and Acclimation Mechanisms. New York: Wiley-Liss, 1990:265-293.
13. Thomashow MF. Molecular genetics of cold acclimation in higher plants. Adv Genet 1990; 28:99-131.
14. Alberdi M, Corcuera LJ. Cold acclimation in plants. Phytochemistry 1991; 30:3177-3184.
15. Lee SP, Chen THH. Molecular biology of plant cold hardiness development. In: Li PH, Christersson L, eds. Advances in Plant Cold Hardiness. Boca Raton: CRC Press, 1993:1-44.
16. Chen THH. Plant adaptation to low temperature stress. Can J Plant Pathol 1994; 16:231-236.
17. Palva ET. Gene expression under low temperature stress. In: Basra AS, ed. Stress-Induced Gene Expression in Plants. Chur, Switzerland: Harwood Acad. Publ, 1994:103-130.
18. Hughes MA, Dunn MA. The molecular biology of plant acclimation to low temperature. J Exp Bot 1996; 47:291-305.
19. Palva ET, Heino P. Molecular mechanism of plant cold acclimation and freezing tolerance. In: Li PH, Chen THH, eds. Plant Cold Hardiness. Molecular Biology, Biochemistry, and Physiology. New York: Plenum Press, 1997:3-14.
20. Larcher W. Physiological Plant Ecology. Ecophysiology and Stress Physiology of Functional Groups, 3rd ed. Berlin, Heidelberg: Springer, 1995.
21. Li PH, Sakai A. Plant Cold Hardiness and Freezing Stress: Mechanisms and Crop Implications, vol 2. New York: Academic Press, 1982.
22. Li PH. Plant Cold Hardiness. New York: Alan R Liss, 1987.
23. Li PH, Christersson L. Advances in Plant Cold Hardiness. Boca Raton: CRC Press, 1993.
24. Li PH, Chen THH. Plant Cold Hardiness. Molecular Biology, Biochemistry, and Physiology. New York: Plenum Press, 1997.
25. Steponkus PL. Role of plasma membrane in freezing injury and cold acclimation. Ann Rev Plant Physiol 1984; 35:543-584.
26. Tao D-L, Li PH. Cell wall and freezing injury: a mini-review. In: Li PH, Christersson L, eds. Advances in Plant Cold Hardiness. Boca Raton: CRC Press, 1993:195-201.
27. Olien CR. Freezing stresses and survival. Ann Rev Plant Physiol 1967; 18:387-408.

28. Pearce RS. Extracellular ice and cell shape in frost-stressed cereal leaves: a low-temperature scanning-electron-microscopy study. Planta 1988; 175:313-324.
29. Pearce RS, Ashworth E. Cell shape and localization of ice in leaves of overwintering wheat during frost stress in the field. Planta 1992; 188:324-331.
30. Brush RA, Griffith M, Mlynarz A. Characterization and quantification of intrinsic ice nucleators in winter rye (*Secale cereale*) leaves. Plant Physiol 1994; 104:725-735.
31. Palta JP. Stress interactions at the cellular and membrane levels. Hort Sci 1990; 25:1377-1381.
32. Steponkus PL. Cold acclimation and freezing injury from a perspective of the plasma membrane. In: Katterman F, ed. Environmental Injury to Plants. San Diego: Academic Press, 1990:1-16.
33. Pihakaski K, Steponkus PL. Freeze-induced phase transitions in the plasma membrane of isolated protoplasts. Physiol Plant 1987; 69:666-674.
34. Uemura M, Steponkus PL. Effect of cold acclimation on the incidence of two forms of freezing injury in protoplasts isolated from rye leaves. Plant Physiol 1989; 91:1131-1137.
35. Uemura M, Steponkus PL. Parallel effects of freezing and osmotic stress on the ATPase activity and protein composition of the plasma membrane of winter rye seedlings. Plant Physiol 1989: 91:961-969.
36. Webb MS, Steponkus PL. Freeze-induced membrane ultrastructural alterations in rye (*Secale cereale*) leaves. Plant Physiol 1993; 101:955-963.
37. Steponkus PL, Lynch DV, Uemura M. The influence of cold acclimation on the lipid composition and cryobehaviour of the plasma membrane of isolated rye protoplasts. Phil Trans R Soc Lond; 1990: B326:571-583.
38. Gordon-Kamm WJ, Steponkus PL. The behavior of the plasma membrane following osmotic contraction of isolated protoplasts: implications in freezing injury. Protoplasma 1984; 123:83-94.
39. Steponkus PL, Lynch DV. freeze–thaw-induced destabilization of the plasma membrane and the effects of cold acclimation. J Bioenerg Biomembr 1989; 21:21-41.
40. Steponkus PL, Uemura M, Webb MS. A contrast of the cryostability of the plasma membrane of winter rye and spring oat. Two species that widely differ in their freezing tolerance and plasma membrane lipid composition. Adv Low-Temp Biol 1993; 2:211-312.
41. Dowgert MF, Steponkus PL. Behavior of the plasma membrane of isolated protoplasts during a freeze–thaw cycle. Plant Physiol 1984; 75:1139-1151.
42. Gordon-Kamm WJ, Steponkus PL. Lamellar-to-hexagonal^{II} phase transitions in the plasma membrane of isolated protoplasts after freeze-induced dehydration. Proc Natl Acad Sci USA 1984; 81:6373-6377.
43. Fujikawa S, Steponkus PL. Freeze-induced alterations in the ultrastructure of the plasma membrane of rye protoplasts isolated from cold-acclimated leaves. Cryobiology 1990; 27:665-666.
44. Webb MS, Uemura M, Steponkus PL. A comparison of freezing injury in oat and rye: Two cereals at the extremes of freezing tolerance. Plant Physiol 1994; 104:467-478.
45. Uemura M, Joseph RA, Steponkus PL. Cold acclimation of *Arabidopsis thaliana*: effect on plasma membrane lipid composition and freeze-induced lesions. Plant Physiol 1995; 109:15-30.
46. Bartolo ME, Wallner SJ, Ketchum. Comparison of freezing tolerance in cultured plant cells and their respective protoplasts. Cryobiology 1987; 24:53-57.
47. Singh J, Iu B, Johnson-Flanagan AM. Membrane alterations in winter rye and *Brassica napus* cells during lethal freezing and plasmolysis. Plant Cell Environ 1987; 10:163-168.
48. Harvey DMR, Pihakaski K. Ultrastructural changes arising from freezing of leaf blade cells of rye (*Secale cereale*): an investigation using freeze-substitution. Physiol Plant 1989; 76:262-270.
49. Palta JP. Plasma membrane ATPase as a key site of perturbation in response to freeze–thaw stress. Curr Top Plant Biochem Physiol 1989; 8: 41-68.

50. Johnson-Flanagan AM, Singh J. Membrane deletion during plasmolysis in hardened and non-hardened plant cells. Plant Cell Environ 1986; 9:299-305.
51. Pearce RS, Willison JHM. A freeze-etch study of the effects of extracellular freezing on cellular membranes of wheat. Planta 1985; 163:304-316.
52. Iswari S, Palta JP. Plasma membrane ATPase activity following reversible and irreversible freezing injury. Plant Physiol 1989; 90:1088-1095.
53. Arora R, Palta JP. A loss in the plasma membrane ATPase activity and its recovery coincides with incipient freeze–thaw injury and postthaw recovery in onion bulb scale tissue. Plant Physiol 1991; 95:846-852.
54. Palta JP, Whitaker BD, Weiss LS. Plasma membrane lipids associated with genetic variability in freezing tolerance and cold acclimation of *Solanum* species. Plant Physiol 1993; 103:793-803.
55. Pihakaski-Maunsbach K, Harvey DMR. X-ray microanalytical (EDX) investigation of potassium distributions in mesophyll cells of non-acclimated and cold-acclimated rye leaves. Plant Cell Environ 1992; 15:585-591.
56. Huner NPA, Palta JP, Li PH, Carter JV. Anatomical changes in leaves of Puma rye in response to growth at cold-hardening temperatures. Bot Gaz 1981; 142:55-62.
57. Griffith M, Brown GN. Cell wall deposits in winter rye *Secale cereale* L. 'Puma' during cold acclimation. Bot Gaz 1982; 143:486-490.
58. Griffith M, Huner NPA, Espelie KE, Kolattukudy PE. Lipid polymers accumulate in the epidermis and mestome sheath cell walls during low temperature development of winter rye leaves. Protoplasma 1985; 125:53-64.
59. Olien CR. Interference of cereal polymers and related compounds with freezing. Cryobiology 1965; 2:47-54.
60. Kindel PK, Liao S-Y, Liske MR, Olien CR. Arabinoxylans from rye and wheat seed that interact with ice. Carbohydrate Res 1989; 187:173-185.
61. Jian LC, Sun LH, Sun DL. Glycoproteins at the cell surface in cold hardy and cold tender wheat (*Triticum aestivum* L.). In: Li PH, ed. Plant Cold Hardiness. New York: Alan R Liss, 1987: 59-66.
62. Aloni R, Griffith M. Functional xylem anatomy in root-shoot junctions of six cereal species. Planta 1991; 184:123-129.
63. Zámecník J, Bieblová, Grospietsch M. Safety zone as a barrier to root-shoot ice propagation. Plant Soil 1994; 167:149-155.
64. Jian L-C, Sun L-H, Wei X-Y. Microtubule cytoskeleton in relation to plant cold hardiness. In: Li PH, Christersson L, eds. Advances in Plant Cold Hardiness. Boca Raton: CRC Press, 1993:125-137.
65. Pihakaski-Maunsbach K, Puhakainen T. Effect of cold exposure on cortical microtubules of rye (*Secale cereale*) as observed by immunocytochemistry. Physiol Plant 1995; 93:563-571.
66. Gusta LV, Fowler DB. Cold resistance and injury in winter cereals. In: Mussel H, Staples RC, eds. Stress Physiology in Crop Plants. New York: John Wiley & Sons, 1979: 159-178.
67. Krol M, Griffith M, Huner NPA. An appropriate physiological control for environmental temperature studies: comparative growth kinetics of winter rye. Can J Bot 1984; 62:1062-1068.
68. White PJ, Cooper HD, Earnshaw MJ, Clarkson DT. Effects of low temperature on the development and morphology of rye (*Secale cereale*) and wheat (*Triticum aestivum*). Ann Bot 1990; 66:559-566.
69. Krol M, Huner NPA. Growth and development at cold hardening temperatures. Pigment and benzoquinone accumulation in winter rye. Can J Bot 1985; 62:716-721.
70. Huner N, Williams J, Krol M, Boese S, Hurry V, Lapointe L, Reynolds T. Photosynthesis, low temperature development and freezing tolerance. Curr Top Plant Biochem Physiol 1989; 8:6-20.
71. Falk S, Maxwell D, Gray G, Rezansoff D, Huner N. Photosynthetic acclimation to low temperature in higher plants and algae. Curr Top Bot Res 1993; 1:281-292.

72. Öquist G, Hurry VM, Huner NPA. Low-temperature effects on photosynthesis and correlation with freezing tolerance in spring and winter cultivars of wheat and rye. Plant Physiol 1993; 101:245-250.
73. Hurry VM, Keerberg O, Pärnik T, Gardeström P, Öquist G. Cold-hardening results in increased activity of enzymes involved in carbon metabolism in leaves of winter rye (*Secale cereale* L.). Planta 1995; 195:554-562.
74. Strand Å, Hurry V, Gustafsson P, Gardeström P. Development of *Arabidopsis thaliana* leaves at low temperatures releases the suppression of photosynthesis and photosynthetic gene expression despite the accumulation of soluble carbohydrates. Plant J 1997; 12:605-614.
75. Hurry VM, Strand Å, Tobiæson M, Gardeström P, Öquist G. Cold hardening of spring and winter wheat and rape results in differential effects on growth, carbon metabolism, and carbohydrate content. Plant Physiol 1995; 109:697-706.
76. Huner NPA, Öquist G, Hurry V, Krol M, Falk A, Griffith M. Photosynthesis, photoinhibition and low temperature acclimation in cold tolerant plants. Photosynth Res 1993; 37:19-39.
77. Öquist G, Huner NPA. Cold-hardening-induced resistance to photoinhibition of photosynthesis in winter rye is dependent upon an increased capacity for photosynthesis. Planta 1993; 189:150-156.
78. Gray GR, Savitch LV, Ivanov AG, Huner NPA. Photosystem II excitation pressure and development of resistance to photoinhibition. II. Adjustment to photosynthetic capacity in winter wheat and winter rye. Plant Physiol 1996; 110:61-71.
79. Hurry V, Tobiæson M, Krömer S, Gardeström P, Öquist G. Mitochondria contribute to increased photosynthetic capacity of leaves of winter rye (*Secale cereale* L.) following cold-hardening. Plant Cell Environ 1995; 18:69-76.
80. Santarius KA. The mechanism of cryoprotection of biomembrane systems by carbohydrates. In: Li PH, Sakai A, eds. Plant Cold Hardiness and Freezing Stress: Mechanisms and Crop Implications. vol 2. New York: Academic Press, 1982:475-486.
81. Santarius KA, Bauer J. Cryopreservation of spinach chloroplast membranes by low-molecular-weight carbohydrates. Cryobiology 1983; 20:83-89.
82. Hincha DK, Schmitt JM. Cryoprotection of thylakoid membranes: I Protection by sugars. In: Dörffling K, Brettschneider B, Tantau H, Pithan K, eds. Cold Adaptation to Cool Climates. Brussels: ECSP-EEC-EAEC, 1994:151-165.
83. Anchordoguy TJ, Rudolph AS, Carpenter JF, Crowe JH. Modes of interaction of cryoprotectants with membrane phospholipids during freezing. Cryobiology 1987; 24:324-331.
84. Koster KL, Lynch DV. Solute accumulation and compartmentation during the cold acclimation of Puma rye. Plant Physiol 1992; 98:108-113.
85. Crespi MD, Zabaleta EJ, Pontis HG, Salerno GL. Sucrose synthase expression during cold acclimation in wheat. Plant Physiol 1991; 96:887-891.
86. Guy CL, Huber JLA, Huber SC. Sucrose phosphate synthase and sucrose accumulation at low temperature. Plant Physiol 1992; 100:502-508.
87. Pontis HG. Fructans and cold stress. J Plant Physiol 1989; 134:148-150.
88. Suzuki M, Nass HG. Fructan in winter wheat, triticale, and fall rye cultivars of varying cold hardiness. Can J Bot 1988; 66:1723-1728.
89. Olien CR, Clark JL. Changes in soluble carbohydrate composition of barley, wheat, and rye during winter. Agron J 1993; 85:21-29.
90. Olien CR, Lester GE. Freeze-induced changes in soluble carbohydrates of rye. Crop Sci 1985; 25:288-290.
91. Olien CR. An adaptive response of rye to freezing. Crop Sci 1984; 24:51-54.
92. Koch KE. Carbohydrate-modulated gene expression in plants. Annu Rev Plant Physiol Plant Mol Biol 1996; 47:509-540.
93. Grierson C, Du J-S, de Torres Zabala M, Beggs K, Smith C, Holdsworth M, Bevan M. Separate *cis* sequences and *trans* factors direct metabolic and developmental regulation of a potato tuber storage protein gene. Plant J 1994; 5:815-826.

94. Ohto M, Hayashi K, Isobe M, Nakamura K. Involvement of Ca^{2+} signalling in the sugar-inducible expression of genes coding for sporamin and α-amylase of sweet potato. Plant J 1995; 7:297-307.
95. Reynolds S, Smith SM. Regulation of expression of the cucumber isocitrate lyase gene in cotyledons upon seed germination and by sucrose. Plant Mol Biol 1995; 29:885-896.
96. Monroy AF, Dhindsa RS. Low-temperature signal transduction: induction of cold-acclimation-specific genes of alfalfa by calcium at 25°C. Plant Cell 1995; 7:321-331.
97. Knight H, Trewavas AJ, Knight MR. Cold calcium signalling in *Arabidopsis* involves two cellular pools and a change in calcium signature after acclimation. Plant Cell 1996; 8:489-503.
98. Chen H-H, Li PH. Potato cold acclimation. In: Li PH, Sakai A, eds. Plant Cold Hardiness and Freezing Stress: Mechanisms and Crop Implications, vol 2. New York: Academic Press, 1982:5-22.
99. Antikainen M, Pihakaski S. Early developments in RNA, protein, and sugar levels during cold stress in winter rye (*Secale cereale*) leaves. Ann Bot 1994; 74:335-341.
100. Perras M, Sarhan F. Energy state of spring and winter wheat during cold hardening. Soluble sugars and adenine nucleotides. Physiol Plant 1984; 60:129-132.
101. Lynch DV, Steponkus PL. Plasma membrane lipid alterations associated with cold acclimation of winter rye seedlings (*Secale cereale* L. cv. Puma). Plant Physiol 1987; 83:761-767.
102. Steponkus PL, Uemura M, Balsamo RA, Arvinte T, Lynch DV. Transformation of the cryobehavior of rye protoplasts by modification of the plasma membrane lipid composition. Proc Natl Acad Sci USA 1988; 85:9026-9030.
103. Uemura M, Steponkus PL. A contrast of the plasma membrane lipid composition of oat and rye leaves in relation to freezing tolerance. Plant Physiol 1994; 104:479-496.
104. Uemura M, Yoshida S. Involvement of plasma membrane alterations in cold acclimation of winter rye seedlings (*Secale cereale* L. cv Puma). Plant Physiol 1984; 75:818-826.
105. Murata N, Ishizaki-Nishizawa O, Higashi S, Hayashi H, Tasaka Y, Nishida I. Genetically engineered alteration in the chilling sensitivity of plants. Nature 1992; 356:710-713.
106. Uemura M, Steponkus PL. Effect of cold acclimation on the lipid composition of the inner and outer membrane of the chloroplast envelope isolated from rye leaves. Plant Physiol 1997; 114:1493-1500.
107. Zhou B-L, Arakawa K, Fujikawa S, Yoshida S. Cold-induced alterations in plasma membrane proteins that are specifically related to the development of freezing tolerance in cold-hardy winter wheat. Plant Cell Physiol 1994; 35:175-182.
108. Larkins BA. Polyribosomes. In: Linskens HF, Jackson JF, eds. Modern Methods of Plant Analysis, New Series, vol 1 - Cell Components. Berlin, Heidelberg: Springer, 1985:331-352.
109. Antikainen M, Pihakaski S. Cold-induced changes in the polysome pattern and protein synthesis in winter rye (*Secale cereale*) leaves. Physiol Plant 1993; 89:111-116.
110. Laroche A, Hopkins WG. Polysomes from winter rye seedlings grown at low temperature. I. Size class distribution, composition, and stability. Plant Physiol 1987; 85:648-654.
111. Perras M, Sarhan F. Polysome metabolism during cold acclimation in wheat. Plant Cell Physiol 1990; 31:1083-1089.
112. Graham D, Patterson BD. Responses of plants to low, nonfreezing temperatures: proteins, metabolism, and cold acclimation. Ann Rev Plant Physiol 1982; 33:347-372.
113. Weiser CJ. Cold resistance and injury in woody plants. Science 1970; 169:1269-1278.
114. Guy CL, Niemi KJ, Brambl R. Altered gene expression during cold acclimation of spinach. Proc Natl Acad Sci USA 1985; 82:3673-3677.
115. Ougham HJ, Howarth CJ. Temperature shock proteins in plants. In: Long SP, Woodward FI, eds. Plants and Temperature. Cambridge: The Company of Biologists Ltd, 1988:259-280.
116. Thomashow MF. Genes induced during cold acclimation in higher plants. Adv Low-Temp Biol 1993; 2:183-210.
117. Lin C, Guo WW, Everson E, Thomashow MF. Cold acclimation in *Arabidopsis* and wheat. A response associated with expression of related genes encoding "boiling-stable" polypeptides. Plant Physiol 1990; 94:1078-1083.

118. Guy CL, Haskell D. Induction of freezing tolerance in spinach is associated with the synthesis of cold acclimation induced proteins. Plant Physiol 1987; 84:872-878.
119. Danyluk J, Sarhan F. Differential mRNA transcription during the induction of freezing tolerance in spring and winter wheat. Plant Cell Physiol 1990; 31:609-619.
120. Welin B, Heino P, Nordin Henriksson K, Palva ET. Plant cold acclimation: possibilities for engineering freezing tolerance. Ag Biotech News and Information 1996; 8:15N-22N.
121. Hincha DK, Heber U, Schmitt JM. Proteins from frost-hardy leaves protect thylakoids against mechanical freeze–thaw damage *in vitro*. Planta 1990; 180:416-419.
122. Lin C, Thomashow MF. A cold-regulated *Arabidopsis* gene encodes a polypeptide having potent cryoprotective activity. Biochem Biophys Res Comm 1992; 183:1103-1108.
123. Hughes MA, Dunn MA, Pearce RS, White AJ, Zhang L. An abscisic-acid-responsive, low temperature barley gene has homology with a maize phospholipid transfer protein. Plant Cell Environ 1992; 15:861-865.
124. Neven LG, Haskell DW, Guy CL, Denslow N, Klein PA, Green LG, Silverman A. Association of 70-kilodalton heat-shock cognate proteins with acclimation to cold. Plant Physiol 1992; 99:1362-1369.
125. Lång V, Palva ET. The expression of a *rab*-related gene, *rab18*, is induced by abscisic acid during cold acclimation process of *Arabidopsis thaliana* (L.) Heynh. Plant Mol Biol 1992; 20:951-962.
126. Wolfraim LA, Langis R, Tyson H, Dhindsa RS. cDNA sequence, expression and transcript stability of a cold acclimation-specific gene, *cas18*, of alfalfa (*Medicago falcata*) cells. Plant Physiol 1993; 101:1275-1282.
127. Zhu B, Chen THH, Li PH. Expression of an ABA-responsive osmotin-like gene during the induction of freezing tolerance in *Solanum commersonii*. Plant Mol Biol 1993; 21:729-735.
128. Lee SP, Chen THH. Molecular cloning of abscisic acid-responsive mRNAs expressed during the induction of freezing tolerance in bromegrass (*Bromus inermis* Leyss) suspension culture. Plant Physiol 1993; 101:1089-1096.
129. Gilmour SJ, Lin C, Thomashow MF. Purification and properties of *Arabidopsis thaliana COR* (Cold-Regulated) gene polypeptides COR15am and COR6.6 expressed in *Escherichia coli*. Plant Physiol 1996; 111:293-299.
130. Uemura M, Gilmour SJ, Thomashow MF, Steponkus PL. Effects of COR6.6 and COR15am polypeptides encoded by *COR* (Cold-Regulated) genes of *Arabidopsis thaliana* on the freeze-induced fusion and leakage of liposomes. Plant Physiol 1996; 111:313-327.
131. Webb MS, Gilmour SJ, Thomashow MF, Steponkus PL. Effects of COR6.6 and COR15am polypeptides encoded by *COR* (Cold-Regulated) genes of *Arabidopsis thaliana* on dehydration-induced phase transitions of phospholipid membranes. Plant Physiol 1996; 111:301-312.
132. Artus NN, Uemura M, Steponkus PL, Gilmour SJ, Lin C, Thomashow MF. Constitutive expression of the cold-regulated *Arabidopsis thaliana COR15a* gene affects both chloroplast and protoplast freezing tolerance. Proc Natl Acad Sci USA 1996; 93:13404-13409.
133. Volger HG, Heber U. Cryoprotective leaf proteins. Biochim Biophys Acta 1975; 412:335-349.
134. Hincha DK, Heber U, Schmitt JM. Freezing ruptures thylakoid membranes in leaves, and rupture can be prevented *in vitro* by cryoprotective proteins. Plant Physiol Biochem 1989; 27:795-801.
135. Sieg F, Schröder W, Schmitt JM, Hincha DK. Purification and characterization of a cryoprotective protein (cryoprotectin) from the leaves of cold-acclimated cabbage. Plant Physiol 1996; 111:215-221.
136. Griffith M, Ala P, Yang DSC, Hon W-C, Moffatt BA. Antifreeze protein produced endogenously in winter rye leaves. Plant Physiol 1992; 100:593-596.
137. Marentes E, Griffith M, Mlynarz A, Brush RA. Proteins accumulate in the apoplast of winter rye leaves during cold acclimation. Physiol Plant 1993; 87:499-507.
138. Hon W-C, Griffith M, Chong P, Yang DSC. Extraction and isolation of antifreeze proteins from winter rye (*Secale cereale* L.) leaves. Plant Physiol 1994; 104:971-980.

139. Antikainen M, Griffith M, Zhang J, Hon W-C, Yang DSC, Pihakaski-Maunsbach K. Immunolocalization of antifreeze proteins in winter rye leaves, crowns, and roots by tissue printing. Plant Physiol 1996; 110:845-857.
140. Hon W-C, Griffith M, Mlynarz A, Kwok YC, Yang DSC. Antifreeze proteins in winter rye are similar to pathogenesis-related proteins. Plant Physiol 1995; 109:879-889.
141. Kader J-C. Lipid-transfer proteins in plants. Annu Rev Plant Physiol Plant Mol Biol 1996; 47:627-654.
142. Urrutia ME, Duman JG, Knight CA. Plant thermal hysteresis proteins. Biochim Biophys Acta 1992; 1121:199-206.
143. Duman JG, Olsen TM. Thermal hysteresis protein activity in bacteria, fungi, and phylogenetically diverse plants. Cryobiology 1993; 30:322-328.
144. Duman JG. Purification and characterization of a thermal hysteresis protein from a plant, the bittersweet nightshade *Solanum dulcamara*. Biochim Biophys Acta 1994; 1206:129-153.
145. DeVries AL. Antifreeze glycopeptides and peptides: interactions with ice and water. Meth Enzymol 1986; 127:293-302.
146. Yang DSC. Protein-ice interactions: antifreeze proteins. In: Epand RM, ed. The Amphipathic Helix. Boca Raton: CRC Press, 1993:369-391.
147. Sicheri F, Yang DSC. Ice-binding structure and mechanism of an antifreeze protein from winter flounder. Nature 1995; 375:427-431.
148. Duman JG. Insect antifreezes and ice-nucleating agents. Cryobiology 1982; 19:613-617.
149. Duman JG, Wu DW, Olsen TM, Urrutia M, Tursman D. Thermal-hysteresis proteins. Adv Low-Temp Biol 1993; 2:131-182.
150. Griffith M, Ewart KV. Antifreeze proteins and their potential use in frozen foods. Biotechnol Adv 1995; 13:375-402.
151. Knight CA, Duman JG. Inhibition of recrystallization of ice by insect thermal hysteresis proteins: a possible cryoprotective role. Cryobiology 1986; 23:256-262.
152. Kurkela S, Franck M. Cloning and characterization of a cold and ABA inducible *Arabidopsis* gene. Plant Mol Biol 1990; 15:137-144.
153. Boothe JG, Sönnichsen FD, de Beus MD, Johnson-Flanagan AM. Purification, characterization, and structural analysis of a plant low-temperature-induced protein. Plant Physiol 1997; 113:367-376.
154. Cutler AJ, Saleem M, Kendall E, Gusta LV, Georges F, Fletcher GL. Winter flounder antifreeze protein improves the cold hardiness of plant tissues. J Plant Physiol 1989; 135:351-354.
155. Georges F, Saleem M, Cutler AJ. Design and cloning of a synthetic gene for the flounder antifreeze protein and its expression in plant cells. Gene 1990; 91:159-165.
156. Hightower R, Baden K, Penzes E, Lund P, Dunsmuir P. Expression of antifreeze proteins in transgenic plants. Plant Mol Biol 1991; 17:1013-1021.
157. Kenward KD, Altschuler M, Hildebrand D, Davies PL. Accumulation of type I fish antifreeze protein in transgenic tobacco is cold-specific. Plant Mol Biol 1993; 23:377-385.
158. Hincha DK, DeVries AL, Schmitt JM. Cryotoxicity of antifreeze proteins and glycoproteins to spinach thylakoid membranes – comparison with cryotoxic sugar acids. Biochim Biophys Acta 1993; 1146:258-264.
159. Wallis JG, Wang, H, Guerra DJ. Expression of a synthetic antifreeze protein in potato reduces electrolyte release at freezing temperatures. Plant Mol Biol 1997; 35:323-330.
160. Stinzi A, Heitz T, Prasad V, Wiedemann-Merdinoglu S, Kauffmann S, Geoffroy P, Legrand M, Fritig B. Plant "pathogenesis-related" proteins and their role in defense against pathogens. Biochimie 1993; 75: 687-706.
161. Collinge DB, Kragh KM, Mikkelsen JD, Nielsen KK, Rasmussen U, Vad K. Plant chitinases. Plant J 1993; 3:31-40.
162. Tronsmo AM, Gregersen P, Hjeljord L, Sandal T, Bryngelsson T, Collinge DB. Cold-induced disease resistance. In: Fritig B, Legrand M, eds. Mechanisms of Plant Defense Responses. Dordrecht, The Netherlands: Kluwer Acad Publ, 1993:369.

163. Gatschet MJ, Taliaferro CM, Porter DR, Anderson MP, Anderson JA, Jackson KW. A cold-regulated protein from bermudagrass crowns is a chitinase. Crop Sci 1996; 36:712-718.
164. Zhu B, Chen THH, Li PH. Analysis of late-blight disease resistance and freezing tolerance in transgenic potato plants expressing sense and antisense genes for an osmotin-like protein. Planta 1996; 198:70-77.
165. Hincha DK, Meins Jr F, Schmitt JM. β-1,3-Glucanase is cryoprotective *in vitro* and is accumulated in leaves during cold acclimation. Plant Physiol 1997; 114:1077-1083.
166. Pihakaski-Maunsbach K, Griffith M, Antikainen M, Maunsbach A. Immunogold localization of glucanase-like antifreeze protein in cold acclimated winter rye. Protoplasma 1996; 191:115-125.
167. Rohringer R, Ebrahim-Nesbat F, Wolf G. Proteins in intercellular washing fluids from leaves of barley (*Hordeum vulgare* L.). J Exp Bot 1983; 34:1589-1605.
168. Ye Z-H, Varner JE. Tissue-specific expression of cell wall proteins in developing soybean tissues. Plant Cell 1991; 3:23-37.
169. Rubinsky B, Arav A, DeVries AL. The cryoprotective effect of antifreeze glycopeptides from antarctic fishes. Cryobiology 1992; 29:69-79.
170. Cloutier Y, Siminovitch D. Correlation between cold- and drought-induced frost hardiness in winter wheat and rye varieties. Plant Physiol 1982; 69:256-258.
171. Siminovitch D, Cloutier Y. Twenty-four-hour induction of freezing and drought tolerance in plumules of winter rye seedlings by desiccation stress at room temperature in the dark. Plant Physiol 1982; 69:250-255.
172. Chen THH, Gusta LV. Abscisic acid-induced freezing resistance in cultured plant cells. Plant Physiol 1983; 73:71-75.
173. Shinozaki K, Yamaguchi-Shinozaki K. Molecular responses to drought and cold stress. Curr Opin Biotechnol 1996; 7:161-167.
174. Close TJ. Dehydrins: A commonalty in the response of plants to dehydration and low temperature. Physiol Plant 1997; 100:291-296.
175. Close TJ, Fenton RD, Yang A, Asghar R, DeMason DA, Crone DE, Meyer NC, Moonan R. Dehydrin: the protein. In: Close TJ, Bray EA, eds. Plant Responses to Cellular Dehydration During Environmental Stress. Rockville: An American Society of Plant Physiologists Series, Current Topics in Plant Physiology, 1993; 10:104-118.
176. Houde M, Danyluk J, Laliberté JF, Rassart É, Dhindsa RS, Sarhan F. Cloning, characterization, and expression of a cDNA encoding a 50-kilodalton protein specifically induced by cold acclimation in wheat. Plant Physiol 1992; 99:1381-1387.
177. Houde M, Dhindsa RS, Sarhan F. A molecular marker to select for freezing tolerance in Gramineae. Mol Gen Genet 1992; 234:43-48.
178. Danyluk J, Houde M, Rassart É, Sarhan F. Differential expression of a gene encoding an acidic dehydrin in chilling sensitive and freezing tolerant gramineae species. FEBS Lett 1994; 344:20-24.
179. Robertson AJ, Weninger A, Wilen RW, Fu P, Gusta LV. Comparison of dehydrin gene expression and freezing tolerance in *Bromus inermis* and *Secale cereale* grown in controlled environments, hydroponics, and the field. Plant Physiol 1994; 106:1213-1216.
180. Houde M, Daniel C, Lachapelle M, Allard F, Laliberté S, Sarhan F. Immunolocalization of freezing-tolerance-associated proteins in the cytoplasm and nucleoplasm of wheat crown tissues. Plant J 1995; 8:583-593.
181. Stanca AM, Crosatti C, Grossi M, Lacerenza NG, Rizza F, Cattivelli L. Molecular adaptation of barley to cold and drought conditions. Euphytica 1996; 92:215-219.
182. Sarhan F, Ouellet F, Vasquez-Tello A. The wheat *wcs120* gene family. A useful model to understand the molecular genetics of freezing tolerance in cereals. Physiol Plant 1997; 101:439-445.
183. Neven LG, Haskell DW, Hofig A, Li Q-B, Guy CL. Characterization of a spinach gene responsive to low temperature and water stress. Plant Mol Biol 1993; 21:291-305.

184. Kazuoka T, Oeda K. Purification and characterization of COR85-oligomeric complex from cold-acclimated spinach. Plant Cell Physiol 1994; 35:601-611.
185. Gilmour SJ, Artus NN, Thomashow MF. cDNA sequence analysis and expression of two cold-regulated genes of *Arabidopsis thaliana*. Plant Mol Biol 1992; 18:13-21.
186. Welin BV, Olson Å, Palva ET. Structure and organization of two closely related low-temperature-induced *dhn/lea/rab*-like genes in *Arabidopsis thaliana* L. Heyhn. Plant Mol Biol 1995; 29: 391-395.
187. Welin BV, Olson Å, Nylander M, Palva ET. Characterization and differential expression of *dhn/lea/rab*-like genes during cold acclimation and drought stress in *Arabidopsis thaliana*. Plant Mol Biol 1994; 26:131-144.
188. Nishida I, Murata N. Chilling sensitivity in plants and cyanobacteria: The crucial contribution of membrane lipids. Annu Rev Plant Physiol Plant Mol Biol 1996; 47:541-568.
189. Murata N, Los DA. Membrane fluidity and temperature perception. Plant Physiol 1997; 115:875-879.
190. Murata N. Molecular species composition of phosphatidylglycerols from chilling-sensitive and chilling-resistant plants. Plant Cell Physiol 1983; 24:81-86.
191. Moon BY, Higashi S-I, Gombos Z, Murata N. Unsaturation of the membrane lipids of chloroplasts stabilizes the photosynthetic machinery against low-temperature photoinhibition in transgenic tobacco plants. Proc Natl Acad Sci USA 1995; 92:6219-6223.
192. Somerville C. Direct tests of the role of membrane lipid composition in low-temperature-induced photoinhibition and chilling sensitivity in plants and cyanobacteria. Proc Natl Acad Sci USA 1995; 92:6215-6218.
193. Wu J, Lightner J, Warwick N, Browse J. Low-temperature damage and subsequent recovery of *fab1* mutant Arabidopsis exposed to 2°C. Plant Physiol 1997; 113:347-356.
194. Kanervo E, Tasaka Y, Murata N, Aro E-M. Membrane lipid unsaturation modulates processing of the photosystem II reaction-center protein D1 at low temperatures. Plant Physiol 1997; 114:841-849.
195. White AJ, Dunn MA, Brown K, Hughes MA. Comparative analysis of genomic sequence and expression of a lipid transfer protein gene family in winter barley. J Exp Bot 1994; 45:1885-1892.
196. Houlston CE, Pearce RS, Dunn MA, Hughes MA. *In situ* localisation of expression of cold-induced barley genes. J Exp Bot 1995; 46 (suppl):67.
197. Thoma S, Kaneko Y, Somerville C. A non-specific lipid transfer protein from *Arabidopsis* is a cell wall protein. Plant J 1993; 3:427-436.
198. Pyee J, Yu H and Kolattukudy PE. Identification of a lipid transfer protein as the major protein in the surface wax of broccoli (*Brassica oleracea*) leaves. Arch Biochem Biophys 1994; 311:460-468.
199. Pyee J and Kolattukudy PE. The gene for the major cuticular wax-associated protein and three homologous genes from broccoli (*Brassica oleracea*) and their expression patterns. Plant J 1995; 7:49-59.
200. Molina A, García-Olmedo F. Developmental and pathogen-induced expression of three barley genes encoding lipid transfer proteins. Plant J 1993; 4:983-991.
201. Brule-Babel AL, Fowler DB. Genetic control of cold hardiness and vernalization requirement in rye. Genome 1989; 32:19-23.
202. Mäntylä E. Molecular mechanisms of cold acclimation and drought tolerance in plants. PhD Thesis. Uppsala: Swedish Univ Agric Sci, 1997.
203. Gilmour SJ, Thomashow MF. Cold acclimation and cold-regulated gene expression in ABA mutants of *Arabidopsis thaliana*. Plant Mol Biol 1991; 17:1233-1240.
204. Nordin K, Vahala T, Palva ET. Differential expression of two related, low-temperature-induced genes in *Arabidopsis thaliana* (L.) Heynh. Plant Mol Biol 1993; 21:641-653.
205. Yamaguchi-Shinozaki K, Shinozaki K. A novel *cis*-acting element in an *Arabidopsis* gene is involved in responsiveness to drought, low-temperature, or high-salt stress. Plant Cell 1994; 6:251-264.

206. Skriver K, Mundy J. Gene expression in response to abscisic acid and osmotic stress. Plant Cell 1990; 2:503-512.
207. Bohnert HJ, Nelson DE, Jensen RG. Adaptations to environmental stresses. Plant Cell 1995; 71099-71111.
208. de Vetten NC, Ferl RJ. Transcriptional regulation of environmentally inducible genes in plants by an evolutionary conserved family of G-box binding factors. Int J Biochem 1994; 26:1055-1068.
209. Dolferus R, De Bruxelles G, Dennis ES, Peacock WJ. Regulation of the *Arabidopsis Adh* gene by anaerobic and other environmental stresses. Ann Bot 1994; 74:301-308.
210. Horwath DP, McLarney BK, Thomashow MF. Regulation of *Arabidopsis thaliana* L. (Heyn) *cor78* in response to low temperature. Plant Physiol 1993; 103:1047-1053.
211. Wilhelm KS, Thomashow MF. *Arabidopsis thaliana cor15b*, an apparent homologue of *cor15a*, is strongly responsive to cold and ABA, but not drought. Plant Mol Biol 1993; 23:1073-1077.
212. Jarillo JA, Capel JC, Leyva A, Martinez-Zapater JN, Salinas J. Two related low-temperature-inducible genes of *Arabidopsis* encode proteins showing high homology to 14-3-3 proteins, a family of putative kinase regulators. Plant Mol Biol 1994; 25:693-704.
213. Capel J, Jarillo JA, Salinas J, Martinez-Zapater JM. Two homologous low-temperature-inducible genes from *Arabidopsis* encode highly hydrophobic proteins. Plant Physiol 1997; 115:569-576.
214. de Bruxelles GL, Peacock WJ, Dennis ES, Dolferus R. Abscisic acid induces the alcohol dehydrogenase gene in *Arabidopsis*. Plant Physiol 1996; 111:381-391.
215. Merlot S, Giraudat J. Genetic analysis of abscisic acid signal transduction. Plant Physiol 1997; 114:751-757.
216. Baker SS, Wilhelm KS, Thomashow MF. The 5'-region of *Arabidopsis thaliana cor15a* has *cis*-acting elements that confer cold-, drought-, and ABA-regulated gene expression. Plant Mol Biol 1994; 24:701-713.
217. Stockinger EJ, Gilmour SJ, Thomashow MF. *Arabidopsis thaliana CBF1* encodes an AP2 domain-containing transcriptional activator that binds to the C-repeat/DRE, a cis-acting DNA regulatory element that stimulates transcription in response to low temperature and water deficit. Proc Natl Acad Sci USA 1997; 94:1035-1040.
218. Jaglo-Ottosen KR, Gilmour SJ, Zarka DG, Schabenberger O, Thomashow MF. *Arabidopsis CBF1* overexpression induces *COR* genes and enhances freezing tolerance. Science 1998; 280:104-106.
219. Olson Å. Molecular responses to cold stress. Regulation of low temperature-induced genes in *Arabidopsis thaliana*. PhD Thesis. Uppsala: Swedish Univ Agric Sci, 1997.
220. Tähtiharju S, Sangwan V, Monroy AF, Dhindsa AF, Borg M. The induction of *kin* genes in cold-acclimating *Arabidopsis thaliana*. Evidence of a role for calcium. Planta 1997; 203:442-447.
221. Harmon AC, Lee J-Y, Yoo B-C, Shao J. Plant membrane-associated protein kinases: proposed great communicators. In: Smallwood M, Knox JP, Bowles DJ, eds. Membranes: Specialized Functions in Plants. Oxford: BIOS Scientific Publ Ltd, 1996:137-150.
222. Hong SW, Jon JH, Kwak JM, Nam HG. Identification of a receptor-like protein kinase gene rapidly induced by abscisic acid, dehydration, high salt, and cold treatments in *Arabidopsis thaliana*. Plant Physiol 1997; 113:1203-1212.
223. Ayres PG, West HM. Stress responses in plants infected by pathogenic and mutualistic fungi. In: Fowden L, Mansfield T, Stoddart J, eds. Plant Adaptation to Environmental Stress. London: Chapman & Hall, 1993:295-311.
224. Tronsmo AM. Predisposing effects of low temperature on resistance to winter stress factors in grasses. Acta Agric Scand 1984; 34:210-220.
225. Tronsmo AM. Induced resistance to biotic stress factors in grasses by frost hardening. In: Kaurin Å, Junttila O, eds. Plant Production in the North. Tromsø: Norwegian Univ Press, 1985: 127-133.

226. Gaudet DA. Progress towards understanding interactions between cold hardiness and snow mold resistance and development of resistant cultivars. Can J Plant Pathol 1994; 16:241-246.
227. Collinge DB, Gregersen PL, Thordal-Christensen H. The induction of gene expression in response to pathogenic microbes. In: Basra AS, ed. Mechanisms of Plant Growth and Improved Productivity. Modern Approaches. New York: Marcel Dekker, Inc, 1994:391-433.
228. Benhamou N, Grenier J, Asselin A, Legrand M. Immunogold localization of β-1,3-glucanases in two plants infected by vascular wilt fungi. Plant Cell 1989; 1:1209-1221.
229. Wubben JP, Eijkelboom CA, De Wit PJGM. Accumulation of pathogenesis-related proteins in the epidermis of tomato leaves infected by *Cladosporium fulvum*. Neth J Plant Pathol 1993; 99 (3 suppl):231-239.
230. Hincha DK. Rapid induction of frost hardiness in spinach seedlings under salt stress. Planta 1994; 194:274-278.
231. Zhu B, Chen THH, Li PH. Expression of three osmotin-like genes in response to osmotic stress and fungal infection in potato. Plant Mol Biol 1995; 28:729-735.
232. Zhu B, Chen THH, Li PH. Activation of two osmotin-like protein genes by abiotic stimuli and fungal pathogen in transgenic potato plants. Plant Physiol 1995; 108:929-937.
233. Yun D-J, D'Urzo MP, Abad L, Takeda S, Salzman R, Chen Z, Lee H, Hasegawa PM, Bressan RA. Novel osmotically induced antifungal chitinases and bacterial expression of an active recombinant isoform. Plant Physiol 1996; 111:1219-1225.
234. Foyer CH, Lopez-Delgado H, Dat JF, Scott IM. Hydrogen peroxide- and glutathione-associated mechanisms of acclimatory stress tolerance and signalling. Physiol Plant 1997; 100:241-254.
235. Bridger GM, Yang W, Falk DE, McKersie BD. Cold acclimation increases tolerance to activated oxygen in winter cereals. J Plant Physiol 1994; 144:235-240.
236. Asada K. Production and action of active oxygen species in photosynthetic tissues. In: Foyer CH, Mullineaux PM, eds. Causes of Photooxidative Stress and Amelioration of Defense Systems in Plants. Boca Raton: CRC Press, 1994:77-104.
237. Inzé D, Van Montagu M. Oxidative stress in plants. Curr Opin Biotechnol 1995; 5:153-158.
238. Kendall EJ, McKersie BD. Free radical and freezing injury to cell membranes of winter wheat. Physiol Plant 1989; 76:86-94.
239. McKersie BD, Chen Y, de Beus M, Bowley SR, Bowler C, Inzé D, D'Halluin K, Botterman J. Superoxide dismutase enhances tolerance of freezing stress in transgenic alfalfa (*Medicago sativa* L.). Plant Physiol 1993; 103:1155-1163.
240. McKersie BD, Bowley SR, Harjanto E, Leprince O. Water-deficit tolerance and field performance of transgenic alfalfa overexpressing superoxide dismutase. Plant Physiol 1996; 111:1177-1181.
241. Wu G, Shortt BJ, Lawrence EB, León J, Fitzsimmons KC, Levine EB, Raskin I, Shah DM. Activation of host defence mechanisms by elevated production of H_2O_2 in transgenic plants. Plant Physiol 1997; 115:427-435.
242. Alscher RG, Donahue JL, Cramer CL. Reactive oxygen species and antioxidants: Relationships in green cells. Physiol Plant 1997; 100:224-233.
243. Huner NPA, Maxwell DP, Gray GR, Savitch LV, Krol M, Ivanov AG, Falk S. Sensing environmental temperature change through imbalances between energy supply and energy consumption: Redox state of photosystem II. Physiol Plant 1996; 98:358-364.
244. Gray GR, Chauvin L-P, Sarhan F, Huner NPA. Cold acclimation and freezing tolerance. A complex interaction of light and temperature. Plant Physiol 1997; 114:467-474.
245. Hincha DK, Sonnewald U, Willmitzer L, Schmitt JM. The role of sugar accumulation in leaf frost hardiness – investigation with transgenic tobacco expressing a bacterial pyrophosphatase or a yeast invertase gene. J Plant Physiol 1996; 147:604-610.
246. Lång V. The role of ABA and ABA-induced gene expression in cold acclimation of *Arabidopsis thaliana*. PhD Thesis. Uppsala: Swedish Univ Agric Sci, 1993.
247. Calkins J, Swanson BT. The distinction between living and dead plant tissue - viability tests in cold hardiness research. Cryobiology 1990; 27:194-211.

Gene expression and cold hardiness in animals

K. B. Storey* and J. M. Storey

Institute of Biochemistry, Departments of Biology and Chemistry, Carleton University, 1125 Colonel By Drive, Ottawa, Ontario, Canada K1S 5B6

1 Introduction

The earth is a cold place. Many polar and alpine environments are perpetually cold and the deep ocean is consistently only 2–4°C. Huge parts of the planet are also seasonally cold and frequently present terrestrial organisms with the challenge of sustaining life at subzero temperatures that are well below the freezing point of body fluids. Nonetheless, life abounds in all of these environments, and organisms have evolved marvelous adaptations to allow them to deal with cold and/or freezing. For poikilothermic organisms seeking to survive when ambient air temperature drops below 0°C, the choices are basically three:

1. use behavioral means to leave the cold environment (e.g. migration) or position themselves in insulated microenvironments where temperature remains above 0°C (e.g. hibernation underwater or deep underground),
2. endure subzero exposures by invoking metabolic adaptations that maintain body fluids in a liquid state (called freeze avoidance), or
3. endure the freezing of extracellular body fluids (called freeze tolerance).

Ecological, physiological and biochemical aspects of freeze avoidance and freeze tolerance in animals have been reviewed many times by ourselves and others.[1–7] Briefly, freeze avoidance involves mechanisms that greatly suppress the temperature at which body fluids crystallize. Taking advantage of the phenomenon of supercooling, the natural tendency of aqueous solutions to cool substantially below their equilibrium freezing points before spontaneously freezing, freeze-avoiding organisms employ mechanisms that stabilize the supercooled state and push the crystallization temperature to a value that is well below the expected environmental minimum that the organism must endure. Polar fish, for example, add specific antifreeze proteins (AFPs) to their body fluids to suppress crystallization temperature by about 2 degrees below the -0.5°C equilibrium freezing point of their blood,

* Corresponding author; e-mail: kbstorey@ccs.carleton.ca

just sufficient to keep the fish from freezing in seawater (full strength seawater at 1000 mOsmol has a freezing point of -1.86°C).[8–10] Terrestrial arthropods (insects, spiders, etc.) have pushed freeze avoidance to much lower limits. Those that winter under the snowpack can typically remain liquid to at least -10 to -15°C with a combination of strategies including the seasonal addition of AFPs (also called thermal hysteresis proteins in insects) to body fluids, elimination of potential ice nucleators from their bodies (e.g. clearing gut contents, seasonal changes in the protein composition of blood), protecting themselves from contact with environmental ice (e.g. with cocoons), and partial dehydration to reduce the free water content of their bodies.[5,7,11] Polar species and those that winter in sites that are not insulated by snow cover have pushed the limits even further, many remaining liquid down to -40 or -50°C. This is achieved with an additional adaptation, the accumulation of high concentrations of sugar or sugar alcohol cryoprotectants that provide colligative suppression of the crystallization temperature; levels of these compounds can often reach about 2 M or 20% of the fresh mass of the animal.[4,5,11]

Although less common than freeze avoidance, freeze tolerance as a mechanism of winter survival occurs in diverse groups of animals including many insects, various intertidal marine invertebrates (e.g. barnacles, mussels, snails), and several species of terrestrially hibernating amphibians and reptiles.[1–7,13] Freeze tolerance is generally defined as the ability to endure the formation of ice in extracellular body fluid compartments while employing mechanisms that defend the liquid state of the cytoplasm. For most species that use this strategy this means tolerating the conversion of 50–65% of total body water into extracellular ice. Physiologically, this has at least three serious consequences on cells and organs:

1. ice can do serious damage to the body tissues, particularly to the integrity of microcapillaries,
2. freezing of blood plasma halts the delivery of oxygen and nutrients to tissues,
3. osmotic effects caused by the "loss" of up to two-thirds of body water into ice crystals result in large reductions in cell volume and great increases in the osmolality and ionic strength of the cytosol.

Freeze-tolerant organisms need adaptations to counter all of these problems. To date, known adaptations include the following:

1. the use of specific ice nucleators (often plasma proteins) to initiate controlled freezing at relatively high subzero temperatures,
2. the ice-binding abilities of AFPs may be used to control the shape of growing crystals and to inhibit recrystallization during long-term freezing,
3. high concentrations of sugar or sugar alcohol cryoprotectants provide colligative limitation of the extent of cell volume reduction and of the percentage of total body water converted to ice,
4. specific membrane protectants (e.g. trehalose, proline) stabilize the lipid bilayer structure against compression stress as cell volume is reduced,
5. well-developed mechanisms of anoxia tolerance maintain cell viability during freeze-induced ischemia, and
6. physiological adjustments regulate the cessation and reactivation of vital signs (breathing, heart beat) during freeze/thaw.

The focus of the present chapter is on one aspect of cold hardiness and freeze tolerance that has only fairly recently come to be appreciated and explored, at least in animal systems. This is the role of gene expression as an adaptive mechanism. From one point of view, it has long been obvious that changes in gene expression must support these phenomena since many examples of changes in the amounts of gene products (proteins) can be cited, particularly during seasonal cold hardening. These include both the specific appearance of new proteins (e.g. AFPs) during cold hardening and the increased synthesis of others (e.g. elevated activities of enzymes needed for cryoprotectant synthesis).[4,10,14–17] Examples of cold-induced downregulation of proteins are also known (e.g. the autumn disappearance of a lipoprotein from stag beetle hemolymph enhances supercooling capacity) as well as examples of protein induction associated with the loss of cold hardiness (e.g. the appearance of cryoprotectant-degrading enzymes in the spring or when diapause is broken).[18,19] All of these are undoubtedly the primary result of changes in gene transcription although controls at translational and posttranslational levels may also be involved.

From another point of view, however, our present understanding of the mechanisms that confer cold hardiness or freeze tolerance is biased by a concentration of effort on the "highly visible" adaptations such as AFPs, ice-nucleating proteins or cryoprotectants, adaptations which provide natural protection against some of the most obvious injuries that occur when unprotected cells are chilled or frozen. However, it is highly likely that many important, but as yet unrecognized, adaptations remain to be discovered that support subzero survival. The question is how to find them. New developments in molecular biology are providing the answer. Modern screening techniques allow broad comparisons to be made of gene expression in control versus stressed states in order to find those genes that are upregulated by cold or freezing; after isolation and sequencing, identification of a stress-induced gene can generally be made via comparison of its DNA or deduced amino acid sequence with those stored in DNA or protein sequence banks. This approach allows an unbiased scan to be made to seek the genes/proteins that are key to the adaptive response to cold or freezing stress and, thereby, to obtain a handle on the full range of metabolic functions that undergo adaptation in order to confer cold hardiness or freeze tolerance upon cells and organisms. The remainder of this chapter will analyze some recent advances in the fields of animal cold hardiness and freeze tolerance that have come from studies of gene expression. This includes new advances in understanding the origin of AFPs, an analysis of cold shock and its role in cold-hardy animals, and the identification of freeze-induced proteins involved in vertebrate freeze tolerance via differential screening of cDNA libraries.

2
Antifreeze proteins

The structures and functions of the antifreeze peptides and proteins of marine teleost fish have been extensively studied and are very well understood.[8–10] The proteins are produced constitutively in polar species and seasonally in temperate species.[9] Four types of fish proteins are known. Antifreeze glycoproteins (AFGPs) occur in antarctic notothenioids and northern cod species and are composed of a tripeptide repeat (Ala-Ala-Thr) with a disaccharide attached to each Thr. The other three lack a carbohydrate component. Type I AFPs are amphiphilic α-helical polypeptides that are rich in alanine (as much as 60 mol%) and occur in flounders and sculpins. Type II AFPs have extensive β-sheet structure and are cysteine-rich with many disulfide bridges; they occur in sea raven, smelt and herring. Type III AFPs have no distinctive features to their amino acid composition and contain a β-sandwich secondary structure; this type occurs in ocean pout and wolffish. Despite their diverse structures, all AFPs act by a similar mechanism, known as adsorption-inhibition. The proteins attach to the growing prism faces of ice crystals to inhibit the further addition of water molecules to the crystal lattice. This action provides noncolligative suppression of the freezing point of body fluids by mechanisms of AFP/ice interactions that have been extensively reviewed elsewhere.[8,10]

A gene-based approach to studies of fish AFPs has recently made advances on several fronts. For example, use of cDNA and genomic clones has allowed examination of the organization and structure of AFP genes. All AFPs appear to be encoded by multigene families and in most cases with an organization of the genes as tandem repeats.[9] Such an arrangement of repeated duplication was probably the fastest way to achieve high expression of the genes in species that were under intense selective pressure to develop high levels of antifreeze output. Indeed, among flounder species, the number of gene copies correlates with the peak plasma AFP levels achieved and with the probable "need" for these proteins in the fish's environment (i.e. higher levels in shallow, versus deeper, water species where contact with ice is greater). A gene-based approach is also proving to be the best way of analyzing the tissue distribution of AFP genes and their function. Fish liver produces the circulating form of type I AFP, but using a liver AFP cDNA probe, several cDNA clones were isolated from a skin cDNA library of winter flounder.[20] Distinct but homologous alanine-rich AFPs were found in skin but the mature polypeptides all lack both the signal and prosequences that are found on liver AFPs, suggesting that the proteins play a role in intracellular freezing protection (notably skin type AFPs are not found in blood). Northern blot analysis also showed that skin-type AFP was strongly expressed in other exterior tissues such as scales, fin and gills.[21] Hence, two multigene AFP families exist in winter flounder, although both contain the same functional amino acid repeats. One produces the secretory liver-type AFPs and the other the nonsecretory AFPs that provide strong antifreeze protection to surface tissues of the fish but are also expressed in lower levels in internal organ.

Other important studies are discovering the evolutionary history of these genes. Fish AFPs and the genes that encode them are relatively new acquisitions that appeared in diverse groups of coldwater and polar fishes an estimated 10–14 million years ago when sea level glaciation began to occur again after an interruption of more than 200 million years. In evolutionary terms, then, the pressure to develop antifreezes was rapid and recent. How was this accomplished? Protein analysis has given no clue as to the origin of AFPs but molecular biology techniques have recently shown that AFPs arose by the recruitment and modification of existing genes. The first example of this came from studies of the cysteine-rich type II AFPs which characterize three distantly related types of fish: sea raven, smelt and herring. Despite phylogenic divergence that occurred long before the Cenozoic glaciation that created the need for AFPs, the proteins of all three fish share sequence identity.[21] Furthermore, they all share structural similarity with lectin-binding proteins and pancreatic stone protein, particularly with the carbohydrate recognition domains of the calcium-dependent lectins.[21–23] However, the AFPs have lost one (smelt, herring) or both (sea raven) of the calcium-binding domains.[21,23] Interestingly, the thermal hysteresis activity of smelt AFP is strongly reduced in the absence of calcium whereas that of sea raven is unaffected.[21] Herring AFP requires Ca^{2+} for normal antifreeze activity (Mn^{2+} can substitute but not other divalent cations) and Ca^{2+} binding converts the protein from a protease-sensitive, inactive apoAFP to a protease-resistant, active AFP.[23] It appears, then, that type II AFPs have arisen independently in phylogenetically distinct fish through the common recruitment and modification of C-type lectin genes.

The AFGPs of antarctic notothenioids and arctic cod present another example of gene evolution. Their AFGPs are nearly identical and consist of a family of isoforms of different sizes that are all composed primarily of a simple glycotripeptide repeat $(\text{Thr-Ala/Pro-Ala})_n$ with galactose-*N*-acetylgalactosamine attached to each Thr.[8] How this odd structure came to be was unknown until studies began to address the structure of the notothenioid AFGP genes. Each gene encodes a large polyprotein precursor containing multiple AFGP molecules. The polyprotein produced by the first AFGP gene to be characterized contained 46 AFGP molecules (44 of isoform 8 and 2 of isoform 7) linked in direct tandem by a 3-residue spacer that was the post-translational cleavage site.[24] The clue to the origin of this strange gene came when the 3' flanking sequence from the termination codon to about 100 nucleotides (nt) downstream was fed into the GenBank database. This indicated about 80% homology with the C terminus of the trypsinogen cDNA of another fish species.[24] Subsequent isolation and sequencing of the AFGP and trypsinogen genes from pancreas of the same notothenioid species (*Dissostichus mawsoni*) showed that the former clearly evolved from the latter. The 5' and 3' ends of the trypsinogen gene were conserved in the AFGP gene (94 and 96% identical) to provide the secretory signal sequence and the 3' untranslated region and then amplification of a 9 nt Thr-Ala-Ala coding element in the trypsinogen progenitor gave rise to the new protein coding region that produces the antifreeze peptides.[24] Not only is the identification of trypsinogen as the progenitor of AFGPs highly significant itself as an excellent example of how an existing gene can evolve into a new one that produces a completely different protein product but it suggests the possible evolutionary origins of various other antifreeze and ice nucleating proteins. As one of the digestive enzymes secreted from the pancreas into the lumen of the intestine, trypsinogen is a constitutive component of the intestinal fluids. These fluids in antarctic fish also contain high levels of AFGPs to inhibit the growth of ice crystals that can enter the fish during ingestion of food. Presumably the need to find a way to prevent intestinal fluid from freezing put a strong evolutionary pressure on genes that encoded proteins found in the intestinal fluid to select and modify any that showed some antifreeze action. Later, the expression of AFGP genes was apparently extended to the liver to provide the plasma antifreeze that is circulated to other extracellular fluid compartments of the fish.

Given the common derivation of type II AFPs from lectins, it might be predicted that the northern cods may have also derived their AFGPs from trypsinogen. Apparently, however, this cannot be the case, for the AFGP gene sequences from cod share no sequence identity with the trypsinogen gene.[25] Futhermore, although the final AFGP products of both species are near identical, the genes of the two fish contain different coding sequences for the functional tripeptide as well as very different intro-exon organization and spacer sequences so that very different processing of the polyprotein precursors must occur.[25] Hence, the arctic cod AFGP gene could not have arisen from the trypsinogen gene and the AFGPs of the two types of fish must, therefore, represent examples of convergent evolution where the same protein product arose from two different genomic origins.

3
Cold-induced gene expression and shock proteins

Temperature change affects the rates of all cellular reactions and the structural properties of all biological macromolecules. It is not surprising, then, that much effort has been put into determining how the proteins of poikilothermic organisms adapt to changing temperatures on acute, seasonal or acclimatory, and evolutionary time scales.[26] With respect to freeze-tolerant or freeze-avoiding species specifically, there has been little or no study of how or whether constitutive proteins and enzymes are altered to allow homeostasis to be maintained in the face of the deep subzero temperatures of winter. The examples of protein/enzyme activity changes in these types of organisms that have been identified are primarily related to changing specific functions in the cold, such as for cryoprotective purposes (e.g. enhanced activities of enzymes in pathways of polyol synthesis),[17] rather than for maintaining homeostasis of cellular metabolism. However, the mechanisms promoting the latter function have been extensively analyzed in other systems, very often using fish models, and are assumed to be of relevance to other situations.[26]

How do changes in gene expression aid a poikilothermic organism in adjusting to a lower environmental temperature on an acclimatory or seasonal basis? First of all, it should be noted that many enzymes/proteins do not show changes in either the amount or type of molecular species during adaptation to a different temperature.[26] Thus, there may be no need to alter the activities of many enzymes that are constitutively present in high activities because these may have no effect on net flux through the pathways in which they participate even if temperature change alters their properties somewhat. The function of other enzymes may be normalized by concurrent changes in other aspects of metabolism such as changes in steady-state substrate levels, stabilization by polyols or other solutes, temperature-dependent changes in cellular pH, or changes in the phospholipid composition of membranes in which transporter proteins are embedded. For those enzymes/proteins whose levels are adjusted for low temperature function, however, several mechanisms exist. In some cases, low temperature leads to adjustments of enzyme concentrations; for example, lactate dehydrogenase-B (LDH-B) activity is higher in heart of northern, cold water versus southern, warm water populations of *Fundulus heteroclitus* due to higher levels of gene transcription and greater amounts of LDH-B mRNA in tissue from cold water fish.[27] Changes in the mixture of allelic variants of an enzyme that are expressed can also provide temperature compensation. Six allelic variants of LDH-B in *F. heteroclitus*, each differing slightly in kinetic properties or thermal stability, appear in different mixtures in fish from different thermal environments.[28] In other instances, expression of different isoforms provides temperature compensation. In carp, for example, muscle function at low temperature can be readjusted by expressing a different mixture of the 28 genes coding for myosin heavy chain isoforms to achieve a higher specific myofibrillar ATPase activity.[29] Acclimation of goldfish to different temperatures also led to the expression of different isoforms of Ca^{2+}-adenosine triphosphatase[30] and similar acclimations induced different isoforms of acetylcholinesterase in trout brain.[31]

Apart from varying the mixture of allelic or isozymic variants, other proteins and enzymes have been identified as being cold-inducible. Some have known functions. For example, cold acclimation of carp results in an 8- to 10-fold increase in the activity of microsomal stearoyl coenzyme A desaturase within 48–60 h, an enzyme that is needed to help restore membrane fluidity at low temperatures by increasing phosphoglyceride unsaturation.[32] In other cases, cold-induced proteins remain unidentified. Thus, *in vitro* translation of the mRNA transcripts isolated from wheat after transfer from 24 to 2°C showed the increased or induced appearance of 71 mRNA species. Eighteen of these appeared only transiently with peak expression after 24 h but levels of 52 mRNA species (and presumably their protein products as well) remained high over the subsequent 4 weeks needed for the plants to acquire freeze tolerance.[33] Of those, 34 were expressed in higher levels in freezing-tolerant winter wheat varieties than in the less tolerant varieties of spring wheat; hence, in this situation, it appears that the acquisition of cold and freeze tolerance may involve the work of many different protein types. What are these proteins?

The first and most thoroughly studied of the protein/gene responses to thermal stress was the heat shock response.[34,35] Exposure of organisms to high temperatures results in a temporary cessation of synthesis of the normal proteins in cells and is replaced instead by production of a conserved suite of proteins known as heat shock proteins (hsps). These typically fall into three molecular weight classes of about 90 kDa, 70 kDa, and small (15–30 kDa) proteins. The heat shock response seems to be ubiquitous in nature and this suggests that the expression of hsps serves a critical homeostatic function. Some hsps appear to be proteases or act to target proteins for proteolysis; their role, then, is to get rid of heat-damaged proteins.[36] Many, however, function as chaperones whose role is to assist in the correct folding, oligomerization and compartmentation of other cellular proteins.[37] Some chaperones are constitutively expressed in cells but others are induced in response to stresses (such as heat) that can cause the unfolding of extant proteins or the misfolding of nascent polypeptides. Heat-induced unfolding of proteins can expose hydrophobic regions that are normally buried in the folded protein and if these regions make contact with similar sites on other unfolded proteins or nascent polypeptides, the result is nonspecific aggregation. Chaperones act to prevent such improper aggregation and promote the proper refolding of proteins.

Cold can also disrupt protein conformation. Low temperature destabilizes hydrophobic interactions but increases the strength of hydrophilic bonds.[38] Because bonds of both these types contribute to many aspects of protein structure (e.g. conformation, subunit association, polymer formation) and function (e.g. binding of substrates or effectors to enzymes or of ligands to receptors), the properties of many proteins can change substantially at low temperature, including numerous instances of cold denaturation.[39] It was perhaps not surprising, then, when studies began to show that hsps induced by heat shock could also enhance survival of cold exposure[40] and that hsps could also be directly induced by cold shock. Cold shock induction of proteins is now well characterized in bacteria where a family of cold shock proteins (csps) with high sequence similarity has been identified.[36,41] Cold shock (e.g. 30 to 10°C transfer) also induces expression of several genes in yeast.[42] In insects, heat shock at 37–41°C elicited synthesis of seven hsps in diapausing pharate larvae of the gypsy moth *Lymantria dispar* (90, 75, 73, 60, 42, 29 and 22 kDa) whereas cold shock (-10 to -20°C) led to production of the two high molecular weight proteins (90, 75 kDa) and the additional synthesis of the 29-kDa protein if recovery was at 25°C, rather than 4°C.[43] Interestingly, whereas normal protein synthesis was suppressed during heat shock and recovery from heat shock in *L. dispar*, synthesis of normal "housekeeping" proteins continued during cold-induced shock protein synthesis; the same phenomenon as been reported in wheat.[33,43] Cold exposure also elicited synthesis of seven different stress proteins in the biting midge *Culicoides variipennis* and induced 72- and 92-kDa proteins in pharate adults of the flesh fly *Sarcophaga crassipalpis*.[44,45] Heat shock mRNA also accumulated in *Drosophila melanogaster* salivary glands during the recovery period after cold exposure.[46] Both heat (32°C) and cold (5°C) also enhanced synthesis of a 65-kDa protein in epidermal cell cultures from the bullfrog *Rana catesbeiana* but only heat exposure stimulated hsp synthesis in a kidney cell line from *Xenopus laevis*.[47] Human cells in culture also responded to transient chilling at 4°C with synthesis of hsp 70 when they were returned to 37°C; activation was attributable to a transcriptional event mediated by activation of the heat shock transcription factor.[48] Similarly, hsp70 mRNA was induced in brown adipose tissue by cold exposure of mice; induction was associated with enhanced binding of heat shock transcription factors to DNA and could be blocked by treatment with adrenergic receptor antagonists.[49]

The above examples of cold-induced synthesis of hsps come largely from organisms that are not naturally cold-tolerant or that were not cold-adapted at the time of the experiment. Clearly, hsps or csps play a key role in helping cells to protect their proteins and preserve integrated metabolic function in the face of a large thermal stress, often outside the normal range of temperatures in which they live. The temperature that constitutes a thermal stress for any given species lies just a few degrees outside the normal body temperature range. Amazingly, hsps can be induced just above 0°C for antarctic fish but not until near 100°C for hyperthermophilic bacteria.[50,51] The threshold induction temperature for hsp synthesis is also an adaptable value; studies with goby fish found that the induction temperature for hsp90 in brain was 36–38°C for 30°C acclimated fish, 28°C for 20°C acclimated fish and 24°C for 10°C acclimated fish.[52] The cold shock response functions similarly. In bacteria and yeast synthesis of csps precedes the reestablishment of normal log-phase growth when cultures are suddenly switched from higher to lower temperatures (e.g. from 30 to 10°C) indicating that csps probably have a role, like hsps, in stabilizing those proteins and processes that are disrupted by the rapid temperature change. In cases where cold shock is more extreme, the synthesis of these proteins still occurs but does not necessarily overcome the damage done by cold exposure. For example, shock proteins were produced by flesh flies *S. crassipalpis* during the recovery period after cold shock but subsequent examination of most pharate adults showed that they were highly damaged.[45] Some csps, like hsps, may be chaperones that help to defend protein structure against thermal denaturation but roles are unknown in many cases.[53] A chaperone called trigger factor is cold-induced in *E. coli* and viability of cultures at 4°C correlates with its levels.[54] However, trigger factor reduced viability of cells at 50°C and, not surprisingly, its levels were negatively correlated with hsp content, indicating that chaperones with different functions are needed at high and low temperature. This is perhaps not surprising since it is not likely that the same proteins and/or cellular functions that need critical attention from chaperones at high temperature will also need it at low.

Whether shock proteins play a role in the survival of subzero temperatures and/or freezing by acclimated animals remains questionable, however. Although there are several examples, cited above, of hsp induction in response to cold shock in cold-sensitive species, studies to date have not found hsp synthesis in response to low temperatures in freeze-tolerant insects. Thus, when brain tissue taken from heat shocked (35°C, 1 h) larvae of the gall fly *Eurosta solidaginis* was incubated *in vitro* with ^{35}S-methionine at 25°C, shock proteins of 94 and 71 kDa were synthesized.[55] However, neither of these proteins was synthesized in comparable incubations when brain tissue was taken from cold (-5°C) or freezing (-20°C) exposed larvae, either autumn-sampled (not yet freeze-tolerant) or winter-sampled (fully freeze-tolerant) larvae.[55] We found the same result when extracts from whole *E. solidaginis* larvae were assessed for cross reaction with antibodies to mammalian hsps (K.B. Storey and D.D. Mosser, unpubl. data). The constitutive hsc70 and the glucose responsive grp78 were present in autumn-collected larvae exposed to 15 or -4°C in the laboratory and also in all eight samples taken at intervals between September and April from larvae held outdoors in their galls over the winter. However, the inducible hsp70 was missing in all cases. Furthermore, similar Western blot analyses showed no change in the expression of any of these three hsps in organs of the freeze-tolerant wood frog in response to either freezing or thawing.[56] Thus, whereas adaptive changes in the levels of constitutive chaperones may be part of the seasonal cold-hardening process (something that remains to be investigated), it does not appear that either cold nor freezing exposure constitute a "shock" for adapted species. It is also interesting that shock proteins are not part of the response to freeze/thaw, which suggests that the changes in the osmolality and ionic strength of the cytosol that are caused by freeze concentration of body fluids are not sufficient to constitute a denaturing stress on protein structure. Given that the apparent roles of most shock proteins are to deal with stress-induced malfolding of proteins, it can be suggested that animals that are tolerant of subzero temperature exposures must have made adaptations that deal with this potential problem. With respect to cold adaptation, there are two possibilities:

1. properties of proteins have been adjusted over evolutionary time so that their structure/function remains stable throughout the subzero range that constitutes their normal experience, or
2. other adaptations ensure protein stability at subzero values (e.g. low molecular weight cryoprotectants, constitutive chaperone-like proteins).

4 Freeze-induced gene expression

The wood frog *Rana sylvatica* is widely used as a model animal by our laboratory and others to analyze the biochemical and physiological adaptations that support natural freezing survival by a vertebrate.[1,2,6] Not only is the phenomenon of natural freeze tolerance extremely interesting in itself, but the mechanisms that underlie freezing survival by individual organs of the frog offer potential solutions for the development of cryopreservation technologies for human organ transplants.

Seasonal and species-specific changes in amounts of selected proteins that support freeze tolerance in frogs have been documented in a number of cases. These include elevated activities of some enzymes involved in the synthesis of glucose as the cryoprotectant by liver, increased numbers of plasma membrane glucose transporters in organs that produce (liver) and take up (muscle) cryoprotectant, and the appearance of proteinaceous ice nucleators in plasma.[56–61] To broaden our view of the types of protein adaptations that support freeze tolerance, we first used ^{35}S-labeled methionine and cysteine to look for both freeze-induced and thaw-induced new protein synthesis in wood frog liver. *In vivo* injection of ^{35}S-labeled amino acids resulted in different patterns of protein labeling in control versus thawed (after 12 h frozen at -1.4°C) frogs, as revealed by isoelectrofocusing (Fig. 1). In particular, proteins with isoelectric points (pI) of about 6.0 were prominent in liver of thawed frogs and labeled proteins in this peak persisted at 5, 10 or 24 h post-injection, representing a greater proportion of the total labeled protein over time. SDS-PAGE analysis of the pI 6.0 peaks from thawed samples revealed the presence of one major low molecular weight protein with a molecular mass of 15 kDa.[62] In another approach we isolated mRNA from liver of control, frozen, and thawed frogs, translated it *in vitro* using ^{35}S-labeled amino acids, and then assessed the resulting protein products. This approach revealed the presence of different populations of mRNA transcripts in all three metabolic situations resulting in the *in vitro* production of both freeze-specific and thaw-specific proteins. Liver showed the largest number of new protein types although heart, kidney, brain and skeletal muscle also showed differential patterns of proteins, indicating different populations of mRNA transcripts in organs of control, frozen and thawed frogs (D. Lobsinger and K. Storey, unpubl. data).

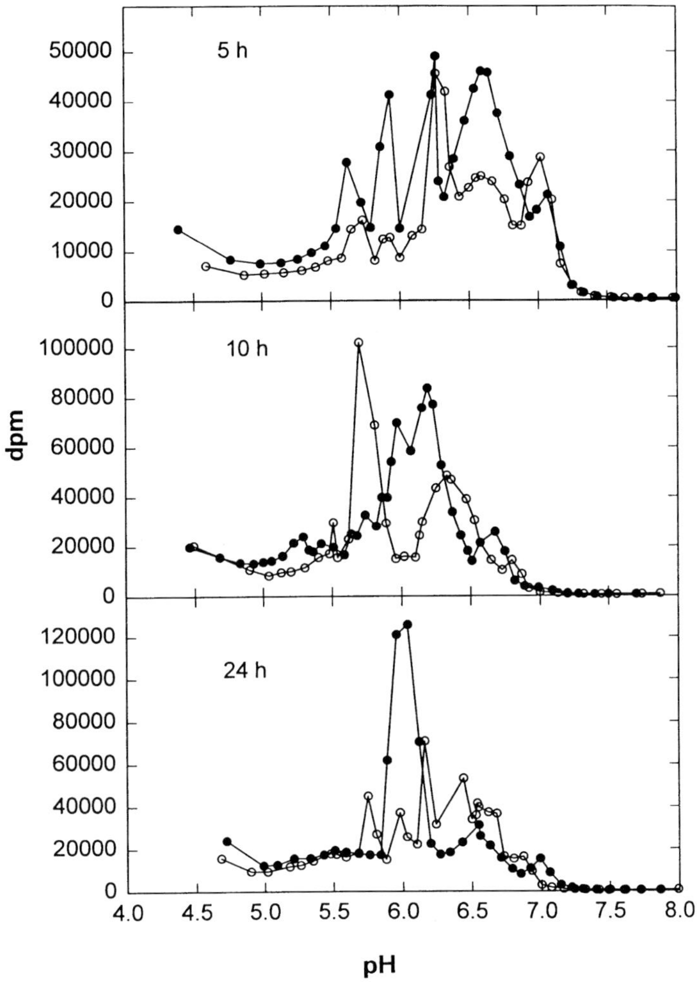

Fig. 1. Isoelectric focusing elution pattern of ^{35}S-labeled proteins produced in liver of wood frogs post-thaw (after 12 h frozen at -1.4°C)
Samples were taken at 5, 10 or 24 h post-injection; disintegrations min^{-1} (dpm) are plotted against measured pH of the fractions for liver extracts from control (*open circles*) and thawed (*filled circles*) frogs (reprinted with permission from ref. 62)

Both *in vivo* labeling and *in vitro* translation techniques, although clearly demonstrating the occurrence of freeze- or thaw-induced new protein synthesis in frog organs, are limited by their inability to identify the proteins involved except after painstaking isolation and purification of sufficiently large amounts of the protein to enable sequencing. A much better and more direct approach to figuring out what protein adaptations are key to freezing survival is to begin from the gene level using differential screening and cloning. Molecular biology techniques have already been applied quite widely to identify genes responsible for cold hardiness in plants,[63] including the genes coding for late embryogenesis abundant protein, dehydrin, hsps, and AFPs.[64–67] Some of these are specifically linked to cold hardening and/or freeze tolerance but others are also upregulated by dehydration or osmotic stresses. Other cold-regulated genes of unknown function are also known; for example, expression of the gene *COR15a* that encodes a chloroplast-targeted polypeptide, enhances *in vivo* freeze tolerance of chloroplasts in *Arabidopsis thaliana*.[68] In another instance, a five-member family of 50-kDa proteins that is cold-induced in wheat and accumulates to 0.9% of total soluble proteins appears to be critical to the freezing protection of cells in the vascular transition zone.[69] Comparable studies with animal systems have been lacking to date but recent work in our laboratory is taking this route. Our first results are uniquely interesting in their identification previously unsuspected protein types as important to freeze tolerance.

Our first studies of freeze-induced gene expression in a freeze-tolerant animal targeted wood frog liver. This central organ must not only express adaptations to protect itself during freezing but since liver is a major site for the synthesis of many plasma proteins, it is also logical to assume that various cryoprotective proteins could also be made and secreted by liver, in much the same way as liver synthesizes the carbohydrate cryoprotectant, glucose, for the entire animal. A cDNA library was produced from liver of frozen wood frogs (24 h at -2.5°C) and screened with ^{32}P-labeled total cDNA probes from liver of control (5°C) versus freezing-exposed frogs.[70] From this, five freezing-responsive cDNA clones representing different genes were isolated. Two of these were identified as homologous with the genes for the α and γ subunits of fibrinogen, respectively, and another showed homology with α-amylase pancreatase.[70] The fourth showed 81–86% homology with the ADP/ATP translocase (AAT), a critically important protein located in the inner mitochondrial membrane that mediates the exchange transport of ADP and ATP.[71] The fifth clone could not be identified by comparison with sequence banks but its cDNA sequence (designated *Fr10*) carried 457 bp and contained a single open reading frame that could encode a small protein of 90 amino acids with a molecular weight of about 10 kDa, named FR10.[72] This novel protein may have a unique function in freeze tolerance. Significantly, the deduced amino acid sequence of FR10 contains a potential nuclear exporting signal sequence (LALVVLVIAISGL) in its highly hydrophobic N-terminal region, similar to the nuclear exporting signal found in the protein kinase A inhibitor protein.

To assess the time course of freeze-induced gene expression, Northern blotting was used; total RNA extracted from livers sampled at different times was separated on formaldehyde agarose gels, blotted onto Nitran membranes and then hybridized against ^{32}P-labeled cDNA probes for the freeze-induced genes. Figure 2 shows the mRNA transcript levels for the fibrinogen genes over the course of 24 h freezing at -2.5°C followed by 24 h thawing recovery at 5°C. The α and γ subunits of fibrinogen were coordinately expressed, transcript accumulation reaching maximal levels of 3-fold greater than controls after 8 h freezing but retaining about 70% of the maximal mRNA levels after 24 h frozen.[70] After thawing, however, transcript levels

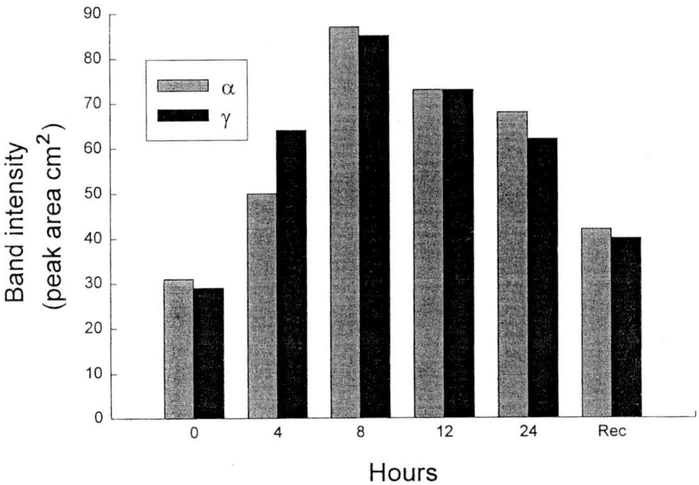

Fig. 2. Histogram showing the relative levels of mRNA transcripts for α and γ subunits of fibrinogen in wood frog liver over a time course of 4-24 h freezing at -2.5°C followed by 24 h thawing recovery (Rec) at 5°C after 24 h freezing
Total RNA isolated from liver samples was analyzed via Northern blots with ^{32}P-labeled cDNA probes for the α and γ subunits; autoradiograms were scanned and band intensities were quantified (reprinted with modification and permission from ref. 70)

fell and had returned to near control values after 24 h. Fibrinogen is composed of two halves each consisting of three subunits (Aα, Bβ and γ) and its structure is highly conserved among vertebrates, ranging from humans to lampreys.[73] The genes for the three subunits are independently transcribed but transcription rates of all three are coordinated.[74] So the similar patterns of freeze-induced gene expression of the two subunits are expected. Although our study did not turn up a freeze-induced clone corresponding to the β subunit, it is highly probable that freezing induced this subunit as well. Freeze-induction of the *Aat* gene had some different characteristics (Fig. 3). Northern blotting showed peak levels of mRNA for this gene (a 4.5-fold increase) after 8 h but levels quickly fell after that, being only about 2.5- and 2-fold higher than controls at 12 and 24 h, respectively, and falling somewhat lower than controls after 24 h thawing.[71] By contrast, the peak accumulation of AAT protein in liver (assessed by immunoblotting with the chicken antibody) was a 2-fold rise over controls which occurred after 24 h freezing and levels remained about 50% higher than controls after 24 h thawing (Fig. 3).[71] The mRNA levels for the unidentified *Fr10* gene showed a third pattern (Fig. 3). A 2-fold rise in mRNA levels occurred within the 1st h of freezing and continued to rise over the entire freezing episode; levels were restored to control values within the 24 h thaw.[72] Hence, the patterns of up- and down-regulation of all three genes were distinctly different and this suggests that (1) the expression of each gene may be under the control of different signals and transcription factors, and (2) the protein products of each gene address different freeze-related metabolic needs. This is also support-

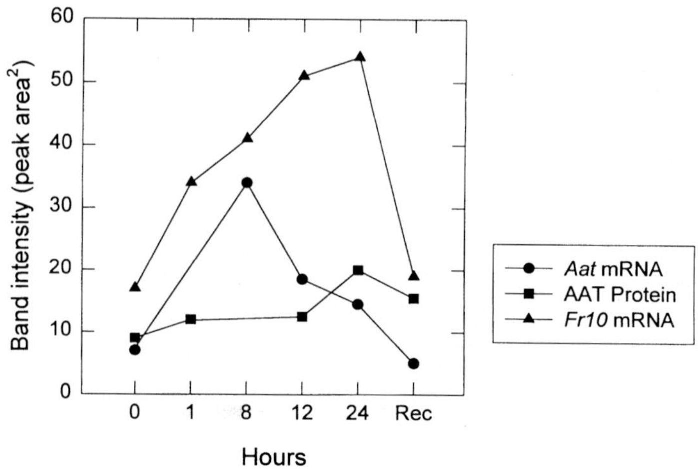

Fig. 3. Changes in *Aat* and *Fr10* mRNA transcript levels and ADP/ATP translocase (AAT) protein content in wood frog liver over a time course of *in vivo* freezing exposure at -2.5°C followed by 24 h thawing at 5°C (Rec); band intensities on Northern blots and immunoblots were scanned and quantified (reprinted with modification and permission from refs. 71,72)

ed by the organ-specific patterns of expression of these three genes in control versus 24 h frozen frogs, as revealed by Northern blotting of organ total mRNA using the liver-specific cDNA clones. Fibrinogen transcripts were strongly upregulated by freezing exposure in frog liver and lung, although total mRNA content was much lower in lung (Fig. 4). Bladder, gut and heart also showed some increase in fibrinogen transcripts, whereas there was no change in the negligible amount of signal detected in brain, kidney or skeletal muscle.[70] By contrast, *Fr10* expression was strongly induced by freezing in liver, lung, gut and bladder of frogs and mRNA levels also rose in brain and heart (Fig. 5).[72] Thus, except for kidney (transcript levels suppressed) and skeletal muscle (no change), *Fr10* induction is widespread in frog organs during freezing, suggesting that its protein product serves a need that is common to many tissues. This contrasts with the results for fibrinogen transcripts which are primarily confined to liver, the organ that is known to produce this plasma protein in all vertebrates. The mRNA transcripts of *Aat* showed a very different pattern of organ expression with strong freeze-induced elevation in liver, lung and bladder but an opposite strong suppression in kidney, heart and gut.[71] The content of AAT protein, as determined by immunoblotting with antibodies to the chicken enzymes, showed the same pattern.

What is critically significant about these results for freeze-tolerant frogs is that none of the liver genes/proteins that were identified by cDNA screening had a prior link with any of the known metabolic adaptations that support natural freeze tolerance. They all fall outside our current ideas about the processes and adaptations that confer freezing survival and, hence, illustrate that there is still much to learn about how cells respond and adapt to survive freezing. Furthermore, all proteins

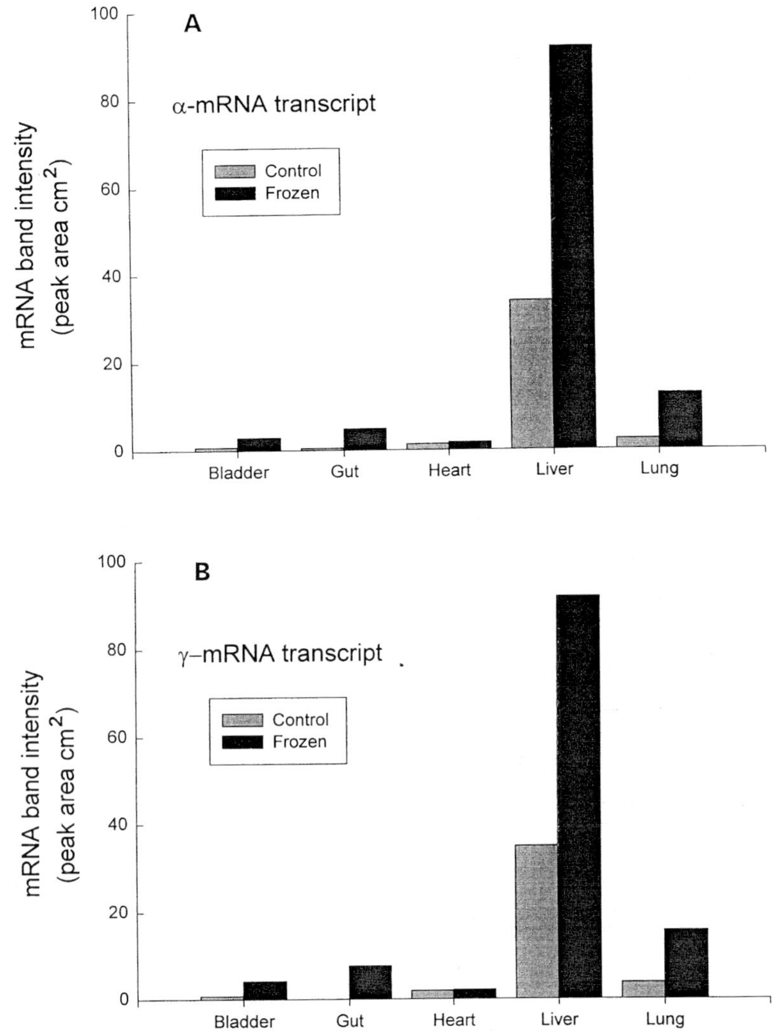

Fig. 4. Differential expression of fibrinogen α (**A**) and γ (**B**) genes in organs of control (5°C acclimated) versus 24 h frozen (-2.5°C) wood frogs
Total RNA isolated from tissue samples was hybridized with ^{32}P-labeled cDNA probes to measure tissue mRNA transcript levels; autoradiograms of Northern blots were scanned and band intensities were analyzed with Imagequant (reprinted with permission from ref. 70)

identified with these recent studies were upregulated in direct response to freezing of the animal, which shows that there are two kinds of adaptive protein responses involved in freeze tolerance – those that are induced seasonally and those that are directly stimulated by freezing. Most previously identified protein adaptations are seasonally induced. For example, the number of plasma membrane glucose trans-

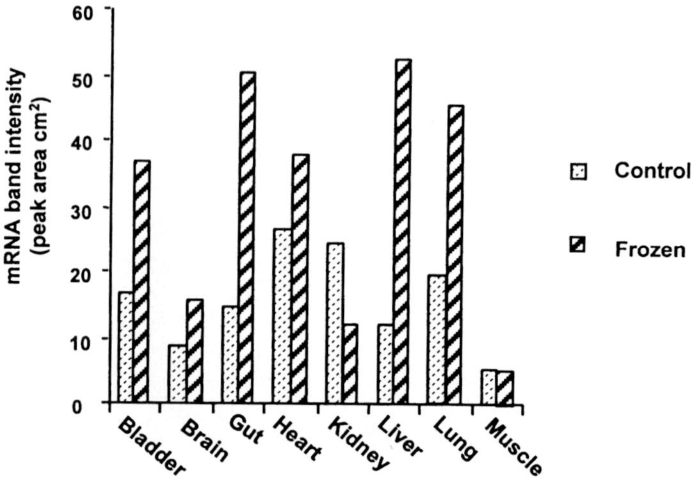

Fig. 5. Differential expression of *Fr10* gene in organs of control (5°C acclimated) versus 24 h frozen (-2.5°C) wood frogs
Total RNA isolated from tissue samples was hybridized with ^{32}P-labeled cDNA probes to measure tissue mRNA transcript levels; autoradiograms of Northern blots were scanned and band intensities were analyzed with Imagequant (reprinted with permission from ref. 72)

porters is 8.5-fold higher in liver of autumn, versus summer, wood frogs (based on cytochalasin B binding sites), an adaptation that would favor high rates of glucose export when the demand for cryoprotectant is triggered by ice forming in body tissues.[58]

What functions might the protein products of these genes serve in freezing survival? Fibrinogen is particularly interesting and significant. Fibrinogen is an acute-phase plasma protein. It is predominantly synthesized in vertebrate liver and secreted into the plasma, with production being stimulated by stresses such as infection, inflammation, and tissue injury.[75] As the final step in the coagulation cascade, thrombin cleaves near the N-termini of the Aα and Bβ chains to release the A and B fibrinopeptides and expose sites for polymerization into the fibrin mesh of a growing blood clot.[76] Hence, we believe that the upregulation of fibrinogen synthesis when frogs freeze may be viewed as a damage repair response, either initiated directly by minor tissue injuries developing as ice propagates through the frog's body or triggered in anticipation of tissue injuries that could become apparent during thawing. In this role, fibrinogen upregulation would address one of the primary causes of freezing damage that have been identified in cryomedical attempts to freeze transplantable tissues and organs. This is the physical damage done by ice expansion within the lumen of microcapillaries that breaks vessel walls and results in the loss of vascular integrity after thawing.[77] Indeed, we have frequently observed hematomas in the leg muscles of frogs after thawing (K.B. Storey and J.M. Storey, unpubl. data), so it can be suggested that some physical damage by ice to tissue/organ vasculature is a common problem for freeze-tolerant frogs. The upregulation of fibrinogen synthesis during freezing could contribute to maximizing the clotting capacity of the blood so that during thawing damaged capillary walls could be quickly and effectively sealed, thereby minimizing internal bleeding injuries. This repair mechanism, coupled with a preventative adaptation that limits the amount of ice that can form within organs (organs dehydrate during freezing, losing up to 25% of their water to extraorgan ice masses),[78] could effectively address the potential problem of physical damage to organs by ice crystals. It will be interesting to determine whether other blood proteins involved in the clotting cascade are also upregulated during freezing. Also interesting about the fibrinogen response was that fibrinogen genes were upregulated in several organs as the result of freezing stress. The protein is typically viewed as liver-specific in mammals[75] and only minute amounts of its transcript were found in extrahepatic organs of control frogs. However, elevated mRNA levels for both α and γ subunits were found in bladder, gut, heart and especially lung during freezing (Fig. 4). This may make a direct contribution to endogenous damage repair in each organ. All these organs may be particularly susceptible to ice damage since bladder, gut and heart are all organs that have a central lumen or chamber within which ice forms whereas lung is a very delicate tissue in frogs.

The role of AAT upregulation during freezing is more problematic. The known function of the ADP/ATP translocase is to catalyze the transmembrane exchange of cytosolic ADP with mitochondrial ATP generated by oxidative phosphorylation.[79] Thus, the protein is a key link between the mitochondrial and cytosolic compartments of aerobic eukaryotic cells. Northern blots showed that AAT was also induced by anoxia in wood frog liver, mRNA levels being greatly enhanced within 1 h under a nitrogen gas atmosphere at 5°C and remaining high after 24 h. This suggests that the freeze-induced elevation of AAT in liver is a response to the ischemia that develops as ice propagates through the frog's body, halting blood circulation and the delivery of oxygen to tissues. Elevated levels of the protein may contribute to minimizing net ATP depletion during freezing or may facilitate the rapid reestablishment of oxidative metabolism when the frog thaws. Wood frogs show good stability of organ adenylates during natural freezing episodes. In liver, for example, total adenylates dropped by only 27% over 3 d of freezing and energy charge fell from 0.86 to 0.70; both effects were soon reversed after thawing.[80] One possible function that the protein could play is in the regulation of the intramitochondrial adenine nucleotide pool size. Under anoxic conditions, cytoplasmic adenine nucleotides can be degraded, decreasing total adenine nucleotide levels.[81] However, the adenine nucleotide pool in the mitochondrial matrix can be preserved by a mechanism that shifts ADP to AMP, which is not subject to further dephosphorylation inside the mitochondria and cannot be transported into the cytosol.[82] Thus, a net movement of nucleotides into the mitochondria, mediated by AAT, can be advantageous in several ways for organisms enduring anoxia stress:

1. it suppresses the activities of ATP-dependent steps in mitochondrial metabolism (e.g. citrulline synthesis, pyruvate carboxylation), shutting down nonessential energy-expensive pathways such as urea synthesis and gluconeogenesis,[82]
2. it enables cells to recover bioenergetic function upon reoxygenation by simply rephosphorylating (rather than resynthesizing) matrix adenine nucleotides, and
3. it may stimulate intramitochondrial gene expression during the anoxia to normoxia transition.

In related studies with turtles, we have found that the genes for other proteins associated with mitochondrial bioenergetics are induced by freezing. While studying the role of gene expression in the anoxia tolerance of turtles we showed the anoxia-induced upregulation in heart of adult turtles (*Trachemys scripta elegans*) of several genes encoded on the mitochondrial genome. These were the genes for NADH-ubiquinone oxidoreductase subunit 5 (*Nad5*) and cytochrome c oxidase subunit 1 (*Cox1*) as well as the mitochondrial WANCY (tryptophan, alanine, asparagine, cysteine, tyrosine) tRNA gene cluster.[83,84] These turtles as adults can endure anoxia for as long as 3 months when submerged in cold water, an adaptation that supports winter hibernation at the bottom of lakes and ponds. During their first winter of life, however, turtles of both the *Trachemys* and related *Chrysemys* genera may encounter freezing conditions.[2] This is because young turtles, after hatching in late summer, stay in their shoreline nests all winter. This response reduces predation on the young turtles but it also exposes them to freezing in their terrestrial nests. Significantly, when we used Northern blotting to assess mRNA levels of the *Cox1* and *Nad5* genes that were upregulated by oxygen deprivation in adult turtles, we found that transcript levels of both genes were also elevated in organs from 20 h frozen hatchling painted turtles, *C. picta marginata*.[83] Levels of mRNA transcript in heart of frozen turtles were 3.5-fold higher for *Cox1* and 70% higher for *Nad5* than controls at 5°C. Both were also elevated in gut and *Nad5* transcript was high in kidney. Levels of the WANCY transcripts increased during freezing in heart, kidney, liver and brain. Thus, the data from both frogs and turtles show that selected proteins involved in mitochondrial bioenergetics are induced by both freezing and anoxia exposures leading to the conclusion that the role of these proteins in freeze-tolerant species is in dealing with the ischemia caused when blood plasma freezes.

The parallel induction of proteins associated with mitochondrial bioenergetics, *Aat* in frogs and *Cox1* and *Nad5* in turtles, by freezing and anoxia stresses also points towards the possible triggers, second messengers and transcription factors that could be involved in mediating freeze-induced gene expression. Freezing imposes multiple stresses on cells. Apart from ischemia, physical damage by ice crystals and dehydration (cell volume decreases and cellular osmolality and ionic strength increases when cell water exits to join extracellular ice masses) must be dealt with. Not only must these concerns be addressed for freezing survival but each of these individual component stresses may be responsible for triggering different parts of the total cryoprotection package. For example, we have demonstrated that dehydration of wood frogs at 5°C elicits the same hyperglycemic response as does freezing, including the same pattern of changes in second messengers, protein kinases and glycogenolytic enzymes in liver.[56,85–87] By contrast, anoxia exposure stimulated none of these. Like the hyperglycemic response, but opposite to *Aat* regulation[71], the expression of fibrinogen genes and of *Fr10* is also linked to a dehydration signal.[70,72] Northern blot analysis compared mRNA transcript levels of freeze-induced genes in livers sampled from wood frogs given experimental anoxia exposure (30 min, 4 or 48 h N_2 gas exposure) or experimental dehydration (20% of total body water lost) and rehydration (40% dehydration followed by rehydration to full water content), both at 5°C. Levels of mRNA transcript for both fibrinogen genes were dramatically elevated when frogs were dehydrated with induced levels similar to those seen in liver after 8 h freezing exposure.[70] By contrast, transcripts were barely detectable on Northern blots of mRNA from livers of anoxic frogs. The identical pattern was seen when *FR10* transcripts were assessed but *Aat* gene expression was the opposite, showing strong induction by anoxia but no change in gene expression during dehydration or rehydration.[71,72] Thus, the stresses imposed by freezing must work through at least two different signal transduction cascades to stimulate the necessary changes in gene expression that initiate freeze-induced adaptive responses. Continuing studies are exploring the signaling systems involved.

5 Conclusions and applications

In conclusion, then, a gene-based approach to studying cold hardiness and freeze tolerance offers multiple ways in which to expand our understanding of these phenomena whether it be deciphering the evolutionary development of proteins, such as AFPs, that confer cold hardiness (a comparable approach could be used to determine the origin of the ice nucleating proteins), or using gene screening methods to identify new protein adaptations that have critical roles in supporting subzero survival. Interesting applications will also arise out of the identification and study of cold- and freeze-induced gene expression. Several studies have already found uses for AFPs in improving the cold or freezing storage of cells and organs.[88–91] Their effects include an ability to block plasma membrane ion channels;[92,93] this action would help to slow or prevent the dissipation of membrane potential difference, which, when lost, leads to multiple serious injuries to cellular metabolism.[94] Transgenic fish (salmon, goldfish) have also been engineered the express AFPs (although at very low levels)[95,96] but the possibility remains of using this method to impart freezing resistance to fish in aquaculture. Furthermore, as noted earlier, one of the critical theoretical results of studies using cDNA screening technology is the ability to search, in an unbiased way, for the full range of protein adaptations that support cold or freezing survival. These types of studies with frogs have already produced several novel findings and implicated the upregulation of selected specific proteins (e.g. fibrinogen) and processes (mitochondrial energetics) that were previously unsuspected as having key roles in freezing survival. This, in turn, suggests that in applied applications of freezing preservation, such as the hyperthermic or freezing storage of tissues and organs for transplant, that attention needs to be paid to stabilizing or protecting the metabolic functions represented by the genes/proteins that are freeze-induced in natural systems.

6 References

1. Storey KB, Storey JM. Freeze tolerance in animals. Physiol Rev 1988; 68:27-84.
2. Storey KB, Storey JM. Natural freeze tolerance in ectothermic vertebrates. Ann Rev Physiol 1992; 54:619-637.

3. Storey KB, Storey JM. Natural freezing survival in animals. Ann Rev Ecol Syst 1996; 27:365-386.
4. Zachariassen KE. Physiology of cold tolerance in insects. Physiol Rev 1985; 65:799-832.
5. Duman JG, Wu DW, Xu L, Tursman D, Olsen TM. Adaptations of insects to subzero temperatures. Quart Rev Biol 66:387-410.
6. Layne JR, Lee RE. Adaptations of frogs to survive freezing. Climate Res 1995; 5:53-59.
7. Lee RE, Costanzo JP, Mugnano JA. Regulation of supercooling and ice nucleation in insects. Eur J Entomol 1996; 93:405-418.
8. Cheng CC, deVries AL. The role of antifreeze glycopeptides and peptides in the freezing avoidance of cold-water fish. In: di Prisco G, ed. Life Under Extreme Conditions. Berlin: Springer, 1991:1-14.
9. Davies PL, Hew CL, Fletcher GL. Fish antifreeze proteins: physiology and evolutionary biology. Can J Zool 1988; 66:2611-2617.
10. Davies PL, Hew CL. Biochemistry of fish antifreeze proteins. FASEB J 1990; 4:2460-2468.
11. Storey KB, Storey JM. Freeze tolerance and freeze avoidance in ectotherms. In: Wang LCH, ed. Advances in Comparative and Environmental Physiology, vol 4. Berlin: Springer, 1989:52-82.
12. Aarset AV. Freezing tolerance in intertidal invertebrates (a review). Comp Biochem Physiol A 1982; 73:571-580.
13. Wharton DA. Cold tolerance strategies of nematodes. Biol Rev 1995; 70:161-185.
14. Horwath KL, Duman JG. Photoperiodic and thermal regulation of antifreeze protein levels in the beetle *Dendroides canadensis*. J Insect Physiol 1983; 29:907-917.
15. Joanisse DR, Storey KB. Enzyme activity profiles in an overwintering population of freeze-tolerant larvae of the gall fly, *Eurosta solidaginis*. J Comp Physiol 1994; 164:247-255.
16. Joanisse DR, Storey KB. Enzyme activity profiles in an overwintering population of freeze-avoiding gall moth larvae, E*piblema scudderiana*. Can J Zool 1994; 72:1079-1086.
17. Storey KB, Storey JM. Biochemical adaptations for winter survival in insects. In: Steponkus PL. Advances in Low Temperature Biology, vol 1. Greenwich CT: JAI Press, 1992:101-140.
18. Neven LG, Duman JG, Beals JM, Castellino FJ. Overwintering adaptations of the stag beetle, *Ceruchus piceus*: removal of ice nucleators in the winter to promote supercooling. J Comp Physiol B 1986; 156:707-716.
19. Yaginuma T, Kobayashi M, Yamashita O. Distinct effects of different low temperatures on the induction of NAD-sorbitol dehydrogenase activity in diapause eggs of the silkworm, *Bombyx mori*. J Comp Physiol B1990; 160:277-285.
20. Gong Z, Ewart KV, Hu Z, Fletcher GL, Hew CL. Skin antifreeze protein genes of the winter flounder, *Pleuronectes americanus*, encode distinct and active polypeptides without the secretory signal and prosequences. J Biol Chem 1996; 271:4106-4112.
21. Ewart KV, Rubinsky B, Fletcher GL. Structural and functional similarity between fish antifreeze proteins and calcium-dependent lectins. Biochem Biophys Res Commun 1992; 185; 335-340.
22. Ng NFL, Hew CL. Structure of an antifreeze polypeptide from the sea raven. Disulfide bonds and similarity to lectin-binding proteins. J Biol Chem 1992; 267:16069-16075.
23. Ewart KV, Yang DSC, Ananthanarayanan VS, Fletcher GL, Hew CL. Ca^{2+}-dependent antifreeze proteins: modulation of conformation and activity by divalent metal ions. J Biol Chem 1996; 271:16627-16632.
24. Chen L, deVries AL, Cheng CHC. Evolution of antifreeze glycoprotein gene from a trypsinogen gene in Antarctic notothenioid fish. Proc Natl Acad Sci 1997; 94:3811-3816.
25. Chen L, deVries AL, Cheng CHC. Convergent evolution of antifreeze glycoproteins in Antarctic notothenioid fish and Arctic cod. Proc Natl Acad Sci 1997; 94:3817-3822.
26. Somero GN. Proteins and temperature. Annu Rev Physiol 1995; 57:43-68.
27. Crawford DL, Powers DA. Evolutionary adaptation to different thermal environments via transcriptional regulation. Mol Biol Evol 1992; 9:806-813.

28. Powers DA, Smith M, Gonzalez-Villasenor I, DiMichelle L, Crawford D, et al. A multidisciplinary approach to the selection/neutralist controversy using the model teleost, *Fundulus heteroclitus*. In: Futuyma D, Antonovics J, eds. Oxford Surveys in Evolutionary Biology, vol 9. Oxford: Oxford Univ Press, 1993:43-107.
29. Goldspink G, Turay L, Hansen E, Ennion S, Gerlach S. Switches in fish myosin genes induced by environmental temperature in muscle of the carp. Symp Soc Exp Biol 1992; 46:139-149.
30. Hwang GC, Watabe S, Hashimoto K. Changes in carp myosin ATPase induced by temperature acclimation. J Comp Physiol B 1990; 160:233-239.
31. Baldwin J, Hochachka PW. Functional significance of isoenzymes in thermal acclimatization: acetylcholinesterases from trout brain. Biochem J 1970; 116:883-887.
32. Tiku PE, Gracey AY, Macartney AI, Beynon RJ, Cossins AR. Cold-induced expression of delta 9-desaturase in carp by transcriptional and posttranslational mechanisms. Science 1996; 271:815-818.
33. Danyluk J, Rassart E, Sarhan F. Gene expression during cold and heat shock in wheat. Biochem Cell Biol 1991; 69:383-391.
34. Schlesinger MJ, Ashburner M, Tissieres A. Heat Shock from Bacteria to Man. Cold Spring Harbor: Cold Spring Harbor Laboratory Press, 1982.
35. Parsell DA, Lindquist S. The function of heat-shock proteins in stress tolerance: degradation and reactivation of damaged proteins. Annu Rev Genet 1994; 27:437-496.
36. Jones PG, Inouye M. The cold-shock response - a hot topic. Mol Microbiol 1994; 11:811-818
37. Ellis RJ, van der Vies SM. Molecular chaperones. Annu Rev Biochem 1991; 60:321-347.
38. Hochachka PW, Somero GN. Biochemical Adaptation. Princeton: Princeton Univ Press, 1984.
39. Franks F, Hatley RHM. Stability of proteins at subzero temperatures: thermodynamics and some ecological consequences. Pure Appl Chem 1991; 63:1367-1380.
40. Komatsu Y, Kaul SC, Iwahashi H, Obuchi K. Do heat shock proteins provide protection against freezing? FEMS Microbiol Lett 1990; 72:159-162.
41. Ray MK, Sitaramamma T, Ghandhi S, Shivaji S. Occurrence and expression of *cspA*, a cold shock gene, in Antarctic psychrotrophic bacteria. FEMS Microbiol Lett 1994; 116:55-60.
42. Munoz-Dorado J, Kondo K, Inouye M, Sone H. Identification of *cis*- and *trans*-acting elements involved in the expression of cold shock-inducible *TIP1* gene of yeast *Saccharomyces cerevisiae*. Nucleic Acids Res 1994; 22: 560-568.
43. Yocum GD, Joplin KH, Denlinger DL. Expression of heat shock proteins in response to high and low temperature extremes in diapausing pharate larvae of the gypsy moth, *Lymantria dispar*. Arch Insect Biochem Physiol 1992; 18:239-249.
44. Nunamaker RA, Dean VC, Murphy KE, Lockwood JA. Stress proteins elicited by cold shock in the biting midge, *Culicoides variipennis sonorensis* Wirth and Jones. Comp Biochem Physiol 1996; 113B: 73-77.
45. Joplin KH, Yocum GD, Denlinger DL. Cold shock elicits expression of heat shock proteins in the flesh fly, *Sarcophaga crassipalpis*. J Insect Physiol 1990; 36:825-834.
46. Petersen NS, Young P, Burton V. Heat shock mRNA accumulation during recovery from cold shock in *Drosophila melanogaster*. Insect Biochem 1990; 20:679-684.
47. Ketola-Pirie, CA, Atkinson BG. Cold- and heat-shock induction of new gene expression in cultured amphibian cells. Can J Biochem Cell Biol 1983; 61:462-471.
48. Liu, AYC, Bian H, Huang LE, Lee YK. Transient cold shock induces the heat shock response upon recovery at 37°C in human cells. J Biol Chem 1994; 269: 14768-14775.
49. Matz JM, Blake MJ, Tatelman HM, Lavoi KP, Holbrook NJ. Characterization and regulation of cold-induced heat shock protein expression in mouse brown adipose tissue. Am J Physiol 1995; 269:R38-R47.
50. Trent JD, Osipiuk J, Pinkau T. Acquired thermotolerance and heat shock in the extremely thermophilic archaebacterium *Sulfolobus* sp. Strain B12. J Bact 1990; 172:1478-1484.
51. Maresca B, Patriarca E, Goldenberg C, Sacco M. Heat shock and cold adaptation in Antarctic fishes: a molecular approach. Comp Biochem Physiol B 1988; 90:623-629.

52. Dietz TJ. Acclimation of the threshold induction temperatures for 70-kDa and 90-kDa heat shock proteins in the fish *Gillichthys mirabilis*. J Exp Biol 1994; 188:333-338.
53. Graumann P, Marahiel MA. Some like it cold: response of microorganisms to cold shock. Arch Microbiol 1996; 166:293-300.
54. Kandror O, Goldberg AL. Trigger factor is induced upon cold shock and enhances viability of *Escherichia coli* at low temperatures. Proc Natl Acad Sci USA 1997; 94:4978-4981.
55. Lee RE, Dommel RA, Joplin KH, Denlinger DL. Cryobiology of the freeze-tolerant gall fly *Eurosta solidaginis*: overwintering energetics and heat shock proteins. Climate Res 1995; 5:61-67.
56. Storey KB, Mosser DD, Douglas DN, Grundy JE, Storey JM. Biochemistry below 0°C: nature's frozen vertebrates. Brazilian J Med Biol Res 1996; 29:283-307.
57. King PA, Rosholt MN, Storey KB. Adaptations of plasma membrane glucose transport facilitate cryoprotectant distribution in freeze tolerant frogs. Am J Physiol 1993; 265:R1036-R1042.
58. King PA, Rosholt MN, Storey KB. Seasonal changes in plasma membrane glucose transport in freeze tolerant wood frogs. Can J Zool 1995; 73:1-9.
59. Mommsen TP, Storey KB. Hormonal effects on glycogen metabolism in isolated hepatocytes of a freeze-tolerant frog. Gen Comp Endocrinol 1992; 87:44-53.
60. Wolanczyk JP, Baust JG, Storey KB. Seasonal ice nucleating activity in the freeze tolerant frog, *Rana sylvatica*. Cryo Lett 1990; 11:143-150.
61. Wolanczyk JP, Storey KB, Baust JG. Nucleating activity in the blood of the freeze tolerant frog, *Rana sylvatica*. Cryobiology 1990; 27:328-335.
62. Storey KB, Storey JM, Churchill TA. De novo protein biosynthesis responses to water stresses in wood frogs: freeze-thaw and dehydration-rehydration. Cryobiology 1997; 34:200-213.
63. Cherry JH (ed.) Biochemical and Cellular Mechanisms of Stress Tolerance in Plants. Berlin: Springer, 1994.
64. Baker J, Steel C, Dure L. Sequence and characterization of 6 Lea proteins and their genes from cotton. Plant Mol Biol 1988; 11:277-291.
65. Close TJ, Kortt AA, Chandler PM. A cDNA-based comparison of dehydration-induced proteins (dehydrins) in barley and corn. Plant Mol Biol 1989; 13:95-108.
66. Neven LG, Haskell DW, Hofig A, Li QB, Guy CL. Characterization of a spinach gene responsive to low temperature and water stress. Plant Mol Biol 1993; 21:291-305.
67. Griffith M, Ala P, Yang DSC, Hon WC, Moffatt BA. Antifreeze protein produced endogenously in winter rye leaves. Plant Physiol 1992; 100:593-596.
68. Artus NN, Uemura M, Steponkus PL, Gilmour SJ, Lin C, Thomashow MF. Constitutive expression of the cold-regulated *Arabidopsis thaliana COR15a* gene affects both chloroplast and protoplast freezing tolerance. Proc Natl Acad Sci USA 1996; 93:13404-13409.
69. Houde M, Daniel C, Lachapelle M, Allard F, Laliberte S, Sarhan F. Immunolocalization of freezing-tolerance-associated proteins in the cytoplasm and nucleoplasm of wheat crown tissues. Plant J 1995; 8:583-593.
70. Cai Q, Storey KB. Freezing-induced genes in wood frog (*Rana sylvatica*): fibrinogen upregulation by freezing and dehydration. Am J Physiol 1997: 272;R1480-R1492.
71. Cai Q, Greenway SC, Storey KB. Differential regulation of the mitochondrial ADP/ATP translocase gene in wood frogs under freezing stress. Biochim Biophys Acta 1997; 1343:69-78.
72. Cai Q, Storey KB. Upregulation of a novel gene by freezing exposure in the freeze-tolerant wood frog (*Rana sylvatica*). Gene 1997; 198:305-312.
73. Weissbach L, Grieninger G. Bipartite mRNA for chicken α-fibrinogen potentially encodes an amino acid sequence homologous to β- and γ-fibrinogens. Proc Natl Acad Sci USA 1990; 87:5198-5202.
74. Roberts LR, Holland LJ. Coordinate transcriptional regulation of the three fibrinogen subunit genes by glucocorticoids in cultured primary liver cells from *Xenopus laevis*. Endocrinology 1993; 132:2563-2570.

75. Huber P, Laurent M, Dalmon J. Human β-fibrinogen gene expression: Upstream sequences involved in its tissue specific expression and its dexamethasone and interleukin 6 stimulation. J Biol Chem 1990; 265:5695-5701.
76. Roberts LR, Nichols LA, Holland LJ. cDNA and amino-acid sequences and organization of the gene encoding the Bβ subunit of fibrinogen from *Xenopus laevis*. Gene 1995; 160:223-228.
77. Rubinsky B, Lee CY, Bastacky J, Onik G. The process of freezing and the mechanism of damage during hepatic cryosurgery. Cryobiology 1987; 27:85-97.
78. Costanzo JP, Lee RE, Wright MR. Cooling rate influences cryoprotectant distribution and organ dehydration in freezing wood frogs. J Exp Zool 1992; 261:373-378.
79. Brandolin G, Le Saux A, Trezeguet V, Lauquin GJM, Vignais PV. Chemical, immunological, enzymatic, and genetic approaches to studying the arrangement of the peptide chain of the ADP/ATP carrier in the mitochondrial membrane. J Bioenerg Biomemb 1993; 25:459-472.
80. Storey KB, Storey JM. Freeze tolerant frogs: cryoprotectants and tissue metabolism during freeze/thaw cycles. Can J Zool 1986; 64:49-56.
81. Dransfield DT, Aprille JR. The influence of hypoxia and anoxia on distribution of adenine nucleotides in isolated hepatocytes. Arch Biochem Biophys 1994; 313:156-165.
82. Aprille JR. Regulation of the mitochondrial adenine nucleotide pool size in liver: mechanism and metabolic role. FASEB J 1988; 2:2547-2556.
83. Cai Q, Storey KB. Anoxia-induced gene expression in turtle heart: up-regulation of mitochondrial genes for NADH-ubiquinone oxidoreductase subunit 5 and cytochrome C oxidase subunit 1. Eur J Biochem 1996; 241:83-92.
84. Cai Q, Storey KB. A novel RNA species from the turtle mitochondrial genome – induction and regulation of transcription and processing under anoxic and freezing stresses. Genome 1997; 40:534-543.
85. Churchill TA, Storey KB. Dehydration tolerance in wood frogs: a new perspective on the development of amphibian freeze tolerance. Am J Physiol 1993; 265:R1324-R1332.
86. Holden CP, Storey KB. Second messenger and cAMP-dependent protein kinase responses to dehydration and anoxia stresses in frogs. J Comp Physiol B 1997; 167:305-312.
87. Holden CP, Storey KB. Signal transduction, second messenger, and protein kinase responses during freezing exposures in the wood frog. Am J Physiol 1996; 271:R1205-R1211.
88. Rubinsky B, Arav B, Fletcher GL. Hypothermic protection – a fundamental property of "antifreeze" proteins. Biochem Biophys Res Commun 1991; 180:566-571.
89. Lee CY, Rubinsky B, Fletcher GL. Hypothermic preservation of whole mammalian organs with "antifreeze" proteins. Cryo-Lett 1992; 13:59-66.
90. Rubinsky B, Arav A, Hong J-S, Lee CY. Freezing of mammalian livers with glycerol and antifreeze proteins. Biochem Biophys Res Commun 1994; 200:732-741.
91. Rubinsky B, Mattioli M, Arav A, Barboni B, Fletcher GL. Inhibition of Ca^{2+} and K^{+} currents by "antifreeze" proteins. Am J Physiol 1992; 262:R542-R545.
92. Negulescu PA, Rubinsky B, Fletcher GL, Machen TE. Fish antifreeze proteins block Ca entry into rabbit parietal cells. Am J Physiol 1992; 263:C1310-1313.
93. Hochachka PW. Defense strategies against hypoxia and hypothermia. Science 1986; 231:234-241.
94. Hew CL, Davies PL, Fletcher G. Antifreeze protein gene transfer in Atlantic salmon. Mol Mar Biol Biotechnol 1992; 1:309-317.
95. Wang R, Zhang P, Gong Z, Hew CL. Expression of the antifreeze protein gene in transgenic goldfish (*Carassius auratus*) and its implication in cold adaptation. Mol Mar Biol Biotechnol 1995; 4:20-26.

Subject index

abscisic acid (ABA) 71, 87, 96, 355
ABA-responsive element 91, 368
acarbose 285
Acari 168
active site 282
Adamussium colbecki 225, 227
adaptive evolution 239
Adorybiotus coronifer 158
ADP/ATP translocase 396, 401
adult life span 191
aeronautic 167
alarm phase 87
alarm reactions 97
algae 6, 8, 24
alkaline phosphatase 41, 265
allochthonous 183
allosteric activators 284
allosteric enzymes 303
allosteric transition 329
alpine 16, 59, 167, 181, 197
alpine lake 16
alpine streams 181
Alteromonas haloplanctis 35, 40, 262, 270, 280
amino acid composition 308
amino acid sequence 239, 302, 308, 319
amino acids 198
aminotransferases 305, 319
Amphibolus nebulosus 159
α-amylase 40, 262, 270, 277
α-amylase 3D structures 280
anaerobiosis 156
anatomical adaptations 354
anhydrobiosis 158
anhydrobiotic organisms 158
Annelida 166
anoxia stress 401
anoxia tolerance 386
antarctic bacteria 16, 257, 277, 305, 319, 335
antarctic fish 219, 240, 297
antarctic lake 16
Anthonomus pomorum 141
antifreeze agents 198, 200
antifreeze protection 47
antifreeze proteins 47, 111, 127, 132, 151, 199, 354, 360, 385
Aphelencus avenae 158
apoplastic abscisic acid 95
apoplastic proteins 361, 363
apoplastic space 129
apoplastic volume 121
Aptenodytes forsteri 247
aptery 191
aquaculture 403
Aquaspirillum arcticum 41, 45
aquatic insects 189
Arabidopsis 368
Arabidopsis thaliana 120
arabinoxylan gums 127
arabinoxylans 354
Araneae 167
Archaeognatha 170
archeal enzyme 323
arctic 59, 167
arginine biosynthesis 319
arginine biosynthetic genes 322
Arrhenius equation 34, 300, 313
Artedidraconidae 222
Arthrobacter 4, 9, 12, 41
Arthrobacter glacialis 35, 45
Arthrobacter globiformis 40, 45
arthropods 190
ascorbic acid 86
aspartate aminotransferase 305
aspartate carbamoyltransferase 319, 325
autochthonous 183
automobilizing elements 337
avoidance of freezing 62
avoidance strategy 81

Bacillus cereus 39, 44
Bacillus psychrophilus 40, 44
Bacillus psychrosaccharolyticus 41
Bacillus subtilis 38

bacterial production 25
bacterial promoters 306
$(\beta/\alpha)_8$-barrel enzymes 282
Bathydraconidae 223
benthic fishes 221
biodiversity 233, 335
black ice 18
body surface 208
Bohr effect 244
Bolitophagus reticulatus 151
bottom dwellers 240
Bovichtidae 240
brachyptery 191
Brassica napus 362
Bromus inermis 121
bud scales 112

cactus 62
calcium binding site 282
calcium channels 88
Ca/Zn metalloprotease 277
capillary suction 123
Carabodes labyrinthicus 157
carbamoyltransferases 319
carbohydrate accumulation 356
catalytic constant k_{cat} 259, 299, 311
catalytic efficiency 259, 299, 311
catalytic mechanism 285
Catharacta maccormicki 248
cell cultures 121
cell dehydration 80
cell injury 79
cell membranes 62
cell surface receptors 84
cell wall compression 124
cell wall matrix 124
cessation of growth 69
Chaenocephalus aceratus 232, 300
Champsocephalus gunnari 223
Channichthyidae 219, 221, 223, 240
channichthyids 231
chemoreceptors 208
chill-intolerant 138
chill-tolerant 58, 138
chilling 349
chilling injury 79
chilling-resistant 79, 80
Chilopoda 169
Chionodraco hamatus 298
Chionodraco myersi 230
chitinases 111, 361
Chlamydomonas nivalis 214
chloride ion 282, 285
Chrysemys marginata 402
ciliates 23
citrate synthase 40, 265
cloud droplets 21
coastal habitat 224
cold acclimation 87, 105, 132, 350, 354
cold-acclimation proteins 44, 359
cold activation 94
cold-active enzymes 257, 277, 297, 319
cold adaptation 59, 188, 198, 200, 259, 297
386
cold cave 9
cold deserts 5
cold hardiness 137, 188, 385, 396
cold-induced gene expression 390
cold-induced genes 61
cold-induced proteins 42
cold injury zone 141
cold-regulated proteins 359
cold-responsive element (C-repeat) 369
cold sensor 46
cold shock 44
cold-shock induction 43
cold-shock proteins 43, 392
cold-shock response 42
cold soils 3, 13
cold tolerance 188
Coleoptera 145, 171
Collembola 143, 155, 170, 197, 200
conifers 66
conjugative plasmid 335
continental shelf 224
Corylus avellana 64
Corynebacterium 9, 35
coryneform bacteria 12
critical photoperiod 69
cryoconite 159, 167, 215
cryoconservation 7
cryopelagic 230
cryopreservation 47
cryoprotectants 6, 132, 149, 357, 386
cryoprotection 360
cryoprotective proteins 396
Cryptococcus 5
Cryptopygus antarcticus 148
crystallization 386
crystallography 278
cuticular permeability 205
cyanobacteria 4, 5, 6, 8, 25
Cygnodraco mawsoni 223
Cytophaga-Flavobacterium 23
cytoplasmic membranes 352
cytoplasmic volume 121
cytorrhysis 105, 126

deep supercooling 105, 114
defoliation 67
dehardening 57, 65, 71
dehydration 154, 159, 365, 402
dehydration-responsive element 92, 369
dehydrative stress 105
dehydrin 83, 365
dehydrogenases 41
Deinococcus 5
Delia radicum 141
denaturation of proteins 39
Dendrobaena octaedra 159
Dendroides canadensis 148
Dermaptera 170
desiccation 154
diaglycerol 94, 96
Diamesa 185, 190
diapause 139, 189
diatoms 25
dielectric poration 127
differential thermal analysis 111
Diplopoda 169
Diplura 170
Diptera 145, 171
Dissostichus 226
Dissostichus mawsoni 298
disulfide bonds 287
DNA replication 338
16S rDNA sequencing 4, 6
dormant buds 111
drought 371
Dubautia menziesii 64
Dytilenchus 158

***E**chiniscus jenningsi* 159
ecological diversity 226, 234
effector proteins 84
egg deposition 227
elastase 263, 268
electrolyte leakage 373
electrospray mass spectrometry 311
embolism 63
enchytraeids 160
Enchytraeus kinkaidi 160
endemic 167, 169, 172, 173
enthalpy 241
enthalpy of activation 300
Entomobrya nivalis 154
Entomobrya sp. 200
entropy of activation 300
environmental constraints 223
enzymes 35, 259, 297, 305, 319
epidermis 208
epifluorescence microscopy 7
episodic frost 63
equilibrium freezing point 106
erythrocytes 244
ethylene 94
ethylene glycol 150
Eudorylaimus 156
eukryal 184, 185
euryzonal 167, 169, 170
extinction 221
extracellular enzymes 259, 263
extracellular freezing 60, 83
extraorgan freezing 111
extrinsic ice nucleation 106

fatty acid 358, 366
fatty acid biosynthesis 37
fatty acid synthetase 37
feeding 227
fibrinogen 396, 400
firn areas 200
fish 388
fish community 227
fish diversity 225
fish enzymes 267
Flexibacter-Bacteroides-Cytophaga 4, 9
flexibility 260, 277, 288
fluorescence microscopy 4
free energy of activation 300
freeze avoidance 105, 137, 198, 385, 390
freeze dehydration 349
freeze-induced gene expression 394
freeze inhibition 199
freeze-susceptible 137
freeze-thaw cycles 5, 6, 9, 13, 128
freeze tolerance 62, 82, 105, 138, 147, 158, 198, 349, 385, 391, 396
freezing 119, 120
freezing damage 400
freezing injury 351, 353
freezing point 122, 198
freezing protection 388
freezing stress 79, 352
freshwater ecosystems 188
frog 395, 400
frost hardening 81, 96
frost hardiness 58
frost resistance 7
frost tolerance 83
frozen sediments 6, 13
fructan 357
functional groups 59

β-**g**alactosidase 41, 265
Gallionella 12

Gastropoda 166
gene expression 89, 269, 368, 385
gene expression systems 345
gene structure 306
gibberellins 70
glacier-dwelling Collembola 197
glacier-fed 183, 187
glaciers 3, 7, 200
glacio-rhithral 186
glass formation 106
glassed solutions 106
Globodera rostochiensis 157
glucanases 111, 361
glucose-6-phosphate dehydrogenase 298
glutamate dehydrogenase 39, 298
glutathione 86
glycerol 83, 150, 356
glycogen phosphorylase 300
glycoproteins 354, 388
glycosyl hydrolases 280, 285
growth rate 34
GTP-binding proteins 86
gut content 214, 215
Gynaephora groenlandica 145

Haemideina maori 145
hardening 57, 65, 355
hatching 233
heat of oxygenation 241
heat-shock proteins 40
heat-shock response 391
Helianthus tuberosus 120
heliochrome 86
heme pocket 246
Hemiptera 171
hemoglobin 239
herbaceous 66, 105
heterogeneous ice nucleation 106, 143
high mountain lakes 18
high-temperature exotherm 112
Histiodraco velifer 222, 230
histological features 208
homeostasis 390
homogeneous ice nucleation 106, 143, 150
hydraulic resistance 123, 124
hydrogen bonding capacity 323
hydrogen exchange 261
hydrolysis 282
hydrophilicity 323
hydrophobic interactions 38
hydrophobicity index 268
Hymenoptera 145, 171
hypokrenal 187
hypokryal 184, 186
Hypsibius klebesbergi 159

ice 7, 13
ice-associated communities 226
ice cover 18
icefish 219, 297
ice formation 80, 350
ice growth 122
ice-nucleating agents 138
ice-nucleating bacteria 148
ice nucleation 61, 105, 124, 351, 386
ice-nucleation-active 106, 132
ice-nucleation proteins 48
ice pressure 127
ice propagation 107, 112
ice recrystallization 362
in-situ hybridization 22
infrared thermography 105
infrared thermometry 122
inhibitors 277
inositol-1,4,5-trisphosphate 96
insect larvae 181
insects 137
interstitial ice growth 123
interstitial water 120, 121
intracellular enzymes 259, 265
intracellular freezing 144, 152, 157
intramitochondrial gene expression 401
intrinsic catalytic efficiency 323
intrinsic nucleation 106
invertase 132
invertebrates 137, 165, 226
ion channels 86
Ips acuminatus 151, 153
Ips typographus 150
irradiance 67
irreversible denaturation 314
ischemia 402
Isopoda 169
Isotoma 'sp. G.' 200
Isotoma saltans 198, 200
Isotomurus palliceps 200

kinetic parameters 299
kryal 184

β-lactamase 41, 263
lactate dehydrogenase 40, 41, 265
Lactococcus lactis 47
lake ice microbial communities 17, 21
larvae 233
larval strategies 190
Lepidoptera 145, 171
Leptinotarsa decemlineata 148

life cycle 168, 189, 227
lipase 41
lipid bilayer 37, 386
lipid fluidity 6
lipid polymers 354
lipid-transfer proteins 367
lipids 36
liposomes 358
Lithobius forficatus 151
long days 67
longitudinal zonation 184, 187
low-temperature exotherm 112
low temperature-induced proteins 360
low temperature-responsive element 368
lumbricid earthworms 159
Lycopersicon esculentum 107

Macrobiotus harmsworthi 159
Mamestra configurata 141
marine isolates 337
mating activities 191
matrix solution 122
Mecoptera 171
Medicago sativa 132
melanism 191
melting point 143
membrane behavior 128
membrane fluidity 36, 79, 200, 366
membrane free energy 127
membrane lipid 36, 81, 366
membrane protectants 386
membrane stability 36
Mesenchytraeus flavus 160
mesophilic host 269
metabolic acidosis 251
metabolic activity 35
metakryal 184, 185
metazoa 24
methane production 4
Michaelis-Menten kinetics 257, 297, 305
microclimatic conditions 202
Micrococcus cryophilus 39
midgut epithelium 212
mitogen-activated protein kinase 94
mobility 260
model systems 120
modeling 72, 119
moisture content 62
molecular adaptations 259
molecular evolution 239
molecular modeling 239, 245, 329
molecular responses 349
monounsaturated fatty acids 37
moraines 3, 5
Moraxella 125, 305
morphological features 197
mountain lake 20
Myzus persicae 141

nanoflagellates 24
negative modification of enthalpy 258
nematodes 156, 166
nesting 227
nitrification 4
nival fauna 165
nonesterified pectins 112
Notothenia neglecta 267
Nototheniidae 222, 240
nototheniids 227
Notothenioidei 220, 240
notothenioids 227
nucleation temperature 106
nutrient requirements 23

oligonucleotid probes 22
Onychiurus 'sp. G' 200
Onychiurus arcticus 155
Opiliones 168
Opuntia fragilis 62
ornithine carbamoyltransferase 319
osmotic loss of water 94
osmotic potential 64, 130
osmotic stress 371
osmoticum 126
ovarian cycle 231
oxidative stress 371
oxygen carriers 239
oxygen free radicals 80
oxygen transport 240

Pagothenia borchgrevinki 230, 241
Panagrolaimus davidi 157
Panonychus ulmi 140
Paraneoptera 170
parthenogenetic 191
pathogen resistance 370
pathogenesis-related proteins 361
pectin-mediated regulation 114
pelagization 225
Pelophila borealis 147
penguins 247
periodic frost 65
periplastic space 126
permafrost 6, 7, 13
permanent ice 20
pest insects 139
pest management 141
Petrova resinella 151

Phaseolus vulgaris 107
phenotypic plasticity 234
phosphatidic acid 83, 86
phosphatidylglycerol 83
phosphoglycerate kinase 265
phosphoinositide 83
phospholipases 86, 94
phospholipid 83, 358
photoinhibition 79
photoperiod 65, 70
photosynthesis 24, 355
photosystem II 89, 93
phyletic diversification 234
Phyllodecta laticollis 147, 151
phylogenetic analysis 250
phytochrome A 68
phytoplankton 25
Picea abies 61
Pieris brassicae 141
pigmentation 8, 9, 191, 208
Pinus sylvestris 59
pit membrane 114
plant acclimation 79
plant hormones 87
plant morphogenesis 89
plant protoplasts 119
plasma membrane 86, 124, 352, 357
plasma membrane dysfunction 94
plasmids 335, 338, 341
plasmolysis 128, 129
plasticity 260
Pleuragramma 226
Pleuragramma antarcticum 222, 241
poikilothermic organisms 390
polar fish 219
polyols 138, 149, 150, 154, 159, 198
polysaccharide 9, 290, 354
polyunsaturated fatty acids 37
Populus tremuloides 115
positive modification of enthalpy 258
postantennal organ 208
prehardening 82
primary production 24
primary structure 250
promoter regions 327
protective dehydration mechanism 160
protective solutes 121
protein crystallography 278
protein diffraction data 279
protein flexibility 38
protein kinases 87, 95
proteinases 41
proteins 95, 387, 388
proton-motive force 93
protoplast model 119
protoplasts 128, 352
Protura 170
Prunus sp. 111
Pseudomonas 4, 8, 35
Pseudomonas fluorescens 35
Pseudomonas fragi 44
Pseudomonas putida 48
Pseudomonas syringae 106, 148
Pseudoscorpiones 168
Psocoptera 171
Psychrobacter uratovorans 39
psychrophilic 3, 6, 8, 12, 34, 290
psychrophilic enzyme 262, 277
psychrophilic transferases 321
psychrotolerant 3, 7, 9, 12, 35
psychrotrophic 3, 6, 34
Pygoscelis adeliae 247
pyrimidine pathway 325
Pytho depressus 151

Rana sylvatica 394
rapid-cooling responses 88
redox agents 86
redox signaling system 89
reducing equivalents 89, 93
refugial-cavatic 169, 170
regularory dimer 329
Rep proteins 344
reproductive strategies 231, 232
rhithral 187
Rhizobium 41
Rhododendron sp. 110
ribonucleases 41
river typologies 184
RNA-binding protein 44
RNA chaperone 43
rolling circle mechanism 344
root effect 244
R state 246
rye protoplasts 128

Saccharomyces cerevisiae 158
salt stress 371
Saltatoria 170
saturated fatty acids 37
Saxifraga caespitosa 65
Secale cereale 120, 350
second freezing 131
second messengers 86
sediments 8, 9
sequencing 325
Shine-Dalgarno-like sequence 306
shoot growth 69

short days 67
signal transduction 84, 369
signaling pathways 79
site-directed mutagenesis 270
Sitobion spp. 141
slow responses 88
slush 18
snow 6, 7, 8, 12
snow algae 8
snow Collembola 197
snowmelt 186
Solanum commersonii 352
Solanum dulcamara 361
Solanum tuberosum 120
spawning 231, 233
specific activity 299
spring-fed stream 183, 187
springs 186
stability of proteins 260
stenothermality 297
stream types 183
Streptococcus lactis 39
stress syndrome 87
substrate binding 282
substrate recognition 289
subtilisin 263, 271
sucrose 72, 356
sucrose synthase 356
sugar alcohol cryoprotectants 198, 386
sugars 357
Sulfolobus solfataricus 306
supercooling 23, 60, 385
supercooling capacity 387
supercooling point 138, 144, 198
surface pools 23
swarming 191
Symphyla 169

Tardigrada 167
taxonomic diversity 234
temperature adaptations 239
temperature optima 23
temperature ranges 187
Tenebrio molitor 152
terminal web 212
terrestrial arthropods 137
terrestrial isolates 336
Tetracanthella wahlgreni 157
thaumatin-like proteins 361
thawing 119, 120
thermal hysteresis 200
thermal hysteresis activity 389
thermal hysteresis proteins 138, 151, 199, 362
thermocouples 107
thermodynamic analysis 239
thermodynamic model 128
thermodynamic parameters 300
thermodynamics 250
thermoinactivation 314
thermolabile 319
thermoregulation 191
thermoresistance 311
thermosensors 46
thermostability 260, 300, 323
Thermotoga maritima 323
Thiobacillus ferroxidans 47
thiobacteria 12
three-dimensional (3D) structures 277
Thripidae 171
tonoplast membrane 129
Trachemys scripta elegans 402
transgenic 68
transient pores 126
transpiration rates 202, 203
trehalose 150, 158, 159, 356
Trematomus 229
Trematomus bernacchii 225, 227, 297
Trematomus hansoni 297
Trematomus newnesi 241
Trichiocampus populi 144
Trichosporon pullulans 45
Trichostrongylus colubriformis 157
trimeric enzyme 323
triosephosphate isomerase 40, 41, 263, 272
Triticum aestivum 350
trophic resources 224
trypsin 263, 268
Turbellaria 165
turtles 402

ultrastructural features 208
unsaturated fatty acids 37
unsaturation 366

ventral tube 206, 210
vertebrate 394, 398
Vibrio 4, 321
Vibrio marinus 39, 41

walled cells 121
water absorption rates 207
water balance 89
water content 199, 202
water evaporation 123
water loss rates 202, 204
water stress 83
white ice 18

winter desiccation 63
woody plants 105

X-ray crystallography 239
X-ray data 278
xylem 354

yeasts 4, 6, 8

zonal vegetation 60
zonation 188
zonobiomes 60

Computer to Film: Saladruck, Berlin
Binding: Buchbinderei Lüderitz & Bauer, Berlin